Material										
Magnesium, extrusion, AZ80X	0.066	35	26	49	21	19	6.5	2.4	12	14.4
Magnesium, sand cast, AZ63-HT	0.066	14	14	40	19	14	6.5	2.4	12	14.4
Monel, wrought, hot-rolled	0.319	50	b	90		40	26/15	9.5	35	7.8
Red brass, cold-rolled	0.316	60	b	75			15	5.6	4	9.8
Red brass, annealed	0.316	15	b	40			15	5.6	50	9.8
Bronze, cold-rolled	0.320	75	b	100			15	6.5	3	9.4
Bronze, annealed	0.320	20	b	50			15	6.5	50	9.4
Titanium alloy, annealed	0.167	135	b	155			14	5.3	13	
Invar, annealed	0.292	42	b	70			21	8.1	41	0.6
Nonmetallic materials										
Douglas fir, green[e]	0.022	4.8	3.4	3.9	0.9		1.6			
Douglas fir, air dry[e]	0.020	8.1	6.4	7.4	1.1		1.9			1.9
Red oak, green[e]	0.037	4.4	2.6	3.5	1.2		1.4			
Red oak, air dry[e]	0.025	8.4	4.6	6.9	1.8		1.8			
Concrete, medium strength	0.087	1.2	1.2	3.0			3.0			6.0
Concrete, fairly high strength	0.087	2.0	2.0	5.0			4.5			6.0

[a] Elastic strength may be represented by proportional limit, yield point, or yield strength at a specified offset (usually 0.2 percent for ductile metals).

[b] For ductile metals (those with an appreciable ultimate elongation), it is customary to assume the properties in compression have the same values as those in tension.

[c] Rotating beam.

[d] Elongation in 8 in.

[e] All timber properties are parallel to the grain.

$1 \text{ MPa} = 10^6 \text{ N/m}^2$

MECHANICS OF MATERIALS

3-0

$$\sigma_n = \bar{S} \cdot \bar{n}_s = S_x l + S_y m + S_z n$$

$$OR = \sigma_x l^2 + \sigma_y m^2 + \sigma_z n^2 + 2\hat{\tau}_{xy} lm + 2\hat{\tau}_{yz} mn + 2\hat{\tau}_{zx} nl$$

$$\tau_{nt}^2 = |\bar{S}|^2 - \sigma_n^2$$

$$S_x = \sigma_x l + \hat{\tau}_{xy} m + \hat{\tau}_{xz} n$$

$$S_y = \tau_{yx} l + \sigma_y m + \hat{\tau}_{yz} m$$

$$S_z = \hat{\tau}_{zx} l + \hat{\tau}_{zy} m + \sigma_z n$$

$$l = \cos \theta_x$$

$$m = \cos \theta_y$$

$$n = \cos \theta_z$$

FOURTH EDITION

MECHANICS OF MATERIALS

ARCHIE HIGDON

Engineering and Technology Education Consultant

EDWARD H. OHLSEN

Department of Engineering Science and Mechanics
Iowa State University

WILLIAM B. STILES

Department of Engineering Mechanics
The University of Alabama

JOHN A. WEESE

School of Engineering
Old Dominion University

WILLIAM F. RILEY

Department of Engineering Science and Mechanics
Iowa State University

JOHN WILEY & SONS

NEW YORK   CHICHESTER   BRISBANE   TORONTO   SINGAPORE

Book designed by Karin Gerdes Kincheloe.
Cover photo by Paul Silverman.
Production supervised by Carolyn Golojuch.
Illustrations by John Balbalis with the assistance of
the Wiley Illustration department.
Manuscript edited by Patricia Brecht under the supervision
of Deborah Herbert.

Library of Congress Cataloging in Publication Data
Main entry under title:

Mechanics of materials.

    Includes index.
    1. Strength of materials. I. Higdon, Archie,
1905–
TA405.M515   1985   620.1'12   84-7583
ISBN 0-471-89044-8

Printed in the United States of America

10 9 8 7

PREFACE

The primary objectives of a course in mechanics of materials are: (1) to develop a working knowledge of the relations between the loads applied to a nonrigid body made of a given material and the resulting deformations of the body, (2) to develop a thorough understanding of the relations between the loads applied to a nonrigid body and the stresses produced in the body, (3) to develop a clear insight into the relations between stress and strain for a wide variety of conditions and materials, and (4) to develop adequate procedures for finding the required dimensions of a member of a specified material to carry a given load subject to stated specifications of stress and deflection. These objectives involve the concepts and skills that form the foundation of all structural and machine design.

The principles and methods used to meet the general objectives are drawn largely from prerequisite courses in mechanics, physics, and mathematics together with the basic concepts of the theory of elasticity and the properties of engineering materials. This book is designed to emphasize the required fundamental principles, with numerous applications to demonstrate and develop logical orderly methods of procedure. Instead of deriving numerous formulas for all types of problems, we have stressed the use of free-body diagrams and the equations of equilibrium, together with the geometry of the deformed body and the observed relations between stress and strain, for the analysis of the force system acting on a body.

This book is designed for a first course in mechanics of deformable bodies. Because of the extensive subdivision into different topics, the book will provide flexibility in the choice of assignments to cover courses of different length and content. The developments of structural applications include the inelastic as well as the elastic range of stress; however, the material is organized so that the book will be found satisfactory for elastic coverage only.

Extensive use has been made of prerequisite course material in statics and calculus. A working knowledge of these subjects is considered an essential prerequisite to a successful study of mechanics of materials as presented in this book.

The illustrative examples and problems have been selected with special attention devoted to problems that require an understanding of the principles of mechanics of materials without demanding excessive time for computational work. A large number of problems are included so that problem assignments may be varied from term to term. Approximately half of the problems are given in SI units.

The answers to about half the problems are included at the end of this book. We feel that the first assignment on a given topic should include some problems for which the answers are given; since the simpler problems are usually reserved for this first assignment, the answers are provided for the first four or five problems of each article and thereafter are given in general for alternate problems. We feel that answers for certain problems should not be given; hence, we have not followed the flat rule of answers for all even-numbered problems. Since the convenient designation of problems for which answers are provided is of great value to those who make up assignment sheets, the problems for which answers are provided are indicated by means of an asterisk after the number.

In general, all given data are assumed to be composed of three significant figures regardless of the number of figures shown. Answers are, therefore, given to three significant figures, unless the number lies between 1 and 2 or any decimal multiple thereof, in which case four significant figures are reported. In Chapter 11, the problems involve material properties, which are quite variable, and possibly stress concentration factors, which are difficult to obtain to more than two significant figures from the graphs; therefore, we hesitate to give results to more than two significant figures. However, in order to avoid confusion, the answers are usually given to three significant figures.

We are grateful for comments and suggestions received from colleagues and from users of the earlier editions of this book. We appreciate also the careful appraisal made by the publisher's consultant and have incorporated those recommended changes which we considered important.

Archie Higdon
Edward H. Ohlsen
William B. Stiles
John A. Weese
William F. Riley

CONTENTS

CHAPTER 6   FLEXURAL LOADING: STRESSES   237

CHAPTER 7   FLEXURAL LOADING: DEFLECTIONS   355

SYMBOLS AND ABBREVIATIONS*

A	area
avg.	average
b	breadth, width
C	compression, Celsius
c	distance from neutral axis or from center of twist to extreme fiber
E	modulus of elasticity in tension or compression
e	eccentricity
Eq.	equation
F	force, or body force, Fahrenheit
f	normal force per unit of length
ft	foot or feet
ft-lb	foot-pound
G	modulus of rigidity (modulus of elasticity in shear)
g	acceleration of gravity

* The symbols and abbreviations used in this book conform essentially with those approved by the International Standards Organization.

h	height, depth of beam
hp	horsepower
I	moment of inertia or second moment of area, interference
$in.$	inch
in.-lb	inch-pound
J	polar moment of inertia of area, joule (N · m)
K	stress concentration factor
k	modulus of spring, factor of safety
kg	kilogram
kip or k	kilopound (1000 lb)
ksi	kilopounds (or kips) per square inch
L	length
lb	pound
M	bending moment
m	meter
max	maximum
min	minimum
N	normal force, newton
N · m	Newton · meter
n	ratio of elastic moduli, number of cycles
P	concentrated load, force
p	fluid pressure
Pa	pascal (N/m^2)
psf	pounds per square foot
psi	pounds per square inch
Q	first moment of area
q	shearing force per unit of length
R	radius, modulus of rupture, resultant force
r	radius of gyration
rad	radian
rpm	revolutions per minute
rps	revolutions per second
S	surface force, resultant stress, stress vector, section modulus ($S = I/c$)
sec	second
sq	square
T	torque, temperature, tension
t	thickness, tangential deviation
U	strain energy
u	strain energy per unit volume
u, v, w	components of displacement
V	shearing force

W	weight, watt $(N \cdot m/s)$
W_k	work
w	load per unit of length
x, y, z	cartesian coordinates
y	deflection
Z	plastic section modulus
α (alpha)	coefficient of thermal expansion
γ (gamma)	shearing strain (unit), specific weight
δ (delta)	total deformation
Δ (delta)	maximum deflection under energy load
ε (epsilon)	normal strain (unit)
η (eta)	efficiency, portion of energy load which is effective
θ (theta)	total angle of twist in radians, slope of deflected beam
μ (mu)	micro (10^{-6})
ν (nu)	Poisson's ratio
ρ (rho)	radius
σ (sigma)	normal stress (unit)
τ (tau)	shearing stress (unit)
ω (omega)	angular velocity
$\circ$	degree
C12 $\times$ 30	a 12-in. rolled channel section with a weight of 30 lb/ft
C305 $\times$ 45	a 305-mm rolled channel section with a mass of 45 kg/m
L6 $\times$ 4 $\times$ 1/2	a rolled angle section with 6-in. and 4-in. legs 1/2 in. thick
L102 $\times$ 89 $\times$ 13	a rolled angle section with 102-mm and 89-mm legs 13 mm thick
S10 $\times$ 35	a 10-in. American standard rolled section (I-beam) with a weight of 35 lb/ft
S127 $\times$ 22	a 127-mm American standard rolled section (I-beam) with a mass of 22 kg/m
W16 $\times$ 100	a 16-in. wide-flange rolled section with a weight of 100 lb/ft
W914 $\times$ 342	a 914-mm wide-flange rolled section with a mass of 342 kg/m
WT8 $\times$ 50	an 8-in. structural T-section (cut from a W-section) with a weight of 50 lb/ft
WT457 $\times$ 171	a 457-mm structural T-section (cut from a W-section) with a mass of 171 kg/m

CHAPTER 1

ANALYSIS OF STRESS—CONCEPTS AND DEFINITIONS

1-1 INTRODUCTION

The primary objective of a course in mechanics of materials is the development of relationships between the loads applied to a nonrigid body and the internal forces and deformations induced in the body. Ever since the time of Galileo Galilei (1564–1642), men of scientific bent have studied the problem of the load-carrying capacity of structural members and machine components, and have developed mathematical and experimental methods of analysis for determining the internal forces and the deformations induced by the applied loads. The experiences and observations of these scientists and engineers of the last three centuries are the heritage of the engineer of today. The fundamental knowledge gained over the last three centuries, together with the theories and analysis techniques developed, permit the modern engineer to design, with complete competence and assurance, structures and machines of unprecedented size and complexity.

The subject matter of this book forms the basis for the solution of three general types of problems:

1. Given a certain function to perform (the transporting of traffic over a river by means of a bridge, conveying scientific instru-

ments to Mars in a space vehicle, the conversion of water power into electric power), of what materials should the machine or structure be constructed, and what should the sizes and proportions of the various elements be? This is the designer's task, and obviously there is no single solution to any given problem.

2.   Given the completed design, is it adequate? That is, does it perform the function economically and without excessive deformation? This is the checker's problem.

3.   Given a completed structure or machine, what is its actual load-carrying capacity? The structure may have been designed for some purpose other than the one for which it is now to be used. Is it adequate for the proposed use? For example, a building may have been designed as an office building, but is later found to be desirable for use as a warehouse. In such a case, what maximum loading may the floor safely support? This is the rating problem.

Since the complete scope of these problems is obviously too comprehensive for mastery in a single course, this book is restricted to a study of individual members and very simple structures or machines. The design courses that follow will consider the entire structure or machine, and will provide essential background for the complete analysis of the three problems.

The principles and methods used to meet the objective stated at the beginning of this chapter depend to a great extent on prerequisite courses in mathematics and mechanics, supplemented by additional concepts from the theory of elasticity and the properties of engineering materials. The equations of equilibrium from *statics* are used extensively, with one major change in the free-body diagrams; namely, most free bodies are isolated by *cutting through* a member instead of removing a pin or some other connection. The forces transmitted by the cut sections are *internal* forces. The intensities of these internal forces (force per unit area) are called *stresses*.

It will frequently be found that the equations of equilibrium (or motion) are not sufficient to determine all the unknown loads or reactions acting on a body. In such cases it is necessary to consider the geometry (the change in size or shape) of the body after the loads are applied. The *deformation* per unit length in any direction or dimension is called *strain*. In some instances, the specified maximum deformation and not the specified maximum stress will govern the maximum load that a member may carry.

Some knowledge of the physical and mechanical properties of materials is required in order to create a design, to properly evaluate a given design or even to write the correct relation between an applied load and the resulting deformation of a loaded member. Essential information will be introduced as required, and more complete information can be obtained from textbooks and handbooks on properties of materials.

1-2 LOAD CLASSIFICATION

Certain terms are commonly used to describe applied loads; their definitions are given here so that the terminology will be clearly understood.

Loads may be classified with respect to time:

1.   A *static load* is a gradually applied load for which equilibrium is reached in a relatively short time.

2.   A *sustained load* is a load that is constant over a long period of time, such as the weight of a structure (called *dead load*). This type of load is treated in the same manner as a static load; however, for some materials and conditions of temperature and stress, the resistance to failure may be different under short-time loading and under sustained loading.

3.   An *impact load* is a rapidly applied load (an energy load). Vibration normally results from an impact load, and equilibrium is not established until the vibration is eliminated, usually by natural damping forces.

4.   A *repeated load* is a load that is applied and removed many thousands of times. The helical springs that close the valves on automobile engines are subjected to repeated loading.

Loads may also be classified with respect to the area over which the load is applied:

1.   A *concentrated load* is a load or force applied at a point. Any load applied to a relatively small area compared with the size of the loaded member is assumed to be a concentrated load; for example, a truck wheel load on the longitudinal members of a bridge.

2.   A *distributed load* is a load distributed along a length or over an area. The distribution may be uniform or nonuniform. The weight of a concrete bridge floor of uniform thickness is an example of a uniformly distributed load.

Loads may be classified with respect to the location and method of application:

1.   A *centric load* is one in which the resultant force passes through the centroid of the resisting section. If the resultant

FIG. 1–1

(a)   (b)   (c)

passes through the centroids of all resisting sections, the loading is termed *axial*. In Fig. 1–1*a*, the line of action of the load *P* passes through the centroid of section *a–a*; therefore, the load is centric for section *a–a*. The force *P* in Fig. 1–1*b* is applied through the plate to the shank of the rivet and is resisted by the force *V* on the shaded section *b–b*. If the thickness *t* is not too large (the usual case for riveted joints), the force *P* can be assumed to be collinear with *V*, thus satisfying the definition of a centric load. The loading illustrated in this figure is called *single shear,* and the force *V* is a shearing force. If the plate *A* was sandwiched between two plates *B* with the rivet extending through all three plates, the rivet would be subjected to two shearing forces. Such loading is called *double shear*. Figure 1–1*c* represents an eyebar used for trusses, tie rods, and linkages. This is an example of axial loading.

2.   A *torsional load* is one that subjects a shaft or some other member to couples that twist the member. If the couples lie in planes transverse to the axis of the member, as in Fig. 1–2, the member is subjected to pure torsion.

3.   A *bending* or *flexural load* is one in which the loads are applied transversely to the longitudinal axis of the member. The applied load may include couples that lie in planes parallel to the axis of the member. A member subjected to bending loads bends

FIG. 1–2

Resultant of
bolt forces
is a couple

or bows along its length. Figure 1–3 illustrates a beam subjected to flexural loading consisting of a concentrated load, a uniformly distributed load, and a couple.

4.　A *combined loading* is a combination of two or more of the previously defined types of loading.

1-3　CONCEPT OF STRESS

In every subject area there are certain fundamental concepts of paramount importance to a satisfactory comprehension of the subject matter. For the subject mechanics of materials, a thorough mastery of the physical significance of stress and strain is paramount. The discussion of stress will be undertaken first, and the study of strain will be taken up in later articles.

The eyebar *AB* from the pile extractor of Fig. 1–4 (used to reclaim steel sheet piles from cofferdams and similar installations) will serve as an example for the introduction of the concept of stress. Although the bar is subjected to dynamic loading in use, an equivalent static load (a static load that will produce the same stresses and strains as the dynamic load) may be used in the design. Since the eybar is a two-force member, the pin reactions at *A* and *B* are collinear with the axis of the bar. A review of two-force members and their application in trusses, as presented in a course in statics, is strongly recommended. When the bar is cut by any plane, such as plane *c–c*, and the free-body diagram of Fig. 1–5 is drawn, equilibrium may be established by placing on the cut section a series of parallel forces (stresses), each of which acts on a unit area. The term *stress* is sometimes used to denote the resultant force on a section; in other instances it is used to indicate the force

FIG. 1–3

FIG. 1–4   *(Courtesy of Vulcan Iron*   FIG. 1–5
           Works Inc., Chicago, Illinois)

per unit area. *In this book stress will always refer to the force per unit area, unless otherwise stated.* The equilibrium equation $\Sigma F_y = 0$ gives the expression

$$\text{applied force} = (\text{stress})(\text{area})$$

It is immediately apparent that the magnitude of the stress on plane c–c is a function of the angle of inclination θ of the plane with respect to the applied load. This fact distinguishes a stress from an ordinary force vector. Since stress is not a vector, the laws of vector addition do not apply to stresses that act on different planes. This need not be cause for concern if, in the application of the equations of equilibrium (or motion), one always replaces a stress with a total force (stress multiplied by the appropriate area), thus reducing the problem to one involving ordinary force vectors. However, stresses that act on a particular plane can be treated as vectors (see Art. 1–5), since they are all associated with the same area.

Because the stress indicated on plane c–c of Fig. 1–5, if evaluated, would convey little information regarding the adequacy of design, it is customary to replace this stress with two components: one perpendicular to the plane, called the *normal stress* (which may be tension T or compression C) and designated by the Greek letter σ; and the other parallel to the plane, called the *shearing stress* and designated by the letter τ. These stresses, when evaluated, may be compared with experimentally determined values for the strength of the material in tension, compression, and shear, thus yielding significant information. Such stresses are indicated in Fig. 1–6a. For this simple case of axial loading, the sense of the normal and shearing stresses has been determined by inspection. Again, the aspect of the plane as an additional characteristic possessed by a stress must be emphasized. In Fig. 1–6b the stresses are replaced by their resultants, the force vectors N and V; therefore, the equations of equilibrium may be applied to this free-body diagram. In the preceding discussion, the assumption is made that the stress is uniformly distributed over the cut section. A necessary (but not sufficient) condition for the distribution to be uniform is that the loading be *centric*. Nonuniform stress distribution under centric loading will be discussed in Art. 1–10. In this book, unless otherwise indicated, a uniform stress distribution is to be assumed for all centric (or assumed centric) loading.

FIG. 1–6

1-4 STRESSES UNDER CENTRIC LOADING

If the equations of equilibrium are applied to the free-body diagram of Fig. 1–6b, the following expressions are obtained:

For $\Sigma F_n = 0$

$$P \cos\theta = N = \frac{\sigma A}{\cos\theta}$$

from which

$$\sigma = \frac{P}{A} \cos^2\theta = \frac{P}{2A} (1 + \cos 2\theta) \qquad (a)$$

and for $\Sigma F_t = 0$

$$P \sin\theta = V = \frac{\tau A}{\cos\theta}$$

from which

$$\tau = \frac{P}{A} \sin\theta \cos\theta = \frac{P}{2A} (\sin 2\theta) \qquad (b)$$

The graph of these expressions in Fig. 1–7 indicates that σ is maximum when θ is 0° or 180°, that τ is maximum when θ is 45° or 135°, and also that $\tau_{max} = \frac{1}{2}\sigma_{max}$. Therefore, the maximum stresses are given by the expressions

$$\sigma_{max} = P/A \qquad (1-1)$$

and

$$\tau_{max} = P/(2A) \qquad (1-2)$$

It must be emphasized that Eq. 1–2 is valid only for centric tensile or compressive loading in one direction. For convenience, the maximum normal stress under axial loading is called the *axial stress*. Note that the normal stress is either maximum or minimum on planes for which the shearing stress

FIG. 1–7

is zero. In Art. 1–7 it is shown that the shearing stress is always zero on the planes of maximum or minimum normal stress.

Laboratory experiments indicate that both shearing and normal stresses under axial loading are important since a brittle material loaded in tension will fail in tension on the transverse plane, whereas a ductile material loaded in tension will fail in shear on the 45° plane.

One observes that when θ of Fig. 1–7 is greater than 90°, the sign of the shearing stress in Eq. *b* changes. The magnitude of the shearing stress for any angle θ, however, is the same as that for 90° + θ. The sign change merely indicates that the shear force vector changes sense, being directed toward the top of the element instead of toward the bottom as in Fig. 1–6. Normal and shearing stresses on planes having aspects θ_1 and 90° + θ_1 are shown in Fig. 1–8.

The equality of shearing stresses on orthogonal planes can be demonstrated by applying the equations of equilibrium to the free-body diagram of a small rectangular block of thickness *dz*, shown in Fig. 1–9. If a shearing force[1] $V_x = \tau_{yx}dx\,dz$ is applied to the top surface of the block, the equation $\Sigma F_x = 0$ will dictate the application of an oppositely directed force V_x to the bottom of the block, thus leaving the block subjected to a clockwise couple; this couple must be balanced by a counterclockwise couple composed of the oppositely directed forces V_y applied to the vertical faces of the block. Finally, application of the equation $\Sigma M_z = 0$ yields the following:

$$\tau_{yx}(dx\,dz)dy = \tau_{xy}(dy\,dz)dx$$

from which

$$\tau_{yx} = \tau_{xy} \qquad\qquad (1\text{–}3)$$

FIG. 1–8   FIG. 1–9

[1] The double subscript on the shearing stress is used to designate both the plane on which the stress acts and the direction of the stress. The first subscript indicates the plane (or rather the normal to the plane), and the second subscript indicates the direction of the stress.

Therefore, *if a shearing stress exists at a point on any plane, there must also exist at this point a shearing stress of the same magnitude on an orthogonal plane.* This statement is also valid when normal stresses are acting on the planes, since the normal stresses occur in collinear but oppositely directed pairs, and thus have zero moment with respect to any axis.

It will be observed that loads on a structure or machine are transmitted to the individual members through some sort of connection, as, for example, the riveted connection of Fig. 1–1*b*, the pins at the ends of the eyebar of Fig. 1–4, and even so simple a detail as the head (or nut) of a bolt (Fig. 1–10) loaded axially in tension. In all these connections, one of the most significant stresses induced is a shearing stress; hence, these members are often said to be subjected to centric shear loading. If the loading is as indicated in Fig. 1–1*b*, the average shearing stress on plane *b–b* is the significant shearing stress (although it is not the maximum stress since the shearing stress cannot be uniformly distributed over the area) and is referred to as the *cross shearing stress*. Another type of shear loading is termed *punching shear,* as, for example, the action of a punch in forming rivet holes in a metal plate, the tendency of building columns to punch through footings, and the tendency of a tensile axial load on a bolt to pull the shank through the head. Under a punching shear load, the significant stress is the average shearing stress on the surface described by the periphery of the punching member and the thickness of the punched member;[2] for example, the shaded cylindrical area of Fig. 1–10.

The following example will serve to illustrate the concepts discussed in Arts. 1–3 and 1–4.

FIG. 1–10

---

[2] This is modified in the case of building columns on concrete footings. Reference should be made to reinforced-concrete design specifications.

Note: In all problems in this textbook the weights of the members will be neglected, unless otherwise specified.

EXAMPLE 1–1

A hoist is made from a 4 × 4-in. wood post *BC* and a 1-in. diameter steel eyebar *AB* as shown in Fig. 1–11*a*. The hoist supports a load *W* of 7000 lb. Determine the axial stresses in members *AB* and *BC* and the cross shearing stress in the 1-in. diameter bolt at *A* which is in double shear.[3]

SOLUTION

The members *AB* and *BC* are two-force members subjected to axial loading. When the equations of equilibrium are applied to the free-body diagram of Fig. 1–11*b*, the following results are obtained:

$$\curvearrowleft + \qquad \Sigma M_C = 0$$

$$7000(5) - (5/13)(F_A)(7) = 0, \qquad F_A = 13{,}000 \text{ lb } T$$

(a)

(b)

(c)

FIG. 1–11

[3] Structural timbers specified as rough sawn will be approximately full size. Cross-sectional dimensions of dressed timbers will be between $\frac{3}{8}$ and $\frac{1}{2}$ in. smaller than nominal. For convenience, all timbers in this book will be considered full size.

$$\sigma_{AB} = \frac{F_A}{A} = \frac{13,000}{(\pi/4)(1)^2} = 16,550 \text{ psi } T \qquad \text{Ans.}$$

Observe that it is not necessary to solve the equation explicitly for the force F_A in the member if only the stress is required. Hence, for the next equation this step will be omitted; thus,

$$(\circlearrowleft+) \qquad \Sigma M_A = 0$$

$$7000(12) - \frac{1}{\sqrt{2}}(16)(\sigma_{BC})(7) = 0$$

$$\sigma_{BC} = \underline{1061 \text{ psi } C} \qquad \text{Ans.}$$

From the free-body diagram of Fig. 1–11c, which is a free-body diagram of the connection at A, and using the value of force F_A as determined above, the cross shearing stress is obtained as follows:

$$(\oplus\rightarrow) \qquad \Sigma F_x = 0$$

$$F_A - 2V = 0$$

$$V = \tau\pi(1)^2/4 = 6500$$

$$\tau = \underline{8280 \text{ psi}} \qquad \text{Ans.}$$

Stress, being the intensity of internal force, has the dimensions of force per unit area (FL^{-2}). Until recently, the commonly used unit for stress in the United States was the pound per square inch, abbreviated as psi. Since metals can sustain stresses of several thousand pounds per square inch, the unit ksi (kips per square inch) is also frequently used (1 ksi = 1000 psi). With the advent of the International System of Units (usually called SI units), units of stress based on the metric system are beginning to be used in the United States and will undoubtedly come into wider use in the next few years.

During the transition, both systems will be encountered, so some problems in this book are given using the English system (pounds and inches) and others with SI units (newtons and meters). For problems with SI units, forces will be given in newtons (N) or kilonewtons (kN), dimensions in meters (m) or millimeters (mm), and masses in kilograms (kg). The SI unit for stress or pressure is a newton per square meter (N/m²), also known as a pascal (Pa). Stress magnitudes normally encountered in engineering applications are expressed in meganewtons per square meter (MN/m²), or megapascals (MPa).

Data will usually be given in either the English or the SI system; a combination of the two systems will not be used for a given problem. When it is necessary to convert units from one system to the other, the following conversion factors will be useful:

1 lb = 4.448 N	1 N = 0.2248 lb
1 in. = 25.40 mm	1 mm = 0.03937 in.
1 ft = 0.3048 m	1 m = 3.281 ft
1 ksi = 6.895 MPa	1 MPa = 0.1450 ksi

[handwritten:] 1 MPa = 10⁶ N/m²

[handwritten:] 1 ksi = 1000 psi

[handwritten:] 1 MPa = 1/7 ksi (Rule of thumb)

Note that one MPa is approximately one-seventh of one ksi. This one-seventh "rule of thumb" is convenient for interpreting the physical significance of answers in SI units for those more accustomed to English units.

Acceleration due to gravity is 9.81 meters per second per second. From Newton's second law, the force due to gravitational attraction on a mass of 1 kg is 9.81 N. Consequently, when a 1-kg mass is supported by a cord in the earth's gravitational field, the tension in the cord is 9.81 N. Similarly the tension in a cable sustaining a mass of m kg is 9.81 m N. A more detailed discussion of SI units is given in App. E.

EXAMPLE 1–2

The crane structure shown in Fig. 1–12a supports an automobile of mass 1800 kg. If member AB is a steel bar 15 mm in diameter,

(a)

(b)

[handwritten:] 1 MPa = .1450 KSI

FIG. 1–12

(a) Determine the force transmitted by member AB.

(b) Determine the axial stress in member AB.

SOLUTION

The tension in the cable at D is $(9.81)(1800)$ or $17,660$ N. From the free-body diagram of Fig. 1–12b, the force in member AB is obtained by summing moments about pin F. Thus,

(a) $$\Sigma M_F = (17,660)(8) - 6F_{AB} = 0$$

Solving for F_{AB} yields

$$F_{AB} = (17,660)(8/6) = 23,547 \text{ N or } \underline{23.5 \text{ kN}}$$ Ans.

(b) $$\sigma = F_{AB}/A = 23,547/(\pi/4)(0.015)^2 = 133.2(10^6) \text{ N/m}^2$$
$$= \underline{133.2 \text{ MPa } T}$$ Ans.

PROBLEMS

> *Note: Unless otherwise specified, all members are assumed to have negligible weight, and all pins used for connections are assumed to be smooth. An asterisk on the problem number indicates that the answer is given in the back of the book.*

1-1*   The loads of Fig. P1–1 are axially applied at sections A, B, C, and D. If the bar has a cross-sectional area of 3 sq in., determine the axial stress in the bar

(a) On a section 10 in. to the right of A.

(b) On a section 60 in. to the right of A.

1-2*   The round bar of Fig. P1–2 has steel, brass, and aluminum sections. Axial loads are applied at cross sections A, B, C, and D. If the allowable

FIG. P1–1

FIG. P1–2

A Steel B Brass C Aluminum D

270 kN → ← 245 kN — 200 kN → ← 225 kN

0.5 m   0.5 m   0.5 m

52.4   *21.3*   *58.1*

axial stresses are 125 MPa in the steel, 70 MPa in the brass, and 85 MPa in the aluminum, determine the required diameters for each of the sections.

1-3* A 2024-T4 aluminum tube with a wall thickness of 0.120 in. and an outside diameter of 1.000 in. is used as a compression member. Determine the axial load on the member when the axial stress is 30 ksi C.

1-4* Axial loads are applied with rigid bearing plates to the concrete structure shown in Fig. P1–4. The cross-sectional areas for the concrete sections are 15,000 mm² for AB, 3000 mm² for BC, and 10,000 mm² for CD. Determine the axial stress in the concrete

(a) On a cross section in the interval AB.

(b) On a cross section in the interval BC.

(c) On a cross section in the interval CD.

1-5* A punch press is used to punch a 1.25-in. diameter hole in a 0.50-in. thick steel plate. Determine the force exerted by the press on the plate if the average punching shear resistance of the steel plate is 50,000 psi.

1-6* Two strips of a plastic material are bonded together, as shown in Fig. P1–6. The average shearing stress in the glue must be limited to 500 kPa.

All loads are in kN

80

D

60   60

Rigid bearing plates

C

85   85

B

120   120

A

80 +120 −130 +240 = 220 kn
804 120 −130 = 90 kn
80 +120 = 200 kn
≈ 270

FIG. P1–4

FIG. P1–6

What length of splice plate is needed if the axial load carried by the joint is 25 kN?

1-7   A brass tube with an outside diameter of 2.00 in. and a wall thickness of 0.375 in. is connected to a steel tube with an inside diameter of 2.00 in. and a wall thickness of 0.250 in. by using a 0.750-in. diameter pin as shown in Fig. P1–7. Determine the cross shearing stress in the pin when the joint is carrying an axial load of 8000 lb.

1-8   If the pin of Problem 1-7 is replaced by a glued joint, determine the length of joint required if the shearing stress in the glue must be limited to 300 psi.

1-9*   A device for determining the shearing strength of wood is shown in Fig. P1–9. The dimensions of the wood specimen are 150 mm by 150 mm by 50 mm. If a load of 50 kN is required to fail the specimen, determine the shearing strength of the wood.

1-10   The steel pipe column shown in Fig. P1–10 has an outside diameter of 4 in. and a wall thickness of 1/2 in. The load imposed on the column by the timber beam is 25,000 lb. Determine

(a)   The average bearing stress at the surface between the column and the steel bearing plate.

(b)   The length of a side of a square bearing plate if the average bearing stress between the steel plate and the wood beam is not to exceed 400 psi.

FIG. P1–7

FIG. P1–9

FIG. P1–10

1-11   The tie rod shown in Fig. P1–11 has a diameter of 40 mm and is used to resist the lateral pressure against the walls of a grain bin. If the tensile stress in the rod is 70 MPa, determine

(a)   The diameter of the washer that must be used if the bearing stress on the wall is not to exceed 3.0 MPa.

(b)   The average punching shear stress in the head of the rod if the head has a diameter of 80 mm and a thickness of 60 mm.

1-12*   The inclined member *AB* of a timber truss is framed into a 4 × 6-in. douglas fir bottom chord, as shown in Fig. P1–12. Determine the axial compressive force in member *AB* when the average shearing stress parallel to the grain in the end of the bottom chord is 150 psi.

1-13*   Member *AD* of the timber truss of Fig. P1–13 is framed into the 100 × 150-mm white oak chord *ABC* as shown in the insert. Determine the dimension *a* that must be used if the average shearing stress parallel to the grain in the end of *AB* is not to exceed 1.50 MPa.

FIG. P1–11

FIG. P1–12

FIG. P1–13

1-14   A lever is attached to the steel gate valve operating shaft of Fig. P1–14 by means of a 1/2 × 1/2 × 3/4-in. key. Determine the average cross shearing stress in the key when a load of 200 lb is applied to the lever.

1-15   A vertical shaft is supported by a thrust collar and bearing plate, as shown in Fig. P1–15. Determine the maximum axial load that can be applied to the shaft if the average punching shear stress in the collar and the average bearing stress between the collar and the plate are limited to 60 and 75 MPa, respectively.

FIG. P1–14

FIG. P1–15

FIG. P1–16

1-16* The tie rod *AB* of Fig. P1–16 has a cross-sectional area of 1.00 sq. in. Pins *A* and *B* have diameters of 1.50 in. Pin *C* has a diameter of 2.00 in. Determine

  (a) The axial stress in tie rod *AB*.

  (b) The cross shearing stress in pin *A*.

  (c) The cross shearing stress in pin *C*.

1-17* The axial stresses are 12 MPa *C* in the wood post *B* and 150 MPa *T* in the steel bar *A* of Fig. P1–17. Determine

  (a) The load *P*.

  (b) The minimum diameter for pin *C* if it is in single shear and the cross shearing stress is limited to 70 MPa.

  (c) The minimum diameter for pin *D* if it is in double shear and the cross shearing stress is limited to 70 MPa.

1-18 The turnbuckles in Fig. P1–18 are tightened until the compression block *D* exerts a force of 25,000 lb on the beam at *B*. Member *D* is a hollow tube with an inside diameter of 1.00 in. and an outside diameter of 2.00 in.

FIG. P1–17

FIG. P1–18

Rods AE and EC each have cross-sectional areas of 2.00 sq. in. The pins at A and C each have a diameter of 1.25 in. Determine

    (a)   The axial stress in member D.

    (b)   The axial stress in rod AE.

    (c)   The cross shearing stress in pin A.

1-19   The pin-connected frame ABC of Fig. P1–19 is subjected to a load P of 450 kN. Member AC is a rectangular bar with a depth of 80 mm and a thickness of 25 mm. Determine

    (a)   The axial stress in member AC.

    (b)   The cross shearing stress in pin B if it has a diameter of 50 mm and is loaded in double shear.

1-20* Member GF of the pin-connected truss of Fig. P1–20 has a cross-sectional area of 2.4 sq. in. Determine the axial stress in member GF.

1-21 For the pin-connected truss of Fig. P1–20, determine

    (a)   The axial stress in member CD if it is a 7/8-in. diameter steel rod.

    (b)   The axial stress in member CF if it is composed of two steel angles having a total cross-sectional area of 1.24 sq. in.

FIG. P1–19

FIG. P1–20

1-22*   Member *GF* of the pin-connected truss of Fig. P1–22 has a cross-sectional area of 4500 mm². Determine the axial stress in member *GF*.

1-23   For the pin-connected truss of Fig. P1–22, determine

   (a)   The axial stress in member *BG* if it is composed of two steel angles having a total cross-sectional area of 2500 mm².

   (b)   The axial stress in member *AB* if it is a 70-mm diameter steel rod.

1-24   For the pin-connected structure of Fig. P1–24, determine the minimum diameter for the pin at joint *D* if the cross shearing stress in the pin must be limited to 7500 psi.

1-25   Member *BE* of the pin-connected structure of Fig. P1–25 is a 25 × 8-mm rectangular bar. The force *P* is applied to the pin at *E* outside members *BE* and *DE* as shown in the insert. The pin at *E* has a diameter of 30 mm. Determine

FIG. P1–22

FIG. P1–24

(a)   The axial stress in member *BE*.

(b)   The maximum cross shearing stress in the pin at *E*.

1-26*   A structural steel bar 1.00 in. in diameter is subjected to an axial tensile load of 12,000 lb. Determine

(a)   The normal and shearing stresses on a plane through the bar that makes an angle of 30° with the direction of the load.

(b)   The maximum normal and shearing stresses in the bar.

1-27*   A structural steel eyebar with a 50 × 150-mm rectangular cross section is subjected to an axial tensile load of 1200 kN. Determine

(a)   The normal and shearing stresses on a plane through the bar that makes an angle of 40° with the direction of the load.

(b)   The maximum normal and shearing stresses in the bar.

1-28   The two parts of the eyebar of Fig. P1–28 are connected with 1/2-in. diameter bolts (one on each side). Specifications for the bolts require that the

FIG. P1–25

Handwritten annotations (left margin):

EM440

$\sigma_n = \frac{P}{A}\cos\theta\cos\theta \quad \theta = 50°$

$\tau_n = \frac{P}{A}\sin\theta\cos\theta$

$\sigma_{max} = \frac{P}{A}$

$\tau_{max} = \frac{P}{2A}$

FIG. P1–28

axial tensile stress not exceed 12.0 ksi and the cross shearing stress not exceed 8.0 ksi. Determine the maximum load P that can be applied to the eyebar without exceeding either specification. The cross-sectional area of the bolts at the root of the threads is 0.126 sq. in.

1-29 The normal stress on plane AB of the rectangular block of Fig. P1–29 is 5.00 MPa C when the load P is applied. If angle θ is 30°, determine

    (a)   The load P.

    (b)   The shearing stress on plane AB.

    (c)   The maximum normal and shearing stresses in the block.

1-30 Specifications for the 3 × 3 × 21-in. block of Fig. P1–30 require that the normal and shearing stresses on plane A–A not exceed 400 psi and 120 psi, respectively. Determine the maximum load P that can be applied without exceeding the specifications.

1-31 Specifications for the wood block of Fig. P1–31 require that the stresses not exceed the following: shear parallel to the grain 0.75 MPa, compression perpendicular to the grain 1.20 MPa, and maximum shear 1.00

Side   End   FIG. P1–29

FIG. P1–30

FIG. P1–31

MPa. Determine the maximum axial load P that can be applied without exceeding the specifications.

1-5   CONCEPT OF STRESS AT A GENERAL POINT IN AN ARBITRARILY LOADED MEMBER

In Arts. 1–3 and 1–4 the concept of stress was introduced by considering the internal force distribution required to satisfy equilibrium in a portion of a bar under centric load. The nature of the force distribution led to uniformly distributed normal and shearing stresses on transverse planes through the bar. In more complicated structural members or machine components the stress distributions will not be uniform on arbitrary internal planes; therefore, a more general concept of the state of stress at a point is needed.

Consider a body of arbitrary shape that is in equilibrium under the action of a system of applied forces. The nature of the internal force distribution at an arbitrary interior point O can be studied by exposing an interior plane through O as shown in Fig. 1–13a. The force distribution required on such an interior plane to maintain equilibrium of the isolated part of the body, in general, will not be uniform; however, any distributed force acting on the small area ΔA surrounding the point of interest O can be replaced by a statically equivalent resultant force ΔF_n through O and a couple ΔM_n. The subscript n indicates that the resultant force and couple are associated with a particular plane through O—namely, the one having an outward normal in the n direction at O. For any other plane through O the values of ΔF and ΔM could be different. Note that the line of action of ΔF_n or ΔM_n may not coincide with the direction of n. If the resultant force ΔF_n is divided by the area ΔA, an average force per unit area (average resultant stress) is obtained. As the area ΔA is made smaller and smaller, the couple ΔM_n vanishes as the force distribution becomes more and more uniform. In the limit a quantity known as the *stress vector*[4] or resultant stress is obtained. Thus,

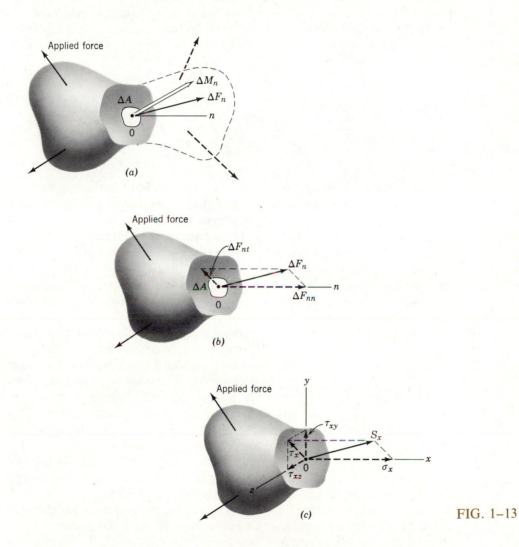

FIG. 1–13

[4] The component of a tensor on a plane is a vector; therefore, on a particular plane, the stresses can be treated as vectors.

$$S_n = \lim_{\Delta A \to 0} \frac{\Delta F_n}{\Delta A}$$

In Art. 1–3 it was pointed out that materials respond to components of the stress vector rather than the stress vector itself. In particular, the components normal and tangent to the internal plane were important. As shown in Fig. 1–13b the resultant force ΔF_n can be resolved into the components ΔF_{nn} and ΔF_{nt}. A normal stress σ_n and a shearing stress τ_n are then defined as

$$\sigma_n = \lim_{\Delta A \to 0} \frac{\Delta F_{nn}}{\Delta A}$$

and

$$\tau_n = \lim_{\Delta A \to 0} \frac{\Delta F_{nt}}{\Delta A}$$

For purposes of analysis it is convenient to reference stresses to some coordinate system. For example, in a Cartesian coordinate system the stresses on planes having outward normals in the x, y, and z directions are usually chosen. Consider the plane having an outward normal in the x direction. In this case the normal and shear stresses on the plane will be σ_x and τ_x, respectively. Since τ_x, in general, will not coincide with the y or z axes, it must be resolved into the components τ_{xy} and τ_{xz}, as shown in Fig. 1–13c.

Unfortunately the state of stress at a point in a material is not completely defined by these three components of the stress vector since the stress vector itself depends on the orientation of the plane with which it is associated. An infinite number of planes can be passed through the point, resulting in an infinite number of stress vectors being associated with the point. Fortunately it can be shown (see Art. 1–9) that the specification of stresses on three mutually perpendicular planes is sufficient to describe completely the state of stress at the point. The rectangular components of stress vectors on planes having outward normals in the coordinate directions are shown in Fig. 1–14. The six faces of the small element are denoted by the directions of their outward normals so that the positive x face is the one whose outward normal is in the direction of the positive x axis. The coordinate axes x, y, and z are arranged as a right-hand system.

The sign convention for stresses is as follows. Normal stresses (indicated by the symbol σ and a single subscript to indicate the plane on which the stress acts) are positive if they point in the direction of the outward normal. Thus, normal stresses are positive if tensile. Shearing stresses are denoted by the symbol τ followed by two subscripts; the first subscript designates the plane on which the shearing stress acts and the second the coordinate axis to which it is parallel. Thus, τ_{xz} is the shearing stress on an x

FIG. 1–14   FIG. 1–15

$$\tau_{xy} = \tau_{yx}$$
$$\tau_{yz} = \tau_{zy}$$
$$\tau_{zx} = \tau_{xz}$$

plane parallel to the z axis. A positive shearing stress points in the positive direction of the coordinate axis of the second subscript if it acts on a surface with an outward normal in the positive direction. Conversely, if the outward normal of the surface is in the negative direction, then the positive shearing stress points in the negative direction of the coordinate axis of the second subscript. The stresses shown on the element in Fig. 1–14 are all positive.

1-6 TWO-DIMENSIONAL OR PLANE STRESS

Before proceeding to a consideration of the general three-dimensional state of stress, we can gain considerable insight into the nature of stress distributions by considering a less complicated state known as two-dimensional or *plane stress*. For this case two parallel faces of the small element of Fig. 1–14 are assumed to be free of stress. For purposes of analysis, let these faces be perpendicular to the z axis. Thus,

$$\sigma_z = \tau_{zx} = \tau_{zy} = 0$$

From Eq. 1–3, however, this also implies that

$$\tau_{xz} = \tau_{yz} = 0$$

The components of stress present for plane stress analysis will be σ_x, σ_y, and $\tau_{xy} = \tau_{yx}$. For convenience this state of stress is usually represented by the two-dimensional sketch shown in Fig. 1–15. The three-dimensional element of which the sketch is a plane projection should be kept in mind at all times. Normal and shearing stresses on an arbitrary plane such as plane A–A in Fig. 1–15 can be obtained by using the free-body diagram method discussed for centric loading in Art. 1–4. The solution to the following example illustrates this method of approach under plane stress conditions.

EXAMPLE 1–3

At a given point in a machine element, the following stresses were evaluated: 8000 psi T and zero shear on a horizontal plane and 4000 psi C on a vertical plane. Determine the stresses at this point on a plane having a slope of 3 vertical to 4 horizontal.

SOLUTION

As an aid to visualization of the data, it is suggested that the differential block of Fig. 1–16a be drawn (a stress picture, not a free-body diagram). The next step is to draw a free-body diagram subjected to ordinary force vectors. Various satisfactory free-body diagrams can be used, such as the wedge-shaped element defined by the three given planes in Fig. 1–16b, in which the shaded area indicates the plane on which the stresses to be evaluated are acting. To this area is assigned arbitrarily the magnitude dA, and the corresponding areas of the horizontal and vertical faces of the element are 0.8 dA and 0.6 dA, respectively. The forces acting on these areas will be as indicated in the diagram. It must be emphasized that *force vectors* act on these areas. For practical purposes, the two-dimensional free-body diagram of

FIG. 1–16   (b)   (c)

Fig. 1–16c will be adequate. Summing forces in the n direction yields the following:

$$\text{(+↘)} \quad \Sigma F_n = 0$$

$$\sigma\, dA + 0.6\, dA\,(4000)(0.6) - 0.8\, dA\,(8000)(0.8) = 0$$

from which

$$\sigma = \underline{3680 \text{ psi } T} \qquad\qquad \text{Ans.}$$

When forces are summed in the t direction, the shearing stress is found to be

$$\tau = \underline{5760 \text{ psi}} \qquad\qquad \text{Ans.}$$

Note that the normal stress should be designated as *tension* or *compression*. The presence of shearing stresses on the horizontal and vertical planes, had there been any, would merely have required two more forces on the free-body diagram: one parallel to the vertical face and one parallel to the horizontal face. Note, however, that the magnitude of the shearing stresses (not the forces) must be the same on any two orthogonal planes (see Art. 1–4) and that the vectors should be directed as in Fig. 1–9 or both vectors reversed, depending on the given data.

PROBLEMS

1-32* Because of internal pressure in a boiler, the stresses at a point in the boiler plate were found to be as shown in the stress picture of Fig. P1–32. Determine the normal and shearing stresses at this point on the inclined plane shown in the figure.

1-33* At a point in a stressed body, there are normal stresses of 95 MPa T on a vertical plane and 150 MPa T on a horizontal plane as shown in Fig. P1–33. Determine the normal and shearing stresses at this point on the inclined plane shown in the figure.

FIG. P1–32   FIG. P1–33

FIG. P1–34   FIG. P1–35

1-34*   The stresses on horizontal and vertical planes at a point are shown in Fig. P1–34. The normal stress on plane $A–A$ at this point is 8000 psi T. Determine

    (a)   The magnitude of the shearing stresses τ_h and τ_v.

    (b)   The magnitude and direction of the shearing stress on the inclined plane $A–A$.

1-35*   The stresses on horizontal and vertical planes at a point are shown in Fig. P1–35. The normal stress on plane AB is 15 MPa T. Determine

    (a)   The normal stress σ_x on the vertical plane.

    (b)   The magnitude and direction of the shearing stress on the inclined plane AB.

1-36   At a point in a structural member, there are stresses on horizontal and vertical planes as shown in Fig. P1–36. Determine the normal and shearing stresses on inclined plane AB.

1-37   At a point in a structural member, there are stresses on horizontal and vertical planes as shown in Fig. P1–37. The normal stress on the inclined plane AB is 25 MPa T. Determine

    (a)   The normal stress σ_x on the vertical plane.

    (b)   The magnitude and direction of the shearing stress on the inclined plane AB.

FIG. P1–36   FIG. P1–37

FIG. P1–38   FIG. P1–39

1-38* At a point in a machine component, the stresses on an inclined plane are σ = 2400 psi T and τ = 600 psi as shown in Fig. P1–38. The normal stress on a vertical plane through the point is zero. Determine

(a) The shearing stresses on horizontal and vertical planes.

(b) The normal stress on a horizontal plane through the point.

1-39* An incomplete free-body diagram for the determination of stresses at a point in a machine component is shown in Fig. P1–39. Complete the diagram and determine the normal and shearing stresses on plane AC. The stresses indicated on the diagram are presented in MPa.

1-40 At a point in a structural member, there are stresses on horizontal and vertical planes as shown in Fig. P1–40. The magnitude of the compressive stress σ_c is three times the magnitude of the tensile stress σ_t. Specifications require that the shearing stress on plane AB not exceed 4000 psi, and the normal stress on plane AB not exceed 7800 psi. Determine the maximum value of stress σ_c that will satisfy the specifications.

1-41 Known stresses at point A in a structural member (see Fig. P1–41) are 150 MPa T and zero shear on plane B–B and 250 MPa C on plane C–C. Determine the stresses on a vertical plane at the point.

1-42 Determine the stresses on a horizontal plane at point A of Prob. 1-41.

FIG. P1–40

FIG. P1–41

1-7 PRINCIPAL STRESSES—PLANE STRESS

As mentioned in Art. 1–3, a thorough understanding of the physical significance of stress is fundamental in mechanics of materials. The concept of stress (internal force per unit area) at a point in a body on a plane through the point must be kept constantly in mind. The fact that the stresses are, in general, different for different planes through any one point has been emphasized previously.

Three important facts about stresses were developed in Art. 1–4:

1. The maximum normal stress for a centric tensile or compressive load in one direction is in the direction of the load on a plane normal to the load, and its magnitude is $\sigma = P/A$.
2. The maximum shearing stress for a centric tensile or compressive load in one direction is on a plane at an angle of 45° to the load, and its magnitude is $\tau = \sigma/2 = P/(2A)$.
3. For all types of loads, a shearing stress at any one point on any plane is accompanied by a shearing stress of equal magnitude on an orthogonal plane through the point.

Once the normal and shearing stresses on orthogonal planes through a point in a body are known, the normal and shearing stresses on any specific oblique plane through the same point can be determined by the free-body diagram method of Art. 1–6. Usually, the critical stresses are the maximum tensile and compressive stresses and the maximum shearing stresses. Therefore, the method of Art. 1–6 is inadequate, unless the slopes of the planes on which the maximum stresses act are known. For a centric load in one direction, the planes on which maximum and minimum normal stresses and maximum shearing stresses act are readily available from the developments of Art. 1–4. As more complicated loads are studied, more general methods for finding the critical stresses become necessary. Two such methods will be presented.

Equations relating the desired normal and shearing stresses σ_n and τ_{nt} on an arbitrary plane through a point oriented at an angle θ with respect to a reference x axis and the known stresses σ_x, σ_y, and $\tau_{xy} = \tau_{yx}$ on the reference planes can be developed using the free-body diagram method of Art. 1–6. Consider the plane stress situation indicated in Fig. 1–17a, where the dotted line A–A represents the trace of any plane through the point (all planes are perpendicular to the plane of zero stress—the plane of the paper). In the following derivation, a counterclockwise angle θ is positive.

Figure 1–17b is a free-body diagram of a wedge-shaped element in which the areas of the faces are dA for the inclined face (plane A–A), $dA \cos \theta$ for the vertical face, and $dA \sin \theta$ for the horizontal face. The force equa-

FIG. 1–17

tion of equilibrium in the n direction gives

$$\sigma_n dA - \tau_{yx}(dA \sin\theta)\cos\theta - \tau_{xy}(dA \cos\theta)\sin\theta - \sigma_x(dA \cos\theta)\cos\theta$$
$$- \sigma_y(dA \sin\theta)\sin\theta = 0$$

from which, since $\tau_{yx} = \tau_{xy}$,

$$\sigma_n = \sigma_x \cos^2\theta + \sigma_y \sin^2\theta + 2\tau_{xy} \sin\theta \cos\theta \qquad (1-4a)$$

or, in terms of the double angle,

$$\sigma_n = \frac{\sigma_x(1 + \cos2\theta)}{2} + \frac{\sigma_y(1 - \cos2\theta)}{2} + \frac{2\tau_{xy}(\sin2\theta)}{2}$$

$$= \frac{\sigma_x + \sigma_y}{2} + \frac{\sigma_x - \sigma_y}{2} \cos2\theta + \tau_{xy} \sin2\theta \qquad (1-4b)$$

For the free-body diagram of Fig. 1–17b, the force equation of equilibrium in the t direction gives

$$\tau_{nt}dA - \tau_{xy}(dA\ \cos\theta)\cos\theta + \tau_{yx}(dA\ \sin\theta)\sin\theta + \sigma_x(dA\ \cos\theta)\sin\theta$$
$$- \sigma_y(dA\ \sin\theta)\cos\theta = 0$$

from which

$$\tau_{nt} = -(\sigma_x - \sigma_y)\sin\theta\ \cos\theta + \tau_{xy}(\cos^2\theta - \sin^2\theta) \qquad (1\text{-}5a)$$

or, in terms of the double angle,

$$\tau_{nt} = -\frac{\sigma_x - \sigma_y}{2}\sin2\theta + \tau_{xy}\cos2\theta \qquad (1\text{-}5b)$$

Equations 1–4 and 1–5 provide a means for determining normal and shearing stresses on any plane whose outward normal is perpendicular to the z axis and is oriented at an angle θ with respect to the reference x axis. When these equations are used, the sign conventions used in their development must be rigorously followed; otherwise, erroneous results will be obtained. These sign conventions can be summarized as follows:

1.   Tensile normal stresses are positive; compressive normal stresses are negative. All of the normal stresses shown on Fig. 1–17 are positive.

2.   A shearing stress is positive if it points in the positive direction of the coordinate axis of the second subscript when it is acting on a surface whose outward normal is in a positive direction. Similarly, if the outward normal of the surface is in a negative direction, then a positive shearing stress points in the negative direction of the coordinate axis of the second subscript. All of the shearing stresses shown on Fig. 1–17 are positive. Shearing stresses pointing in the opposite directions would be negative.

3.   An angle measured counterclockwise from the reference x axis is positive. Conversely, angles measured clockwise from the reference x axis are negative.

4.   The (n, t, z) axes have the same order as the (x, y, z) axes. Both sets of axes form a right-hand coordinate system.

Maximum and minimum values of σ_n will occur at the values of θ for which $d\sigma_n/d\theta$ is equal to zero. Differentiation of σ_n with respect to θ yields

$$\frac{d\sigma_n}{d\theta} = -(\sigma_x - \sigma_y)\sin2\theta + 2\tau_{xy}\cos2\theta \qquad (a)$$

Setting Eq. a equal to zero and solving gives

$$\tan2\theta_p = \frac{\tau_{xy}}{(\sigma_x - \sigma_y)/2} \tag{1-6}$$

Note that the expression for $d\sigma_n/d\theta$ from Eq. a is numerically twice the value of the expression for τ_{nt} from Eq. 1–5b. Consequently, the shearing stress is zero on planes experiencing maximum and minimum values of normal stress. Planes free of shear stress are known as *principal planes*. The normal stresses occurring on principal planes are known as *principal stresses*. The values of θ_p from Eq. 1–6 give the orientation of two principal planes. A third principal plane for the plane stress state has an outward normal in the z direction. For a given set of values of σ_x, σ_y, and τ_{xy}, there are two values of $2\theta_p$ differing by 180° and, consequently, two values of θ_p 90° apart. This proves that the principal planes are normal to each other.

When τ_{xy} and $(\sigma_x - \sigma_y)$ have the same sign, $\tan2\theta_p$ is positive and one value of $2\theta_p$ is between 0° and 90°, with the other value 180° greater, as shown in Fig. 1–17c. Consequently, one value of θ_p is between 0° and 45°, and the other one is 90° greater. In the first case, both $\sin2\theta_p$ and $\cos2\theta_p$ are positive, and in the second case both are negative. When these functions of $2\theta_p$ are substituted into Eq. 1–4b, two in-plane principal stresses σ_{p1} and σ_{p2} are found to be

$$\sigma_{p1, p2} = \frac{\sigma_x + \sigma_y}{2} + \frac{\sigma_x - \sigma_y}{2} \left[\pm \frac{(\sigma_x - \sigma_y)/2}{\sqrt{\left(\dfrac{\sigma_x - \sigma_y}{2}\right)^2 + \tau_{xy}^2}} \right]$$

$$+ \tau_{xy} \left[\pm \frac{\tau_{xy}}{\sqrt{\left(\dfrac{\sigma_x - \sigma_y}{2}\right)^2 + \tau_{xy}^2}} \right]$$

which reduces to

$$\sigma_{p1, p2} = \frac{\sigma_x + \sigma_y}{2} \pm \sqrt{\left(\frac{\sigma_x - \sigma_y}{2}\right)^2 + \tau_{xy}^2} \tag{1-7}$$

Equation 1–7 gives the two principal stresses in the xy plane, and the third one is $\sigma_{p3} = \sigma_z = 0$. Equation 1–6 gives the angles θ_p and $\theta_p + 90°$ between the x (or y) plane and the mutually perpendicular planes on which the principal stresses act. When $\tan2\theta_p$ is positive, θ_p is positive, and the rotation is counterclockwise from the x and y planes to the planes on which the two principal stresses act. When $\tan2\theta_p$ is negative, the rotation is clockwise. Note that one value of θ_p will always be between positive and negative 45° (inclusive) with the other value 90° greater. The numerically greater principal stress will act on the plane that makes an angle of 45° or less with the plane of the numerically larger of the two given normal stresses. This statement can be confirmed in any case by substitution of the value of θ_p from Eq. 1–6 into Eq. 1–4b to obtain the corresponding principal stress.

Note that if one or both of the principal stresses from Eq. 1–7 is negative, the algebraic maximum stress can have a smaller absolute value than the "minimum" stress.

The *maximum in-plane shearing stress* τ_p occurs on planes located by values of θ where $d\tau_{nt}/d\theta$ is equal to zero. Differentiation of Eq. 1–5b yields

$$\frac{d\tau_{nt}}{d\theta} = -(\sigma_x - \sigma_y)\cos2\theta - 2\tau_{xy}\sin2\theta$$

When $d\tau_{nt}/d\theta$ is equated to zero, the value of θ_τ is given by the expression

$$\tan2\theta_\tau = -\frac{(\sigma_x - \sigma_y)/2}{\tau_{xy}} \tag{b}$$

Comparison of Eqs. b and 1–6 reveals that the two tangents are negative reciprocals. Therefore, the two angles $2\theta_p$ and $2\theta_\tau$ differ by 90°, and θ_p and θ_τ are 45° apart. This means that the planes on which the maximum in-plane shearing stresses occur are 45° from the principal planes. The maximum in-plane shearing stresses are found by substituting values of angle functions obtained from Eq. b in Eq. 1–5b. The results are

$$\tau_p = -\frac{\sigma_x - \sigma_y}{2}\left[\mp\frac{(\sigma_x - \sigma_y)/2}{\sqrt{\left(\dfrac{\sigma_x - \sigma_y}{2}\right)^2 + \tau_{xy}^2}}\right]$$

$$+ \tau_{xy}\left[\pm\frac{\tau_{xy}}{\sqrt{\left(\dfrac{\sigma_x - \sigma_y}{2}\right)^2 + \tau_{xy}^2}}\right]$$

which reduces to

$$\tau_p = \pm\sqrt{\left(\frac{\sigma_x - \sigma_y}{2}\right)^2 + \tau_{xy}^2} \tag{1–8}$$

Equation 1–8 has the same magnitude as the second term of Eq. 1–7.

A useful relation between the principal stresses and the maximum in-plane shearing stress is obtained from Eqs. 1–7 and 1–8 by subtracting the values for the two in-plane principal stresses and substituting the value of the radical from Eq. 1–8. The result is

$$\tau_p = (\sigma_{p1} - \sigma_{p2})/2$$

or, in words, the maximum value of τ_{nt} is equal in magnitude to one-half the difference between the two in-plane principal stresses.

In general, when stresses act in three directions it can be shown (see Art. 1–9) that there are three orthogonal planes on which the shearing stress is zero. These planes are known as the principal planes, and the stresses acting on them (the principal stresses) will have three values: one maximum, one minimum, and a third stress between the other two. The maximum shearing stress, τ_{max} on any plane that could be passed through the point, is one-half the difference between the maximum and minimum principal stresses and acts on planes that bisect the angles between the planes of the maximum and minimum normal stresses.

$$\tau_{max} = \frac{\sigma_{max} - \sigma_{min}}{2} \qquad (1-9)$$

When a state of plane stress exists, one of the principal stresses is zero. If the values of σ_{p1} and σ_{p2} from Eq. 1–7 have the same sign, then the third principal stress, σ_{p3} equals zero, will be either the maximum or the minimum normal stress. Thus, the maximum shearing stress may be $(\sigma_{p1} - \sigma_{p2})/2$, $(\sigma_{p1} - 0)/2$, or $(\sigma_{p2} - 0)/2$, depending on the relative magnitudes and signs of the principal stresses. These three possibilities are illustrated in Fig. 1–18, in which one of the two orthogonal planes on which the maximum shearing stress acts is hatched for each example.

The direction of the maximum shearing stress can be determined by drawing a wedge-shaped block with two sides parallel to the planes having the maximum and minimum principal stresses, and with the third side at an angle of 45° with the other two sides. The direction of the maximum shearing stress must oppose the larger of the two principal stresses.

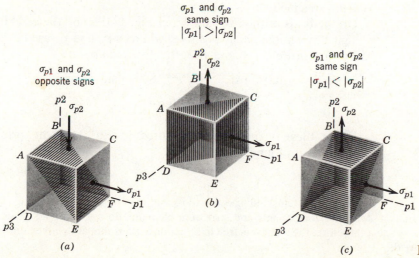

FIG. 1–18

Another useful relation between the principal stresses and the normal stresses on the orthogonal planes, shown in Fig. 1–17, is obtained by adding the values for the two principal stresses as given by Eq. 1–7. The result is

$$\sigma_{p1} + \sigma_{p2} = \sigma_x + \sigma_y \qquad\qquad (c)$$

or, in words, *for plane stress, the sum of the normal stresses on any two orthogonal planes through a point in a body is a constant or invariant.*

In the preceding discussion, "maximum" and "minimum" stresses were considered as algebraic quantities, and it has already been pointed out that the minimum algebraic stress may have a larger magnitude than the maximum stress. However, in the application to engineering problems (which includes the problems in this book), the term *maximum* will always refer to the largest absolute value (largest magnitude).

The application of the formulas and procedures developed in this article are illustrated by the following examples.

EXAMPLE 1–4
At a point in a structural member subjected to plane stress there are stresses on horizontal and vertical planes through the point, as shown in Fig. 1–19a.

   (a)   Determine the principal stresses and the maximum shearing stress at the point.

   (b)   Locate the planes on which these stresses act and show the stresses on a complete sketch.

SOLUTION
   (a)   First define the x–y and n–t (if needed) directions if they have not been specified.[5]

On the basis of the axes shown in Fig. 1–19a and the established sign conventions, σ_x is positive, whereas σ_y and τ_{xy} are negative. For use in Eqs. 1–6 and 1–7, the given values are

$$\sigma_x = +10,000 \text{ psi}, \qquad \sigma_y = -8000 \text{ psi}, \qquad \text{and}$$

$$\tau_{xy} = -4000 \text{ psi}$$

When these values are substituted in Eq. 1–7, the principal stresses are found to be

---

[5] Coordinate systems and axes are tools that may be used to advantage by the analyst. Machine components and structural elements do not come inscribed with x–y axes; therefore, the analyst is free to select directions that are convenient for his work.

FIG. 1–19

(a)

(b)

(c)

$$\sigma_p = \frac{10,000 - 8000}{2} \pm \sqrt{\left(\frac{10,000 + 8000}{2}\right)^2 + (-4000)^2}$$

$$= 1000 \pm 9850$$

$$\sigma_{p1} = \underline{10,850 \text{ psi}} = \underline{10,850 \text{ psi } T} \qquad\qquad \text{Ans.}$$

$$\sigma_{p2} = \underline{-8850 \text{ psi}} = \underline{8850 \text{ psi } C} \qquad\qquad \text{Ans.}$$

$$\sigma_{p3} = \sigma_z = \underline{0} \qquad\qquad \text{Ans.}$$

Since σ_{p1} and σ_{p2} are of opposite sign, the maximum shearing stress is

$$\tau_{max} = \frac{10,850 - (-8850)}{2} = \underline{9850 \text{ psi}} \qquad\qquad \text{Ans.}$$

(b) When the given data are substituted in Eq. 1–6, the results are

$$\tan 2\theta_p = -4000/9000 = -0.4444$$

from which

$$\frac{\tau_{xy}}{\frac{(\sigma_x - \sigma_y)}{2}}$$

$$2\theta_p = -23.96°$$

and

$$\theta_p = -11.98° \text{ or } 12° \text{ ↓}$$

The required sketch is shown in Fig. 1–19b or c.

From Eq. c, the sum of the normal stresses on any two orthogonal planes is a constant for plane stress. Planes *CB* and *DE* of Fig. 1–19 are orthogonal, and the normal stresses on them are obviously equal. Therefore,

$$2\sigma_n = \sigma_x + \sigma_y = 10,000 - 8000 = 2000$$

and

$$\sigma_n = 1000 \text{ psi on the planes of maximum shear}$$

EXAMPLE 1–5

At a point in a structural member subjected to plane stress there are stresses on horizontal and vertical planes through the point, as shown in Fig. 1–20a.

(a) Determine the principal stresses and the maximum shearing stress at the point.

(b) Locate the planes on which these stresses act and show the stresses on a complete sketch.

SOLUTION

(a) On the basis of the axes shown in Fig. 1–20a and the established sign conventions, σ_x is positive, σ_y is positive, and τ_{xy} is negative. For use in Eqs. 1–6 and 1–7, the given values are

$$\sigma_x = +100 \text{ MPa}, \qquad \sigma_y = +80 \text{ MPa}, \qquad \text{and}$$

$$\tau_{xy} = -40 \text{ MPa}$$

When these values are substituted in Eq. 1–7, the principal stresses are found to be

$$\sigma_p = \frac{100 + 80}{2} \pm \sqrt{\left(\frac{100 - 80}{2}\right)^2 + (-40)^2}$$

$$= 90 \pm 41.23$$

FIG. 1–20

$$\sigma_{p1} = \underline{131.2 \text{ MPa } T} \qquad\qquad \text{Ans.}$$

$$\sigma_{p2} = \underline{48.8 \text{ MPa } T} \qquad\qquad \text{Ans.}$$

$$\sigma_{p3} = \sigma_z = \underline{0} \qquad\qquad \text{Ans.}$$

Since σ_{p1} and σ_{p2} have the same sign, the maximum shearing stress is

$$\tau_{max} = (131.2 - 0)/2 = \underline{65.6 \text{ MPa}} \qquad \text{Ans.}$$

(b) When the given data are substituted in Eq. 1–6, the results are

$$\tan 2\theta_p = -40/10 = -4$$

from which

$$2\theta_p = -75.97°$$

and

$$\theta_p = -37.985° \text{ or } 38°$$

The maximum shearing stress occurs on the plane making an angle of 45° with the planes of maximum and minimum normal stress—in this case, 131.2 MPa and zero. The complete sketch is given in Fig. 1–20b where the upper (wedge-shaped) block is the orthographic projection of the lower block. The three-dimensional Fig. 1–20c is presented as an aid to the visualization of Fig. 1–20b.

PROBLEMS

1-43* The stresses shown in Fig. P1–43 act at a point in a machine component. Determine the normal and shearing stresses on plane AB by using Eqs. 1–4 and 1–5.

1-44* The stresses shown in Fig. P1–44 act at a point in a structural member. Determine the normal and shearing stresses on plane AB by using Eqs. 1–4 and 1–5.

1-45* The stresses shown in Fig. P1–45 act at a point in a structural member. Determine the normal and shearing stresses on plane AB by using Eqs. 1–4 and 1–5.

1-46* The stresses shown in Fig. P1–46 act at a point in a machine component. Determine the normal and shearing stresses on plane AB by using Eqs. 1–4 and 1–5.

1-47 The stresses shown in Fig. P1–47 act at a point in a machine compo-

FIG. P1–43   FIG. P1–44

FIG. P1–45

6 ksi

15 ksi

18 ksi

65°

B

A

FIG. P1–46

120 MPa

64°

B

20 MPa

80 MPa

A

nent. Determine the normal and shearing stresses on plane *AB* by using Eqs. 1–4 and 1–5.

1-48   The stresses shown in Fig. P1–48 act at a point in a structural member. Determine the normal and shearing stresses on plane *AB* by using Eqs. 1–4 and 1–5.

1-49*   The stresses shown in Fig. P1–49 act at a point in a machine compo-

12 ksi

A

25 ksi

52°

10 ksi

B

FIG. P1–47

125 MPa

B

75 MPa

65 MPa

55°

A

FIG. P1–48

σ_t

τ_{nt}

σ_n

15°

10 ksi

9 ksi

18 ksi

FIG. P1–49

FIG. P1–50

nent. Determine the normal stresses σ_n and σ_t and the shearing stress τ_{nt} at the point.

1-50   The stresses shown in Fig. P1–50 act at a point in a structural member. Determine the normal stresses σ_n and σ_t and the shearing stress τ_{nt} at the point.

1-51*   Determine and show on a sketch the principal and maximum shearing stresses at the point indicated in Fig. P1–43.

1-52*   Determine and show on a sketch the principal and maximum shearing stresses at the point indicated in Fig. P1–44.

1-53   Determine and show on a sketch the principal and maximum shearing stresses at the point indicated in Fig. P1–45.

1-54   Determine and show on a sketch the principal and maximum shearing stresses at the point indicated in Fig. P1–46.

1-55*   Determine and show on a sketch the principal and maximum shearing stresses at the point indicated in Fig. P1–47.

1-56   Determine and show on a sketch the principal and maximum shearing stresses at the point indicated in Fig. P1–48.

1-57*   Determine and show on a sketch the principal and maximum shearing stresses at the point indicated in Fig. P1–49.

1-58   Determine and show on a sketch the principal and maximum shearing stresses at the point indicated in Fig. P1–50.

1-59*   At a point in a stressed body, a normal stress of 16 ksi T and an unknown negative shearing stress exist on a vertical plane. A normal stress of 24 ksi C exists on a horizontal plane. The magnitude of the maximum shearing stress at the point is 25 ksi. Determine and show on a sketch the principal stresses at the point and the unknown shearing stresses on the vertical and horizontal planes.

1-60   At a point in a stressed member, a normal stress of 64 MPa C and an unknown positive shearing stress exist on a horizontal plane. One principal stress at the point is 8 MPa C, and the maximum shearing stress has a magnitude of 95 MPa. Determine and show on a sketch the unknown stresses on the horizontal and vertical planes, and the principal stresses at the point.

1-61   The principal compressive stress on a vertical plane through a point in a wooden block is equal to four times the principal compressive stress on a horizontal plane. The plane of the grain is 30° clockwise from the vertical plane. If the normal and shearing stresses on the plane of the grain must not exceed 250 psi C and 120 psi shear, determine the maximum permissible compressive stress on the horizontal plane.

Problems 1-62 through 1-71 are problems in plane stress in which certain stresses or angles are given. Determine the unknown quantities for each problem and draw a sketch indicating the stresses and planes on which they act. Note that in some problems there might be more than one possible value of θ_p depending on the sign of τ_{xy}.

	Units	σ_x	σ_y	τ_{xy}	σ_{p1}	σ_{p2}	τ_{max}	θ_p
1-62*	ksi	4	12	−6	–	–	–	–
1-63	MPa	−85	40	−40	–	–	–	–
1-64	ksi	6	−10	–	−13	–	–	–
1-65*	MPa	20	−80	–	100	–	–	–
1-66*	ksi	10	–	–	16	–	15	–
1-67	MPa	75	–	–	130	–	80	–
1-68*	ksi	–	–	−12	3	–	–	−30°
1-69	MPa	–	–	21	55	–	–	20°
1-70	ksi	–	−8	−12	–	−14	–	–
1-71*	MPa	–	28	42	–	8	–	–

1-8   MOHR'S CIRCLE FOR PLANE STRESS

The equations for finding the principal stresses and the maximum shearing stresses at a point in a stressed member were developed in Art. 1–7. The German engineer Otto Mohr (1835–1918) developed a useful pictorial or graphic interpretation of these equations.[6] This method, commonly called Mohr's circle,[7] involves the construction of a circle in such a manner that the coordinates of each point on the circle represent the normal and shearing stresses on one plane through the stressed point, and the angular position of the radius to the point gives the orientation of the plane. The proof that normal and shearing components of stress on an arbitrary plane through a point can be represented as points on a circle follows from the general expressions for normal and shearing stresses derived in Art. 1–7. Recall Eqs. 1–4b and 1–5b and rewrite in the following form:

$$\sigma_n - \frac{\sigma_x + \sigma_y}{2} = \frac{\sigma_x - \sigma_y}{2} \cos2\theta + \tau_{xy} \sin2\theta$$

$$\tau_{nt} = -\frac{\sigma_x - \sigma_y}{2} \sin2\theta + \tau_{xy} \cos2\theta$$

Squaring both equations, adding, and simplifying yield

$$\left(\sigma_n - \frac{\sigma_x + \sigma_y}{2}\right)^2 + \tau_{nt}^2 = \left(\frac{\sigma_x - \sigma_y}{2}\right)^2 + \tau_{xy}^2$$

This is the equation of a circle in terms of the variables σ_n and τ_{nt}. The circle is centered on the σ axis at a distance $(\sigma_x + \sigma_y)/2$ from the τ axis, and the radius of the circle is given by

$$R = \sqrt{\left(\frac{\sigma_x - \sigma_y}{2}\right)^2 + \tau_{xy}^2}$$

Normal stresses are plotted as horizontal coordinates, with tensile stresses (positive) plotted to the right of the origin and compressive stresses (negative) plotted to the left. Shearing stresses are plotted as vertical coordinates, with those tending to produce a clockwise rotation of the stress element plotted above the σ axis, and those tending to produce a counterclockwise rotation of the stress element plotted below the σ axis. The method for

[6] The particular case for plane stress was presented by the German engineer K. Culmann (1829–1881) about sixteen years before Mohr's paper.

[7] Mohr's circle is also used as a pictorial interpretation of the formulas for obtaining maximum and minimum second moments of areas and maximum products of inertia. See *Engineering Mechanics,* Higdon and Stiles, Prentice-Hall, Englewood Cliffs, N.J., 1976. Mohr's circle is also used for plane strain, three-dimensional stress, and maximum and minimum moments of inertia of mass.

interpreting the sign to be associated with a particular shear stress value obtained from a Mohr's circle analysis will be illustrated in the discussion and example problems that follow.

The Mohr's circle for any point subjected to plane stress can be drawn when the stresses on two mutually perpendicular planes through the point are known. Consider, for example, the stress situation of Fig. 1–17a with σ_x greater than σ_y and plot in Fig. 1–21 the points representing the given stresses. The coordinates of point V of Fig. 1–21 are the stresses on the vertical plane through the stressed point of Fig. 1–17a, and point H is determined by the stresses on the horizontal plane through the point. Line VH is the diameter of Mohr's circle for the stresses at the point. Because τ_{yx} is equal to τ_{xy}, point C, the center of the circle, is on the σ axis. Line CV on Mohr's circle represents the plane (the vertical plane of Fig. 1–17a) through the stressed point from which the angle θ is measured. The coordinates of each point on the circle represent σ_n and τ_{nt} for one particular plane through the stressed point, the abscissa representing σ_n and the ordinate representing τ_{nt}. To demonstrate this statement, draw any radius CF in Fig. 1–21 at an angle 2θ counterclockwise from radius CV. From the figure, it is apparent that

FIG. 1–21

$$OF' = OC + CF \cos(2\theta_p - 2\theta)$$

and since CF equals CV, the above equation reduces to

$$OF' = OC + CV \cos2\theta_p \cos2\theta + CV \sin2\theta_p \sin2\theta$$

Referring to Fig. 1–21, note that

$$CV \cos2\theta_p = CV' = (\sigma_x - \sigma_y)/2$$

$$CV \sin2\theta_p = VV' = \tau_{xy}$$

and

$$OC = (\sigma_x + \sigma_y)/2 = \sigma_{avg}$$

Therefore,

$$OF' = OC + CV' \cos2\theta + VV' \sin2\theta$$

$$= \frac{\sigma_x + \sigma_y}{2} + \frac{\sigma_x - \sigma_y}{2} \cos2\theta + \tau_{xy} \sin2\theta$$

This expression is identical to Eq. 1–4b of Art. 1–7. Therefore, OF' is equal to σ_n. In a similar manner,

$$F'F = CF \sin(2\theta_p - 2\theta)$$

$$= CV \sin2\theta_p \cos2\theta - CV \cos2\theta_p \sin2\theta$$

$$= V'V \cos2\theta - CV' \sin2\theta$$

$$= \tau_{xy} \cos2\theta - \frac{\sigma_x - \sigma_y}{2} \sin2\theta = \tau_{nt}$$

which is Eq. 1–5b of Art. 1–7.

Since the horizontal coordinate of each point on the circle represents a particular value of σ_n, the maximum normal stress is represented by OD, and its value is

$$\sigma_{p1} = OD = OC + CD = OC + CV$$

$$= \frac{\sigma_x + \sigma_y}{2} + \sqrt{\left(\frac{\sigma_x - \sigma_y}{2}\right)^2 + \tau_{xy}^2}$$

which agrees with Eq. 1–7.

Likewise, the vertical coordinate of each point on the circle represents a particular value of τ_{nt} (called the in-plane shearing stress), which means that the maximum in-plane shearing stresses are represented by CA and CB, and their value is

$$\tau_p = CA = CV = \sqrt{\left(\frac{\sigma_x - \sigma_y}{2}\right)^2 + \tau_{xy}^2}$$

which agrees with Eq. 1–8. As noted in Art. 1–7, if the two nonzero principal stresses have the same sign, the maximum shearing stress at the point will not be in the plane of the applied stresses.

The angle $2\theta_p$ from CV to CD is counterclockwise or positive, and its tangent is

$$\tan 2\theta_p = \frac{\tau_{xy}}{(\sigma_x - \sigma_y)/2}$$

which is Eq. 1–6. From the derivation of Eq. 1–7, the angle between the vertical plane and one of the principal planes was θ_p. In obtaining the same equation from Mohr's circle, the angle between the radii representing these same two planes is $2\theta_p$. In other words, all angles in Mohr's circle are twice the corresponding angles for the actual stressed body. The angle from the vertical plane to the horizontal plane in Fig. 1–17a is 90°, but in Fig. 1–21 the angle between line CV (which represents the vertical plane) and line CH (which represents the horizontal plane) in Mohr's circle is 180°.

The results obtained from Mohr's circle have been shown to be identical with the equations derived from the free-body diagram of Art. 1–7. Consequently, Mohr's circle provides an extremely useful aid in the solution for, and visualization of, the stresses on various planes through a point in a stressed body in terms of the stresses on two mutually perpendicular planes through the point. Although Mohr's circle can be drawn to scale and used to obtain values of stresses and angles by direct measurements on the figure, it is probably more useful as a pictorial aid to the analyst who is performing analytical determinations of stresses and their directions at the point.

When the state of stress at a point is specified by means of a sketch of a small element, the procedure for drawing and using Mohr's circle to obtain specific stress information can be briefly summarized as follows:

1. Choose a set of x–y reference axes.
2. Identify the stresses σ_x, σ_y and $\tau_{xy} = \tau_{yx}$ and list them with the proper sign.
3. Draw a set of σ–τ coordinate axes with σ and τ positive to the right and upward, respectively.
4. Plot the point $(\sigma_x, -\tau_{xy})$ and label it point V (vertical plane).
5. Plot the point (σ_y, τ_{yx}) and label it point H (horizontal plane).
6. Draw a line between V and H. This establishes the center C and the radius R of Mohr's circle.

7. Draw the circle.
8. An extension of the radius between C and V can be identified as the x axis or the reference line for angle measurements (i.e., $\theta = 0°$).

By plotting points V and H as $(\sigma_x, -\tau_{xy})$ and (σ_y, τ_{yx}), respectively, shear stresses that tend to rotate the stress element clockwise will plot above the σ axis, while those tending to rotate the element counterclockwise will plot below the σ axis. The use of a negative sign with one of the shearing stresses $(\tau_{xy}$ or $\tau_{yx})$ is required for plotting purposes, since for a given state of stress, the shearing stresses $(\tau_{xy} = \tau_{yx})$ have only one sign (both are positive or both are negative). The use of the negative sign at point V on Mohr's circle brings the direction of angular measurements 2θ on Mohr's circle into agreement with the direction of angular measurements θ on the stress element.

Once the circle has been drawn, the normal and shearing stresses on an arbitrary inclined plane AB having an outward normal n that is oriented at an angle θ with respect to the reference x axis can be obtained from the coordinates of a point F (see Fig. 1–21) on the circle that is located at angular position 2θ from the reference axis through point V. The coordinates of point F must be interpreted as stresses σ_n and $-\tau_{nt}$. Other points on Mohr's circle that provide stresses of interest are

1. Point D that provides the principal stress σ_{p1}.
2. Point E that provides the principal stress σ_{p2}.
3. Point A that provides the maximum in-plane shearing stress $-\tau_p$ and the accompanying normal stress σ_{avg} that acts on the plane.

A negative sign must be used when interpreting shearing stresses τ_{nt} and τ_p obtained from the circle since a shearing stress tending to produce a clockwise rotation of the stress element is a negative shearing stress when a right-hand n–t coordinate system is used.

Problems of the type presented in Art. 1–6 as well as those of Art. 1–7 can readily be solved by this semigraphic method, as will be illustrated in the following example.

EXAMPLE 1–6
At a point in a structural member subjected to plane stress there are stresses on horizontal and vertical planes through the point, as shown in Fig. 1–22a. Determine and show on a sketch

(a) The principal stresses and maximum shearing stress at the point.
(b) The normal and shearing stresses on plane A–A through the point.

SOLUTION

Mohr's circle is constructed from the given data as follows: on a set of coordinate axes (Fig. 1–22b) plot point V (representing the stresses on the vertical plane) at $(8, -4)$ because the stresses on the vertical plane are 8 ksi T and 4 ksi (counterclockwise) shear. Likewise, point H (representing the stresses on the horizontal plane) has the coordinates $(-6, 4)$. Draw line HV, which is a diameter of Mohr's circle, and note that the center of the circle is at $(1, 0)$. The radius of the circle is

$$CV = \sqrt{7^2 + 4^2} = 8.06 \text{ ksi}$$

(a) The principal stresses and the maximum shearing stress at the point are

$$\sigma_{p1} = OD = 1 + 8.06 = \underline{9.06 \text{ ksi } T} \qquad\qquad \text{Ans.}$$

$$\sigma_{p2} = OE = 1 - 8.06 = -7.06 = \underline{7.06 \text{ ksi } C} \qquad \text{Ans.}$$

$$\sigma_{p3} = \sigma_z = \underline{0} \qquad\qquad\qquad\qquad\qquad \text{Ans.}$$

(b)

(a)

(c)

(d)

FIG. 1–22

Since σ_{p1} and σ_{p2} have opposite signs, the maximum shearing stress is

$$\tau_p = \tau_{max} = CA = CB = \underline{8.06 \text{ ksi}} \qquad \text{Ans.}$$

The principal planes are represented by lines CD and CE, where

$$\tan 2\theta_p = 4/7 = 0.5714$$

which gives

$$2\theta_p = 29.75°\text{↰}$$

Since the angle $2\theta_p$ is counterclockwise, the principal planes are counterclockwise from the vertical and horizontal planes of the stress block, as shown in Fig. 1–22c. To determine which principal stress acts on which plane, note that as the radius of the circle rotates counterclockwise, the end of the radius CV moves from V to D, indicating that as the initially vertical plane rotates through 14.88°, the stresses change to 9.06 ksi T and zero. Note also that the end H or radius CH moves to E, indicating that as the initially horizontal plane rotates through 14.88°, the stresses change to 7.06 ksi C and zero. The required sketch is Fig. 1–22c.

(b) Plane $A–A$ is 45° counterclockwise from the vertical plane; therefore, the corresponding radius of Mohr's circle is 90°, 2θ, counterclockwise from the line CV, and is shown as CF on Fig. 1–22b. The coordinates of the point F are seen to be (5, 7), which means that the stresses on plane $A–A$ are 5.00 ksi T and 7.00 ksi shear in a direction to produce a clockwise couple on a stress element having plane $A–A$ as one of its faces. This shear stress would be defined as a negative shear stress since to produce a clockwise couple, it must be directed in the negative t direction associated with plane $A–A$ (see Fig. 1–22d). Note that Fig. 1–22d does not show all of the stresses acting on the element.

PROBLEMS

Note: The bodies in the following problems are assumed to be in a state of plane stress. For interpretation of signs on shearing stresses, the horizontal (x) and vertical (y) directions are positive to the right and upward, respectively.

1-72*   Use Mohr's circle to solve Prob. 1-32.

1-73*   Use Mohr's circle to solve Prob. 1-33.

1-74   Use Mohr's circle to solve Prob. 1-36,

1-75*   Use Mohr's circle to solve Probs. 1-43 and 1-51.

1-76*   Use Mohr's circle to solve Probs. 1-44 and 1-52.

1-77   Use Mohr's circle to solve Probs. 1-45 and 1-53.

1-78   Use Mohr's circle to solve Probs. 1-46 and 1-54.

1-79   Use Mohr's circle to solve Probs. 1-47 and 1-55.

1-80   Use Mohr's circle to solve Probs. 1-48 and 1-56.

1-81*   Use Mohr's circle to solve Probs. 1-49 and 1-57.

1-82   Use Mohr's circle to solve Probs. 1-50 and 1-58.

1-83*   Use Mohr's circle to solve Prob. 1-62.

1-84   Use Mohr's circle to solve Prob. 1-63.

1-85   Use Mohr's circle to solve Prob. 1-64.

1-86*   Use Mohr's circle to solve Prob. 1-65.

1-87   Use Mohr's circle to solve Prob. 1-40.

1-88   Use Mohr's circle to solve Prob. 1-61.

1-89*   At a point in a machine component, the normal stresses are 160 MPa T on a vertical plane and 40 MPa C on a horizontal plane. A negative shearing stress of 50 MPa acts on the vertical plane at the point. Determine and show on a sketch

(a)   The stresses on a plane 30° clockwise from the vertical plane.

(b)   The principal and maximum shearing stresses at the point.

1-90*   At a point in a structural member, the normal stresses are 5000 psi C on a vertical plane and 4000 psi T on a horizontal plane. A negative shearing stress of 1000 psi acts on the horizontal plane at the point. Determine and show on a sketch

(a)   The stresses on a plane 60° counterclockwise from the vertical plane through the point.

(b)   The principal and maximum shearing stresses at the point.

1-91   At a point in a stressed body, the normal stresses are 160 MPa T on a vertical plane and 240 MPa C on a horizontal plane. An unknown negative shearing stress acts on the vertical plane. The maximum shearing stress at

FIG. P1–93

the point has a magnitude of 250 MPa. Determine the principal stresses at the point and the shearing stress on the vertical plane.

1-92   At a point in a stressed body, the normal stresses are 25 ksi T on a vertical plane and 15 ksi T on a horizontal plane. An unknown positive shearing stress acts on a horizontal plane. The maximum shearing stress at the point has a magnitude of 16.5 ksi. Determine the principal stresses at the point and the shearing stress on the vertical plane.

1-93*   At a point in a stressed body, the nonzero principal stresses are oriented as shown in Fig. P1–93. Determine and show on a sketch

   (a)   The stresses on plane $a–a$.

   (b)   The stresses on horizontal and vertical planes at the point.

   (c)   The maximum shearing stress at the point.

1-94*   At a point in a stressed body, the nonzero principal stresses are oriented as shown in Fig. P1–94. Determine and show on a sketch

   (a)   The stresses on plane $a–a$.

   (b)   The stresses on horizontal and vertical planes at the point.

   (c)   The maximum shearing stress at the point.

1-95   At a point in a stressed member, a normal stress of 10 ksi C and an unknown positive shearing stress acts on a horizontal plane. One principal

FIG. P1–94

stress at the point is 2 ksi C. The maximum shearing stress at the point has a magnitude of 13 ksi. Determine the unknown stresses on the vertical and horizontal planes and the unknown principal stresses.

1-96 At a point in a stressed body, there is a normal stress of 75 MPa T and an unknown negative shearing stress on a horizontal plane. One principal stress at the point is 200 MPa T. The maximum in-plane shearing stress at the point has a magnitude of 85 MPa. Determine the unknown stresses on the vertical and horizontal planes, the unknown principal stresses, and the maximum shearing stress at the point.

1-9 GENERAL STATE OF STRESS AT A POINT

The general state of stress at a point was previously illustrated in Fig. 1–14. Expressions for the stresses on any oblique plane through the point in terms of stresses on the reference planes can be developed with the aid of the free-body diagram of Fig. 1–23, where the n axis is normal to the oblique (shaded) face. The areas of the faces of the element are dA for the oblique face and $dA \cos\theta_x$, $dA \cos\theta_y$, and $dA \cos\theta_z$ for the x, y, and z faces, respectively.[8] The resultant force F on the oblique face is $S\,dA$, where S is the *resultant stress* (stress vector—see Art. 1–5) on the area and is equal to $\sqrt{\sigma_n^2 + \tau_{nt}^2}$.

FIG. 1–23

[8] These relationships can be established by considering the volume of the tetrahedron in Fig. 1–23. Thus, $V = (1/3)\,dn\,dA = (1/3)\,dx\,dA_x = (1/3)\,dy\,dA_y = (1/3)\,dz\,dA_z$. But $dn = dx \cos\theta_x = dy \cos\theta_y = dz \cos\theta_z$; therefore, $dA_x = dA \cos\theta_x$, $dA_y = dA \cos\theta_y$, $dA_z = dA \cos\theta_z$.

The forces on the x, y, and z faces are shown as three components, the magnitude of each being the product of the area by the appropriate stress. If we use l, m, and n for $\cos\theta_x$, $\cos\theta_y$, and $\cos\theta_z$, respectively, the three force equations of equilibrium in the x, y, and z directions are

$$F_x = S_x dA = \sigma_x dA \, l + \tau_{yx} dA \, m + \tau_{zx} dA \, n$$

$$F_y = S_y dA = \sigma_y dA \, m + \tau_{zy} dA \, n + \tau_{xy} dA \, l$$

$$F_z = S_z dA = \sigma_z dA \, n + \tau_{xz} dA \, l + \tau_{yz} dA \, m$$

from which the three orthogonal components of the resultant stress are

$$S_x = \sigma_x l + \tau_{yx} m + \tau_{zx} n$$

$$S_y = \tau_{xy} l + \sigma_y m + \tau_{zy} n \qquad (a)$$

$$S_z = \tau_{xz} l + \tau_{yz} m + \sigma_z n$$

The normal component σ_n of the resultant stress S equals $S_x l + S_y m + S_z n$; therefore, from Eq. a, the following equation for the normal stress on any oblique plane through the point is obtained:

$$\sigma_n = \sigma_x l^2 + \sigma_y m^2 + \sigma_z n^2 + 2\tau_{xy} lm + 2\tau_{yz} mn + 2\tau_{zx} nl \qquad (1\text{--}10)$$

The shearing stress τ_{nt} on the oblique plane can be obtained from the relation $S^2 = \sigma_n^2 + \tau_{nt}^2$. For a given problem, the values of S and σ_n will be obtained from Eqs. a and 1–10.

A principal plane was previously defined as a plane on which the shearing stress τ_{nt} is zero. The normal stress σ_n on such a plane was defined as a principal stress σ_p. If the oblique plane of Fig. 1–23 is a *principal plane,* then $S = \sigma_p$ and $S_x = \sigma_p l$, $S_y = \sigma_p m$, $S_z = \sigma_p n$. When these components are substituted into Eq. a, the equations can be rewritten to produce the following homogeneous linear equations in l, m, and n:

$$(\sigma_p - \sigma_x)l - \tau_{yx} m - \tau_{zx} n = 0$$

$$(\sigma_p - \sigma_y)m - \tau_{zy} n - \tau_{xy} l = 0 \qquad (b)$$

$$(\sigma_p - \sigma_z)n - \tau_{xz} l - \tau_{yz} m = 0$$

This set of equations has a nontrivial solution only if the determinant of the coefficients of l, m, and n is equal to zero. Thus,

$$\begin{vmatrix} (\sigma_p - \sigma_x) & -\tau_{yx} & -\tau_{zx} \\ -\tau_{xy} & (\sigma_p - \sigma_y) & -\tau_{zy} \\ -\tau_{xz} & -\tau_{yz} & (\sigma_p - \sigma_z) \end{vmatrix} = 0$$

Expansion of the determinant yields the following cubic equation for determining the principal stresses:

$$\sigma_p^3 - (\sigma_x + \sigma_y + \sigma_z)\sigma_p^2 + (\sigma_x\sigma_y + \sigma_y\sigma_z + \sigma_z\sigma_x - \tau_{xy}^2 - \tau_{yz}^2 - \tau_{zx}^2)\sigma_p$$
$$- (\sigma_x\sigma_y\sigma_z - \sigma_x\tau_{yz}^2 - \sigma_y\tau_{zx}^2 - \sigma_z\tau_{xy}^2 + 2\tau_{xy}\tau_{yz}\tau_{zx}) = 0 \quad (1\text{--}11)$$

For given values of σ_x, σ_y, . . ., τ_{zx}, Eq. 1–11 gives three values for the principal stresses σ_{p1}, σ_{p2}, σ_{p3}. By substituting these values for σ_p, in turn, into Eq. b and using the relation

$$l^2 + m^2 + n^2 = 1 \qquad (c)$$

three sets of direction cosines may be determined for the normals to the three principal planes. The foregoing discussion verifies the existence of three mutually perpendicular principal planes for the most general state of stress.

In developing equations for maximum and minimum normal stresses, the special case will be considered in which $\tau_{xy} = \tau_{yz} = \tau_{zx} = 0$. No loss in generality is introduced by considering this special case since it involves only a reorientation of the reference x, y, z axes to coincide with the principal directions. Since the x, y, z planes are now principal planes, the stresses σ_x, σ_y, σ_z become σ_{p1}, σ_{p2}, and σ_{p3}. Solving Eq. a for the direction cosines yields

$$l = S_x/\sigma_{p1} \qquad m = S_y/\sigma_{p2} \qquad n = S_z/\sigma_{p3}$$

By substituting these values into Eq. c, the following equation is obtained:

$$\frac{S_x^2}{\sigma_{p1}^2} + \frac{S_y^2}{\sigma_{p2}^2} + \frac{S_z^2}{\sigma_{p3}^2} = 1 \qquad (d)$$

The plot of Eq. d is the ellipsoid shown in Fig. 1–24. It can be observed that the magnitude of σ_n is everywhere less than that of S (since $S^2 = \sigma_n^2 + \tau_{nt}^2$)

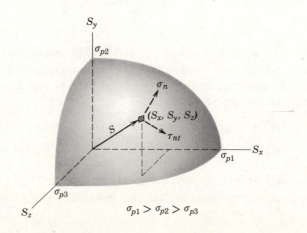

$\sigma_{p1} > \sigma_{p2} > \sigma_{p3}$

FIG. 1–24

except at the intercepts, where S is σ_{p1}, σ_{p2}, or σ_{p3}. Therefore, it can be concluded that two of the principal stresses (σ_{p1} and σ_{p3} of Fig. 1–24) are the maximum and minimum normal stresses at the point. The third principal stress is intermediate in value and has no particular significance. The discussion above demonstrates that the set of principal stresses includes the maximum and minimum normal stresses at the point.

Continuing with the special case where the given stresses σ_x, σ_y, and σ_z are principal stresses, we can develop equations for the maximum shearing stress at the point. The resultant stress on the oblique plane is given by the expression

$$S^2 = S_x^2 + S_y^2 + S_z^2$$

Substitution of values for S_x, S_y, and S_z from Eq. a, with zero shearing stresses, yields the expression

$$S^2 = \sigma_x^2 l^2 + \sigma_y^2 m^2 + \sigma_z^2 n^2 \tag{e}$$

Also, from Eq. 1–10,

$$\sigma_n^2 = (\sigma_x l^2 + \sigma_y m^2 + \sigma_z n^2)^2 \tag{f}$$

Since $S^2 = \sigma_n^2 + \tau_{nt}^2$, an expression for the shearing stress τ_{nt} on the oblique plane is obtained from Eqs. e and f as

$$\tau_{nt} = \sqrt{\sigma_x^2 l^2 + \sigma_y^2 m^2 + \sigma_z^2 n^2 - (\sigma_x l^2 + \sigma_y m^2 + \sigma_z n^2)^2} \tag{1–12}$$

The planes on which maximum and minimum shearing stresses occur can be obtained from Eq. 1–12 by differentiating with respect to the direction cosines l, m, and n. One of the direction cosines in Eq. 1–12 (n for example) can be eliminated by solving Eq. c for n^2 and substituting into Eq. 1–12. Thus,

$$\tau_{nt} = \{(\sigma_x^2 - \sigma_z^2)l^2 + (\sigma_y^2 - \sigma_z^2)m^2$$
$$+ \sigma_z^2 - [(\sigma_x - \sigma_z)l^2 + (\sigma_y - \sigma_z)m^2 + \sigma_z]^2\}^{1/2} \tag{g}$$

By taking the partial derivatives of Eq. g, first with respect to l and then with respect to m and equating to zero, the following equations are obtained for determining the direction cosines associated with planes having maximum and minimum shearing stresses:

$$l[\tfrac{1}{2}(\sigma_x - \sigma_z) - (\sigma_x - \sigma_z)l^2 - (\sigma_y - \sigma_z)m^2] = 0 \tag{h}$$

$$m[\tfrac{1}{2}(\sigma_y - \sigma_z) - (\sigma_x - \sigma_z)l^2 - (\sigma_y - \sigma_z)m^2] = 0 \tag{i}$$

One solution of these equations is obviously $l = m = 0$. Then from Eq. c, $n = \pm 1$. Solutions different from zero are also possible for this set of equations.

Consider first that $m = 0$; then from Eq. h, $l = \pm\sqrt{1/2}$ and from Eq. c, $n = \pm\sqrt{1/2}$. Also if $l = 0$, then from Eq. i, $m = \pm\sqrt{1/2}$ and from Eq. c, $n = \pm\sqrt{1/2}$. Repeating the above procedure by eliminating l and m in turn from Eq. g yields other values for the direction cosines that make the shearing stresses maximum or minimum. All the possible solutions are listed in Table 1–1. In the last line of the table the planes corresponding to the direction cosines in the column above are shown shaded. Note that in each case only one of the two possible planes is shown.

TABLE 1–1 Direction Cosines for Shearing Stresses

| | Minimum | | | Maximum | | |
	1	2	3	4	5	6
l	± 1	0	0	$\pm\sqrt{1/2}$	$\pm\sqrt{1/2}$	0
m	0	± 1	0	$\pm\sqrt{1/2}$	0	$\pm\sqrt{1/2}$
n	0	0	± 1	0	$\pm\sqrt{1/2}$	$\pm\sqrt{1/2}$

Substituting values for the direction cosines from column 4 of Table 1–1 into Eq. 1–12, with σ_x, σ_y, σ_z replaced with σ_{p1}, σ_{p2}, σ_{p3}, yields

$$\tau_{max} = \sqrt{\tfrac{1}{2}\sigma_{p1}^2 + \tfrac{1}{2}\sigma_{p2}^2 + 0 - (\tfrac{1}{2}\sigma_{p1} + \tfrac{1}{2}\sigma_{p2})^2}$$

$$= \frac{\sigma_{p1} - \sigma_{p2}}{2}$$

Similarly, using the values of the cosines from columns 5 and 6 gives

$$\tau_{max} = \frac{\sigma_{p1} - \sigma_{p3}}{2} \quad \text{and} \quad \tau_{max} = \frac{\sigma_{p2} - \sigma_{p3}}{2}$$

Of these three possible results, the largest magnitude will be the maximum stress; hence, the expression for the maximum shearing stress is

$$\tau_{max} = \frac{\sigma_{max} - \sigma_{min}}{2} \tag{1–13}$$

which verifies the statement in Art. 1–7 regarding maximum shearing stress.

PROBLEMS

1-97   Demonstrate that Eq. 1–11 reduces to Eq. 1–7 for the state of plane stress.

1-98*   At a point in a stressed body, the known stresses are $\sigma_x = 40$ MPa T, $\sigma_y = 20$ MPa C, $\sigma_z = 20$ MPa T, $\tau_{xy} = +40$ MPa, $\tau_{yz} = 0$, and $\tau_{zx} = 0$. Determine

   (a)   The normal and shearing stresses on a plane whose outward normal is oriented at angles of 57°, 57°, and 50.4° with the x, y, and z axes, respectively.

   (b)   The principal stresses and the maximum shearing stress at the point.

1-99*   At a point in a stressed body, the known stresses are $\sigma_x = 14$ ksi T, $\sigma_y = 12$ ksi T, $\sigma_z = 10$ ksi T, $\tau_{xy} = +4$ ksi, $\tau_{yz} = -4$ ksi, and $\tau_{zx} = 0$. Determine

   (a)   The normal and shearing stresses on a plane whose outward normal is oriented at angles of 40°, 60°, and 66.2° with the x, y, and z axes, respectively.

   (b)   The principal stresses and the maximum shearing stress at the point.

1-100   At a point in a stressed body, the known stresses are $\sigma_x = 60$ MPa T, $\sigma_y = 90$ MPa T, $\sigma_z = 60$ MPa T, $\tau_{xy} = +120$ MPa, $\tau_{yz} = +75$ MPa, and $\tau_{zx} = +90$ MPa. Determine

   (a)   The normal and shearing stresses on a plane whose outward normal is oriented at angles of 70.5°, 48.2°, and 48.2° with the x, y, and z axes, respectively.

   (b)   The principal stresses and the maximum shearing stress at the point.

1-101   At a point in a stressed body, the known stresses are $\sigma_x = \sigma_y = \sigma_z = 15$ ksi T, $\tau_{xy} = +6$ ksi, $\tau_{yz} = 0$, $\tau_{zx} = +8$ ksi. Determine

   (a)   The normal and shearing stresses on a plane whose outward normal makes equal angles with the x, y, and z axes.

   (b)   The principal stresses and the maximum shearing stress at the point.

1-102*   At a point in a stressed body, the known stresses are $\sigma_x = 72$ MPa T, $\sigma_y = 32$ MPa C, $\sigma_z = 0$, $\tau_{xy} = +21$ MPa, $\tau_{yz} = 0$, and $\tau_{zx} = +21$ MPa. Determine

   (a) The normal and shearing stresses on a plane whose outward normal makes equal angles with the x, y, and z axes.

   (b) The principal stresses and the maximum shearing stress at the point.

1-103* At a point in a stressed body, the known stresses are $\sigma_x = 60$ MPa T, $\sigma_y = 40$ MPa T, $\sigma_z = 20$ MPa T, $\tau_{xy} = +40$ MPa, $\tau_{yz} = +20$ MPa, and $\tau_{zx} = +30$ MPa. Determine

   (a) The principal stresses and the maximum shearing stress at the point.

   (b) The orientation of the plane on which the maximum tensile stress acts.

1-104 At a point in a stressed body, the known stresses are $\sigma_x = 18$ ksi T, $\sigma_y = 12$ ksi T, $\sigma_z = 6$ ksi T, $\tau_{xy} = +12$ ksi, $\tau_{yz} = -6$ ksi, and $\tau_{zx} = +9$ ksi. Determine

   (a) The principal stresses and the maximum shearing stress at the point.

   (b) The orientation of the plane on which the maximum tensile stress acts.

1-105* At a point in a stressed body, the known stresses are $\sigma_x = 100$ MPa T, $\sigma_y = 100$ MPa C, $\sigma_z = 80$, MPa T, $\tau_{xy} = +50$ MPa, $\tau_{yz} = -70$ MPa, and $\tau_{zx} = -64$ MPa. Determine

   (a) The normal and shearing stresses on a plane whose outward normal is oriented at angles of 40°, 60°, and 66.2° with the x, y, and z axes, respectively.

   (b) The principal stresses and the maximum shearing stress at the point.

   (c) The orientation of the plane on which the maximum tensile stress acts.

1-106 At a point in a stressed body, the known stresses are $\sigma_x = 18$ ksi C, $\sigma_y = 15$ ksi C, $\sigma_z = 12$ ksi C, $\tau_{xy} = -15$ ksi, $\tau_{yz} = +12$ ksi, and $\tau_{zx} = -9$ ksi. Determine

   (a) The normal and shearing stresses on a plane whose outward normal is oriented at angles of 35°, 75°, and 59.2° with the x, y, and z axes, respectively.

   (b) The principal stresses and the maximum shearing stress at the point.

   (c) The orientation of the plane on which the maximum compressive stress acts.

1-10 STRESS CONCENTRATION

In the foregoing articles, it was assumed that the average stress, as determined by the expression σ or τ equals P/A, is the significant or critical stress. For many problems, this is true; for other problems, however, the maximum normal stress on a given section may be considerably greater than the average; for certain combinations of loading and material, the maximum rather than the average is the important stress. If there exists in the structural or machine element a discontinuity that interrupts the stress path (called a stress trajectory),[9] the stress at the discontinuity may be considerably greater than the nominal (average, in the case of centric loading) stress on the section; thus, there is a *stress concentration* at the discontinuity. This is illustrated in Fig. 1–25 in which a type of discontinuity is shown in the upper figure and the approximate distribution of normal stress on a transverse plane is shown in the accompanying lower figure. The ratio of the maximum stress to the nominal stress on the section is known as the *stress concentration factor*. Thus, the expression for the maximum normal stress in a centrically loaded member becomes

$$\sigma = K(P/A)$$

where A is either the gross or the net area (area at the reduced section) depending on the value used for K, the stress concentration factor. Curves for stress concentration factors will be found in Chap. 11. It is important that

FIG. 1–25   *(a)*   *(b)*   *(c)*

[9] A stress trajectory is a line everywhere parallel to the maximum normal stress.

the user of such curves (or tables of factors) ascertain whether the factors are based on the gross or net section.

A classic example of the solution of a problem involving a localized redistribution of stress occurs in the case of a small circular hole in a wide plate under uniform unidirectional tension.[10] The theory of elasticity solution is expressed in terms of a radial stress σ_r, a tangential stress σ_θ, and a shearing stress $\tau_{r\theta}$, as shown in Fig. 1–26. The equations are

$$\sigma_r = \frac{\sigma}{2}\left(1 - \frac{a^2}{r^2}\right) - \frac{\sigma}{2}\left(1 - \frac{4a^2}{r^2} + \frac{3a^4}{r^4}\right)\cos2\theta$$

$$\sigma_\theta = \frac{\sigma}{2}\left(1 + \frac{a^2}{r^2}\right) + \frac{\sigma}{2}\left(1 + \frac{3a^4}{r^4}\right)\cos2\theta$$

$$\tau_{r\theta} = \frac{\sigma}{2}\left(1 + \frac{2a^2}{r^2} - \frac{3a^4}{r^4}\right)\sin2\theta$$

On the boundary of the hole (at $r = a$) these equations reduce to

$$\sigma_r = 0$$

$$\sigma_\theta = \sigma(1 + 2\cos2\theta)$$

$$\tau_{r\theta} = 0$$

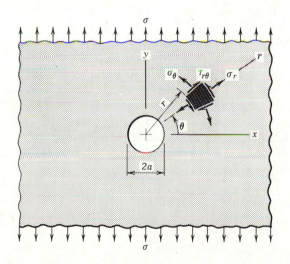

FIG. 1–26

[10] This solution was obtained by G. Kirsch; see *Z. Ver. deut. Ing.*, Vol. 42, 1898. See also A. P. Boresi and P. P. Lynn, *Elasticity in Engineering Mechanics*, Prentice-Hall, Englewood Cliffs, N.J., 1974, pp. 304–309.

At $\theta = 0°$, the tangential stress σ_θ equals 3σ where σ is the uniform tensile stress in the plate in regions far removed from the hole. Thus, the stress concentration factor associated with this type of discontinuity is 3.

The localized nature of a stress concentration can be evaluated by considering the distribution of the tangential stress σ_θ along the x axis ($\theta = 0°$). Here

$$\sigma_\theta = \frac{\sigma}{2}\left(2 + \frac{a^2}{r^2} + \frac{3a^4}{r^4}\right)$$

At a distance $r = 3a$ (one hole diameter from the hole boundary) this equation yields $\sigma_\theta = 1.074\sigma$. Thus, the stress that began as three times the nominal at the boundary of the hole has decayed to a value only 7 percent greater than the nominal at a distance of one diameter from the hole. This rapid decay is typical of the redistribution of stress in the neighborhood of a discontinuity.

Stress concentration is not significant in the case of static loading of a ductile material (defined in Art. 3–2) because the material will yield inelastically in the region of high stress and, with the accompanying redistribution of stress, equilibrium may be established and no harm done. However, if the load is an impact or repeated load, instead of the action described above, the material may fracture. Also, if the material is brittle, even a static load may cause fracture. Therefore, in the case of impact or repeated loading on any material or static loading on a brittle material, the presence of stress concentration should not be ignored. Before we leave the subject of stress concentration, it should be noted that in regions of support and load application, the stress distribution varies from the nominal (defined as the stress obtained from elementary theories of stress distribution—uniform for centric loading). This fact was discussed in 1864 by Barre de Saint-Venant (1797–1886), a French mathematician. Saint-Venant observed that although localized distortions in such regions produced stress distributions different from the theoretical distributions, these localized effects disappeared at some distance (the implication being that the distance is not of great magnitude[11]) from such locations. This statement is known as *Saint-Venant's principle* and is constantly used in engineering design. A more complete discussion of stress concentration is included in Chap. 11.

---

[11] For example, it can be shown mathematically that the localized effect of a concentrated load on a beam may disappear at a section slightly greater than the depth of the beam away from the load. See *Theory of Elasticity*, S. Timoshenko and J. N. Goodier, Third Edition, McGraw-Hill, New York, 1970.

CHAPTER 2

ANALYSIS OF STRAIN— CONCEPTS AND DEFINITIONS

2-1 INTRODUCTION

The state of stress at an arbitrary point in a body resulting from a system of loads was discussed in Chapter 1. The relationships between stresses on planes having different aspects at a point were developed using equilibrium considerations. No assumptions involving deformations or materials used in fabricating the body were made; therefore, the results are valid for an idealized rigid body or for a real deformable body. In the design of structural elements or machine components, the deformations experienced by the body as a result of the applied loads often represent as important a design consideration as the stresses previously discussed. For this reason the nature of the deformations experienced by a real deformable body as a result of internal force or stress distributions will be studied, and methods to measure or compute deformations will be established.

2-2 DISPLACEMENT, DEFORMATION, AND THE CONCEPT OF STRAIN

When a system of loads is applied to a machine component or structural element, individual points of the body generally move. This movement of a point with respect to some convenient reference system of axes is a vector quantity known as a *displacement*. In some instances displacements are associated with a translation and/or rotation of the body as a whole. Since the size or shape of the body is not changed by this type of displacement (referred to as a rigid-body displacement), they have no significance in the analysis of deformations. When displacements are induced by an applied load or a temperature change, individual points of the body move relative to each other; therefore, the size and/or shape of the body is altered. The change in any dimension associated with these load- or temperature-induced displacements is known as a *deformation* and will be designated by the Greek letter δ.

Under general conditions of loading, deformations will not be uniform throughout the body. Some line segments will experience extensions while others will experience contractions. Different segments (of the same length) along the same line may experience different amounts of extension or contraction. Similarly, angle changes between line segments may vary with position and orientation in the body. This nonuniform nature of load-induced deformations requires an analysis of deformations similar to the analysis of internal forces developed in Chapter 1.

Strain is a quantity used to provide a measure of the intensity of a deformation (deformation per unit length) just as stress is used to provide a measure of the intensity of an internal force (force per unit area). In Chapter 1 two types of stresses were defined: normal stresses and shearing stresses. This same classification is used for strains. *Normal strain,* designated by the letter ε, is used to provide a measure of the elongation or contraction of an arbitrary line segment in a body during deformation. *Shearing strain,* designated by the letter γ, is used to provide a measure of angular distortion (change in angle between two lines that are orthogonal in the undeformed state). The deformation or strain may be the result of a change in temperature, of a stress, or of other physical phenomena such as grain growth or shrinkage. In this book only strains resulting from changes in temperature or stress are considered. The term *strain* is sometimes used to denote total deformation δ; however, *in this book, strain will always refer to deformation per unit length for normal strains or changes in angle for shearing strains.*

2-3   NORMAL AND SHEARING STRAINS

The deformation (change in length and width) of a simple bar under an axial load (see Fig. 2–1a) can be used to illustrate the idea of a normal strain. The average axial strain (a normal strain, hereafter called *axial* strain) ε_{avg} over the length of the bar is obtained by dividing the axial deformation δ_n by the length L of the bar. Thus,

$$\varepsilon_{\text{avg}} = \delta_n/L \qquad\qquad (a)$$

FIG. 2–1

(a)   *(b)*

In those cases in which the deformation is nonuniform along the length of the bar (a long bar hanging under its own weight, for example), the average axial strain given by Eq. *a* may be significantly different from the true axial strain at an arbitrary point P along the bar. The true axial strain at a point can be determined by making the length over which the axial deformation is measured smaller and smaller. In the limit a quantity defined as the axial strain at the point, $\varepsilon(P)$, is obtained. This limit process is indicated by the expression

$$\varepsilon(P) = \lim_{L \to 0} \frac{\delta_n}{L} = \frac{d\delta_n}{dL} \qquad (b)$$

In a similar manner a deformation involving a change in shape (distortion) can be used to illustrate a shearing strain. An average shearing strain γ_{avg} associated with two reference lines that are orthogonal in the undeformed state (two edges of the element shown in Fig. 2–1*b*) can be obtained by dividing the shearing deformation δ_s (displacement of the top edge of the element with respect to the bottom edge) by the distance L between these two edges. The shearing strain is defined here for small deformations only (i.e., tan $\gamma = \gamma$). Thus,

$$\gamma_{\text{avg}} = \frac{\delta_s}{L} \qquad (c)$$

Again, for those cases in which the deformation is nonuniform, the shearing strain at a point, $\gamma(P)$, associated with two orthogonal reference lines is obtained by measuring the shearing deformation as the size of the element is made smaller and smaller. In the limit

$$\gamma(P) = \lim_{L \to 0} \frac{\delta_s}{L} = \frac{d\delta_s}{dL} \qquad (d)$$

Since shearing strain is the tangent of the angle of distortion, which for small angles is equal to the angle in radians, an equivalent expression for shearing strain that is sometimes useful for calculations is

$$\gamma(P) = \frac{\pi}{2} - \theta' \qquad (e)$$

In this expression θ' is the angle in the deformed state between the two initially orthogonal reference lines.

Equations a through e indicate that both normal and shearing strains are dimensionless quantities; however, normal strains are frequently expressed in units of in./in. or microin./in. while shearing strains are expressed in radians or microradians. The symbol μ is frequently used to indicate micro (10^{-6}).

From the definition of normal strain given by Eq. a or b it is evident that normal strain is positive when the line elongates and negative when the line contracts. In general, if the axial stress is tensile, the axial deformation will be an elongation. Therefore, positive normal strains are referred to as _tensile strains_. The reverse will be true for compressive axial stresses; therefore, negative normal strains are referred to as _compressive strains_. From Eq. e it is evident that shearing strains will be positive if the angle between the reference lines decreases. If the angle increases, the shearing strain is negative. Positive and negative shearing strains are not given special names. Normal and shearing strains for most engineering materials in the elastic range (see Art. 3–3) seldom exceed values of 0.2 percent, which is equivalent to 0.002 in./in. or 0.002 rad.

PROBLEMS

2-1* A 1/2-in. round structural steel rod is loaded in tension to fracture. The 8.00-in. gage length of the rod was marked off in 1.00-in. lengths before loading. After the rod broke, the strain in the 2.00-in. length containing the fracture was found to be 0.450 in./in., and the total elongation of the other six inches was found to be 1.50 in. Determine the average strain for the 8.00-in. gage length.

2-2* A structural steel bar was loaded in tension to fracture. The 200-mm gage length of the rod was marked off in 25-mm lengths before loading. After the rod broke, the 25-mm segments were found to have lengthened to 30.0, 30.5, 31.5, 34.0, 44.5, 32.0, 31.0, and 30.0 mm, consecutively. Compare the average strain over the 200-mm gage length with the maximum average strain over any 50-mm length.

2-3* A rigid bar C is supported by two steel rods as shown in Fig. P2–3.

FIG. P2–3   FIG. P2–4

There is no strain in the vertical bars before load P is applied. After load P is applied, the axial strain in bar B is 0.0006 in./in. Determine

(a)   The axial strain in bar A.

(b)   The axial strain in bar A if there is a 0.001-in. clearance in the connection between A and C before the load is applied.

2-4*   A rigid steel plate A is supported by three rods as shown in Fig. P2–4. There is no strain in the rods before the load P is applied After load P is applied, the axial strain in rod C is 0.0024 m/m. Determine

(a)   The axial strain in rods B.

(b)   The axial strain in rods B if there is a 0.15-mm clearance in the connection between A and C before the load P is applied.

2-5   The load P produces an axial strain in the brass post B of Fig. P2–5 of 0.0014 in./in. Determine

(a)   The axial strain in the aluminum alloy rod A.

(b)   The axial strain in the aluminum alloy rod A if there is a 0.005-in. clearance in the connection between A and C in addition to the 0.009-in. clearance between B and C.

EM440

2-6   The load P produces an axial strain in the steel post D of Fig. P2–6 of 0.0075 m/m. Determine

(a)   The axial strain in the aluminum rod C.

(b)   The axial strain in the aluminum alloy rod C if there is a 0.10-mm clearance in the connection at E in addition to the 0.09-mm clearance between B and D before the load P is applied.

$E = \frac{\delta}{E} = .0075$
E FOR STEEL ≈ 30×10³ ksi
ASSUME NO BENDING
OR BUCKLING
TENTION IN C
CONTINUOUS

2-7*   The axial strain in a suspended bar of varying cross section (due to its weight) is given by the expression $\varepsilon = (\gamma y)/(3E)$ where γ is the specific weight of the material, y is the distance from the free (bottom) end of the bar,

FIG. P2–5

FIG. P2–6

and E is a material constant. Determine the average axial strain over the length L of the bar in terms of γ, L, and E.

2-8* A steel cable is used to support an elevator cage at the bottom of a 600-m deep mineshaft. A uniform axial strain of 260 μm/m is produced in the cable by the gravitational force on the mass of the cage (weight of the cage). At each point the gravitational force on the mass of the cable (weight of the cable) produces an additional axial strain that is proportional to the length of the cable below the point. If the total axial strain at a point at the upper end of the cable is 500 μm/m, determine the total elongation of the cable.

2-9 A steel rod is subjected to a nonuniform heating that produces an extensional (axial) strain that is proportional to the square of the distance from the unheated end ($\varepsilon = kx^2$).

(a) Derive an expression for the change in length of the rod.

(b) If the strain is 0.001 in./in. at the midpoint of a 100-in. rod, determine the change in length of the rod.

(c) Determine the average and maximum strains in the rod of part (b).

2-10 Mutually perpendicular axes in an unstressed member were found to be oriented at 89.54° when the member was stressed. Determine the shearing strain associated with these axes in the stressed member.

2-11* A thin rectangular plate is uniformly deformed as shown in Fig. P2–11. Compute the shearing strain at P associated with the two edges (PQ and PR) that were orthogonal in the undeformed plate.

2-12 A thin triangular plate is uniformly deformed as shown in Fig. P2–12. Compute the shearing strain at P associated with the two edges (PQ and PR) that were orthogonal in the undeformed plate.

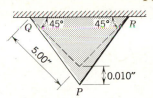

2-13* The thin triangular plate shown in Fig. P2–13 is uniformly deformed such that $\varepsilon_{PQ} = +0.0030$ in./in., $\varepsilon_{QR} = 0$, and $\varepsilon_{PR} = +0.0050$ in./in. Compute the shearing strain at P associated with the two edges (PQ and PR) that were orthogonal in the undeformed plate.

2-14 The thin rectangular plate shown in Fig. P2–14 is uniformly deformed such that $\varepsilon_{PQ} = +0.0040$ m/m, $\varepsilon_{PR} = +0.0020$ m/m, and $\varepsilon_{PS} = +0.0060$ m/m. Compute the shearing strain at P associated with the two edges (PQ and PR) that were orthogonal in the undeformed plate.

2-4 THE STATE OF STRAIN AT A POINT

The material of Art. 2–3 serves to convey the concept of strain as a unit deformation, but it is inadequate for other than one-directional loading. The extension of the concept to biaxial loading is essential because of the important role played by the strain in experimental methods of stress evaluation. In many practical problems involving the design of structural or machine elements, the configuration and loading are too complicated to permit stress determination solely by mathematical analysis; hence, this technique is supplemented by laboratory measurements.

Strains can be measured by several methods but, except for the simplest cases, stresses cannot be obtained directly. Consequently, the usual procedure in experimental stress analysis is to measure strains and calculate the state of stress from the stress-strain equations presented in Art. 3–3.

The complete state of strain at an arbitrary point P in a body under load can be determined by considering the deformation associated with a small

FIG. P2–13

FIG. P2–14

FIG. 2–2

(a)

(b)

volume of material surrounding the point. For convenience the volume is normally assumed to have the shape of a rectangular parallelepiped with its faces oriented perpendicular to the reference axes x, y, and z in the undeformed state, as shown in Fig. 2–2a. Since the element of volume is very small, deformations are assumed to be uniform; therefore, parallel planes remain plane and parallel and straight lines remain straight in the deformed element, as shown in Fig. 2–2b. The final size of the deformed element is determined by the lengths of the three edges dx', dy', and dz'. The distorted shape of the element is determined by the angles θ'_{xy}, θ'_{yz}, and θ'_{zx} between faces.

The Cartesian components of strain at the point can be expressed in terms of the deformations by using the definitions of normal and shearing strain presented in Art. 2–3. These are the strain components associated with the Cartesian components of stress discussed in Art. 1–5 and shown in Fig. 1–13. Thus,

$$\varepsilon_x = \frac{dx' - dx}{dx} = \frac{d\delta_x}{dx} \qquad \gamma_{xy} = \frac{\pi}{2} - \theta'_{xy}$$

$$\varepsilon_y = \frac{dy' - dy}{dy} = \frac{d\delta_y}{dy} \qquad \gamma_{yz} = \frac{\pi}{2} - \theta'_{yz} \qquad (2\text{--}1a)$$

$$\varepsilon_z = \frac{dz' - dz}{dz} = \frac{d\delta_z}{dz} \qquad \gamma_{zx} = \frac{\pi}{2} - \theta'_{zx}$$

In a similar manner the normal strain component associated with a line oriented in an arbitrary n direction and the shearing strain component associated with two arbitrary orthogonal lines oriented in the n and t directions in the undeformed element are given by

$$\varepsilon_n = \frac{dn' - dn}{dn} = \frac{d\delta_n}{dn} \qquad \gamma_{nt} = \frac{\pi}{2} - \theta'_{nt} \qquad (2-1b)$$

Alternative forms of Eq. 2–1, which will be useful in later developments, are

$$dx' = (1 + \varepsilon_x)dx \qquad \theta'_{xy} = \frac{\pi}{2} - \gamma_{xy}$$

$$dy' = (1 + \varepsilon_y)dy \qquad \theta'_{yz} = \frac{\pi}{2} - \gamma_{yz}$$

$$(2-2)$$

$$dz' = (1 + \varepsilon_z)dz \qquad \theta'_{zx} = \frac{\pi}{2} - \gamma_{zx}$$

$$dn' = (1 + \varepsilon_n)dn \qquad \theta'_{nt} = \frac{\pi}{2} - \gamma_{nt}$$

2-5 TWO-DIMENSIONAL OR PLANE STRAIN

The method of relating the components of strain associated with a Cartesian coordinate system to the normal and shearing strains associated with other orthogonal directions will be illustrated by considering the *two-dimensional or plane strain* case. If the xy plane is taken as the reference plane, then for conditions of plane strain $\varepsilon_z = \gamma_{zx} = \gamma_{zy} = 0$. Consider Fig. 2–3 in which the shaded rectangle represents the small unstrained element of material having the configuration of a rectangular parallelepiped. When the body is subjected to a system of loads, the element assumes the shape indicated by the dashed

FIG. 2–3

lines in Fig. 2–3. The dimensions of the deformed element are given in terms of strains obtained from Eq. 2–2.

An expression for normal strain in the n direction, ε_n, can be obtained by applying the law of cosines to the triangle $OC'B'$ shown in Fig. 2–3. Thus,

$$(OB')^2 = (OC')^2 + (C'B')^2 - 2(OC')(C'B') \cos \left(\frac{\pi}{2} + \gamma_{xy} \right)$$

or in terms of strains

$$[(1 + \varepsilon_n)dn]^2 = [(1 + \varepsilon_x)dx]^2 + [(1 + \varepsilon_y)dy]^2$$
$$- 2[(1 + \varepsilon_x)dx][(1 + \varepsilon_y)dy][-\sin \gamma_{xy}] \qquad (a)$$

Substituting $dx = dn \cos \theta$ and $dy = dn \sin \theta$ into Eq. a yields

$$(1 + \varepsilon_n)^2(dn)^2 = (1 + e_x)^2(dn)^2(\cos^2\theta) + (1 + \varepsilon_y)^2(dn)^2(\sin^2\theta)$$
$$+ 2(dn)^2(\sin \theta)(\cos \theta)(1 + \varepsilon_x)(1 + \varepsilon_y)(\sin \gamma_{xy}) \qquad (b)$$

Since the strains are small, it follows that $\varepsilon^2 \ll \varepsilon$, $\sin \gamma \approx \gamma$, and so forth; hence, all second-degree terms such as ε^2, $\gamma\varepsilon$, and the like, can be neglected as Eq. b is expanded to become

$$1 + 2\varepsilon_n = (1 + 2\varepsilon_x) \cos^2\theta + (1 + 2\varepsilon_y) \sin^2\theta + 2\gamma_{xy}\sin \theta \cos \theta$$

which reduces to

$$\varepsilon_n = \varepsilon_x \cos^2\theta + \varepsilon_y \sin^2\theta + \gamma_{xy} \sin \theta \cos \theta \qquad (2\text{–}3a)$$

or in terms of the double angle

$$\varepsilon_n = \frac{\varepsilon_x + \varepsilon_y}{2} + \frac{\varepsilon_x - \varepsilon_y}{2} \cos 2\theta + \frac{\gamma_{xy}}{2} \sin 2\theta \qquad (2\text{–}3b)$$

An expression for the shearing strain γ_{nt} can be obtained by applying the law of sines to the triangle $OC'B'$ (shown in Fig. 2–3) to obtain the rotations ϕ_1 and ϕ_2 associated with the n and t directions. Thus,

$$\frac{(1 + \varepsilon_n) \, dn}{\sin \left(\frac{\pi}{2} + \gamma_{xy} \right)} = \frac{(1 + \varepsilon_y) \, dy}{\sin(\theta + \phi_1 - \gamma_1)} \qquad (c)$$

Since the strains are small,

$$\frac{(1 + \varepsilon_y) \, dy}{(1 + \varepsilon_n) \, dn} \approx (1 + \varepsilon_y - \varepsilon_n) \sin \theta$$

and

$$\frac{\sin(\theta + \phi_1 - \gamma_1)}{\sin\left(\dfrac{\pi}{2} + \gamma_{xy}\right)} \approx \sin\theta + (\phi_1 - \gamma_1)\cos\theta$$

Hence, Eq. c can be reduced to

$$(\varepsilon_y - \varepsilon_n)\sin\theta = (\phi_1 - \gamma_1)\cos\theta \qquad (d)$$

Substituting Eq. 2–3a into Eq. d and solving for ϕ_1 yields

$$\phi_1 = -(\varepsilon_x - \varepsilon_y)\sin\theta\cos\theta - \gamma_{xy}\sin^2\theta + \gamma_1 \qquad (e)$$

The rotation ϕ_2 associated with the t direction is obtained by substituting $\pi/2 + \theta$ for θ in Eq. e. Thus,

$$\phi_2 = (\varepsilon_x - \varepsilon_y)\sin\theta\cos\theta - \gamma_{xy}\cos^2\theta + \gamma_1$$

The shearing strain γ_{nt} is equal to the difference between the rotations ϕ_1 and ϕ_2. Therefore,

$$\gamma_{nt} = -2(\varepsilon_x - \varepsilon_y)\sin\theta\cos\theta + \gamma_{xy}(\cos^2\theta - \sin^2\theta) \qquad (2\text{–}4a)$$

or in terms of the double angle

$$\gamma_{nt} = -(\varepsilon_x - \varepsilon_y)\sin 2\theta + \gamma_{xy}\cos 2\theta \qquad (2\text{–}4b)$$

The similarity between Eqs. 2–3 and 2–4 for plane strain and Eqs. 1–4 and 1–5 for plane stress in Art. 1–7 indicates that all the relationships developed for plane stress can be applied to plane strain by substituting ε_x for σ_x, ε_y for σ_y, and $\gamma_{xy}/2$ for τ_{xy}. Thus, from Eqs. 1–6, 1–7, and 1–8, expressions are obtained for determining the in-plane principal directions, the in-plane principal strains, and the maximum in-plane shearing strain

$$\tan 2\theta_p = \frac{\gamma_{xy}}{\varepsilon_x - \varepsilon_y} \qquad (2\text{–}5)$$

$$\varepsilon_{p1}, \varepsilon_{p2} = \frac{\varepsilon_x + \varepsilon_y}{2} \pm \sqrt{\left(\frac{\varepsilon_x - \varepsilon_y}{2}\right)^2 + \left(\frac{\gamma_{xy}}{2}\right)^2} \qquad (2\text{–}6)$$

$$\gamma_p = 2\sqrt{\left(\frac{\varepsilon_x - \varepsilon_y}{2}\right)^2 + \left(\frac{\gamma_{xy}}{2}\right)^2} \qquad (2\text{–}7)$$

In the previous equations, normal strains that are tensile and shearing strains that decrease the angle between the faces of the element at the origin of coordinates (see Fig. 2–3) are positive.

FIG. 2–4

$$\epsilon_{p1} > 0 > \epsilon_{p2} \qquad\qquad \epsilon_{p1} > \epsilon_{p2} > 0 \qquad\qquad 0 > \epsilon_{p1} > \epsilon_{p2}$$

When a state of plane strain exists, Eq. 2–6 gives the two in-plane principal strains while the third principal strain, ε_{p3}, is $\varepsilon_z = 0$. An examination of Eqs. 2–6 and 2–7 indicates that the maximum in-plane shearing strain is the difference between the in-plane principal strains, but this may not be the maximum shearing strain at the point. The maximum shearing strain at the point may be $(\varepsilon_{p1} - \varepsilon_{p2})$, $(\varepsilon_{p1} - 0)$, or $(0 - \varepsilon_{p2})$, depending on the relative magnitudes and signs of the principal strains. The lines associated with the maximum shearing strain bisect the angles between lines experiencing maximum and minimum normal strains. The three possibilities are illustrated in Fig. 2–4.

The pictorial or graphic representation of Eqs. 1–4 and 1–5, known as Mohr's circle for stress, can be used with Eqs. 2–3 and 2–4 to yield a Mohr's circle for strain. The equation for the strain circle obtained from the equation for the stress circle by using a change in variables is

$$\left(\varepsilon_n - \frac{\varepsilon_x + \varepsilon_y}{2}\right)^2 + \left(\frac{\gamma_{nt}}{2}\right)^2 = \left(\frac{\varepsilon_x - \varepsilon_y}{2}\right)^2 + \left(\frac{\gamma_{xy}}{2}\right)^2$$

FIG. 2–5

The variables in this equation are ε_n and $\gamma_{nt}/2$. The circle is centered on the ε axis at a distance $(\varepsilon_x + \varepsilon_y)/2$ from the origin and has a radius

$$R = \sqrt{\left(\frac{\varepsilon_x - \varepsilon_y}{2}\right)^2 + \left(\frac{\gamma_{xy}}{2}\right)^2}$$

Mohr's circle for the strains of Fig. 2–3 (with $\varepsilon_x > \varepsilon_y$) is given in Fig. 2–5. It is apparent that the sign convention for shearing strain needs to be extended to cover the construction of Mohr's circle. Observe that for positive shearing strain (indicated in Fig. 2–3), the edge of the element parallel to the x axis tends to rotate counterclockwise while the edge parallel to the y axis tends to rotate clockwise. For Mohr's circle construction, the clockwise rotation will be designated positive and the counterclockwise designated negative. This is consistent with the sign convention for shearing stresses given in Art. 1–8.

EXAMPLE 2–1

The strain components at a point in a body under a state of plane strain are $\varepsilon_x = +1200\mu$, $\varepsilon_y = -600\mu$, and $\gamma_{xy} = +900\mu$. Determine the principal strains and the maximum shearing strain at the point. Show the principal strain deformations and the maximum shearing strain distortion on a sketch.

SOLUTION

The given data are substituted in Eqs. 2–5, 2–6, and 2–7 to yield the in-plane principal strains at the point, their orientations, and the maximum in-plane shearing strain.

$$\tan 2\theta_p = \frac{900}{1200 + 600}$$

$$(\varepsilon_{p1}, \varepsilon_{p2})(10^6) = \frac{1200 - 600}{2} \pm \sqrt{\left(\frac{1200 + 600}{2}\right)^2 + \left(\frac{900}{2}\right)^2}$$

$$\gamma_p(10^6) = 2\sqrt{\left(\frac{1200 + 600}{2}\right)^2 + \left(\frac{900}{2}\right)^2}$$

from which

$$\varepsilon_{p1} = \underline{+1306\mu}$$

$$\varepsilon_{p2} = \underline{-706\mu}$$

$$\varepsilon_{p3} = \underline{0} \qquad\qquad \text{Ans.}$$

$$\theta_p = \underline{13.3°}$$

$$\gamma_p = \underline{+2012\mu} = \gamma_{max}$$

FIG. 2–6

(a)

(b)

The required sketch is given in Fig. 2–6a. The given data can also be used to construct a Mohr's circle for the state of strain at the point. The Mohr's circle solution to the problem is shown in Fig. 2–6b.

PROBLEMS

> *Note: In the following problems, the bodies are assumed to be in a state of plane strain ($\varepsilon_z = \gamma_{zx} = \gamma_{zy} = 0$). The symbol μ is used to represent micro (10^{-6}).*

2-15* The strain components at a point are $\varepsilon_x = +1000\mu$, $\varepsilon_y = -800\mu$, and $\gamma_{xy} = -800\mu$. Determine the principal strains and the maximum shearing strain at the point. Show the principal strain deformations and the maximum shearing strain distortion on a sketch.

2-16* Solve Prob. 2-15 if the strain components at the point are $\varepsilon_x = +1200\mu$, $\varepsilon_y = +960\mu$, and $\gamma_{xy} = -960\mu$.

2-17 Solve Prob. 2-15 if the strain components at the point are $\varepsilon_x = +300\mu$, $\varepsilon_y = +900\mu$, and $\gamma_{xy} = +1500\mu$.

2-18 Solve Prob. 2-15 if the strain components at the point are $\varepsilon_x = -750\mu$, $\varepsilon_y = -390\mu$, and $\gamma_{xy} = +900\mu$.

2-19* The strain components at a point are $\varepsilon_x = -600\mu$, $\varepsilon_y = +1200\mu$, and $\gamma_{xy} = +2000\mu$.

    (a) Determine the principal strains and the maximum shearing strain at the point.

FIG. P2–19   FIG. P2–20

(b)   Determine the strain components ε_n, ε_t, and γ_{nt} if the x–y and n–t axes are oriented as shown in Fig. P2-19.

(c)   Sketch Mohr's circle for the state of strain at this point.

(d)   Prepare a sketch that illustrates the deformed shapes of the rectangles shown in Fig. P2-19.

2-20   Solve Prob. 2-19 if the strain components at the point are $\varepsilon_x = -800\mu$, $\varepsilon_y = +640\mu$, $\gamma_{xy} = -960\mu$, and the x–y and n–t axes are oriented as shown in Fig. P2-20.

2-21*   Solve Prob. 2-19 if the strain components at the point are $\varepsilon_x = +1200\mu$, $\varepsilon_y = +200\mu$, and $\gamma_{xy} = -1600\mu$. The x–y and n–t axes are oriented as shown in Fig. P2-19 except the angle θ is 64°.

2-22   Solve Prob. 2-19 if the strain components at the point are $\varepsilon_x = +720\mu$, $\varepsilon_y = +1500\mu$, and $\gamma_{xy} = -1200\mu$. The x–y and n–t axes are oriented as shown in Fig. P2-20 except the angle θ is 38°.

Note: In Probs. 2-23 through 2-31, certain strains and angles are given. Determine the unknown quantities for each problem and prepare a sketch showing the principal strain deformations and the maximum shearing strain distortions. In some problems there may be more than one possible value of θ_p depending on the sign of γ_{xy}.

	ε_x	ε_y	γ_{xy}	ε_{p1}	ε_{p2}	γ_{max}	θ_p
2-23*	$+400\mu$	$+1200\mu$	-1200μ	–	–	–	–
2-24	-680μ	$+320\mu$	-640μ	–	–	–	–
2-25*	$+480\mu$	-800μ	–	$+1040\mu$	–	–	–
2-26	$+300\mu$	-1200μ	–	$+1500\mu$	–	–	–
2-27*	$+800\mu$	–	–	$+1280\mu$	–	$+2400\mu$	–
2-28	$+450\mu$	–	–	$+780\mu$	–	$+960\mu$	–
2-29*	–	–	-1800μ	$+225\mu$	–	–	$-30°$
2-30	–	–	$+840\mu$	$+1100\mu$	–	–	$20°$
2-31*	–	-640μ	-1920μ	–	-1120μ	–	–

2-6 PRINCIPAL STRAINS—PLANE STRESS

In most experimental work involving strain measurement, the strains are measured on a free (unstressed) surface of a member where a state of plane stress exists. If the outward normal to the surface is taken as the z-direction, then $\sigma_z = \tau_{zx} = \tau_{zy} = 0$. Since this state of stress offers no restraint to out-of-plane deformations, a normal strain ε_z develops in addition to the in-plane strains ε_x, ε_y, and γ_{xy}. The shearing strains γ_{zx} and γ_{zy} remain zero; therefore, the normal strain ε_z is a principal strain. In Art. 2–5, expressions were developed for the plane strain case relating the in-plane principal strains (ε_{p1} and ε_{p2}) and their orientations to the in-plane strains ε_x, ε_y, and γ_{xy}. For the plane stress case, which involves the normal strain ε_z in addition to the in-plane strains, similar expressions are needed.

As an illustration of the effects of an out-of-plane displacement on the deformation (change in length) of a line segment originally located in the xy plane, consider line AB of Fig. 2–7. As a result of the loads imposed on the member, the line AB is displaced and extended into the line $A'B'$. The displacements associated with point A' are u in the x-direction, v in the y-direction, and w in the z-direction. Point B' displaces $u + du$ in the x-direction, $v + dv$ in the y-direction, and $w + dw$ in the z-direction. The deformation δ_{AB} is obtained from the original length of the line and the displacements du, dv, and dw of point B' with respect to A'. Thus,

$$(L + \delta_{AB})^2 = (L + du)^2 + (dv)^2 + (dw)^2$$

and after squaring both sides, the result is

$$L^2 + 2L\delta_{AB} + \delta_{AB}^2 = L^2 + 2L(du) + (du)^2 + (dv)^2 + (dw)^2$$

If the deformations are small, the second degree terms can be neglected; hence,

$$\delta_{AB} = du$$

FIG. 2–7

This indicates that the normal strain along AB (δ_{AB} divided by L) is not affected by the presence of the out-of-plane displacements. In fact, none of the in-plane strains are affected; therefore, Eqs. 2–3 and 2–4 and their Mohr's circle representation are valid not only for the plane strain case but also for the plane stress case present when strain measurements are made on a free surface.

Electrical resistance strain gages have been developed to provide accurate measurements of normal strain. The resistance gage may be a wire gage consisting of a length of 0.001-in. diameter wire arranged as shown in Fig. 2–8a and cemented between two pieces of paper, or it may be an etched foil conductor mounted on epoxy or Bakelite backing (see Fig. 2–8b). The wire or foil gage is cemented to the material for which the strain is to be determined. As the material is strained, the wires are lengthened or shortened; this changes the electrical resistance of the gage. The change in resistance is measured by means of a Wheatstone bridge, which may be calibrated to directly read strain. Figure 6–1 shows SR-4 wire strain gages mounted on an aluminum beam. Shearing strains are more difficult to measure directly than

FIG. 2–8

normal strains and are often obtained by measuring normal strains in two or three different directions. The shearing strain γ_{xy} can be computed from the normal strain data by using Eq. 2–3a. For example, consider the most general case of three arbitrary normal strain measurements as shown in Fig. 2–9. From Eq. 2–3a

$$\varepsilon_a = \varepsilon_x \cos^2\theta_a + \varepsilon_y \sin^2\theta_a + \gamma_{xy} \sin \theta_a \cos \theta_a$$

$$\varepsilon_b = \varepsilon_x \cos^2\theta_b + \varepsilon_y \sin^2\theta_b + \gamma_{xy} \sin \theta_b \cos \theta_b$$

$$\varepsilon_c = \varepsilon_x \cos^2\theta_c + \varepsilon_y \sin^2\theta_c + \gamma_{xy} \sin \theta_c \cos \theta_c$$

From the measured values of ε_a, ε_b, and ε_c and a knowledge of the gage orientations θ_a, θ_b, and θ_c with respect to the reference x axis, the values of ε_x, ε_y, and γ_{xy} can be determined by simultaneous solution of the three equations. In practice the angles θ_a, θ_b, and θ_c are selected to simplify the calculations. Multiple-element strain gages used for this type of measurement are known as strain rosettes. Several rosette configurations marketed commercially are shown in Fig. 2–10.

In this book the angles used to identify the normal strain directions of the various elements of a rosette will always be measured counterclockwise from the reference x axis. Once ε_x, ε_y, and γ_{xy} have been determined, Eqs. 2–5, 2–6, and 2–7 or the corresponding Mohr's circle can be used to determine the in-plane principal strains, their orientations, and the maximum in-plane shearing strain at the point.

In Art. 3–3 it will be shown that for plane stress

$$\varepsilon_z = \varepsilon_{p3} = -\frac{\nu}{1 - \nu} (\varepsilon_x + \varepsilon_y) \tag{2–8}$$

FIG. 2–9

FIG. 2–10

where ν is Poisson's ratio (a property of the material used in fabricating the member), which is defined in Art. 3–2. For the case of plane stress this out-of-plane principal strain is important, since the maximum shearing strain at the point may be $(\varepsilon_{p1} - \varepsilon_{p2})$, $(\varepsilon_{p1} - \varepsilon_{p3})$, or $(\varepsilon_{p3} - \varepsilon_{p2})$, depending on the relative magnitudes and signs of the principal strains at the point.

Strain-measuring transducers such as electrical resistance strain gages measure the average normal strain along some gage length and not strain at a point. So long as the gage length is kept small, errors associated with such measurements can be kept within acceptable limits. The following examples illustrate the application of Eq. 2–3a and Mohr's circle to principal strain and maximum shearing strain determinations under conditions of plane stress.

EXAMPLE 2–2

A strain rosette, composed of three resistance gages making angles of 0°, 60°, and 120° with the x axis, was mounted on the free surface of a material for which Poisson's ratio is 1/3. Under load the following strains were measured:

$$\varepsilon_0 = \varepsilon_x = +1000\mu, \qquad \varepsilon_{60} = -650\mu, \qquad \varepsilon_{120} = +750\mu$$

Determine the principal strains and the maximum shearing strain. Show the directions of the in-plane principal strains on a sketch.

SOLUTION

The given data are substituted in Eq. 2–3b to yield the following (in terms of μ):

$$\varepsilon_x = +1000 \text{ (measured)}$$

$$-650 = \frac{1000 + \varepsilon_y}{2} + \frac{1000 - \varepsilon_y}{2}(\cos 120°) + \frac{\gamma_{xy}}{2}(\sin 120°)$$

$$= \frac{1000 + \varepsilon_y}{2} + \frac{1000 - \varepsilon_y}{2}\left(-\frac{1}{2}\right) + \frac{\gamma_{xy}}{2}\left(\frac{\sqrt{3}}{2}\right)$$

$$750 = \frac{1000 + \varepsilon_y}{2} + \frac{1000 - \varepsilon_y}{2}(\cos 240°) + \frac{\gamma_{xy}}{2}(\sin 240°)$$

$$= \frac{1000 + \varepsilon_y}{2} + \frac{1000 - \varepsilon_y}{2}\left(-\frac{1}{2}\right) + \frac{\gamma_{xy}}{2}\left(-\frac{\sqrt{3}}{2}\right)$$

from which $\varepsilon_y = -266.7$ and $\gamma_{xy} = -1616.6$. These values are then used to construct the Mohr's circle of Fig. 2–11. The extensions of the radii to the plotted points ε_x, ε_y are designated x and y as an aid in determining the principal directions. Since γ_{xy} is negative, the angle between the edges of the

FIG. 2–11

element at the origin increases; hence, the edge parallel to the x axis rotates clockwise ($\gamma_{xy}/2$ is plotted positive), and the edge parallel to the y axis rotates counterclockwise ($\gamma_{xy}/2$ is plotted negative). The in-plane principal strains obtained from the circle are $\varepsilon_{p1} = 1394\mu$ and $\varepsilon_{p2} = -660\mu$. The principal strain perpendicular to the surface, $\varepsilon_z = \varepsilon_{p3}$, is found from Eq. 2–8 as follows:

$$\varepsilon_z = \varepsilon_{p3} = -\frac{1/3}{1 - (1/3)} (1000\mu - 267\mu) = -367\mu$$

Since ε_{p3} is less compressive than the in-plane principal strain ε_{p2}, the maximum shearing strain at the point is the maximum in-plane shearing strain γ_p, obtained from the circle, as

$$\gamma_{\max} = \gamma_p = \pm(1394\mu + 660\mu) = \pm2054\mu = \underline{\pm2050\mu} \qquad \text{Ans.}$$

The principal strains are

$$\varepsilon_{p1} = \varepsilon_{\max} = \underline{+1394\mu}$$

$$\varepsilon_{p3} = \varepsilon_{\text{int}} = \underline{-367\mu} \qquad \text{Ans.}$$

$$\varepsilon_{p2} = \varepsilon_{\min} = \underline{-660\mu}$$

as shown in the accompanying sketch.

PROBLEMS

> *Note: In the following plane stress problems, the symbol μ is used to represent micro (10^{-6}). For the determination of ε_z, assume Poisson's ratio is 0.3.*

2-32* At a point on the free surface of a machine part, the known strains are $\varepsilon_x = +500\mu$, $\varepsilon_y = -300\mu$, and $\gamma_{xy} = +600\mu$. Determine the principal strains and the maximum shearing strain at the point. Show the principal strain deformations and the maximum shearing strain distortion on a sketch.

2-33* At a point on the free surface of a machine part, the known strains are $\varepsilon_x = +600\mu$, $\varepsilon_y = +480\mu$, and $\gamma_{xy} = -480\mu$. Use Mohr's circle to determine the principal strains and the maximum shearing strain at the point. Show the principal strain deformations and the maximum shearing strain distortion on a sketch.

2-34* An investigator measured average normal strains at a point on the free surface of a stressed body with the strain rosette shown in Fig. P2–34. The strain values obtained from the gages were: $\varepsilon_a = +1100\mu$, $\varepsilon_b = +1000\mu$, and $\varepsilon_c = -100\mu$. Determine the principal strains and the maximum shearing strain at the point.

2-35 The strain rosette shown in Fig. P2–35 was used to obtain the following normal strain data at a point on the free surface of a machine part: $\varepsilon_a = +900\mu$, $\varepsilon_b = -450\mu$, and $\varepsilon_c = +400\mu$. Determine

    (a)   The strain components ε_x, ε_y, and γ_{xy}.

    (b)   The principal strains and the maximum shearing strain at the point.

FIG. P2–34

FIG. P2–35

FIG. P2–36

FIG. P2–37

2-36*   The strain rosette shown in Fig. P2–36 was used to obtain the following normal strain data at a point on the free surface of a machine part: $\varepsilon_a = +950\mu$, $\varepsilon_b = +225\mu$, and $\varepsilon_c = +500\mu$. Determine

(a)   The strain components ε_x, ε_y, and γ_{xy} at the point.

(b)   The principal strains and the maximum shearing strain at the point.

2-37   The strain rosette shown in Fig. P2–37 was used to obtain the following normal strain data at a point on the free surface of a machine part: $\varepsilon_a = -200\mu$, $\varepsilon_b = -450\mu$, and $\varepsilon_c = -875\mu$. Determine

(a)   The strain components ε_x, ε_y, and γ_{xy} at the point.

(b)   The principal strains and the maximum shearing strain at the point.

2-38*   The strain rosette shown in Fig. P2–38 was used to obtain the following normal strain data at a point on the free surface of a stressed body: $\varepsilon_a = +850\mu$, $\varepsilon_b = +1300\mu$, and $\varepsilon_c = -50\mu$. Determine

(a)   The strain components ε_x, ε_y, and γ_{xy} at the point.

(b)   The principal strains and the maximum shearing strain at the point.

2-39   The strain rosette shown in Fig. P2–39 was used to obtain the following normal strain data at a point on the free surface of a stressed body: $\varepsilon_a = +350\mu$, $\varepsilon_b = +300\mu$, and $\varepsilon_c = +950\mu$. Determine

(a)   The strain components ε_x, ε_y, and γ_{xy} at the point.

(b)   The principal strains and the maximum shearing strain at the point.

2-40*   Average normal strains measured at angles of 30°, 60° and 135° counterclockwise from the horizontal x axis at a point on the free surface of a stressed body are $+800\mu$, $+300\mu$, and -400μ, respectively. Determine the principal and maximum shearing strains at the point.

FIG. P2–38

FIG. P2–39

2-41 Average normal strains measured at angles of 15°, 75°, and 135° counterclockwise from the horizontal x axis at a point on the free surface of a stressed body are -400μ, -800μ, and -1200μ, respectively. Determine the principal and maximum shearing strains at the point.

2-42* Average normal strains measured at angles of 15°, 45°, and 120° counterclockwise from the horizontal x axis at a point on the free surface of a stressed body are $+600\mu$, -200μ, and $+100\mu$, respectively. Determine the principal and maximum shearing strains at the point.

2-43 Use Mohr's circle to determine the strains that would be indicated by the strain gages shown in Fig. P2–43 if the two in-plane principal strains at the point are $\varepsilon_{p1} = 800\mu$, $\varepsilon_{p2} = 250\mu$ and the principal strain ε_{p1} is located 30° clockwise from the x axis.

2-44* Use Mohr's circle to determine the strains that would be indicated by the strain gages shown in Fig. P2–44 if the two in-plane principal strains at the point are $\varepsilon_{p1} = +1000\mu$, $\varepsilon_{p2} = -400\mu$ and the principal strain ε_{p1} is located 15° counterclockwise from the x axis.

2-45 At a point on the free surface of a machine part, the strain rosette shown in Fig. P2–43 was used to obtain the following normal strain data: $\varepsilon_a = +1200\mu$, $\varepsilon_b = +400\mu$, and $\varepsilon_c = +500\mu$. Determine the principal strains and the maximum shearing strain at the point. Show the principal strain deformations and the maximum shearing strain distortion on a sketch.

FIG. P2–43

FIG. P2–44

CHAPTER 3

MATERIAL PROPERTIES AND STRESS-STRAIN RELATIONSHIPS

3-1 INTRODUCTION

The satisfactory performance of a structure frequently is determined by the amount of deformation or distortion that can be permitted. A deflection of a few thousandths of an inch might make a boring machine useless, whereas the boom on a dragline might deflect several inches without impairing its usefulness. It is often necessary to relate the loads on a structure, or on a member in a structure, to the deflection the loads will produce. Such information can be obtained by plotting diagrams showing loads and deflections for each member and type of loading in a structure, but such diagrams will vary with the dimensions of the members, and it would be necessary to draw new diagrams each time the dimensions were varied. A more useful diagram is one showing the relation between the stress and strain. Such diagrams are called stress-strain diagrams.

FIG. 3–1 Hydraulic testing machine set up for a tension test. The load on the specimen is indicated on one of the large dials at upper right. Inset shows an extensometer (for measuring strains) attached to the specimen.

3-2 STRESS-STRAIN DIAGRAMS

Data for stress-strain diagrams are usually obtained by applying an axial load to a test specimen and measuring the load and deformation simultaneously. A testing machine (Fig. 3–1) is used to strain the specimen and to measure the load required to produce the strain. The stress is obtained by dividing the load by the initial cross-sectional area of the specimen. The area will change somewhat during the loading, and the stress obtained using the initial area is obviously not the exact stress occurring at higher loads. It is the stress most commonly used, however, in designing structures. The stress obtained by dividing the load by the actual area is frequently called the *true stress* and is useful in explaining the fundamental behavior of materials. Strains are usually relatively small in materials used in engineering structures, often less than 0.001, and their accurate determination requires special measuring equipment. Normal strain can be obtained by measuring the deformation δ in a length L and dividing δ by L. Instruments for measuring the deformation δ are called strain gages, or sometimes extensometers or compressometers, and obtain the desired accuracy by multiplying levers, dial indicators, beams of light, or other means. The electrical resistance strain gage described in Art. 2–6 is widely used for this type of measurement.

True strain, like true stress, is computed on the basis of the actual length of the test specimen during the test and is used primarily to study the fundamental properties of materials. The difference between nominal stress and strain, computed from initial dimensions of the specimen, and true stress and strain is negligible for stresses usually encountered in engineering structures, but sometimes the difference becomes important with larger stresses and strains.

A more complete discussion of the experimental determination of stress and strain will be found in various books on experimental stress analysis.[1,2]

Figures 3–2a and b show tensile stress-strain diagrams for structural steel (a low-carbon steel) and for a magnesium alloy. These diagrams will be used to explain a number of properties useful in the study of mechanics of materials.

The initial portion of the stress-strain diagram for most materials used in engineering structures is a straight line. The stress-strain diagrams for some materials, such as gray cast iron and concrete, show a slight curve even at very small stresses, but it is common practice to draw a straight line to average the data for the first part of the diagram and neglect the curvature. The proportionality of load to deflection was first recorded by Robert Hooke

---

[1] *Handbook of Experimental Stress Analysis,* M. Hetenyi, Wiley, New York, 1950.

[2] *Experimental Stress Analysis,* J. W. Dally and W. F. Riley, McGraw-Hill, New York, 1978.

FIG. 3–2 Stress-strain diagrams for structural steel and a magnesium alloy.

who observed in 1678, *"Ut tensio sic vis"* (as the stretch so the force); this is frequently referred to as Hooke's law. Thomas Young, in 1807, suggested what amounts to using the ratio of stress to strain to measure the stiffness of a material. This ratio is called *Young's modulus* or the *modulus of elasticity* and is the slope of the straight line portion of the stress-strain diagram. Young's modulus is written as

$$E = \sigma/\varepsilon \quad \text{or} \quad G = \tau/\gamma \tag{3-1}$$

where E is used for normal stress and strain and G (sometimes called the modulus of rigidity) is used for shearing stress and strain. The maximum stress for which stress and strain are proportional is called the *proportional limit* and is indicated by the ordinates at points A on Fig. 3–2a or b.

For points on the stress-strain curve beyond the proportional limit (such as point C on Fig. 3–2c), other quantities such as the *tangent modulus* and the *secant modulus* are used as measures of the stiffness of a material. The tangent modulus E_t is defined as the slope of the stress-strain diagram at a particular stress level. Thus, the tangent modulus is a function of the stress (or strain) for stresses greater than the proportional limit. For stresses less than the proportional limit, the tangent modulus is the same as Young's modulus. The secant modulus E_s is the ratio of the stress to the strain at any point on the diagram. Young's modulus E, the tangent modulus E_t, and the secant modulus E_s are all illustrated in Fig. 3–2c.

The action is said to be *elastic* if the strain resulting from loading disappears when the load is removed. The *elastic limit* is the maximum stress for which the material acts elastically. For most materials it is found that the stress-strain diagram for unloading is approximately parallel to the loading portion (see line BC in Fig. 3–2b). If the specimen is again loaded, the stress-strain diagram will usually follow the unloading curve until it reaches a stress a little less than the maximum stress attained during the initial loading, at which time it will start to curve in the direction of the initial loading curve. As indicated in Fig. 3–2b, the proportional limit for the second loading is greater than that for the initial loading. This phenomenon is called *strain hardening* or *work hardening*.

When the stress exceeds the elastic limit (or proportional limit for practical purposes), it is found that a portion of the deformation remains after the load is removed. The deformation remaining after an applied load is removed is called *plastic deformation*. Plastic deformation independent of the time duration of the applied load is known as *slip*. *Creep* is plastic deformation that continues to increase under a constant stress. In many instances creep continues until fracture occurs; however, in other instances the rate of creep decreases and approaches zero as a limit. Some materials are much more susceptible to creep than are others, but most materials used in engineering exhibit creep at elevated temperatures. The total strain is thus made up of elastic strain, possibly combined with plastic strain that results from slip, creep, or both. When the load is removed, the elastic portion of the strain is recovered, but the plastic part (slip and creep) remains as permanent set.[3]

---

[3] In some instances a portion of the strain that remains immediately after the stress is removed may disappear after a period of time. This reduction of strain is sometimes called *recovery*.

A precise value for the proportional limit is difficult to obtain, particularly when the transition of the stress-strain diagram from a straight line to a curve is gradual. For this reason, other measures of stress that can be used as a practical elastic limit are required. The yield point and the yield strength for a specified offset are frequently used for this purpose.

The *yield point* is the stress at which there is an appreciable increase in strain with no increase in stress, with the limitation that, if straining is continued, the stress will again increase. This latter specification indicates that there is a kink or "knee" in the stress-strain diagram, as indicated in Fig. 3–2*a*. The yield point is easily determined without the aid of strain-measuring equipment because the beam of the testing machine drops (or if a dial indicator is used, the needle halts) at the yield point. Unfortunately, few materials possess this property, the most common examples being the low-carbon steels.

The *yield strength* is defined as the stress that will induce a specified permanent set, usually 0.05 to 0.3 percent, which is equivalent to a strain of 0.0005 to 0.003. The yield strength is particularly useful for materials with no yield point. The yield strength can be conveniently determined from a stress-strain diagram by laying off the specified offset (permanent set) on the strain axis as *OC* in Fig. 3–2*b* and drawing a line *CB* parallel to *OA*. The stress indicated by the intersection of *CB* and the stress-strain diagram is the yield strength.

The maximum stress, based on the original area, developed in a material before rupture is called the *ultimate strength* of the material, and the term may be modified as the ultimate tensile, compressive, or shearing strength of the material. Ductile materials undergo considerable plastic tensile or shearing deformation before rupture. When the ultimate strength of a ductile material is reached, the cross-sectional area of the test specimen starts to decrease or neck down, and the resultant load that can be carried by the specimen decreases. Thus, the stress based on the original area decreases beyond the ultimate strength of the material, although the true stress continues to increase until rupture.

Many engineering structures are designed so that the stresses are less than the proportional limit, and Young's modulus provides a simple and convenient relationship between the stress and strain. When the stress exceeds the proportional limit, the deformation becomes partially plastic, and no simple relation exists between the stress and strain. Various empirical equations have been proposed relating the stress and strain beyond the proportional limit. The following equation, known as the Ramberg-Osgood[4] formula, is one that has been used in plastic analysis:

[4] "Description of Stress-Strain Curves by Three Parameters," W. Ramberg and W. R. Osgood, *NACA* TN 902, 1943.

$$\varepsilon = \varepsilon_0 \left[\frac{\sigma}{\sigma_0} + \frac{3}{7} \left(\frac{\sigma}{\sigma_0} \right)^n \right] \tag{3-2}$$

in which ε and σ are the strain and stress to be related, and ε_0, σ_0, and n are experimentally determined constants. Corresponding values of stress and strain can also be read from stress-strain diagrams. A stress-strain diagram similar to the one in Fig. 3–2d is frequently assumed for mild steel or other material with similar properties in order to simplify the calculations. In this book, material with such a stress-strain diagram will be designated *elasto-plastic,* and the proportional limit and yield points will be assumed to be the same. For mild steel the plastic strain that occurs at the yield point with no increase in stress is 15–20 times the elastic strain at the proportional limit.

Strength and stiffness are not the only properties of interest to a design engineer. Another important property is *ductility,* defined as the capacity for plastic deformation in tension or shear. This property controls the amount of cold forming to which a material may be subjected. The forming of automobile bodies, the bending of concrete-reinforcing bars, and the manufacture of fencing and other wire products all require ductile materials. Ductility is also an important property of materials used for fabricated structures. Under static loading the presence of stress concentration in the region of rivet holes or welds may be ignored, since the ductility permits considerable plastic action to take place in the region of high stresses, with a resulting redistribution of stress and the establishment of equilibrium. Two commonly used quantitative indices of ductility are the *ultimate elongation* (expressed as a percentage elongation of the gage length at rupture) and the *reduction of cross-sectional area* at the section where rupture occurs (expressed as a percentage of the original area).

The property indicating the resistance of a material to failure by creep is known as the *creep limit* and is defined as the maximum stress for which the plastic strain will not exceed a specified amount during a specified time interval at a specified temperature. This property is important when designing parts to be fabricated with polymeric materials (commonly known as plastics) and when designing metal parts that will be subjected to high temperatures and sustained loads (for example, the turbine blades in a turbojet engine).

A material loaded in one direction will undergo strains perpendicular to the direction of the load as well as parallel to it. The ratio of the lateral or perpendicular strain (ε_{lat} or ε_t) to the longitudinal or axial strain (ε_{long} or ε_a) for a uniaxial state of stress is called Poisson's ratio, after Simeon D. Poisson, who identified the constant in 1811. Poisson's ratio is a constant for stresses below the proportional limit and has a value between 1/4 and 1/3 for most metals. The symbol ν is used for Poisson's ratio, which is given by the equation

$$\nu = -\frac{\varepsilon_{lat}}{\varepsilon_{long}} = -\frac{\varepsilon_t}{\varepsilon_a}$$

The ratio $\nu = -\varepsilon_t/\varepsilon_a$ is valid only for a uniaxial state of stress. It will be shown in Art. 3–3 that Poisson's ratio is related to E and G by the formula

$$E = 2(1 + \nu)G$$

The properties discussed in this article are primarily concerned with static or continuous loading. Properties that are important when considering energy or impact loads and repeated loads are discussed in Art. 3–4 and Chapters 11 and 12. A more complete discussion of properties of materials is contained in several excellent properties books.[5,6]

PROBLEMS

3-1* At the proportional limit, an 8-in. gage length of a 1/2-in. diameter alloy bar has elongated 0.012 in., and the diameter has been reduced 0.00025 in. The total axial load carried was 4800 lb. Determine the following properties of the material:

    (a)   The modulus of elasticity.

    (b)   Poisson's ratio.

    (c)   The proportional limit.

3-2* At the proportional limit, a 200-mm gage length of a 15-mm diameter alloy bar has elongated 0.90 mm and the diameter has been reduced 0.022 mm. The total axial load carried was 58.4 kN. Determine the following properties of this material:

    (a)   The modulus of elasticity.

    (b)   Poisson's ratio.

    (c)   The proportional limit.

3-3* A 100-kip axial load is slowly applied to a 1 by 4 by 90-in. rectangular bar. When loaded, the 4-in. side measures 3.9986 in., and the length has increased 0.09 in. Determine Poisson's ratio, Young's modulus, and the modulus of rigidity of the material.

[5] *The Structure and Properties of Materials, Vol. III,* "Mechanical Behavior," H. W. Hayden, W. G. Moffatt, and John Wulff, Wiley, New York, 1965.

[6] *Elements of Material Science,* L. H. VanVlack, second edition, Addison-Wesley, Reading, Mass., 1964.

3-4 A 6 × 50-mm flat alloy bar elongates 1.22 mm in a length of 1.50 m under a total axial load of 35.0 kN. The proportional limit of the material is 300 MPa. Determine

    (a) The axial stress in the bar.

    (b) The modulus of elasticity of the material.

    (c) The change in the two lateral dimensions if Poisson's ratio for the material is 0.32.

3-5* A 1.50-in. diameter rod 20 ft long elongates 0.48 in. under a load of 53 kips. The diameter of the rod decreases 0.001 in. during the loading. Determine the following properties of the material:

    (a) Poisson's ratio.

    (b) Youngs modulus.

    (c) The modulus of rigidity.

3-6 A 40-mm diameter rod 6 m long elongates 12 mm under a load of 240 kN. The diameter of the rod decreases 0.028 mm during the loading. Determine the following properties of the material:

    (a) Poisson's ratio.

    (b) Young's modulus.

    (c) The modulus of rigidity.

3-7 A tensile test specimen having a diameter of 0.505 in. and a gage length of 2.000 in. was tested to fracture. Load and deformation data obtained during the test were as follows:

Load (lb)	Change in Length (in.)	Load (lb)	Change in Length (in.)
0	0	12,600	0.0600
2,200	0.0008	13,200	0.0800
4,300	0.0016	13,900	0.1200
6,400	0.0024	14,300	0.1600
8,200	0.0032	14,500	0.2000
8,600	0.0040	14,600	0.2400
8,800	0.0048	14,500	0.2800
9,200	0.0064	14,400	0.3200
9,500	0.0080	14,300	0.3600
9,600	0.0096	13,800	0.4000
10,600	0.0200	13,000	Fracture
11,800	0.0400		

Determine

(a) The modulus of elasticity.

(b) The proportional limit.

(c) The ultimate strength.

(d) The yield strength (0.05-percent offset).

(e) The yield strength (0.20-percent offset).

(f) The fracture stress.

(g) The true fracture stress if the final diameter of the specimen at the location of the fracture was 0.425 in.

(h) The tangent modulus at a stress level of 46,000 psi.

(i) The secant modulus at a stress level of 46,000 psi.

3-8 A tensile test specimen having a diameter of 11.28 mm and a gage length of 50 mm was tested to fracture. Load and deformation data obtained during the test were as follows:

Load (kN)	Change in Length (mm)	Load (kN)	Change in Length (mm)
0	0		
7.6	0.02	43.8	1.50
14.9	0.04	45.8	2.00
22.2	0.06	48.3	3.00
28.5	0.08	49.7	4.00
29.9	0.10	50.4	5.00
30.6	0.12	50.7	6.00
32.0	0.16	50.4	7.00
33.0	0.20	50.0	8.00
33.3	0.24	49.7	9.00
36.8	0.50	47.9	10.00
41.0	1.00	45.1	Fracture

Determine

(a) The modulus of elasticity.

(b) The proportional limit.

(c) The ultimate strength.

(d) The yield strength (0.05-percent offset).

(e) The yield strength (0.20-percent offset).

(f) The fracture stress.

(g)   The true fracture stress if the final diameter of the specimen at the location of the fracture was 9.50 mm.

(h)   The tangent modulus at a stress level of 315 MPa.

(i)   The secant modulus at a stress level of 315 MPa.

3-3   GENERALIZED HOOKE'S LAW FOR ISOTROPIC MATERIALS

Hooke's law (see Eq. 3–1) may be extended to include biaxial and triaxial states of stress frequently encountered in engineering applications. Consider, for example, the differential cube of Fig. 3–3 subjected to *biaxial* principal stresses in the elastic range (obeying Hooke's law). The shaded square indicates the unstrained configuration of the cube. Under the action of the stress σ_x, the cube extends in the x direction and contracts in the y direction to the configuration indicated by the dotted lines (deformations greatly exaggerated). Then, under the action of the stress σ_y superimposed on σ_x, the cube assumes the outline shown by the hatched dashed lines. If the material is isotropic, Young's modulus has the same value for all directions and the final deformation in the x direction is

V = POISSON Ratio

$$\delta_x = \varepsilon_x dx = \frac{\sigma_x}{E}\, dx - \nu\, \frac{\sigma_y}{E}\, dx$$

and the three principal strains are

$$\varepsilon_x = (\sigma_x - \nu\sigma_y)/E$$

$$\varepsilon_y = (\sigma_y - \nu\sigma_x)/E \qquad\qquad (3\text{–}5)$$

$$\varepsilon_z = -\nu(\sigma_x + \sigma_y)/E$$

$$\varepsilon_x = \frac{\sigma_x}{E} - \nu\frac{\sigma_y}{E} - \nu\frac{\sigma_z}{E}$$

FIG. 3–3

The analysis above is readily extended to triaxial principal stresses, and the expressions for strain become

$$\varepsilon_x = [\sigma_x - \nu(\sigma_y + \sigma_z)]/E$$

$$\varepsilon_y = [\sigma_y - \nu(\sigma_x + \sigma_z)]/E \qquad (3\text{-}6)$$

$$\varepsilon_z = [\sigma_z - \nu(\sigma_x + \sigma_y)]/E$$

In these expressions, tensile stresses and strains are considered positive, and compressive stresses and strains are considered negative.

When Eqs. 3–5 are solved for the stresses in terms of the strains, they give

$$\sigma_x = \frac{E}{1 - \nu^2} (\varepsilon_x + \nu\varepsilon_y)$$

$$\sigma_y = \frac{E}{1 - \nu^2} (\varepsilon_y + \nu\varepsilon_x) \qquad (3\text{-}7)$$

Equations 3–7 can be used to calculate normal stresses from measured or computed normal strains (see Art. 2–6). When Eqs. 3–6 are solved for stresses in terms of strains, they give

$$\sigma_x = \frac{E}{(1 + \nu)(1 - 2\nu)} [(1 - \nu)\varepsilon_x + \nu(\varepsilon_y + \varepsilon_z)]$$

$$\sigma_y = \frac{E}{(1 + \nu)(1 - 2\nu)} [(1 - \nu)\varepsilon_y + \nu(\varepsilon_z + \varepsilon_x)] \qquad (3\text{-}8)$$

$$\sigma_z = \frac{E}{(1 + \nu)(1 - 2\nu)} [(1 - \nu)\varepsilon_z + \nu(\varepsilon_x + \varepsilon_y)]$$

In order to simplify the physical concept, the derivation of the previous equations was based on a cube of material subjected to principal stresses (see Fig. 3–3). However, as demonstrated in Art. 2–6, normal strains are unaffected by the presence of displacements perpendicular to the normal strain direction (such as those produced by shearing strains); therefore, Eqs. 3–5 through 3–8 are valid when shearing stresses exist on the faces of the cube.

Torsion test specimens are used to study material behavior under pure shear, and it is observed that a shearing stress produces only a single corresponding shearing strain. Thus, Hooke's law extended to shearing stresses is simply

$$\tau = G\gamma \qquad (3\text{-}9)$$

Since $\tau_{max} = G\gamma_{max}$ and the maximum shearing strain for the case of plane stress may be dependent on the out-of-plane strain $\varepsilon_z = \varepsilon_{p3}$, Eq. 2–8 of Art. 2–6 is frequently used to determine the third principal strain from measured in-plane strain data. This equation is derived by substituting Eq. 3–7 into Eq. 3–5 as follows:

$$\varepsilon_z = -\frac{\nu(\sigma_x + \sigma_y)}{E} = -\frac{\nu}{E}\frac{E}{1 - \nu^2}[(\varepsilon_x + \nu\varepsilon_y) + (\varepsilon_y + \nu\varepsilon_x)]$$

$$= -\frac{\nu}{(1 - \nu)(1 + \nu)}[(1 + \nu)\varepsilon_x + (1 + \nu)\varepsilon_y]$$

$$\varepsilon_z = -\frac{\nu}{1 - \nu}(\varepsilon_x + \varepsilon_y) \tag{2–8}$$

It is important to note that the three elastic constants E, G, and ν are not independent. A useful relationship between these constants can be developed as follows: when only shearing stresses (no normal stresses) act on a pair of orthogonal (x, y) faces of an element (see Fig. 3–4), it is said to be loaded in *pure shear,* and Mohr's circles for the stresses and strains at the point have their centers at the origins of coordinates, as shown in Figs. 3–4c and d. Applying Eq. 3–7 to Fig. 3–4b and Eq. 3–9 to Fig. 3–4a yields

$$\sigma_p = \frac{E}{1 - \nu^2}(\varepsilon_p - \nu\varepsilon_p) = \tau_{max}$$

However,

$$\tau_{max} = G\gamma_{max} = G(2\varepsilon_p)$$

(a)   (b)   (c)   (d)

FIG. 3–4

Therefore,

$$\frac{E}{1 + \nu} \, \varepsilon_p = 2G\varepsilon_p$$

or

$$E = 2(1 + \nu)G \qquad \text{(3-4)}$$

Solve for G.

Equation 3–4 can also be derived by using the concepts of elastic strain energy discussed in Art. 3–4, (refer to Prob. 3-19).

Equation 3–4 substituted in Eq. 3–9 yields an alternate form of generalized Hooke's law for shearing stress and strain (isotropic material); thus,

$$\tau_{xy} = G\gamma_{xy} = \frac{E}{2(1 + \nu)} \, \gamma_{xy}$$

$$\tau_{yz} = G\gamma_{yz} = \frac{E}{2(1 + \nu)} \, \gamma_{yz} \qquad \text{(3-10)}$$

$$\tau_{zx} = G\gamma_{zx} = \frac{E}{2(1 + \nu)} \, \gamma_{zx}$$

Since Eq. 3–7 is so widely used in experimental stress determination, the following example is offered to illustrate two methods of application.

EXAMPLE 3–1

Assume the strains of Exam. 2–2 were measured on a model constructed of steel for which the modulus of elasticity and Poisson's ratio may be taken as $30(10^6)$ psi and 0.30, respectively. Determine the principal and maximum shearing stresses, and show them on a complete sketch.

SOLUTION

The principal strains from Exam. 2–2 are

$$\varepsilon_{max} = +1394(10^{-6}) \qquad \varepsilon_{min} = -660(10^{-6}) \qquad \gamma_{max} = 2054(10^{-6})$$

From Eq. 3–7,

$$\sigma_{max} = \frac{30(10^6)}{1 - 0.09} \, [1394 + 0.3(-660)](10^{-6}) = \frac{30}{0.91} \, (1196)$$

$$= 39,400 \text{ psi } T \qquad \text{Ans.}$$

$$\sigma_{min} = \frac{30}{0.91} \, [-660 + 0.3(1394)] = 7980 \text{ psi } C \qquad \text{Ans.}$$

From Eq. 3–10,

$$\tau_{max} = \frac{30(10^6)}{2(1 + 0.3)} (2054)(10^{-6}) = \underline{23,700 \text{ psi}} \qquad \text{Ans.}$$

Note that the value for τ_{max} checks that given by Eq. 1–9; thus,

$$\tau_{max} = \frac{\sigma_{p1} - \sigma_{p2}}{2} = \frac{39,400 + 7,980}{2} = 23,690 \text{ psi}$$

The directions of the principal stresses will be the same as those for the principal strains; hence, from Fig. 2–11, the directions are obtained for the sketch of Fig. 3–5.

ALTERNATE SOLUTION
Additional computed strains from Exam. 2–2 are

$$\varepsilon_y = -266.7(10^{-6}) \qquad \gamma_{xy} = -1616.6(10^{-6})$$

From Eq. 3–7,

$$\sigma_x = \frac{30(10^6)}{1 - 0.09} [1000 + 0.3(-266.7)](10^{-6}) = \frac{30}{0.91} (920) = 30,330 \text{ psi}$$

$$\sigma_y = \frac{30}{0.91} [-266.7 + 0.3(1000)] = 1098 \text{ psi}$$

and from Eq. 3–10,

$$\tau_{xy} = \frac{30(10^6)}{2(1 + 0.3)} (-1616.6)(10^{-6}) = -18,650 \text{ psi}$$

The Mohr's circle of Fig. 3–6 is constructed from these data and the following results are obtained:

FIG. 3–5

FIG. 3–6

$$\sigma_{max} = \underline{39.4 \text{ ksi } T} \qquad \text{Ans.}$$

$$\sigma_{min} = \underline{7.98 \text{ ksi } C} \qquad \text{Ans.}$$

$$\tau_{max} = \underline{23.7 \text{ ksi}} \qquad \text{Ans.}$$

$$\theta_p = \underline{25.96°} \quad \text{Ans.}$$

The complete answer is given in Fig. 3–5.

PROBLEMS

Note: In the following problems, when the material is named but the properties are not given, refer to Table A–1 in Appendix A for the properties. Use Eq. 3–4 to determine Poisson's ratio when data from Table A–1 are used.

3-9*   At a point on the surface of a plane stressed region, measured strains reduce to the following: $\varepsilon_x = +800\mu$, $\varepsilon_y = -200\mu$, and $\gamma_{xy} = -200\mu$. The material on which the strains were measured is 6061-T6 wrought aluminum alloy. Determine and show on a sketch

(a) The stresses σ_x, σ_y, and τ_{xy} at the point.

(b) The principal stresses and the maximum shearing stress at the point.

3-10* At a point on the surface of a plane stressed region, measured strains reduce to the following: $\varepsilon_x = +900\mu$, $\varepsilon_y = -100\mu$, and $\gamma_{xy} = +400\mu$. The material on which the strains were measured has the following properties: $E = 210$ GPa and $\nu = 0.30$. Determine and show on a sketch

(a) The stresses σ_x, σ_y, and τ_{xy} at the point.

(b) The principal stresses and the maximum shearing stress at the point.

3-11 On the free surface of a cold-rolled red brass machine component, the following strains were measured by means of a $(0–45–90)°$ rosette:

$$\varepsilon_0 = +500\mu, \qquad \varepsilon_{45} = +450\mu, \qquad \varepsilon_{90} = +250\mu$$

Determine

(a) The principal strains and the maximum shearing strain at the point. Prepare a sketch showing all of these strains.

(b) The principal stresses and the maximum shearing stress at the point. Prepare a sketch showing all of these stresses.

3-12 At a point on the free surface of a steel ($E = 210$ GPa and $G = 84$ GPa) machine component, the strains measured at angles of 0°, 45°, and 90° with respect to the x axis were $+1000\mu$, $+1100\mu$, and $+400\mu$, respectively. Determine

(a) The principal strains and the maximum shearing strain at the point.

(b) The principal stresses and the maximum shearing stress at the point.

(c) The normal stress on a plane whose outward normal is oriented 60° counterclockwise from the x axis.

3-13* The strain rosette shown in Fig. P3–13 was used to obtain the following strain data at a point on the free surface of a steel ($E = 30 \times 10^6$ psi and $\nu = 0.30$) machine part:

$$\varepsilon_a = +600\mu, \qquad \varepsilon_b = +500\mu, \qquad \varepsilon_c = -200\mu$$

Determine the principal stresses and the maximum shearing stress at the point by first computing the principal strains and the maximum shearing strain at the point and using these strains to compute the stresses. Show all of the stresses on a sketch.

3-14* Solve Prob. 3-13 by first computing σ_x, σ_y, and τ_{xy} and using these stresses in Mohr's circle to determine the principal stresses and the maxi-

FIG. P3–13   FIG. P3–15

mum shearing stress at the point.

3-15   The strain rosette shown in Fig. P3–15 was used to obtain the following strain data at a point on the free surface of an aluminum alloy ($E = 70$ GPa and $\nu = 0.33$) machine part:

$$\varepsilon_a = +2000\mu, \qquad \varepsilon_b = +1500\mu, \qquad \varepsilon_c = -1300\mu$$

Determine the principal stresses and the maximum shearing stress at the point by first computing the principal strains and the maximum shearing strain at the point and using these strains to compute the stresses. Show all of the stresses on a sketch.

3-16   Solve Prob. 3-15 by first computing σ_x, σ_y, and τ_{xy} and using these stresses in Mohr's circle to determine the principal stresses and the maximum shearing stress at the point.

3-17*   At a point on the free surface of an aluminum alloy ($E = 10 \times 10^6$ psi and $\nu = 1/3$) machine component, the three strain gages shown in Fig. P3–17 were used to obtain the following strain data:

$$\varepsilon_a = +800\mu, \qquad \varepsilon_b = +960\mu, \qquad \varepsilon_c = +800\mu$$

Determine

   (a)   The maximum normal strain at the point.
   (b)   The maximum shearing strain at the point.
   (c)   The maximum tensile stress at the point.
   (d)   The maximum shearing stress at the point.
   (e)   The principal stress directions.

3-18   At a point on the free surface of a steel alloy ($E = 200$ GPa and $\nu = 0.30$) machine part, the three strain gages shown in Fig. P3–18 were used to obtain the following strain data:

$$\varepsilon_a = +600\mu, \qquad \varepsilon_b = +960\mu, \qquad \varepsilon_c = +300\mu$$

FIG. P3–17

FIG. P3–18

Determine

 (a) The maximum normal strain at the point.

 (b) The maximum shearing strain at the point.

 (c) The maximum tensile stress at the point.

 (d) The maximum shearing stress at the point.

 (e) The principal stress directions.

3-4 STRAIN ENERGY

A significant topic involving stress and strain is the strain energy concept, which is not only important to an understanding of the behavior of materials under dynamic loading, but also has useful applications in the solution of certain types of problems involving static loading.

The concept of strain energy is illustrated by Fig. 3–7, in which (*a*) represents a bar of uniform cross section subjected to a *slowly applied* axial load P and held at the upper end by a support assumed to be rigid. From the

FIG. 3–7

load-deformation diagram (Fig. 3–7b) the work done in elongating the bar an amount δ_2 is

$$W_k = \int_0^{\delta_2} P \, d\delta \qquad (a)$$

where P is some function of δ. The work done on the bar must equal the change in energy of the material,[7] and this energy change, because it involves the strained configuration of the material, is termed *strain energy U*. If δ is expressed in terms of axial strain ($\delta = L\varepsilon$) and P in terms of axial stress ($P = A\sigma$), Eq. *a* becomes

$$W_k = U = \int_0^{\varepsilon_2} (\sigma)(A)(L)d\varepsilon = AL \int_0^{\varepsilon_2} \sigma \, d\varepsilon \qquad (b)$$

where σ is a function of ε (see Fig. 3–7c). If Hooke's law applies,

$$\varepsilon = \sigma/E \qquad d\varepsilon = d\sigma/E$$

and Eq. *b* becomes

$$U = \left(\frac{AL}{E}\right) \int_0^{\sigma_1} \sigma \, d\sigma$$

or

$$U = AL \left(\frac{\sigma_1^2}{2E}\right) \qquad (c)$$

Equation *c* gives the *elastic strain energy* (which is, in general, recoverable[8]) for axial loading of a material obeying Hooke's law. The quantity in parentheses, $\sigma_1^2/(2E)$, is the elastic strain energy u in tension or compression per unit volume, or strain energy intensity, for a particular value of σ. For shear loading the expression would be identical except that σ would be replaced by τ and E by G.

The integral $\int \sigma d\varepsilon$ of Eq. *b* represents the area under the stress-strain curve and, if evaluated from zero to the elastic limit (for practical purposes, the proportional limit), yields a property known as the *modulus of resilience*. The modulus is defined as the maximum strain energy per unit volume that a material will absorb without inelastic deformation. Customary units are inch-pounds per cubic inch or newton-meters per cubic meter. The area under the entire stress-strain curve from zero to rupture gives the property

---

[7] Known as *Clapeyron's theorem*, after the French engineer B. P. E. Clapeyron (1799–1864).

[8] Elastic hysteresis is neglected here as an unnecessary complication.

known as the *modulus of toughness* and denotes the energy per unit volume necessary to rupture the material. The values of these moduli for the steel of Fig 3–2a are

$$u_R = (35,000)(0.001,17)/2 = 22 \text{ in.-lb/cu. in.}$$

and

$$u_T = 57,000(0.3) = 17,000 \text{ in.-lb/cu. in.,}$$

where the 57,000 is an estimated average value of stress for a rectangle having approximately the same area as that under the upper curve of Fig. 3–2a.

The concept of elastic strain energy can be extended to include biaxial and triaxial loadings by writing the expression for strain energy intensity u as $\sigma\varepsilon/2$ and adding the energies due to each of the stresses. Since energy is a positive scalar quantity, the addition is the arithmetic sum of the energies. For a system of triaxial principal stresses σ_x, σ_y, σ_z, the total elastic strain energy intensity is

$$u = (1/2)\sigma_x\varepsilon_x + (1/2)\sigma_y\varepsilon_y + (1/2)\sigma_z\varepsilon_z \qquad (d)$$

When the expressions for strains in terms of stresses from Eq. 3–6 of Art. 3–3 are substituted into Eq. d, the result is

$$u = \frac{1}{2E}\{\sigma_x[\sigma_x - \nu(\sigma_y + \sigma_z)] + \sigma_y[\sigma_y - \nu(\sigma_z + \sigma_x)]$$
$$+ \sigma_z[\sigma_z - \nu(\sigma_x + \sigma_y)]\}$$

from which

$$u = \frac{1}{2E}[\sigma_x^2 + \sigma_y^2 + \sigma_z^2 - 2\nu(\sigma_x\sigma_y + \sigma_y\sigma_z + \sigma_z\sigma_x)] \qquad (e)$$

If σ_x, σ_y, σ_z are not principal stresses, the strain energy due to the shearing stresses on the x, y, z planes must be added to Eq. d or e to obtain the total elastic strain energy intensity; thus,

$$u = \frac{1}{2}(\sigma_x\varepsilon_x + \sigma_y\varepsilon_y + \sigma_z\varepsilon_z + \tau_{xy}\gamma_{xy} + \tau_{yz}\gamma_{yz} + \tau_{zx}\gamma_{zx}) \qquad (3\text{–}11a)$$

$$u = \frac{1}{2E}[\sigma_x^2 + \sigma_y^2 + \sigma_z^2 - 2\nu(\sigma_x\sigma_y + \sigma_y\sigma_z + \sigma_z\sigma_x)]$$
$$+ \frac{1}{2G}(\tau_{xy}^2 + \tau_{yz}^2 + \tau_{zx}^2) \qquad (3\text{–}11b)$$

The concepts discussed and expressions developed in this article will be found useful in the development of certain beam deflection expressions in Chapter 7 and a theory of failure in Chapter 9. There are other practical applications in structural engineering.

PROBLEMS

Note: In the following problems, when the material is named but the properties are not given, refer to Table A–1 in Appendix A for the properties. Use Eq. 3–4 to determine Poisson's ratio when data from Table A–1 are used.

3-19 Develop Eq. 3–4 by writing the elastic strain energy intensity expressions for the element of Fig. 3–4a in terms of the shearing stress τ_{xy} and for the element of Fig. 3–4b in terms of the principal stresses σ_p and noting that the strain energy intensity can have only one value for a particular state of stress at a point.

3-20* Determine the elastic strain energy intensity at the point of Exam. 1–4 in Chapter 1 by

(a) Using the stresses given in the problem statement.

(b) Using the principal stresses.

The material is structural steel.

3-21* Determine the elastic strain energy intensity at the point of Exam. 1–5 in Chapter 1 by

(a) Using the stresses given in the problem statement.

(b) Using the principal stresses.

The moduli of elasticity and rigidity for the material are 200 GPa and 80 GPa, respectively.

3-22 Determine the elastic strain energy intensity at the point of Exam. 1–6 in Chap. 1 by

(a) Using the stresses given in the problem statement.

(b) Using the principal stresses.

(c) Using τ_p and σ_{avg}.

The material is structural steel.

3-23*   Determine the elastic strain energy intensity at the point of Prob. 1-52 in Chapter 1 by

   (a)   Using the stresses given in the problem statement.

   (b)   Using the principal stresses as given in the answer section.

   (c)   Using τ_p and σ_{avg}.

The moduli of elasticity and rigidity for the material are 210 GPa and 84 GPa, respectively.

3-24   Determine the elastic strain energy intensity at the point of Exam. 3–1 by

   (a)   Using the values for the principal stresses and strains.

   (b)   Using the values for the stresses and strains associated with the x, y planes (note ε_x from Exam. 2–2 is $+1000\mu$).

3-25*   Determine the elastic strain energy intensity at the point of Prob. 3-9 by

   (a)   Using the stresses and strains associated with the x, y planes.

   (b)   Using the principal stresses.

Refer to the answer section for stress values.

3-26   Determine the elastic strain energy intensity at the point of Prob. 3-10 by

   (a)   Using the stresses and strains associated with the x, y planes.

   (b)   Using the principal stresses.

Refer to the answer section for stress values.

3-5 TEMPERATURE EFFECTS

Most engineering materials when unrestrained expand when heated and contract when cooled. The strain due to a 1° temperature change is designated by α and is known as the coefficient of thermal expansion. The strain due to a temperature change of $\Delta T°$ is

$$\varepsilon_T = \alpha(\Delta T) \qquad (a)$$

The coefficient of thermal expansion is approximately constant for a considerable range of temperatures (in general, the coefficient increases with an increase of temperature). For a homogeneous, isotropic material, the coefficient applies to all dimensions (all directions). Values of the coefficient of expansion for several materials are included in App. A.

   When a member is restrained (free movement prevented) while a temperature change takes place, stresses (referred to as *thermal stresses*) are induced in the member. If the action is elastic, thermal stresses are easily computed by (1) assuming that the restraining influence has been removed and the member permitted to expand or contract freely, (2) applying forces that cause the member to assume the configuration dictated by the restraining influence. For example, the bar AB of Fig. 3–8a is securely fastened to rigid supports at the ends. If the end B is assumed to be cut loose from the wall and the temperature drops, the end will move to B' a distance δ_T, as indicated in Fig. 3–8b. Next, a force P, shown in Fig. 3–8c, is applied to B of sufficient magnitude to move end B through a distance δ_P so that the length of the bar is again L, the distance between the walls. Since the walls do not move, $\delta_T = \delta_P$, where $\delta_T = \alpha L \Delta T$, and $\delta_P = L\varepsilon$; if Hooke's law applies, $\varepsilon = \sigma/E$, where σ is the induced axial stress.

In case a stress exists before the temperature change takes place (the rod of Fig. 3–8a might have been tightened by means of a turnbuckle or nuts at the ends), the stress due to the temperature change may be added algebraically to the original stress by using the *principle of superposition*. The principle states that stresses (at a point on a given plane) due to different loads may be computed separately and added algebraically, provided that the sum of the stresses does not exceed the proportional limit of the material and that the structure remains stable. In explanation of this last limitation, consider a member with a length considerably greater than the lateral dimension subjected to axial compression loads. This member may successfully carry either of two applied loads, and the sum of the axial stresses for both loads may be within the proportional limit; yet, upon application of both loads, the member may collapse (this action is discussed further in Chapter 10).

The effects of temperature change under more complicated states of stress (plane stress or the three-dimensional state of stress) can be evaluated by observing that an arbitrary normal strain ε_n will be composed of two

FIG. 3–8

parts: namely, a strain ε_σ associated with the stresses and a strain ε_T associated with the temperature change. Thus,

$$\varepsilon_n = \varepsilon_\sigma + \varepsilon_T \qquad\qquad (b)$$

In cases in which the material is completely restrained, the strains due to the stresses must be the negative of the strains due to the temperature change, as the total strain must be zero. The strains resulting from stresses are given by generalized Hooke's law (Eq. 3–6). The strains resulting from a temperature change are given by Eq. *a*. When Eqs. 3–6 and *a* are substituted in Eq. *b*, the stress-strain-temperature equations for a linear elastic isotropic material are obtained. The results are as follows:

$$\varepsilon_x = \frac{1}{E} \left[\sigma_x - \nu(\sigma_y + \sigma_z) \right] + \alpha(\Delta T)$$

$$\varepsilon_y = \frac{1}{E} \left[\sigma_y - \nu(\sigma_z + \sigma_x) \right] + \alpha(\Delta T) \qquad\qquad (3\text{–}12)$$

$$\varepsilon_z = \frac{1}{E} \left[\sigma_z - \nu(\sigma_x + \sigma_y) \right] + \alpha(\Delta T)$$

Often it is more convenient to have equations for stress in terms of strain. When Eqs. 3–12 are solved for stresses in terms of strains, they yield

$$\sigma_x = \frac{E}{(1 + \nu)(1 - 2\nu)} \left[(1 - \nu)\varepsilon_x + \nu(\varepsilon_y + \varepsilon_z) \right] - \frac{E\alpha(\Delta T)}{(1 - 2\nu)}$$

$$\sigma_y = \frac{E}{(1 + \nu)(1 - 2\nu)} \left[(1 - \nu)\varepsilon_y + \nu(\varepsilon_z + \varepsilon_x) \right] - \frac{E\alpha(\Delta T)}{(1 - 2\nu)} \qquad (3\text{–}13)$$

$$\sigma_z = \frac{E}{(1 + \nu)(1 - 2\nu)} \left[(1 - \nu)\varepsilon_z + \nu(\varepsilon_x + \varepsilon_y) \right] - \frac{E\alpha(\Delta T)}{(1 - 2\nu)}$$

Many of the problems considered later in this text can be classified as plane stress problems. For the case of plane stress in the xy plane ($\sigma_z = \tau_{zx} = \tau_{zy} = 0$), Eqs. 3–12 reduce to

$$\varepsilon_x = \frac{1}{E} (\sigma_x - \nu\sigma_y) + \alpha(\Delta T)$$

$$\varepsilon_y = \frac{1}{E} (\sigma_y - \nu\sigma_x) + \alpha(\Delta T) \qquad\qquad (3\text{–}14)$$

$$\varepsilon_z = -\frac{\nu}{E} (\sigma_x + \sigma_y) + \alpha(\Delta T)$$

In a similar manner, Eqs. 3–13 reduce to

$$\sigma_x = \frac{E}{1 - \nu^2}(\varepsilon_x + \nu\varepsilon_y) - \frac{E\alpha(\Delta T)}{1 - \nu}$$

$$\sigma_y = \frac{E}{1 - \nu^2}(\varepsilon_y + \nu\varepsilon_x) - \frac{E\alpha(\Delta T)}{1 - \nu}$$

(3–15)

Additional effects that an increase (decrease) in temperature can produce include a decrease (increase) in modulus of elasticity, a decrease (increase) in yield strength, a decrease (increase) in tensile strength, an increase (decrease) in ductility, and an increase (decrease) in creep rate. The changes in material properties for most engineering materials (polymer materials excepted) are small for a temperature change of a few hundred degrees Fahrenheit near room temperature.

PROBLEMS

3-27* The Golden Gate Bridge has a span of 4200 ft. Determine the change in length of a steel longitudinal floor support beam (that must be provided for with expansion joints) due to a temperature change of 50 °F. Use $\alpha = 6.5$ (10^{-6})/°F.

3-28* A large cement kiln has a length of 125 m and a diameter of 3.5 m. Determine the change in length and diameter of the structural steel shell caused by an increase in temperature of 125 °C. Use $\alpha = 11.9$ (10^{-6})/°C.

3-29* Determine the movement of the pointer of Fig. P3–29 with respect to the scale zero, when the temperature increases 100 °F. The coefficients of thermal expansion are 6.6 (10^{-6})/°F for the steel and 12.5 (10^{-6})/°F for the aluminum.

3-30 Determine the horizontal movement of point A in Fig. P3–30 due to a temperature drop of 45 °C. Assume that member AE has an insignificant coefficient of thermal expansion. The coefficients of thermal expansion are 11.9 (10^{-6})/°C for the steel and 22.5 (10^{-6})/°C for the aluminum alloy.

3-31* A steel surveyor's tape 1/2-in. wide by 1/64-in. thick is exactly 100-ft long at 70 °F and under a pull of 10 lb. What correction should be introduced if the tape is used to make a 100-ft measurement at a temperature of 100 °F and under a pull of 20 lb. Use $E = 30$ (10^6) psi and $\alpha = 6.5$ (10^{-6})/°F.

3-32 A steel tie rod containing a turnbuckle has its ends attached to rigid walls. During the summer when the temperature is 30 °C, the turnbuckle is

FIG. P3–29

FIG. P3–30

tightened to produce a stress in the rod of 15 MPa. Determine the stress in the rod in the winter when the temperature is $-10\,°C$. Use $\alpha = 11.9\,(10^{-6})/°C$ and $E = 200$ GPa.

3-33* A 20-ft section of steel railroad rail has a cross-sectional area of 12 sq. in. Both ends of the rail are tight against adjacent rails that for this problem, can be assumed to be rigid. The rail is supported against lateral movement. The coefficient of thermal expansion for the steel is $6.5\,(10^{-6})/°F$. For an increase in temperature of $100\,°F$, determine

   (a) The maximum normal stress in the rail.

   (b) The maximum shearing stress in the rail.

   (c) The internal force on a cross section of the rail.

3-34* A 6-m long by 50-mm diameter rod of aluminum alloy ($E = 70$ GPa, $\alpha = 22.5(10^{-6})/°C$, and $\nu = 0.346$) is attached at the ends to supports that yield to permit a change in length of 1.00 mm in the rod when stressed. When the temperature is $35\,°C$, there is no stress in the rod. After the temperature of the rod drops to $-20\,°C$, determine

   (a) The maximum normal stress in the rod.

   (b) The change in diameter of the rod.

3-35 A steel tire is to be heated and placed on a 72.00-in. diameter locomotive drive wheel. At 80 °F, the tire has an inside diameter of 71.94 in. The coefficient of thermal expansion α for the steel is $6.5(10^{-6})/°F$. Determine

   (a) The temperature increase required for the inside diameter of the tire to be 72.04 in. to facilitate placing the tire on the wheel.

   (b) The maximum normal stress in the tire when it has cooled to -20 °F. Assume that the diameter of the wheel does not change.

3-36*   A prismatic bar, free of stress at room temperature, is fastened to rigid walls at its ends. One end of the bar is heated to 60 °C above room temperature while the other end is maintained at room temperature. The change in temperature ΔT along the bar is proportional to the square of the distance from the unheated end. Determine

(a)   The axial stress induced in the bar in terms of the coefficient of thermal expansion α and the modulus of elasticity E of the bar.

(b)   The axial stress in the bar if it is made of an aluminum alloy that has a coefficient of thermal expansion of 22.5 $(10^{-6})/$°C and a modulus of elasticity of 70 GPa.

3-37   A rectangular plate is mounted between rigid supports that prevent motion in both the x and y directions. The plate is free to expand in the z direction. Determine the strain in the z direction resulting from a change in temperature ΔT in terms of the thermal (α) and elastic (E and ν) properties of the plate material. Assume that the plate does not buckle.

3-6   DESIGN LOADS, WORKING STRESSES, AND FACTOR OF SAFETY

As mentioned in Art. 1–1, the designer is required to select a material and properly proportion a member therefrom to perform a specified function without failure. Failure is defined as the state or condition in which a member or structure no longer functions as intended. Several types of failure warrant discussion at this point. *Elastic failure* is characterized by excessive elastic deformation—a bridge may deflect elastically under traffic to an extent that may result in discomfort to the vehicle passengers; a close-fitting machine part may deform sufficiently under load to prevent proper operation of the machine. When a structure is designed to avoid elastic failure, the stiffness of the material, indicated by Young's modulus, is the significant property. *Slip failure* is characterized by excessive plastic deformation due to slip. Yield strength, yield point, and proportional limit are used as indices of strength with respect to failure by slip for members subjected to static loads. *Creep failure* is characterized by excessive plastic deformation over a long period of time under constant stress. For machines or structures that are to be subjected to relatively high stress, high temperature, or both over a long period of time, creep is a design consideration, and the creep limit is the strength index to be used. The creep limit normally decreases as the temperature increases. *Failure by fracture* is a complete separation of the material. The ultimate strength of a material is the index of resistance to failure by fracture under static loads in which creep is not involved.

Most design problems involve many unknown variables. The load that the structure or machine must carry is usually estimated; the actual load may vary considerably from the estimate, especially when loads at some future time must be considered. Since testing usually damages the material, the properties of the material used in the structure cannot be evaluated directly but are normally determined by testing specimens of a similar material. Furthermore, the actual stresses that will exist in a structure are unknown because the calculations are based on assumptions about the distribution of stresses in the material. Because of these and other unknown variables, it is customary to design for the load required to produce failure, which is larger than the estimated actual load, or to use a working (or design) stress below the stress required to produce failure.

Figure 3–9 is a photograph showing the tail section of a Boeing 707 transport plane being prepared for a vibration test. Such tests can be used to measure the ultimate load-carrying capacity of the structure.

A _working stress_ or _allowable stress_ is defined as the maximum stress permitted in the design computation. The _factor of safety_ may be defined as the ratio of a failure-producing load to the estimated actual load. Factor of safety may also be defined as the ratio of the strength of a material to the maximum computed stress in the material. The latter factor of safety may be used to determine an allowable (or working) stress by dividing the strength of a material by the factor of safety; however, because actual sections of

Vibration test of tail section of airplane. (_Courtesy Boeing Airplane Co., Seattle, Washington_)

FIG. 3–9

structural and machine elements, in general, are not the same as the minimum computed sections, the actual (computed) factor of safety is the strength divided by the maximum computed stress based on actual sections of members in the design.

Factor of safety is also defined as the ratio of the strain energy per unit volume required to cause failure (slip or fracture) to the maximum computed strain energy per unit volume in a material under dynamic loading.

It will be shown in Chapter 6 that the factor of safety based on load may be greater than that based on stress. In other words, the term is meaningless without a statement of the criterion on which it is based. In this book, unless otherwise indicated, the term *factor of safety* will denote the *ratio of the strength of the material to the maximum computed stress (unit strain energy for energy loading)*. For a given design, the factor of safety based on the ultimate strength is, in general, quite different from that based on the elastic strength (proportional limit, yield point, or yield strength); hence, the term *strength* is ambiguous without a qualifying statement regarding the pertinent mode of failure. The following example illustrates the use of a factor of safety.

EXAMPLE 3–2

The improvised system of cables in Fig. 3–10a is used to remove the pile Q. A tractor with a maximum drawbar pull of 70 kN is available to apply the

FIG. 3–10

force P. Cable A is a 25-mm diameter iron hoisting cable with a breaking load of 130 kN. Cables B, C, and D are parts of a continuous 20-mm diameter carbon steel aircraft cable with a breaking strength of 220 kN. Cable E is a 23-mm diameter steel hoisting cable with a nominal area of 380 mm^2 and an ultimate tensile strength, based on the nominal area, of 500 MPa. The minimum factor of safety for each cable, based on fracture, is to be 2.0. Determine

(a) The maximum allowable force that can be exerted on the pile.
(b) The factor of safety for each of the five cables when the load in part (a) is applied.

SOLUTION

(a) The two free-body diagrams needed to solve the problem are shown in Fig. 3–10b and c. The maximum allowable tension in each cable can be obtained from the strength of the cable and the factor of safety because for axial loading the factor of safety based on load is the same as that based on stress. The allowable tensions are

$$A = 130/2 = 65 \text{ kN}$$

$$B = C = D = 220/2 = 110 \text{ kN}$$

$$E = 380(10^{-6})(500)(10^6)/2 = 95\,(10^3)\text{N} = 95 \text{ kN}$$

The maximum tractor pull 70 kN is greater than the allowable force in cable A; therefore, the maximum tractor pull cannot be utilized without exceeding the specifications. If the force in A is assumed to be 65 kN, the equations of equilibrium applied to Fig. 3–10b give

$$\frac{5}{13} B + \frac{15}{17} 65 - \frac{4}{5} C = 0$$

and

$$\frac{12}{13} B - \frac{8}{17} 65 - \frac{3}{5} C = 0$$

from which $B = 116.0$ kN and $C = 127.5$ kN. Since the maximum allowable load on cables B and C is only 110 kN, it is apparent that the force in A must be less than the allowable load. The force in B is less than that in C; therefore, cable C is the controlling member of Fig. 3–10b. The equations of equilibrium for this free-body diagram with $C = 110$ kN give the results

$$A = 56.1 \text{ kN} \quad \text{and} \quad B = 100.1 \text{ kN}$$

The equations of equilibrium for Fig. 3–10c are

$$\frac{4}{5} C - \frac{24}{25} E = 0$$

and

$$\frac{3}{5} C + \frac{7}{25} E - D = 0$$

When $C = 110$ kN is substituted in these equations, the results are

$$D = E = 91.7 \text{ kN}$$

Both these forces are less than the maximum allowable loads in the corresponding cables and are therefore safe loads. The strength of cable C is thus the limiting factor. The maximum pull on the pile is the force in cable D and is

$$\underline{91.7 \text{ kN}} \qquad \text{Ans.}$$

(b)   The factor of safety in each of the cables is

$$k_A = 130/56.1 = \underline{2.32} \diagup \qquad \text{Ans.}$$

$$k_B = 220/100.1 = \underline{2.20} \qquad \text{Ans.}$$

$$k_C = 220/110 = \underline{2.00} \qquad \text{Ans.}$$

$$k_D = 220/91.7 = \underline{2.40} \qquad \text{Ans.}$$

$$k_E = 190/91.7 = \underline{2.07} \qquad \text{Ans.}$$

As noted previously, failure may be due to excessive elastic deformation. To guard against such failures, maximum deformations may be specified, and for many designs the limiting specification is the maximum deformation rather than the maximum stress.

PROBLEMS

> *Note: Unless otherwise specified, all members are assumed to have negligible weight, and all pins used for connections are assumed to be smooth.*

3-38* A structural steel bar (see App. A) that is supporting an axial tensile load of 32,000 lb has a cross-sectional area of 2 sq. in. Determine

(a) The factor of safety with respect to failure by slip.

(b) The factor of safety with respect to failure by fracture.

3-39* A steel tension member is to be fabricated by using two angles fastened back to back. The elastic and ultimate strengths of the steel are 250 and 450 MPa, respectively. The member is to carry an axial load of 290 kN with a minimum factor of safety of 2 with respect to failure by slip. Determine the cross-sectional area needed and select the lightest pair of angles from App. C that will be acceptable. Determine the actual factor of safety for the angles chosen.

3-40* The load W in Fig. P3–40 is supported by two tie rods. Rod A is made of annealed red brass, has a cross-sectional area of 0.18 sq. in., and has a working stress specification of 12 ksi. Rod B is made of annealed bronze, has a cross-sectional area of 0.10 sq. in., and has a working stress specification of 17 ksi. Determine

(a) The maximum safe load W that can be supported.

(b) The factors of safety with respect to failure by slip based on the working stress specifications.

(c) The factor of safety with respect to failure by fracture for each of the rods when the load of part (a) is being supported.

3-41 The tie rod AB of Fig. P3–41 has a cross-sectional area of 950 mm². The cross-sectional area of pin C is 1850 mm². Both the tie rod and the pin are made of steel, for which the elastic strengths are 250 MPa in tension and 150 MPa in shear, and the ultimate strengths are 450 MPa in tension and 270 MPa in shear. Determine

(a) The axial stress in rod AB.

(b) The cross shearing stress in the pin at C.

(c) The factor of safety with respect to failure by slip for rod AB.

(d) The factor of safety with respect to failure by fracture for pin C.

FIG. P3–40

FIG. P3–41

(3-42)*   The 1/2-in. diameter rod *CD* and the 1/2-in. diameter rivet *B* (which is in single shear) of Fig. P3–42 are made of aluminum alloy 2014-T4. Determine

(a) The axial stress in rod *CD*.

(b) The cross shearing stress in the rivet at *B*.

(c) The factor of safety with respect to failure by slip for rod *CD*.

(d) The factor of safety with respect to failure by fracture for the rivet at *B*.

3-43   The force *P* in Fig. P3–43 is supported by a 6-mm carbon steel aircraft cable having a length *L* and a breaking strength of 36 kN. Determine the location of the force *P* (distance *a* in terms of *L*) and the maximum value of *P* that can be supported with a factor of safety of 2 with respect to failure by fracture.

FIG. P3–42   FIG. P3–43

FIG. P3–44   FIG. P3–45

3-44   Member *BE* of the pin-connected frame of Fig. P3–44 is a 1 × 3/8-in. rectangular bar made of a 0.4-percent carbon steel that has elastic and ultimate strengths in tension of 53 and 84 ksi, respectively. The pin at *E* has a diameter of 1.25 in. and is made of a hardened steel that has an ultimate strength in shear of 45 ksi. The force *P* is applied to the pin outside members *BE* and *DE*. Determine

(a)   The axial stress in member *BE*.

(b)   The maximum cross shearing stress in the pin at *E*.

(c)   The factor of safety with respect to failure by slip for member *BE*.

(d)   The factor of safety with respect to failure by fracture for pin *E*.

3-45*   Member *CE* of the pin-connected frame of Fig. P3–45 is a hollow tube that has a cross-sectional area of 1200 mm² and is made of an aluminum alloy that has elastic and ultimate strengths in compression of 280 and 470 MPa, respectively. The pin at *C* has a diameter of 20 mm and is made of an aluminum alloy that has an ultimate strength in shear of 280 MPa. The cable is attached to the pins and passes over a smooth pulley at *D*. Determine

(a)   The axial stress in tube *CE*.

(b)   The maximum cross shearing stress in the pin at *C*.

(c)   The factor of safety with respect to failure by slip for member *CE*.

(d)   The factor of safety with respect to failure by fracture for pin *C*.

CHAPTER 4

AXIAL LOADING APPLICATIONS AND PRESSURE VESSELS

4-1 INTRODUCTION

The problem of determining internal forces and deformations at all points within a body subjected to external forces is extremely difficult when the loading or geometry of the body is complicated. A refined analytical method of analysis that attempts to obtain general solutions to such problems is known as the *theory of elasticity*. The number of problems solved by such methods has been limited; therefore practical solutions to most design problems are obtained by what has become known as the *mechanics of materials approach*. With this approach, real structural elements are analyzed as idealized models subjected to simplified loadings and restraints. The resulting solutions are approximate, since they consider only the effects that significantly affect the magnitudes of stresses, strains, and deformations.

In Chapters 1 and 2, the concepts of stress and strain were developed and a discussion of material behavior in Chapter 3 led to the development of equations relating stress to strain. In the remaining chapters of the book, the stresses and deformations produced in a wide variety of structural members by axial, torsional, and flexural loadings will be considered. The "mechanics of material analyses," as presented here, are somewhat less rigorous than the "theory of elasticity approach," but experience indicates that the results obtained are quite satisfactory for most engineering problems.

4-2   DEFORMATIONS IN AXIALLY LOADED BARS

When a bar of uniform cross section is axially loaded by forces applied at the ends (two-force member), the axial strain along the length of the bar is assumed to have a constant value, and the elongation (or contraction) of the bar resulting from the axial load P may be expressed as $\delta = \varepsilon L$ by definition. If Hooke's law applies, Eq. 3–5 of Art. 3–3 reduces to $\varepsilon = \sigma/E$, where σ is the axial stress P/A induced by the load. Thus, the axial deformation may be expressed in terms of stress, or load, as follows:

$$\delta = \varepsilon L = \frac{\sigma L}{E} \tag{4-1a}$$

or

$$\delta = \frac{PL}{AE} \tag{4-1b}$$

The first form will frequently be found convenient in elastic problems in which limiting axial stress and axial deformation are both specified. The stress corresponding to the specified deformation can be obtained from Eq. 4–1a and compared to the specified working stress, the smaller of the two values being used to compute the unknown load or cross-sectional area. In general, Eq. 4–1a is the preferred form when the problem involves the determination or comparison of stresses. If a bar is subjected to a number of axial loads applied at different points along the bar, or if the bar consists of parts having different cross-sectional areas or parts composed of different materials, then the change in length of each part can be computed by using Eq. 4–1b. These individual changes in length of the various parts of the bar can then be added algebraically to give the total change in length of the complete bar.

For those cases in which the axial force or the cross-sectional area varies continuously along the length of the bar, Eq. 4–1 is not valid. In Art. 2–3 the axial strain at a point for the case of nonuniform deformation was defined as $\varepsilon = d\delta/dL$. Thus, the increment of deformation associated with a differential element of length $dL = dx$ may be expressed as $d\delta = \varepsilon dx$. If Hooke's law applies, the strain may again be expressed as $\varepsilon = \sigma/E$, where $\sigma = P_x/A_x$. The subscripts indicate that both the applied load P and the cross-sectional area A may be functions of position x along the bar. Thus,

$$d\delta = \frac{P_x\,dx}{A_x E} \tag{a}$$

Integrating Eq. a yields the following expression for the total elongation (or contraction) of the bar:

$$\delta = \int_0^L d\delta = \int_0^L \frac{P_x\,dx}{A_x E} \tag{4-2}$$

Equation 4–2 gives acceptable results for tapered bars, provided the angle between the sides of the bar does not exceed 20°.

EXAMPLE 4–1
The rigid yokes B and C of Fig. 4–1a are securely fastened to the 2-in. square steel ($E = 30,000$ ksi) bar AD. Determine

(a) The maximum normal stress in the bar.
(b) The change in length of the complete bar.

SOLUTION
Since bar AD is subjected to a number of axial loads applied at different points along the bar, the different sections AB, BC, and CD of the bar will transmit different levels of load. An internal force (load) diagram, such as the one shown in Fig. 4–1b, provides a pictorial representation of the levels of internal force in each of the sections and serves as an aid for stress and deformation calculations.

The load diagram is easily constructed by isolating different portions of the bar with imaginary cuts and computing the internal force that must be

(a)

(b)

FIG. 4–1

transmitted by the cross section exposed by the cut to maintain equilibrium of the isolated portion of the bar. For example, an imaginary cut anywhere in section AB will expose a cross section that must transmit 82 kips to maintain equilibrium of the portions of the bar on either side of the cut.

(a) Examination of the load diagram indicates that the maximum load transmitted by any section of the bar is 82 kips; therefore, since the bar has a uniform cross section,

$$\sigma_{max} = \frac{P_{max}}{A} = \frac{82}{4} = \underline{20.5 \text{ ksi } T} \qquad \text{Ans.}$$

(b) Since different parts of the bar are transmitting different levels of load, Eq. 4–1 must be used to determine the changes in length associated with each of the different parts of the bar. These individual values are then added algebraically to give the change in length of the complete bar. Thus,

$$\delta_{AD} = \sum \frac{PL}{AE} = \delta_{AB} + \delta_{BC} + \delta_{CD}$$

$$= \frac{82(8)(12)}{4(30,000)} - \frac{12(5)(12)}{4(30,000)} + \frac{45(4)(12)}{4(30,000)}$$

$$= 0.0656 - 0.0060 + 0.0180$$

$$= \underline{0.0776 \text{ in. increase}} \qquad \text{Ans.}$$

PROBLEMS

4-1* Specifications for a steel ($E = 30,000,000$ psi) machine part require that the axial stress not exceed 20,000 psi and that the total elongation not exceed 0.015 in. If the part is a bar 2 in. in diameter and 3 ft long, determine the greatest permissible axial load to which it may be subjected.

4-2* A structural tension member of aluminum alloy 6061-T6 has a rectangular cross section 25 × 75 mm and is 2 m long. Determine the maximum permissible axial load that may be applied if the total elongation of the member is not to exceed 4 mm and the minimum factor of safety with respect to failure by fracture is to be 3. The modulus of elasticity, elastic strength, and ultimate strength of the aluminum are 70 GPa, 270 MPa, and 310 MPa, respectively.

4-3* A steel bar 10 ft long has a 2-in. diameter over one-half its length and a

1.5-in. diameter over the other half. The modulus of elasticity and elastic strength for the steel are 30×10^6 psi and 53,000 psi, respectively.

(a) How much will the bar elongate under a tensile load of 30 kips?

(b) How much will a bar of uniform cross section, which has the same weight as the bar of part (a), elongate under the action of the 30-kip load?

(c) What are the factors of safety with respect to failure by slip for the two bars of parts (a) and (b)?

4-4. A steel rod, which has a diameter of 25 mm and a length of 1.50 m, is attached to the end of a bronze tube, which has an internal diameter of 25 mm, a wall thickness of 10 mm, and a length of 3.00 m. The moduli of elasticity and elastic strengths for the steel and the bronze are 200 and 100 GPa and 250 and 140 MPa, respectively. Determine

$,45 mm$

$\frac{\pi}{4}\left(OD^2 - ID^2\right)$

(a) The load required to stretch the assembly 3.00 mm.

(b) The factor of safety with respect to failure by slip when the load of part (a) is applied.

4-5* The tension member shown in Fig. P4-5 consists of a steel pipe A, which has an outside diameter of 6 in. and an inside diameter of 4.5 in., and a solid aluminum alloy bar B, which has an outside diameter of 4 in. The moduli of elasticity for the steel and aluminum are 29×10^6 psi and 10.6×10^6 psi, respectively. Determine the overall elongation of the member under the action of the loads indicated on the figure.

4-6 The compression member shown in Fig. P4–6 consists of a solid aluminum bar A, which has an outside diameter of 100 mm, a brass tube B, which has an outside diameter of 150 mm and an inside diameter of 100 mm, and a steel pipe C, which has an outside diameter of 200 mm and an inside diameter of 125 mm. The moduli of elasticity of the aluminum, brass, and steel are

FIG. P4–5

FIG. P4–6

FIG. P4–7

FIG. P4–8

73, 100, and 210 GPa, respectively. Determine the overall shortening of the member under action of the indicated loads.

4-7   A steel (E = 30,000 ksi) bar of rectangular cross section consists of uniform and tapered sections as shown in Fig. P4–7. The width of the tapered section varies linearly from 5 in. at the top to 2 in. at the bottom. The bar has a constant thickness of 1/2 in. Determine the elongation of the bar resulting from application of the 30,000-lb load. Neglect the weight of the bar.

4-8*   An aluminum alloy (E = 70 GPa) bar of circular cross section consists of uniform and tapered sections as shown in Fig. P4–8. The diameter of the tapered section varies linearly from 50 mm at the top to 20 mm at the bottom. The uniform section at the top is hollow and has a wall thickness of 10 mm. Determine the elongation of the bar resulting from application of the 75-kN load. Neglect the weight of the bar.

FIG. P4–10

FIG. P4–11

4-9 A bar of uniform cross section and homogeneous material hangs vertically while suspended from one end. Determine, in terms of W, L, A, and E, the extension of the bar due to its own weight W.

4-10* Determine the extension, due to its own weight, of the homogeneous conical bar of Fig. P4–10. Express the result in terms of L, E, and the specific weight γ of the material. Assume that the taper of the bar is slight enough for the assumption of a uniform axial stress distribution over a cross section to be valid.

4-11 Determine the extension, due to its own weight, of the homogeneous bar of Fig. P4–11. Express the result in terms of L, E, and the specific weight γ of the material. Assume that the taper of the bar is slight enough for the assumption of a uniform axial stress distribution over a cross section to be valid.

4-3   DEFORMATIONS IN A SYSTEM OF AXIALLY LOADED BARS

It is sometimes necessary to determine axial deformations and strains in a loaded system of pin-connected deformable bars (two-force members). The problem is approached through a study of the geometry of the deformed system, from which the axial deformations δ of the various bars in the system are obtained. Suppose, for example, one is interested in the axial deformations of bars AB, BC, and BD of Fig. 4–2 in which the solid lines represent the unstrained (unloaded) configuration of the system and the dashed lines represent the configuration due to a force applied at B. From

FIG. 4–2

the Pythagorean theorem, the axial deformation in bar AB is

$$\delta_{AB} = \sqrt{(L + y)^2 + x^2} - L$$

Transposing the last term and squaring both sides gives

$$\delta_{AB}^2 + 2L\delta_{AB} + L^2 = L^2 + 2Ly + y^2 + x^2$$

If the displacements are small (the usual case for stiff materials and elastic action), the terms involving the squares of the displacements may be neglected; hence,

$$\delta_{AB} \approx y$$

In a similar manner,

$$\delta_{BD} \approx x$$

The axial deformation in bar BC is

$$\delta_{BC} = \sqrt{(R \cos \theta - x)^2 + (R \sin \theta + y)^2} - R$$

Transposing the last term and squaring both sides gives

$$\delta_{BC}^2 + 2R\delta_{BC} + R^2 = R^2 \cos^2\theta - 2Rx \cos \theta \\ + x^2 + R^2 \sin^2\theta + 2Ry \sin \theta + y^2$$

Neglecting small second-degree terms and noting that $\sin^2\theta + \cos^2\theta = 1$, one obtains

$$\delta_{BC} \approx y \sin \theta - x \cos \theta$$

or in terms of the deformations of the other two bars,

$$\delta_{BC} \approx \delta_{AB} \sin \theta - \delta_{BD} \cos \theta$$

The geometric interpretation of this equation is indicated by the shaded right triangles of Fig. 4–3.

FIG. 4–3

The general conclusion that may be drawn from the above discussion is that, *for small displacements,* the axial deformation in any bar may be assumed equal to the component of the displacement of one end of the bar (relative to the other end) taken in the direction of the unstrained orientation of the bar. Any rigid members of the system will change orientation or position but will not be deformed in any manner. For example, if bar *BD* of Fig. 4–2 was rigid and subjected to a small downward rotation, point *B* could be assumed to be displaced vertically through a distance *y*, and δ_{BC} would be equal to $y \sin \theta$.

PROBLEMS

Note: For the problems in this article, assume small deformations unless otherwise noted.

4-12* The rigid bar *DC* of Fig. P4–12 is horizontal under no load, and bars *A* and *B* are unstressed. When the load *P* is applied, the strain in bar *B* is found to be 0.0015. Determine

(a) The factor of safety with respect to failure by slip for bar *A*.

(b) The load *P*.

(c) The vertical displacement (deflection) of point *C*.

FIG. P4–12

The following data apply:

Bar	Young's Modulus, psi	Yield Strength for 0.2% Offset, psi	Cross-Sectional Area, sq. in.
A (steel)	$30(10)^6$	50,000	2.0
B (brass)	$15(10)^6$	40,000	1.0

4-13*   The rigid bar CD of Fig. P4–13 is horizontal under no load, and bars A and B are unstressed. When the load P is applied, the strain in bar B is found to be 0.0012. Determine

    (a)   The factor of safety with respect to failure by slip for bar A.

    (b)   The load P.

    (c)   The vertical displacement (deflection) of point D.

The following data apply:

Bar	Young's Modulus, GPa	Yield Strength for 0.2% Offset, MPa	Cross-Sectional Area, mm²
A (aluminum alloy)	73	280	900
B (steel)	210	430	1200

4-14   Bar A of Fig. P4–14 is made of an aluminum alloy with a proportional limit of 55.0 ksi and a modulus of elasticity of 10.5×10^3 ksi. Bars B are made of structural steel with a yield point of 36 ksi and a modulus of elastic-

FIG. P4–13

FIG. P4–14   FIG. P4–17

ity of 29×10^3 ksi. All bars have rectangular cross sections 2.5 by 1.0 in. and are unstressed when P is zero. When the load P is applied, the strain in bar A is found to be 0.0010. Determine

    (a)   The axial stresses in the bars.

    (b)   The load P.

    (c)   The displacement of pin C.

4-15*   Solve Problem 4-14 when there is a clearance of 0.04 in. between the pin at C and bar A. The load is applied to pin C, which fits snuggly in bars B and moves downward 0.04 in. before contacting bar A.

4-16   Solve Problem 4-14 when there is a clearance of 0.06 in. between the pin at C and bar A. The load is aplied to pin C, which fits snuggly in bars B and moves downward 0.06 in. before contacting bar A.

4-17*   Three bars, each 50 mm wide by 25 mm thick by 4 m long, are connected and loaded as shown in Fig. P4–17. Bar A is made of Monel, which has a proportional limit of 400 MPa and a modulus of elasticity of 180 GPa. Bar B is made of a Magnesium alloy that has a proportional limit of 100 MPa and a modulus of elasticity of 40 GPa. Bar C is made of structural steel that has a yield point of 240 MPa and a modulus of elasticity of 200 GPa. When the load P is applied, the strain in bar B is found to be 0.0010. Determine

    (a)   The axial stresses in the bars.

    (b)   The load P.

    (c)   The displacement of the pin at D.

4-18 Solve Problem 4-17 when there is an initial clearance of 1.75 mm between the pin at D and bar B.

4-19* Solve Problem 4-17 when there is an initial clearance of 3.00 mm between the pin at D and bar B.

4-20 Rod AC of Fig. P4–20 is 8 ft long, has a cross-sectional area of 1.00 sq. in., and is made of an aluminum alloy that has a modulus of elasticity of 10.8×10^6 psi. Rod BC is 12 ft long, has a cross-sectional area of 0.62 sq. in., and is made of structural steel that has a modulus of elasticity of 29×10^6 psi. Determine

    (a)   The axial stresses in the rods.

    (b)   The lengthening of each of the rods.

    (c)   The horizontal and vertical displacements of point C.

4-21 Bar AB and pipe strut BC of Fig. P4–21 are made of structural steel that has a modulus of elasticity of 200 GPa. The cross-sectional areas are 650 mm^2 and 925 mm^2 for AB and BC, respectively. Determine

    (a)   The axial stresses in bar AB and strut BC.

    (b)   The lengthening or shortening of bar AB and strut BC.

    (c)   The horizontal and vertical displacements of point B.

4-4 STATICALLY INDETERMINATE AXIALLY LOADED MEMBERS

In the work so far, it has always been possible to find the internal forces in any member of a structure by means of the equations of equilibrium; that is, the structures were statically determinate. In any structure, if the number of unknown forces and distances exceeds the number of independent equations

FIG. P4–20   FIG. P4–21

of equilibrium that are applicable, the structure is said to be *statically inde-terminate,* and it is necessary to write additional equations involving the geometry of the deformations in the members of the structure. The following outline of procedure will be helpful in the analysis of problems involving statically indeterminate structures.

1. Draw a free-body diagram.
2. Note the number of unknowns involved (magnitudes and po-sitions).
3. Recognize the type of force system on the free-body diagram and note the number of independent equations of equilibrium available for this system.
4. If the number of unknowns exceeds the number of equilib-rium equations, a deformation equation must be written for each extra unknown.
5. When the number of independent equilibrium equations and deformation equations equals the number of unknowns, the equa-tions can be solved simultaneously. Deformations and forces must be related in order to solve the equations simultaneously. Hooke's law and the definitions of stress and strain can be used when all stresses are less than the corresponding proportional limits. If some of the stresses exceed the proportional limits of the materials, the stress-strain diagrams can be used to relate the loads and deformations or simplifying assumptions (such as per-fect plasticity, see Fig. 3–2d) can be made to obtain satisfactory results.

It is recommended that a displacement diagram be drawn showing deforma-tions to assist in obtaining the correct deformation equation. The displace-ment diagram should be as simple as possible (a line diagram), with the deformations indicated with exaggerated magnitudes and clearly dimen-sioned. Loading members (members assumed to be rigid), especially, should be indicated by single lines. Note that an equilibrium equation and the corre-sponding deformation equation *must be compatible*; that is, when a tensile force is assumed for a member in the free-body diagram, a tensile deforma-tion must be indicated for the same member in the deformation diagram. If the diagrams are compatible, a negative result will indicate that the assump-tion was wrong; however, the magnitude of the result will be correct.

    In all the following problems and examples, the loading members, pins, and supports are assumed to be rigid, and the mechanism is so constructed that the force system is coplanar. The procedure outlined above is illustrated in the following examples.

EXAMPLE 4–2

Bar *CD* in the pin-connected structure of Fig. 4–4*a* is to be considered rigid and is horizontal before loads are applied to the structure. Member *A* is an aluminum alloy bar with a modulus of elasticity of 75 GPa, a proportional limit of 270 MPa, and a cross-sectional area of 1000 mm². Member *B* is a structural steel bar with a modulus of elasticity of 200 GPa, a yield point of 250 MPa, and a cross-sectional area of 500 mm². Determine the axial stresses in bars *A* and *B* and the deflection of point *D* when a load of 150 kN is applied to the structure at point *D*.

SOLUTION

A free-body diagram of member *CD* and portions of members *A* and *B* is shown in Fig. 4–4*b*. The free-body diagram contains four unknown forces: C_x, C_y, F_A, and F_B; therefore, since only three equilibrium equations are available, the problem is statically indeterminate. As the load *P* is applied to member *CD*, it will tend to rotate clockwise about pin *C* and produce deformations in members *A* and *B* as shown in Fig. 4–4*c*. The extensions shown in Fig. 4–4*c* are compatible with the tensile forces shown in members *A* and *B* in Fig. 4–4*b*. The unknown reaction at *C* is not needed to complete the solution of the problem and can be eliminated from further consideration by

FIG. 4–4

summing moments about pin C. The equilibrium and deformation equations needed to solve for F_A and F_B are

$$\left(\!\!\!\!+\right) \quad \Sigma M_C = 0 = P(5) - F_A(4) - F_B(1.5)$$

or

$$4F_A + 1.5F_B = 150(5) = 750 \tag{a}$$

and from similar triangles in Fig. 4–4c

$$\frac{\delta_A}{4} = \frac{\delta_B}{1.5} = \frac{F_A L_A}{4A_A E_A} = \frac{F_B L_B}{1.5 A_B E_B}$$

$$= \frac{F_A(2)}{4(1000)(10^{-6})(75)(10^9)} = \frac{F_B(1.5)}{1.5(500)(10^{-6})(200)(10^9)}$$

or

$$F_A = 1.5 F_B \tag{b}$$

Solving Eqs. a and b simultaneously yields

$$F_A = 150 \text{ kN} \qquad F_B = 100 \text{ kN}$$

The stresses in the two bars are

$$\sigma_A = \frac{F_A}{A_A} = \frac{150(10^3)}{1000(10^{-6})} = 150(10^6) \text{ N/m}^2 = \underline{150 \text{ MPa } T} \qquad \text{Ans.}$$

$$\sigma_B = \frac{F_B}{A_B} = \frac{100(10^3)}{500(10^{-6})} = 200(10^6) \text{ N/m}^2 = \underline{200 \text{ MPa } T} \qquad \text{Ans.}$$

These results indicate that stresses in both bars are below the proportional limits in the materials; therefore, the assumption of linear elastic behavior used in the development of Eq. 4–1b is valid.

The deflection of point D can be determined by observing in Fig. 4–4c that

$$\delta_D = \frac{5}{4} \delta_A = \frac{5(150)(10^3)(2)}{4(1000)(10^{-6})(75)(10^9)}$$

$$= 5.00(10^{-3}) \ m = \underline{5.00 \text{ mm downward}} \qquad \text{Ans.}$$

EXAMPLE 4–3

The rigid plate C in Fig. 4–5a is fastened to the 0.50-in. diameter steel rod A and to the aluminum pipe B. The other ends of A and B are fastened to rigid

FIG. 4–5

(a)

(b)

(c)

(d)

supports. When the force P is zero, there are no stresses in A and B. Rod A is made of low-carbon steel assumed to be elastoplastic (see Fig. 3–2d) with a proportional limit and yield point of 40 ksi ($E = 30,000$ ksi). Pipe B has a cross-sectional area of 2.0 sq. in. and is made of an aluminum alloy with the stress-strain diagram shown in Fig. 4–5b. A load P of 30 kips is applied to C as shown. Determine the axial stresses in A and B and the displacement of plate C.

SOLUTION
Figure 4–5c is a free-body diagram of plate C and portions of the members A

and B. The free-body diagram contains two unknown forces, and only one equation of equilibrium is available; therefore, the problem is statically indeterminate. As the load is applied to the plate C, it moves downward an amount δ, which represents the total deformation in members A and B; see Fig. 4–5d. This observation provides a displacement equation that can be used with the equilibrium equation to solve the problem. The two equations are

$$(\uparrow+) \qquad \Sigma F_y = 0 = P_A + P_B - 30$$

or

$$(\pi/4)(0.50)^2\sigma_A + 2\sigma_B = 30 \qquad (a)$$

and

$$\delta_A = \delta_B$$

or

$$10\varepsilon_A = 20\varepsilon_B$$

In order to solve these equations, the stresses and strains must be related. If the stresses are less than the corresponding proportional limits, Hooke's law can be used. If Hooke's law is assumed to be valid, the strain equation becomes

$$\frac{10\sigma_A}{30(10^3)} = \frac{20\sigma_B}{10(10^3)}$$

from which

$$\sigma_A = 6\sigma_B \qquad (b)$$

The value of $E_B = 10(10^3)$ ksi is obtained as the slope of the stress-strain diagram in Fig. 4–5b. The simultaneous solution of Eqs. a and b gives $\sigma_A = 56.6$ ksi T and $\sigma_B = 9.45$ ksi C.

These results indicate that bar A is stressed beyond the proportional limit of the steel and that Hooke's law does not apply. Since the material is assumed to be perfectly plastic beyond the proportional limit, the stress in bar A must be 40 ksi. When this value is substituted in equilibrium Eq. a, the equation becomes

$$(\pi/16)(40) + 2\sigma_B = 30$$

from which

$$\sigma_B = \underline{11.07 \text{ ksi } C} \qquad \text{and} \qquad \sigma_A = \underline{40 \text{ ksi } T} \qquad \text{Ans.}$$

The stress in B is less than the proportional limit for aluminum; therefore, Hooke's law can be used to determine the movement of the plate, which is

$$\delta = \delta_B = 20(11.07)/(10^4) = \underline{0.0221 \text{ in. downward}} \qquad \text{Ans.}$$

If the stress in B had been more than the proportional limit of the aluminum, it would have been necessary to determine the strain from the curve in Fig. 4–5b. If the material in bar A were not perfectly plastic, a trial-and-error solution would be required, since the stresses in both A and B depend on the movement of C.

EXAMPLE 4–4

The steel post B of Fig. 4–6a is 2 by 2 in. square, and the aluminum alloy bar C is 1.5 in. wide by 1 in. thick. Bar A and the bearing block on B are to be

FIG. 4–6   (b)   (c)

considered rigid; all connections are treated as smooth pins; the temperature is assumed to remain constant; the modulus of elasticity of the aluminum alloy is taken as (10^7) psi. Under no load, the clearance between A and B is 0.002 in. Determine the maximum permissible load P that may be applied as shown if the axial stresses are not to exceed 15,000 psi for the steel and 18,000 psi for the aluminum alloy.

SOLUTION

The free-body diagram of body A in Fig. 4–6b shows five unknown forces; however, one of the axial stresses, either σ_B or σ_C, will equal the allowable value (which is given), and the other one will be less than the allowable stress (or equal to it in an exceptional case). The force system is coplanar, for which three independent equations of equilibrium are available. With four unknowns and three equations, the system is statically indeterminate, and an additional equation must be obtained from the deformations of B and C. The displacement diagram of Fig. 4–6c shows the positions of the members, after the load P is applied, as dashed lines. The deformations of members B and C are related by the equation (from similar triangles)

$$\frac{\delta_B + 0.002}{4} = \frac{\delta_C}{12}$$

or

$$3\delta_B + 0.006 = \delta_C$$

In terms of strain $(\delta = L\varepsilon)$, the relation is

$$3(12)\varepsilon_B + 0.006 = 24\varepsilon_C$$

Since the stresses are limited to values less than the proportional limit (see App. A), Hooke's law $(\sigma = E\varepsilon)$ applies, and the equation becomes

$$\frac{36\sigma_B}{30(10^6)} + 0.006 = \frac{24\sigma_C}{10(10^6)}$$

or

$$\sigma_B + 5000 = 2\sigma_C$$

When σ_B equals its allowable value of 15,000 psi, σ_C becomes 10,000 psi as determined from this strain relation. It is apparent that higher values for σ_C would yield a stress in the steel above the specified limit. Therefore, the axial stresses are 15,000 psi in the steel and 10,000 psi in the aluminum. From Fig. 4–6b and the equilibrium equation $\Sigma M_D = 0$,

$$4(15,000)4 + 12(10,000)(3/2) - 12P = 0$$

from which

$$P = \underline{35{,}000 \text{ lb as shown}} \qquad\qquad \text{Ans.}$$

EXAMPLE 4–5

If the mechanism of Exam. 4–4 is loaded with a force P of 10,000 lb as shown and the temperature increases 20° F, what are the axial stresses in members B and C? For the coefficients of thermal expansion per degree F, use $13(10^{-6})$ for the aluminum and $6.5(10^{-6})$ for the steel.

SOLUTION

The free-body diagram in Fig. 4–6b is valid for this problem. Following the suggestion of Art. 3–5, the thermal deformation is first permitted to take place by removing all restraints (bar A is imagined to be removed). The deformations of B and C due to the change of temperature are indicated as δ_T values in Fig. 4–7a. Loads are then applied to B and C, which change their lengths so that the rigid bar A can be connected to them, as indicated in Fig.

FIG. 4–7

4–7b. The deformations due to the forces are indicated as δ_P values. The deformation equation is

$$\frac{\delta_{BP} - \delta_{BT} + 0.002}{4} = \frac{\delta_{CP} + \delta_{CT}}{12}$$

or

$$3\delta_{BP} - 3\delta_{BT} + 0.006 = \delta_{CP} + \delta_{CT}$$

When deformations are replaced by their values in terms of strains, the equation becomes

$$3(12)\varepsilon_{BP} - 3(12)\varepsilon_{BT} + 0.006 = 24\varepsilon_{CP} + 24\varepsilon_{CT}$$

When the strains due to stress are replaced by their values in terms of stress (if Hooke's law is assumed to be valid) and the strains due to temperature change by their numerical values, the displacement equation becomes

$$\frac{36\sigma_B}{30(10^6)} - \frac{36(6.5)20}{(10^6)} + \frac{6}{(10^3)} = \frac{24\sigma_C}{10(10^6)} + \frac{24(13.0)20}{(10^6)}$$

or

$$\sigma_B = 2\sigma_C + 4100 \qquad\qquad (c)$$

From Fig. 4–6b and the equilibrium equation $\Sigma M_D = 0$,

$$4\sigma_B(4) + (3/2)\sigma_C(12) - 10{,}000(12) = 0 \qquad\qquad (d)$$

The simultaneous solution of Eqs. c and d gives

$$\sigma_C = \underline{1088 \text{ psi } T} \qquad\qquad\qquad\text{Ans.}$$

and

$$\sigma_B = \underline{6280 \text{ psi } C} \qquad\qquad\qquad\text{Ans.}$$

These stresses are well within the proportional limits of the materials; therefore, Hooke's law is valid.

The use of colored pencils may be helpful in the construction of the deformation diagram. Use one color for temperature deformations and another for load deformations.

Fig. 4–7a indicates that with no applied load ($P = 0$), the bar A might have occupied some position such as that shown dashed (marked A') and would therefore have induced compression in both bars B and C. In the solution given, the assumption was made that the force P was of sufficient magnitude to induce tension in bar C. The positive sign on the value for σ_C

indicates that the assumption is correct *if the deformation diagram and the free-body diagram are compatible*; that is, when a tensile deformation is assumed in the deformation diagram, a tensile stress must be indicated on the free-body diagram. This very important consideration must not be overlooked.

PROBLEMS

> *Note: Unless otherwise specified, all members are assumed to have negligible weight, and all pins used for connections are assumed to be smooth. The temperature remains constant unless otherwise specified.*

4-22* Nine 1-in. diameter steel ($E = 30,000$ ksi) reinforcing bars are used in the short concrete ($E = 3000$ ksi) pier shown in Fig. P4–22. A load of 200,000 lb is applied to the pier through a rigid capping plate. Determine

    (a)   The stresses in the concrete and in the steel rods.

    (b)   The shortening of the pier.

4-23* The $150 \times 150 \times 500$-mm oak ($E = 12$ GPa) block shown in Fig. P4–23 was reinforced by bolting two $50 \times 150 \times 500$-mm steel ($E = 200$ GPa) plates to opposite sides of the block. If the stresses in the wood and the steel

FIG. P4–22

FIG. P4–23

FIG. P4–24

FIG. P4–25

are to be limited to 32 MPa and 150 MPa, respectively, determine

   (a)   The maximum compressive load that can be applied to the rein-
         forced block.

   (b)   The shortening of the block when the load of part (a) is applied.

4-24   The assembly shown in Fig. P4–24 consists of a steel bar A ($E =$
30,000 ksi and $A = 1.00$ in.^{2}), a rigid bearing plate C that is securely fastened
to bar A, and a bronze bar B ($E = 15,000$ ksi and $A = 4.00$ in.^{2}). The elastic
strengths of the steel and bronze are 62 ksi and 75 ksi, respectively. A
clearance of 0.015 in. exists between the bearing plate C and bar B before the
assembly is loaded. After a load of 75,000 lb is applied to the bearing plate,
determine

   (a)   The axial stresses in bars A and B.

   (b)   The factors of safety with respect to failure by slip for each of the
         members.

   (c)   The vertical displacement of bearing plate C.

4-25*   The assembly shown in Fig. P4–25 consists of a steel cylinder A
($E = 210$ GPa) that has an inside diameter of 160 mm and a wall thickness of
15 mm, a rigid bearing plate C, and an aluminum alloy bar B ($E = 73$ GPa)
that has an outside diameter of 50 mm. The elastic strengths of the steel and
aluminum are 430 MPa and 330 MPa, respectively. An axial load of 2500 kN
is applied to the bearing plate as shown in Fig. P4–25. Determine

   (a)   The axial stresses in members A and B.

   (b)   The factors of safety with respect to failure by slip for each of the
         members.

   (c)   The horizontal displacement of the bearing plate.

4-26   Solve Problem 4–24 if the temperature of bar A increases 100 °F and
the temperature of bar B increases 50 °F after the load P is applied. The
coefficients of thermal expansion are 6.6 $(10^{-6})/°F$ and 9.4$(10^{-6})/°F$ for the
steel and bronze, respectively.

FIG. P4–28

FIG. P4–30

4-27*   Solve Problem 4–25 if the temperature of cylinder A increases 50 °C and the temperature of bar B increases 25 °C after the load P is applied. The coefficients of thermal expansion are 11.9 (10^{-6})/°C and 22.5 (10^{-6})/°C for the steel and aluminum, respectively.

4-28   Bars A and B of the pin-connected structure of Fig. P4–28 are connected to fixed supports at the top and to a rigid bar CD at the bottom. Bar CD is horizontal under no load, and bars A and B are unstressed. After the 30-kip load is applied to the structure, determine

(a)   The axial stresses in bars A and B.

(b)   The factors of safety with respect to failure by slip for each of the bars.

(c)   The vertical displacement of point D.

The following data apply:

Bar	Young's Modulus, ksi	Yield Strength for 0.2% Offset, ksi	Cross-Sectional Area, in.²
A (aluminum alloy)	10,600	41	2.00
B (steel)	29,000	36	2.50

4-29   Solve Problem 4-28 if a clearance of 0.025 in. exists in the connection at the bottom of bar B before the 30-kip load is applied.

4-30*   Bar CD of the pin-connected structure of Fig. P4–30 is to be considered rigid and is horizontal when the structure is not loaded. Bars *A* and *B* are unstressed before the load *P* is applied. After the 24-kN load *P* is applied, determine

(a) The axial stresses in bars *A* and *B*.

(b) The factor of safety with respect to failure by slip for each of the bars.

(c) The vertical displacement of point *C*.

The following data apply:

Bar	Young's Modulus, GPa	Yield Strength for 0.2% Offset, MPa	Cross-Sectional Area, mm²
A (steel)	200	500	150
B (brass)	100	400	200

4-31 Solve Problem 4-30 if the temperature of bar *A* increases 100 °C and the temperature of bar *B* increases 50 °C after the 24-kN load is applied. The coefficients of thermal expansion are 11.9 (10⁻⁶)/°C for the steel and 17.6 (10⁻⁶)/°C for the brass.

4-32* Bar *A* of Fig. P4–32 is a steel rod ($E = 30,000$ ksi and $A = 1.25$ in.²). Member *B* is a brass post ($E = 15,000$ ksi and $A = 4.00$ in.²). Determine the maximum permissible value for the load *P* if the allowable axial stresses are 30 ksi for the steel and 20 ksi for the brass.

4-33 Bar *B* of the pin-connected system of Fig. P4–33 is made of an alumi-

FIG. 4–32

FIG. P4–33

FIG. P4–34

num alloy (E = 70 GPa), and bar A is made of a hardened carbon steel (E = 210 GPa). The cross-sectional areas are 1200 mm² for A and 300 mm² for B. Bar CDE is to be considered rigid. When the system is unloaded, there is a 0.125-mm clearance in the connection at D, which must be taken up before bar A will be stressed. Determine the maximum permissible load that may be applied downward at a point 150 mm to the right of D if the allowable axial stresses are 215 MPa for the steel and 135 MPa for the aluminum alloy.

4-34 In Fig. P4–34, bar B is made of steel (E = 30,000 ksi) and bar A is made of an aluminum alloy (E = 10,600 ksi). Each bar has a cross-sectional area of 1.50 sq. in. Determine the maximum permissible value for the load P if the allowable axial stresses are 20 ksi for the steel and 10 ksi for the aluminum alloy.

4-35*   Bar BF of Fig. P4–35 is made of steel (E = 210 GPa) and bar CE is

FIG. P4–35

FIG. P4–36   FIG. P4–37

Rigid washer

made of aluminum alloy ($E = 73$ GPa). The cross-sectional areas are 1000 mm² for BF and 1500 mm² for CE. As a result of a misalignment of the pinholes at A, B, and C, a force of 45 kN must be applied upward at D, after pins A and B are in place, to permit insertion of pin C. Determine the axial stress in bar CE when the force P is removed with all pins in place.

4-36* The pin-connected structure of Fig. P4–36 occupies the position shown under no load. When the loads D ($=16$ kips) and E (unknown) are applied, the rigid bar C becomes horizontal. Bar A is made of an aluminum alloy ($E = 10,600$ ksi) and bar B is made of red brass ($E = 15,000$ ksi). The cross-sectional areas are 0.75 sq. in. for A and 2.25 sq. in. for B. Determine the weight of E and the maximum axial stress in the system.

4-37 A high-strength steel bolt passes through a brass sleeve as shown in Fig. P4–37. After the unit is assembled, the nut is tightened 1/4 turn (0.03125 in.) and the temperature is raised 50 °F. Determine the axial stresses that develop in the bolt and in the sleeve under these conditions. The following data apply:

Member	Cross-Sectional Area, in.²	Young's Modulus, ksi	Coefficient of Thermal Expansion, /°F
Bolt	0.785	30,000	6.6 (10⁻⁶)
Sleeve	1.767	15,000	9.8 (10⁻⁶)

4-38* The two faces of the clamp, shown in Fig. P4–38, are 250 mm apart when the two stainless-steel bolts connecting them are unstretched. A force P is applied to separate the faces of the clamp so that an aluminum alloy bar with a length of 250.50 mm can be inserted as shown. After the load P is removed, the temperature is raised 100 °C. Determine the axial stresses in the bolts and in the bar, and the distance between the faces of the clamps.

FIG. P4–38   FIG. P4–39

40 mm → |← 250 mm →| ← 40 mm

Aluminum alloy bar

P ←   → P

The following data apply:

Member	Cross-Sectional Area, mm²	Young's Modulus, GPa	Coefficient of Thermal Expansion, /°C
Bolt	115 (each)	190	17.3 (10^{-6})
Bar	625	73	22.5 (10^{-6})

4-39*   A load P will be supported by a structure consisting of a rigid plate A, two aluminum alloy bars B ($E = 10{,}600$ ksi), and a stainless steel bar C ($E = 28{,}000$ ksi), as shown in Fig. P4–39. The bars are unstressed when the structure is assembled at room temperature (72 °F). Each bar has a cross-sectional area of 2.00 sq. in. The coefficients of thermal expansion are 12.5 (10^{-6})/°F for the aluminum and 9.6 (10^{-6})/°F for the stainless steel. Determine the axial stresses in the bars after a 40-kip load is applied and the temperature is increased to 250 °F.

4-40   All members of the structure of Fig. P4–40 are elastic. The deflection of the center of the steel beam C with respect to the pins at the ends is given by the expression $\delta = F/28$ where δ is in mm and F is the resultant of the force P and the reaction of member A at the center of the beam in kN. Before loading, beam C just touches the rigid bearing block on top of member A. Member A is an aluminum alloy ($E = 70$ GPa and elastic strength = 270 MPa) and member B is annealed red brass ($E = 100$ GPa and elastic strength = 100 MPa). The cross-sectional areas are 2500 mm² for post A and 400 mm² each for bars B. Determine the maximum permissible load P that may be applied with a factor of safety of 2.5 with respect to failure by slip.

FIG. P4–40

FIG. P4–41

4-41* Bar AB in the system of Fig. P4–41 is to be considered rigid. Bar C is made of an elastoplastic steel (see Fig. 3–2d) and post D is made of cold-rolled brass. Determine the load P that will produce an axial stress of 30 MPa in post D. The following data apply:

Member	Young's Modulus, GPa	Elastic Strength, MPa	Cross-Sectional Area, mm^2
C (steel)	200	240	600
D (brass)	100	410	2000

4-42 Determine the vertical displacement of point A of Fig. P4–41 when the load P is 90 kN. See Problem 4-41 for material properties and cross-sectional areas.

4-43* The pin-connected structure of Fig. P4–43 consists of a cold-rolled bronze bar A and two elastoplastic steel (see Fig. 3–2d) bars B. Determine the axial stresses in the bars and the displacement of pin C when the load P is 225 kips. The following data apply:

Bar	Young's Modulus, ksi	Elastic Strength, ksi	Cross-Sectional Area, in.^{2}
A (bronze)	15,000	60	3.00
B (steel)	30,000	36	2.50

FIG. P4–43

FIG. P4–46

4-44   Solve Problem 4-43 when there is a clearance of 0.05 in. between the pin at C and bar A that must be closed before bar A carries any load.

4-45   Determine the axial stresses in the bars of Problem 4-43 if the temperature increases 100°F after the 225-kip load is applied. The coefficients of thermal expansion for the bronze and steel are 9.4 (10⁻⁶)/°F and 6.6 (10⁻⁶)/°F, respectively.

4-46*   The mechanism of Fig. P4–46 consists of an elastoplastic steel (see Fig. 3–2d) rod A, a cold-rolled brass rod B, and a rigid bar C. Determine the axial stress induced in rod A by advancing the nut 2.5 mm at the top of rod B. The following data apply:

Bar	Young's Modulus, GPa	Elastic Strength, MPa	Cross-Sectional Area, mm²
A (steel)	200	250	300
B (brass)	100	350	720

4-47   Members A, B, and C of the mechanism of Fig. P4–47 are made of low-carbon steel (see Fig. 3–2d), concrete, and annealed red brass, respectively. Bell cranks D and E are to be considered rigid. Determine the maximum permissible value of the load P if a minimum factor of safety of 2 with respect to failure by slip must be maintained in members, A, B, and C. The following data apply:

FIG. P4–47

Member	Young's Modulus, ksi	Elastic Strength, ksi	Cross-Sectional Area, in.^{2}
A (steel)	30,000	36	0.50
B (concrete)	4,000	3	4.00
C (brass)	15,000	15	1.50

4-48* Determine the axial deformations in rods A and C of Fig. P4–47 if the load P is increased until the axial stress in member B is 3 ksi. See Problem 4-47 for material properties and cross-sectional areas.

4-49 In Fig. P4–49, the steel bars A and C are pinned to the rigid member D, and the rigid yoke E is securely attached to bar C and the cold-rolled

FIG. P4–49

FIG. P4–50

brass bar B. The cross-sectional areas are 2500 mm² for A, 2500 mm² for B, and 1250 mm² for C. Use 200 GPa and 100 GPa for Young's modulus of steel and brass, respectively. The axial stresses are not to exceed 140 MPa in the steel and 210 MPa in the brass, and the deflection at E is not to exceed 1.25 mm downward. Determine the maximum allowable value for P.

4-50* In Fig. P4–50, bars A and B are connected by rigid links to a fixed support at the top and to a rigid bar CD at the bottom. Determine the axial stresses in bars A and B when the load P is 40 kips. The following data apply:

Bar	Material	Cross-Sectional Area, in.²	Proportional Limit, ksi	Young's Modulus, ksi
A	Aluminum alloy (see Fig. 4–5b)	2.00	33	10,000
B	Elastoplastic steel (see Fig. 3–2d)	2.50	36	30,000

4-51 Solve Problem 4-50 when the load is increased to 60 kips.

4-52* Determine the deflection at point C of Fig. P4–50 when the load P is increased to 65 kips. See Problem 4-50 for material properties and cross-sectional areas.

4-5 THIN-WALLED PRESSURE VESSELS

A pressure vessel is described as *thin-walled* when the ratio of the wall thickness to the radius of the vessel (an approximately cylindrical or spherical shape is implied) is so small that the distribution of normal stress on a plane perpendicular to the surface of the shell is essentially uniform through-out the thickness of the shell. Actually, this stress varies from a maximum value at the inside surface to a minimum value at the outside surface of the shell, but it can be shown by application of Eq. 4–6 in Art. 4–6 (see Prob. 4-69) that if the ratio of the wall thickness to the inner radius of the vessel is less than 0.1, the maximum normal stress is less than 5 percent greater than the average. Boilers, gas storage tanks, pipelines, metal tires, and hoops are normally analyzed as thin-walled elements. Gun barrels, certain high-pressure vessels in the chemical processing industry, and cylinders and piping for heavy hydraulic presses need to be treated as thick-walled vessels.

Problems involving thin-walled vessels subjected to liquid (or gas) pressure p are readily solved with the aid of free-body diagrams of sections of the vessels together with the fluid contained therein. In the following subsections, spherical, cylindrical, and other thin shells of revolution are considered.

Spherical Pressure Vessels   A typical thin-walled spherical pressure vessel used for gas storage is shown in Fig. 4–8. If the weights of the gas

Hortonsphere for gas storage in Superior, Wisconsin. (*Courtesy of Chicago Bridge and Iron Co., Chicago, Ill.*)   FIG. 4–8

FIG. 4–9

(a)   **(b)**

and vessel are negligible (a common situation), symmetry of loading and geometry requires that stresses on sections that pass through the center of the sphere be equal. Thus, on the small element shown in Fig. 4–9a,

$$\sigma_x = \sigma_y = \sigma_n$$

Furthermore, there are no shearing stresses on any of these planes since there are no loads to induce them. The normal stress component in a sphere is known as a meridional or axial stress and is commonly denoted as σ_m or σ_a.

The free-body diagram shown in Fig. 4–9b can be used to evaluate the stress $\sigma_x = \sigma_y = \sigma_n = \sigma_a$ in terms of the pressure p, and the radius r and thickness t of the spherical vessel. The force R is the resultant of the internal forces that act on the cross-sectional area of the sphere which is exposed by passing a plane through the center of the sphere. The force P is the resultant of the fluid forces acting on the fluid remaining within the hemisphere. From a summation of forces in the x direction,

$$R - P = 0$$

or

$$2\pi r t \sigma_a = p\pi r^2$$

from which

$$\sigma_a = \frac{pr}{2t} \tag{4–3}$$

Cylindrical Pressure Vessels   A typical thin-walled cylindrical pressure vessel used for gas storage is shown in Fig. 4–10. Normal stresses,

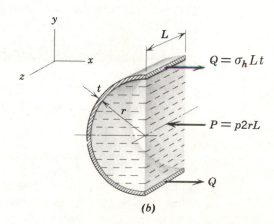

Cylindrical tanks for gas storage in Northlake, Illinois. (*Courtesy of Chicago Bridge and Iron Co., Chicago, Ill.*)   FIG. 4–10

such as those shown on the small element of Fig. 4–11a, are easy to evaluate by using appropriate free-body diagrams. The normal stress component on a transverse plane is known as an axial or meridional stress and is commonly denoted as σ_a or σ_m. The normal stress component on a longitudinal plane is known as a hoop, tangential, or circumferential stress and is denoted as σ_h, σ_t, or σ_c. There are no shearing stresses on transverse or longitudinal planes.

(a)   (b)

FIG. 4–11

The free-body diagram used for axial stress determinations is similar to Fig. 4–9b, which was used for the sphere, and the results are the same. The free-body diagram used for the hoop stress determination is shown in Fig. 4–11b. The force P is the resultant of the fluid forces acting on the fluid remaining within the portion of the cylinder isolated by the longitudinal plane and two transverse planes. The forces Q are the resultant of the internal forces on the cross-sectional area exposed by a longitudinal plane containing the axis of the cylinder. For simplicity, the pressure and stresses in the z direction have been omitted. From a summation of forces in the x direction,

$$2Q - P = 0$$

or

$$2\sigma_h L t = p 2 r L$$

from which

$$\sigma_h = \frac{pr}{t} \qquad (4\text{–}4a)$$

Also (see Eq. 4–3),

$$\sigma_a = \frac{pr}{2t} \qquad (4\text{–}4b)$$

The previous analysis of the stresses in a cylindrical vessel subjected to uniform internal pressure indicates that the stress on a longitudinal plane is twice the stress on a transverse plane. Consequently, a longitudinal joint needs to be twice as strong as a transverse (or girth) joint. As there are no shearing stresses on the longitudinal and transverse planes, they are principal planes, and σ_h and σ_a are principal stresses. The hoop stress σ_h is the maximum principal stress, and the axial stress σ_a is the intermediate principal stress. The third principal stress is zero for a point at the outside surface of the shell and is equal to $-p$ for a point at the inside surface. Therefore, the maximum shearing stress will not occur on a plane perpendicular to the surface of the vessel. The maximum shearing stress occurs at the inside surface of the vessel on planes inclined 45° to both the radial and tangential (hoop) directions and parallel to the longitudinal (axial) axis of the cylinder.

Thin Shells of Revolution   The discussion in the two previous subsections was limited to thin-walled cylindrical and spherical vessels under uniform internal pressure. The theory can be extended, however, to include other shapes and other loading conditions. Consider, for example, the thin shell of revolution shown in Fig. 4–12. Such shells are generated by rotating a plane curve, called the meridian, about an axis lying in the plane of the

FIG. 4–12

curve. Shapes that can be formed in this manner include the sphere, hemisphere, torus (doughnut), cylinder, cone, and ellipsoid. In shells of revolution, the two unknown principal stresses are a meridional stress σ_m that acts on a plane perpendicular to the meridian and a tangential stress σ_t that acts on a plane perpendicular to a parallel. Both of these stresses are shown on the small element of Fig. 4–12. The two stresses can be evaluated by using two equilibrium equations.

The small element of Fig. 4–12 is shown enlarged in Fig. 4–13. The element has a uniform thickness, is subjected to an internal pressure p, and has different curvatures in two orthogonal directions. The resultant forces on the various surfaces are shown on the diagram. Summing forces in the n direction gives

$$P - 2F_m \sin d\theta_m - 2F_t \sin d\theta_t = 0$$

FIG. 4–13

from which

$$p(2r_t d\theta_t)(2r_m d\theta_m) = 2\sigma_m(2tr_t d\theta_t) \sin d\theta_m + 2\sigma_t(2tr_m d\theta_m) \sin d\theta_t$$

and since, for small angles, $\sin d\theta \approx d\theta$, the above equation becomes

$$\frac{\sigma_m}{r_m} + \frac{\sigma_t}{r_t} = \frac{p}{t} \tag{4-5}$$

Since Eq. 4–5 contains two unknown stresses, an additional independent equation is needed. Such an equation can be obtained by considering equilibrium of a portion of the vessel above or below the parallel that passes through the point of interest. The following example illustrates the application of Eq. 4–5.

EXAMPLE 4-6
A pressure vessel of 1/4-in. steel plate has the shape of a paraboloid closed by a thick flat plate, as shown in Fig. 4–14a. The equation of the generating parabola is $y = x^2/4$, where x and y are in inches. Determine the axial and hoop stresses in the shell at a point 16 in. above the bottom due to an internal gas pressure of 250 psi gage.

SOLUTION
The stress σ_a, which must be tangent to the shell, will be determined first with the aid of the free-body diagram of Fig. 4–14b, which represents a thin slice of the vessel and gas, with the forces perpendicular to the slice omitted. From the equation of the parabola, the radius x and the slope of the shell

FIG. 4–14   (a)   (b)

dy/dx at $y = 16$ in. are determined to be 8 in. and 4/1, respectively. Summing forces in the y direction gives

$$-\int_A dP + \int dF \cos \alpha = 0$$

$$-p\pi x^2 + \sigma_a 2\pi x t \cos \alpha = 0$$

Substituting the given data yields

$$-250(\pi)(8^2) + \sigma_a(2\pi)(8)(1/4)(4/\sqrt{17}) = 0$$

from which

$$\sigma_a = 4123 = \underline{4120 \text{ psi } T} \qquad \text{Ans.}$$

In order to find σ_h from Eq. 4–5, the radii r_1 and r_2 at the point must be determined. The radius of curvature of the shell in the xy plane r_1 is determined from the expression

$$r_1 = \frac{[1 + (dy/dx)^2]^{1.5}}{d^2y/dx^2} = \frac{(1 + 4^2)^{1.5}}{1/2} = 140.19 \text{ in.}$$

and the perpendicular radius r_2 is found from the geometry of Fig. 4–14 as

$$r_2 = 8(\sqrt{17}/4) = 8.246 \text{ in.}$$

Then, from Eq. 4–5

$$p/t = (\sigma_a/r_1) + (\sigma_h/r_2)$$

$$250(4) = (4123/140.19) + (\sigma_h/8.246)$$

from which

$$\sigma_h = 8003 = \underline{8000 \text{ psi } T} \qquad \text{Ans.}$$

PROBLEMS

4-53* Determine the maximum normal and shearing stresses in a 12-in. diameter basketball that has a 1/16-in. wall thickness after it has been inflated to a gage pressure of 20 psi.

4-54* A spherical pressure vessel 12 m in diameter contains gas at a pressure of 350 kPa. Determine the force transmitted across a 200-mm length of the joint uniting the two halves of the sphere.

FIG. P4–58   FIG. P4–61

4-55   Determine the maximum normal and shearing stresses in a 40-in. diameter cylindrical compressed-air tank with hemispherical ends that has a 1/4-in. wall thickness if it is being subjected to an internal pressure of 120 psi.

4-56   Determine the thickness of steel plate required for a 2-m diameter cylindrical boiler with hemispherical ends if the tensile and shear stresses in the plate are not to exceed 85 MPa and 40 MPa, respectively, when the internal pressure in the boiler is 1500 kPa.

4-57*   At Hoover Dam, water may be discharged under a maximum head of 600 ft (pressure approximately 260 psi). What thickness of plate would be required for a 5-ft diameter pressure pipe if the working tensile stress is limited to 13,000 psi because of the welded joints?

4-58*   The cylindrical pressure tank of Fig. P4–58 is made of 10-mm steel plate. The normal stress at point A on plane $B–B$ (which is perpendicular to the surface of the plate at A) is 80 MPa T. Determine the air pressure in the tank.

4-59   A water tank 30 ft in diameter and 50 ft high is made of 1/2-in. steel plate. Determine the maximum normal stress in the tank when it is filled to a level of 40 ft. Assume that water weighs 62.4 lb per cu ft.

4-60   A cylindrical pressure vessel 1.50 m in diameter is constructed by wrapping a 15-mm thick steel plate into a spiral and butt-welding the edges of the plate. The butt-welded seams form an angle of 30° with a transverse plane through the cylinder. Determine the normal stress perpendicular to the weld and the shearing stress parallel to the weld when the internal pressure in the vessel is 1500 kPa.

FIG. P4–62   FIG. P4–63

4-61*   A cylindrical pressure tank is fabricated by butt-welding 3/4-in. plate with a spiral seam, as shown in Fig. P4–61. The pressure in the tank is 400 psi, and an axial load of 30,000 lb is applied to the end of the tank through a rigid bearing plate. Determine the normal stress perpendicular to the weld and the shearing stress parallel to the weld.

4-62   The strains measured on the outside surface of the cylindrical pressure vessel of Fig. P4–62 are $\varepsilon_1 = +720\mu$ and $\varepsilon_2 = +330\mu$, where the angle θ is unknown. The diameter of the vessel is 500 mm and the thickness of the plate is 3 mm. The properties of the steel are as follows: $E = 210$ GPa, $\nu = 0.30$, elastic strength $= 360$ MPa, and ultimate strength $= 580$ MPa. Determine

(a)   The stresses σ_1 and σ_2 in the shell.

(b)   The principal stresses at the outside surface.

(c)   The pressure in the vessel.

(d)   The factor of safety with respect to failure by fracture.

4-63*   A hemispherical tank of radius r and thickness t is supported by a flange, as shown in the cross section of Fig. P4–63, and is filled with a fluid having a specific weight γ. Determine, in terms of γ, r, and t, the stresses σ_m and σ_t at a depth $y = r/2$.

4-64   The thin-walled spherical water tank shown in Fig. P4–64 is completely full of water and open to the atmosphere at the top. Determine the stresses σ_m and σ_t at a point on the equator of the sphere in terms of γ, r, and t, where γ is the specific weight of the water, r is the radius of the sphere, and t is the wall thickness of the sphere.

4-65   A steam dome in a boiler is generated by revolving one-quarter of an ellipse $(a = 2b = r)$ about its minor axis as shown in Fig. P4–65. Determine, in terms of p, r, and t, the meridional and tangential stresses σ_m and σ_t at the top of the dome and at a point $x = 0.75\,r$ from the axis of the vessel for a uniform internal pressure p.

FIG. P4–64

FIG. P4–65

4-66* The conical water tank shown in Fig. P4–66 was fabricated from 1/8-in. steel plate. When the tank is completely full of water (specific weight $\gamma = 62.4$ lb per cu ft), determine the axial and hoop stresses σ_a and σ_h at a point in the wall 8 ft below the apex of the cone.

4-67* A reducer in a pipeline (see Fig. P4–67) has a wall thickness of 4 mm. Assume that the flange bolts at the 150-mm end take all the end thrust (no compression in the flange at the 75-mm end). The pipeline is under a static water pressure of 420 kPa (gage). Determine the axial and hoop stresses σ_a and σ_h at a point midway between the flanges.

4-68 An internal pressure p is applied to the thin-walled toroidal shell (pressurized doughnut) shown in Fig. P4–68. Determine the axial and hoop stresses σ_a and σ_h at points A and B on the horizontal plane of symmetry of the shell in terms of p, R, r, and t. Hint: The hoop stress σ_h at point A can be

FIG. P4–66   **FIG. P4–67**

FIG. P4–68

(handwritten top view annotations):
TOP VIEW

(handwritten right-side annotations):

$r_m = r$

$r_t @ A = -(R-r)$

$r_t @ B = (R+r)$

$\left(\frac{1}{r_m}\right)\left(\frac{1}{r_t}\right)$ = GAUSSIAN CURVATURE

(+) - WEAK

0 - DEVELOPABLE SURFACE

(−) - STRONG

determined by using a free body consisting of one quarter of the shell, such as the part shown cross-hatched in Fig. P4–68.

4-6 THICK-WALLED CYLINDRICAL PRESSURE VESSELS

The problem of determining the tangential stress σ_t and the radial stress σ_r at any point in the wall of a thick-walled cylinder, in terms of the pressures applied to the cylinder, was solved by the French elastician G. Lame in 1833. The results can be applied to a wide variety of design situations involving cylindrical pressure vessels, hydraulic cylinders, piping systems, shrink- and press-fit applications, etc.

Consider a thick-walled cylinder having an inner radius a and an outer radius b, as shown in Fig. 4–15a. The cylinder is subjected to an internal pressure p_i and an external pressure p_o. For purposes of analysis, the thick-walled cylinder can be considered to consist of a series of thin rings. A typical ring located at a radial distance ρ from the axis of the cylinder and having a thickness $d\rho$ is shown by the dashed lines in Fig. 4–15a. As a result of the internal and external pressure loadings, a radial stress σ_r would develop at the interface between rings located at radial position ρ, while a slightly different radial stress $(\sigma_r + d\sigma_r)$ would develop at radial position $(\rho + d\rho)$. These stresses would be uniformly distributed over the inner and outer surfaces of the ring, as shown in Fig. 4–15b. Shearing stresses would not develop on the inner and outer surfaces of the ring, since the pressure loadings do not tend to force the rings to rotate with respect to one another.

In Art. 4–5, which dealt with thin-walled pressure vessels, it was shown that a tangential or hoop component of stress develops when a pressure difference exists between the inner and outer surfaces of a thin shell or ring. The planes on which these tangential stresses act can be exposed by considering only a small part of a ring, as shown shaded in Fig. 4–15b. Since the ring is assumed to be thin, the tangential stress σ_t can be considered to be uniformly distributed through the thickness of the ring. A relationship between radial stress σ_r and tangential stress σ_t can be obtained from equilib-

(handwritten right margin notes):

@ PT A

$\sigma_m t \cdot 2\pi(R-r) = q \, P\pi(R^2 - (R-r)^2)$

$\sigma_m =$

$\dfrac{\sigma_m}{r} + \dfrac{\sigma_t}{(R-r)} = \dfrac{P}{t}$

$\sigma_t =$

AT PT B

$\sigma_m t \cdot 2\pi(R+r) = P\pi(R+r)^2 - R^2)$

$\dfrac{\sigma_m}{r} + \dfrac{\sigma_t}{(R+r)} = \dfrac{P}{t}$

FIG. 4–15

(a)

(b)

(c)

rium considerations. A free-body diagram of a small part of a ring, such as the one shown in Fig. 4–15c for the shaded part of Fig. 4–15b, is useful for this determination. The axial stress σ_a, which may be present in the cylinder, has been omitted from this diagram since it does not contribute to equilibrium in the radial or tangential directions. The free body is assumed to have a length dl along the axis of the cylinder.

From a summation of forces in the radial direction

$$(\sigma_r + d\sigma_r)(\rho + d\rho)d\theta \, dl - \sigma_r\rho \, d\theta \, dl - 2\sigma_t \, d\rho \, dl \sin\frac{d\theta}{2} = 0 \qquad (a)$$

By neglecting higher-order terms and noting that for small angles $\sin d\theta/2 \approx d\theta/2$, Eq. a can be reduced to

$$\rho \frac{d\sigma_r}{d\rho} + \sigma_r - \sigma_t = 0 \qquad (b)$$

Equation b cannot be integrated directly, since both σ_r and σ_t are functions of radial position ρ. In previous instances, when such statically indeterminate situations were encountered, the problem was solved by considering deformations of the structure.

For the case of a thick-walled cylinder, the axial strain ε_a at any point in the wall of the cylinder can be expressed in terms of σ_a, σ_r, and σ_t by using generalized Hooke's Law (see Eq. 3–6). Thus,

$$\varepsilon_a = [\sigma_a - \nu(\sigma_r + \sigma_t)]/E \qquad (c)$$

The assumption normally made concerning strains in the thick-walled cylinder, which has been verified by careful measurement, is that axial strain is uniform. This means that plane transverse cross sections before loading remain plane and parallel after the internal and external pressures are applied. So far as the axial stress σ_a is concerned, two cases are of interest in a wide variety of design applications. In the first case, the axial loads induced by the pressure are not carried by the walls of the cylinder ($\sigma_a = 0$). This situation arises in gun barrels and in many types of hydraulic cylinders where pistons carry the axial loads. In the second case, the walls of the cylinder carry the loads. This situation occurs in pressure vessels with various types of end closures or heads. In this second case, in regions of the cylinder away from the ends, it has been found that the axial stress is uniformly distributed over the cross section. Hence ε_a, σ_a, E, and ν are constant for the two cases being considered; therefore, it follows from Eq. c that

$$\sigma_r + \sigma_t = (\sigma_a - E\varepsilon_a)/\nu = 2C_1 \qquad (d)$$

The constant is taken as $2C_1$ for convenience in the following derivation.

When the value for σ_t from Eq. d is substituted into Eq. b, this latter equation may be written as

$$\rho\,\frac{d\sigma_r}{d\rho} + 2\sigma_r = 2C_1 \qquad (e)$$

If Eq. e is multiplied by ρ, the terms before the equal sign can be expressed as

$$\frac{d}{d\rho}\,(\rho^2\sigma_r) = 2C_1\rho$$

Integrating yields

$$\rho^2\sigma_r = C_1\rho^2 + C_2$$

where C_2 is a constant of integration. Thus,

$$\sigma_r = C_1 + \frac{C_2}{\rho^2} \tag{f}$$

The tangential stress σ_t is then obtained from Eq. d as

$$\sigma_t = C_1 - \frac{C_2}{\rho^2} \tag{g}$$

Values for the constants C_1 and C_2 in Eqs. f and g can be determined by using the known values for the pressures at the inside and outside surfaces of the cylinder. These values, commonly referred to as boundary conditions, are

$$\sigma_r = -p_i \quad \text{at} \quad \rho = a$$

$$\sigma_r = -p_o \quad \text{at} \quad \rho = b$$

The minus signs indicate that the pressures (normally considered as positive quantities) produce compressive normal stresses at the surfaces on which they are applied. Substituting the boundary conditions into Eq. f yields

$$C_1 = \frac{a^2 p_i - b^2 p_o}{b^2 - a^2}$$

$$C_2 = -\frac{a^2 b^2 (p_i - p_o)}{b^2 - a^2}$$

The desired expressions for σ_r and σ_t are obtained by substituting these values for C_1 and C_2 into Eqs. f and g. Thus,

$$\sigma_r = \frac{a^2 p_i - b^2 p_o}{b^2 - a^2} - \frac{a^2 b^2 (p_i - p_o)}{(b^2 - a^2)\rho^2}$$

$$\sigma_t = \frac{a^2 p_i - b^2 p_o}{b^2 - a^2} + \frac{a^2 b^2 (p_i - p_o)}{(b^2 - a^2)\rho^2} \tag{4-6}$$

Radial and circumferential deformations δ_ρ and δ_c play important roles in all press-fit or shrink-fit problems. The change in circumference δ_c of the thin ring shown in Fig. 4–15a, when the pressures p_i and p_o are applied to the cylinder, may be expressed in terms of the radial displacement δ_ρ of a point on the ring as

$$\delta_c = 2\pi\delta_\rho$$

The circumferential deformation δ_c may also be expressed in terms of the tangential strain ε_t as

$$\delta_c = \varepsilon_t c$$

where $c = 2\pi\rho$ is the circumference of the ring. Thus,

$$\delta_\rho = \varepsilon_t \rho$$

For most applications involving shrink fits, the axial stress $\sigma_a = 0$. The tangential strain ε_t can then be expressed in terms of the radial stress σ_r and the tangential stress σ_t by using the generalized Hooke's law. Thus,

$$\varepsilon_t = (\sigma_t - \nu\sigma_r)/E$$

The radial displacement of a point in the wall is then obtained in terms of the radial and tangential stresses present at the point as

$$\delta_\rho = (\sigma_t - \nu\sigma_r)\rho/E \qquad (4\text{--}7)$$

As mentioned previously, Eqs. 4–6 and 4–7 can be used to compute stresses and deformations in a wide variety of design situations involving pressure vessels, hydraulic cylinders, and so on. Reduced forms of these equations are used with sufficient frequency to warrant consideration of the following special cases.

CASE 1   (INTERNAL PRESSURE ONLY)

If the loading is limited to an internal pressure p_i ($p_o = 0$), Eqs. 4–6 reduce to

$$\sigma_r = \frac{a^2 p_i}{b^2 - a^2}\left(1 - \frac{b^2}{\rho^2}\right)$$

$$\sigma_t = \frac{a^2 p_i}{b^2 - a^2}\left(1 + \frac{b^2}{\rho^2}\right)$$

Examination of these equations indicates that σ_r is always a compressive stress, while σ_t is always a tensile stress. In addition, σ_t is always larger than σ_r and is maximum at the inside surface of the cylinder. A typical distribution of the stress σ_t is shown in Exam. 4–7. Substituting the values of σ_r and σ_t from the previous two equations into Eq. 4–7 yields the deformation equation applicable for this special case ($p_o = 0$ and $\sigma_a = 0$). Thus,

$$\delta_\rho = \frac{a^2 p_i}{(b^2 - a^2)E\rho}[(1 - \nu)\rho^2 + (1 + \nu)b^2]$$

Handwritten margin notes:

ρ – RADIAL

c – CIRCUMFERENCE

$\sigma_{max} = \dfrac{a^2 p_i}{b^2 - a^2}\left(1 + \dfrac{b^2}{a^2}\right)$

$b = 150\,mm$

TUBE

$\sigma_{max} = \dfrac{a^2(p_i) - b^2 p_o}{b^2 - a^2} +$

$\dfrac{a^2(b^2)(p_i - p_o)}{(b^2 - a^2)a^2}$

$p_o = 65.625$

$\sigma_{max} = \dfrac{b^2(p_o)}{c^2 - b^2}\left(1 + \dfrac{c^2}{b^2}\right)$

$c = 222.5\,mm$

$t = c - b = 72.5\,mm$

CASE 2   (EXTERNAL PRESSURE ONLY)

If the loading is limited to an external pressure p_o ($p_i = 0$), Eqs. 4–6 reduce to

$$\sigma_r = -\frac{b^2 p_o}{b^2 - a^2}\left(1 - \frac{a^2}{\rho^2}\right)$$

$$\sigma_t = -\frac{b^2 p_o}{b^2 - a^2}\left(1 + \frac{a^2}{\rho^2}\right)$$

In this case, both σ_r and σ_t are always compressive. The tangential stress is always larger than the radial stress and assumes its maximum value at the inner surface of the cylinder. Substituting the values of σ_r and σ_t from the previous two equations into Eq. 4–7 yields the deformation equation applicable for this special case ($p_i = 0$ and $\sigma_a = 0$). Thus,

$$\delta_\rho = -\frac{b^2 p_o}{(b^2 - a^2)E\rho}[(1 - \nu)\rho^2 + (1 + \nu)a^2]$$

CASE 3   (EXTERNAL PRESSURE ON A SOLID CIRCULAR CYLINDER)

Radial and tangential stresses σ_r and σ_t and the radial displacement δ_ρ for this special case can be obtained from the expression developed for case 2 by letting the hole in the cylinder vanish ($a = 0$). Thus,

$$\sigma_r = -p_o$$

$$\sigma_t = -p_o$$

and

$$\delta_\rho = -(1 - \nu)p_o\rho/E$$

The minus sign in these equations indicates that the stresses are both compressive and the radius of the cylinder is reduced when the external pressure p_o is applied. For this case the stresses are independent of radial position ρ and have a constant magnitude equal to the applied pressure.

4-7   SHRINK-FITTING

In the fabrication of gun barrels and in numerous other applications, a laminated thick-walled cylinder is frequently formed by shrinking an outer cylinder (called a hoop, jacket, or shell) over an inner cylinder (frequently called the cylinder or tube). The purpose of this fabricating procedure is to increase the load-carrying capabilities of the assembly by developing an initial state of compressive tangential stress in the inner cylinder and tensile tangential stress in the outer cylinder. When the stresses induced by an internal pressure are superimposed on this initial state of stress, the resultant stress distribution makes more efficient use of the material than the distribution that would exist without the shrink-fit-induced initial stresses (see Exam. 4–7).

The determination of stresses in such a laminated cylinder requires the solution of a statically indeterminate problem that makes use of deformations. Consider the laminated thick-walled cylinder shown in Fig. 4–16 that has been formed by heating the jacket until it could be slipped over the tube. The heating of the jacket is required for assembly, since the jacket has an inside diameter that is slightly smaller than the outside diameter of the tube. The dotted sections in Fig. 4–16 illustrate jacket and tube configurations (greatly exaggerated) before they were fitted together. As the jacket cools after assembly, an interfacial contact pressure p_s develops and produces stresses and deformations in both the jacket and the tube. The initial difference in diameters between the jacket and the tube at the interface is known as the interference I. The interference is related to radial deformations of the jacket and tube at the interface by the expression

$$I = 2\delta = 2|\delta_1| + 2|\delta_2| \qquad (a)$$

The magnitudes of the radial deformations $|\delta_1|$ and $|\delta_2|$ can be determined by using the deformation expressions developed previously in Art. 4–6 for special cases 1 and 2. Thus,

$$|\delta_1| = \frac{b^2 p_s}{(c^2 - b^2)Eb}\left[(1 - v)b^2 + (1 + v)c^2\right]$$

$$|\delta_2| = \frac{b^2 p_s}{(b^2 - a^2)Eb}\left[(1 - v)b^2 + (1 + v)a^2\right]$$

FIG. 4–16

Substituting these values into Eq. *a* yields

$$I = \frac{4b^3 p_s(c^2 - a^2)}{E(c^2 - b^2)(b^2 - a^2)} \tag{4-8}$$

or

$$p_s = \frac{EI(c^2 - b^2)(b^2 - a^2)}{4b^3(c^2 - a^2)} \tag{4-9}$$

Equation 4–8 provides a means for establishing the interference that can be tolerated without exceeding maximum allowable stress levels in the jacket or the tube. This type of information is required, for example, when bearing races are being designed for press-fit or shrink-fit applications. Equation 4–9 provides a means for establishing the interfacial pressure resulting from a known amount of interference. Once this pressure is known, shrink-fit stress levels can be established in both the jacket and the tube. Load-induced stresses in the laminated cylinder will add to these fabrication-induced values.

EXAMPLE 4–7

A laminated thick-walled hydraulic cylinder was fabricated by shrink-fitting a steel jacket having an outside diameter of 300 mm onto a steel tube with an inside diameter of 100 mm and an outside diameter of 200 mm. The interference was 0.15 mm. Assume E = 200 GPa.

(a) Determine the interfacial contact pressure.
(b) Determine the maximum tensile stress in the laminated cylinder resulting from the shrink fit.
(c) Determine the maximum tensile stress in the laminated cylinder after an internal pressure of 200 MPa is applied.
(d) Compare the stress distribution in the laminated cylinder with the stress distribution in a uniform cylinder for an internal pressure of 200 MPa.

SOLUTION

(a) The interfacial contact pressure (shrink pressure) can be determined by using Eq. 4–9. Thus,

$$p_s = \frac{200(10^9)(0.00015)(0.150^2 - 0.100^2)(0.100^2 - 0.050^2)}{4(0.100^3)(0.150^2 - 0.050^2)}$$

$$= 35.16(10^6) \text{ N/m}^2 = \underline{35.2 \text{ MPa}} \qquad \text{Ans.}$$

(b) The maximum tensile stress resulting from this shrink pressure will be the tangential stress σ_t that occurs at the inside surface of the jacket. This stress can be determined by using the reduced form of Eq. 4–6 for the case of internal pressure only. Thus,

$$\sigma_t = \frac{(0.100^2)(35.16)(10^6)}{0.150^2 - 0.100^2}\left(1 + \frac{0.150^2}{0.100^2}\right)$$

$$= 91.4(10^6) \text{ N/m}^2 = \underline{91.4 \text{ MPa } T} \qquad \text{Ans.}$$

(c) Stresses in the laminated cylinder after the internal pressure is applied will result from both the shrink pressure and the applied pressure. Again, the maximum tensile stress will be a tangential stress; however, it may occur either at the inside surface of the jacket or at the inside surface of the tube, depending on the magnitudes of the stresses associated with the two pressure loadings. Reduced forms of Eq. 4–6 can be used for each of the loadings.

For the tube,

$$\sigma_t = \sigma_t(\text{tube interference}) + \sigma_t(\text{laminated cylinder due to } p_i)$$

$$= -\frac{(0.100^2)(35.16)(10^6)}{0.100^2 - 0.050^2}\left(1 + \frac{0.050^2}{0.050^2}\right) + \frac{(0.050^2)(200)(10^6)}{0.150^2 - 0.050^2}\left(1 + \frac{0.150^2}{0.050^2}\right)$$

$$= -93.8(10^6) + 250(10^6)$$

$$= 156.2(10^6) \text{ N/m}^2 = \underline{156.2 \text{ MPa } T}$$

For the jacket,

$$\sigma_t = \sigma_t(\text{jacket interference}) + \sigma_t(\text{laminated cylinder due to } p_i)$$

$$= \frac{(0.100^2)(35.16)(10^6)}{0.150^2 - 0.100^2}\left(1 + \frac{0.150^2}{0.100^2}\right) + \frac{(0.050^2)(200)(10^6)}{0.150^2 - 0.050^2}\left(1 + \frac{0.150^2}{0.100^2}\right)$$

$$= 91.4(10^6) + 81.3(10^6)$$

$$= 172.7(10^6) \text{ N/m}^2 = \underline{172.7 \text{ MPa } T} = \sigma_{max} \qquad \text{Ans.}$$

(d) Radial and tangential stress distributions in the laminated cylinder and in a uniform cylinder for an internal pressure of 200 MPa are shown in Fig. 4–17.

FIG. 4–17

(a) Uniform cylinder

(b) Laminated cylinder

PROBLEMS

 4-69 Demonstrate that for a cylindrical vessel with a shell thickness $(b-a)$ of one-tenth the radius a, subjected only to internal pressure, the error involved in computing the hoop tension σ_t by the method of Art. 4–5 instead of Eq. 4–6 is approximately 5 percent.

4-70* A steel cylinder with an inside diameter of 8 in. and an outside diameter of 16 in. is subjected to an internal pressure of 15,000 psi. Determine the maximum tensile stress and the maximum shearing stress in the cylinder.

4-71* A steel cylinder with an inside diameter of 200 mm and an outside diameter of 300 mm is subjected to an internal pressure of 75 MPa. Determine

    (a) The maximum tensile stress in the cylinder.

    (b) The radial and tangential stresses at a point midway in the cylinder wall.

4-72 A thick-walled cylindrical pressure vessel with an inside diameter of 4 in. and an outside diameter of 8 in. has a design specification that limits the maximum shearing stress to 24,000 psi. Determine the maximum internal pressure that may be applied to the vessel.

4-73* A steel cylinder with an inside diameter of 50 mm and an outside diameter of 250 mm is subjected to an internal pressure of 85 MPa and an external pressure of 30 MPa. Determine the circumferential force per unit length carried by a circular element having an inside diameter of 100 mm and a thickness of 25 mm.

4-74 A laminated thick-walled cylinder was fabricated by shrinking a steel ($E = 30,000$ ksi) sleeve having an inside diameter of 7.992 in. and an outside diameter of 12.00 in. onto a steel tube with an inside diameter of 4.00 in. and an outside diameter of 8.000 in. Determine

    (a) The maximum internal pressure that can be applied to the assembly if the tangential stress at the inner surface of the cylinder is not to exceed 30,000 psi.

    (b) The maximum shearing stress in the cylinder after the application of this internal pressure.

4-75 A laminated cylinder consists of a jacket with an inside diameter of 250 mm and an outside diameter of 300 mm shrunk onto a tube with an inside diameter of 150 mm. The initial shrinkage pressure is 28 MPa. The cylinder is then subjected to an internal pressure of 200 MPa. Determine

    (a) The initial interference if the cylinder is made of steel ($E = 200$ GPa).

    (b) The maximum tensile stresses in the jacket and in the tube.

4-76* A jacket is to be shrunk onto a tube having an inside diameter of 4 in. and an outside diameter of 8 in. Determine the minimum thickness for the jacket if the maximum tensile stresses in the jacket and the tube are not to

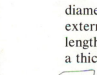

$a = 2"$
$b = 4"$

$\hat{\tau}_{max} = 24000$ psi

FIND P_i

@ $r = 2"$ σ_θ AT max (+)
    σ_r AT max (−)

SINCE $\tau_{r\theta} = 0$

σ_θ, σ_r become σ_{P_1} & σ_{P_2}

$2\tau_{max} = \sigma_\theta - \sigma_r$

σ_θ @ $P_0 = 0$
σ_r @ $r = a$ $\sigma_r = -P_i$
$\sigma_\theta = 30$ ksi
LET $a = 1.9$
$\sigma_\theta = 28.49$ ksi

LET $b = 4.1$
$\sigma_\theta = 23.62$ ksi

exceed 30 ksi when the laminated cylinder is subjected to an internal pressure of 25 ksi.

4-77   A steel cylinder ($E = 200$ GPa) having an internal diameter of 150 mm and an external diameter of 200 mm is shrunk over a tube having an internal diameter of 100 mm. If the laminated cylinder is designed so that the maximum tensile stress in the tube will equal the maximum tensile stress in the jacket after an internal pressure of 165 MPa is applied to the assembly, determine

(a)   The initial shrinkage pressure between the tube and the jacket.

(b)   The interference required to produce this shrinkage pressure.

4-78*   A laminated cylinder consists of a jacket with an internal diameter of 8 in. and an external diameter of 12 in. shrunk over a tube with an internal diameter of 4 in. When the laminated cylinder is subjected to an internal pressure of 20 ksi, the maximum tensile stress in the tube is 20 ksi. Determine

(a)   The maximum tensile stress in the jacket.

(b)   The initial shrinkage pressure.

4-79*   A thick-walled cylinder with an inside diameter of 150 mm was designed with a minimum wall thickness to withstand an internal pressure of 105 MPa and to develop a maximum tensile stress of 175 MPa. If the internal pressure must be doubled, what is the minimum thickness for a shrink-fitted jacket that will prevent the maximum tensile stress in the assembly from exceeding 175 MPa?

4-80   A steel cylinder ($E = 30,000$ ksi) with an inside diameter of 4 in. was designed for a maximum tensile stress of 20 ksi when subjected to an internal pressure of 12 ksi. The internal pressure must now be increased to 16 ksi without increasing the maximum tensile stress in the cylinder. If the cylinder is to be strengthened by winding with square, high-strength-steel wire 0.10 in. on a side, determine

(a)   The initial tensile force that must be applied to the wire as it is wrapped onto the cylinder.

(b)   The tensile stress in the wire when the internal pressure of 16 ksi is applied to the cylinder.

CHAPTER 5

TORSIONAL LOADING

5-1 INTRODUCTION

In the design of machinery (and some structures) the problem of transmitting a torque (a couple) from one plane to a parallel plane is frequently encountered. The simplest device for accomplishing this function is a circular shaft such as that connecting an electric motor with a pump, compressor, or other mechanism. For example, the shaft AB of Fig. 5–1 transmits a torque from the driving motor A to the coupling B. A modified free-body diagram (the weight and bearing reactions are not shown because they contribute no useful information to the torsion problem) of a typical installation is shown in Fig. 5–2. The resultant of the electromagnetic forces applied to the armature A of the motor is a couple resisted by the resultant of the bolt forces (another couple) acting on the flange coupling B, and the circular shaft transmits the torque from armature to coupling. The problem is the determination of significant stresses in and the deformation of the shaft.

As a point of departure, a segment of the shaft between transverse planes a–a and b–b of Fig. 5–2 will be studied. In so doing, the complicated stress patterns at the locations of torque-applying devices may be avoided, and Saint-Venant's principle applies. A free-body diagram of the segment of

FIG. 5–1   A 1000-Hp. 180-rpm motor coupled to ball and tube mills. (*Photo courtesy of Electric Machinery Mfg. Co., Minneapolis, Minn.*)

FIG. 5–2

FIG. 5–3

the shaft is shown in Fig. 5–3 with the torque applied by the armature indicated on the left end as T. The resisting torque T_r, indicated at the right end, is the resultant of the differential forces dF acting on the transverse plane b–b. The force dF is equal to $\tau\,dA$ where τ is the shearing stress *on the transverse plane* at a distance ρ from the shaft center, and dA is a differential area. *In keeping with the notation outlined in Art. 1–5, the shearing stress should be designated $\tau_{z\theta}$ to indicate that it acts on the z face in the direction of increasing θ.* For the elementary theory of torsion of circular sections, however, the shearing stress on any transverse plane is always perpendicular to the radius to the point, and the stress will be referred to as τ or τ_c for convenience. If the shaft is in equilibrium, the summation of moments about the axis of the shaft indicates that

$$T = T_r = \int_{\text{area}} \rho\tau\,dA \qquad (5\text{–}1)$$

The law of variation of the shearing stress on the transverse plane must be known before the integral of Eq. 5–1 can be evaluated. The approach to the problem of stress distribution will be prefaced by a brief reference to the past.

In 1784 C. A. Coulomb, a French engineer, developed experimentally the relationship between the applied torque and the angle of twist for circular bars.[1] A. Duleau, another French engineer, in a paper published in 1820[1]

---

[1] From *History of Strength of Materials,* S. P. Timoshenko, McGraw-Hill, New York, 1953.

FIG. 5–4

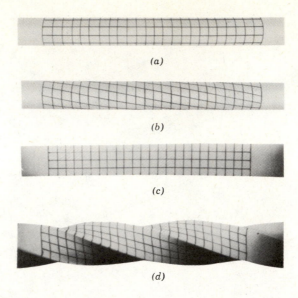

(a)

(b)

(c)

(d)

derived analytically the same relationship by making the assumption that *a plane section before twisting remains plane after twisting and a diameter remains a straight line*. Visual examination of twisted models indicates that these assumptions are apparently correct for circular sections either solid or hollow (provided the hollow section is circular and symmetrical with respect to the axis of the shaft), but incorrect for any other shape. Compare, for example, the distortions of the rubber models with circular and rectangular cross sections shown in Fig. 5–4.

5-2   TORSIONAL SHEARING STRAIN

If the assumptions noted above are made, the distortion of the shaft will be as indicated in Fig. 5–5a, where points B and D on a common radius in a plane section move to B' and D' in the same plane and still on the same radius. The angle θ is called the *angle of twist*. The surface ABB' of Fig. 5–5a is shown developed in Fig. 5–5b, and a differential cube of the material at B assumes at B' the familiar distortion due to shearing stress. At this point the assumption will be made that *all longitudinal elements have the same length* (which limits the results to straight shafts of constant diameter). From Fig. 5–5, the following strain relations may be written:

$$\tan \gamma_c = \frac{BB'}{L} = \frac{c\theta}{L}$$

and

FIG. 5-5

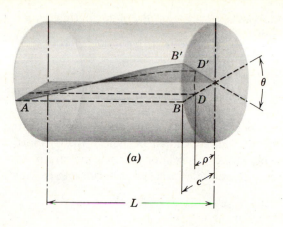

(a)

$\sigma = \dfrac{E}{A}$

$\sigma_H = \dfrac{pr}{t}$

$\sigma_L = \dfrac{pr}{2t}$

$T = \displaystyle\int_A \rho \mathcal{T} \, dA$

Radius = R
Outer Fiber = c
    R = c = ρ

Angle of twist = ρ

$\tan \, \&_c = \dfrac{d}{L} = \tan \, \theta$

$d = \gamma L$

$d = r\theta$

$\boxed{\gamma L = r\theta}$

$\boxed{\gamma = \gamma_G}$

(b)

$$\tan \gamma = \frac{DD'}{L} = \frac{\rho\theta}{L}$$

or, *if the strain is small,*

$$\gamma_c = \frac{c\theta}{L} \quad \text{and} \quad \gamma = \frac{\rho\theta}{L} \qquad (5\text{--}2)$$

From Eq. 5-2,

$$\frac{\gamma_c L}{c} = \frac{\gamma L}{\rho}$$

and as the longitudinal elements have the same length, the expression reduces to

$$\gamma = \frac{\gamma_c}{c} \rho \qquad (5\text{--}3)$$

which indicates that the shearing strain ($\gamma_{z\theta}$) is proportional to the distance from the axis of the shaft. Note that Eq. 5–3 is valid for elastic or inelastic action and for homogeneous or heterogeneous materials, provided the

VALID FOR
ELASTIC
& INELASTIC
BEHAVIOR

strains are not too large ($\tan \gamma \approx \gamma$). Problems in this book will be assumed to satisfy this requirement.

5-3 TORSIONAL SHEARING STRESS—THE ELASTIC TORSION FORMULA

If the assumption is now made that Hooke's law applies (the accompanying limitation is that the stresses must be below the proportional limit of the material), shearing stress τ is related to shearing strain γ by the expression $\tau = G\gamma$ and Eq. 5–3 can be expressed in terms of stresses as

$$\tau = \frac{\tau_c}{c} \rho \qquad (5\text{--}4)$$

When Eq. 5–4 is substituted into Eq. 5–1, the result is

$$T_r = \frac{\tau_c}{c} \int_{\text{area}} \rho^2 dA = \frac{\tau}{\rho} \int_{\text{area}} \rho^2 dA \qquad (a)$$

The integral in Eq. a is the polar second moment of area J (called the polar moment of inertia), with respect to the longitudinal axis of the shaft (see Appendix B). Equation a can then be written as

$$T_r = \frac{\tau J}{\rho}$$

or in terms of the unknown shearing stress as

$$\tau = \frac{T\rho}{J} \qquad (5\text{--}5)$$

Equation 5–5 is known as the *elastic torsion formula*, in which τ is *the shearing stress on a transverse plane at a distance ρ from the axis of the shaft,* and T is the resisting torque, which, in general, is obtained from a free-body diagram and an equilibrium equation. Note that the transverse plane is a plane of maximum shearing stress. Note also that Eq. 5–5 applies only for linearly elastic action in homogeneous and isotropic materials.

5-4 TORSIONAL DISPLACEMENTS

Frequently the amount of twist in a shaft (or structural element) is of paramount importance, making the computation of the angle of twist a necessary accomplishment of the machine or structural designer. The fundamental approach to the problem, involving members satisfying the limitations of Art. 5–3, is provided by the following equations:

$$\gamma = \frac{\rho\theta}{L} \quad \text{or} \quad \gamma = \frac{\rho d\theta}{dL} \qquad \textit{ELASTIC} \quad (5\text{--}2)$$
$$\textit{f INELASTIC}$$

$$\tau = \frac{T\rho}{J} \qquad (5\text{--}5)$$

$$\textit{ONLY ELASTIC REGION}$$

$$G = \frac{\tau}{\gamma} \qquad (3\text{--}9)$$

The second form of Eq. 5–2 is useful when the torque or the cross section varies as a function of position along the length of the shaft. Equation 5–2 is valid for both elastic and inelastic action. Equation 5–5 is the elastic torsion formula that provides the shearing stress τ on a transverse plane at a distance ρ from the axis of the shaft. Equation 3–9 is Hooke's law for shearing stresses. The last two expressions are limited to stresses below the proportional limit of the material (elastic action). The three equations can be combined to give several different relationships; for example,

$$\theta = \frac{\gamma L}{\rho} = \frac{\tau L}{\rho G} \qquad (5\text{--}6a)$$

or

$$\theta = \frac{TL}{JG} \qquad (5\text{--}6b)$$

The first form of Eq. 5–6 will often be most useful in dual-specification problems (limiting θ and τ are specified).

The angle of twist determined from the above expressions is for a length of shaft of constant diameter sufficiently removed from sections where pulleys, couplings, or other mechanical devices are attached to make Saint-Venant's principle applicable. However, for practical purposes, it is customary to neglect local distortion at all connections and compute angles as though there were no discontinuities, and as though the torques applied were distributed over the cross sections in exactly the same manner in which the stresses given by the torsion formula are distributed. The following examples illustrate the application of the principles developed.

EXAMPLE 5–1
The steel shaft of Fig. 5–6a is in equilibrium under the torques shown. Determine

FIG. 5–6

(a)

(b)

(c)

(d)

(e)

(f)

(g)

(a) The maximum shearing stress in the shaft.

(b) The angle of twist of end *B* of the 6-in. segment with respect to end *A*.

(c) The angle of twist of end *C* with respect to end *A*.

SOLUTION

(a) In general, free-body diagrams should be drawn in order to evaluate correctly the resisting torque. Such diagrams are shown in Figs. 5–6*b* and *c*, where in part *b* the shaft is cut by any transverse plane through the 6-in. segment, and T_6 is the resisting torque on this section. Similarly, in Fig. 5–6*c* the plane is passed through the 4-in. section, and T_4 is the resisting torque on this section. A torque diagram, such as the one shown in Fig. 5–6*d*, provides a pictorial representation of the levels of torque being transmitted by each of the sections and serves as an aid for stress and deformation calculations. The location of the maximum shearing stress is not apparent; hence, the stress must be checked at both sections. Thus, from Fig. 5–6*b* or from Fig. 5–6*d*,

$$\Sigma M = 0, \qquad T_6 = 20 - 5 = 15 \text{ ft-kips} = 15{,}000(12) \text{ in.-lb}$$

and

$$\tau_{max} = \frac{Tc}{J} = \frac{15{,}000(12)(3)}{(\pi/2)(3^4)} = \frac{40{,}000}{3\pi}$$

From Fig. 5–6*c* or from Fig. 5–6*d*,

$$\Sigma M = 0, \qquad T_4 = 5 \text{ ft-kips} = 5000(12) \text{ in.-lb}$$

and

$$\tau_{max} = \frac{5000(12)(2)}{(\pi/2)(2^4)} = \frac{15{,}000}{\pi} > \frac{40{,}000}{3\pi}$$

Therefore, the maximum shearing stress is

$$15,000/\pi = \underline{4770 \text{ psi}} \qquad \text{Ans.}$$

This stress is less than the shearing proportional limit of any steel; hence, the torsion formula applies. Note that had the larger torque been carried by the smaller section, the maximum stress would obviously occur in the small section, and only one stress determination would have been required.

(b)   As an aid to visualizing the distortion of the shaft, the segments AB and BC and the torques acting on them are drawn separately in Figs. 5–6e and f with the distortions greatly exaggerated. As the resultant torque of 15 ft-kips twists segment AB through the angle θ_{AB}, points b and d move to b' and d' respectively, and segment BC may be considered a rigid body rotating through the same angle, point c moving to c', after which the torque of 5 ft-kips acting on BC twists this part of the shaft back through the angle θ_{BC}, point c' moving back to c''. The resultant distortion for the entire shaft is shown in Fig. 5–6g.

When computing distortions, it is recommended that the two fundamental relationships of Eqs. 5–2 and 5–5 plus Hooke's law be used; thus,

$$\gamma = \frac{c\theta}{L} = \frac{\tau}{G} = \frac{Tc}{JG}$$

When this expression is applied to Fig. 5–6e, the twist in segment AB is obtained as

$$\theta_{AB} = \frac{(20 - 5)(12)(9)(12)}{(\pi/2)(3^4)(12)(10^3)} = \frac{40}{\pi(10^3)} = \underline{12.73(10^{-3}) \text{ rad}} \quad \text{Ans.(b)}$$

(c)   From Figs. 5–6f and g,

$$\theta_{AC} = \theta_{AB} - \theta_{BC} = \frac{40}{\pi(10^3)} - \frac{5(12)(5)(12)}{(\pi/2)(2^4)(12)(10^3)}$$

$$= \frac{40 - 37.5}{\pi(10^3)} = \underline{0.796(10^{-3}) \text{ rad}} \qquad \text{Ans.(c)}$$

EXAMPLE 5–2

The solid circular tapered shaft of Fig. 5–7 is subjected to end torques applied in transverse planes. Determine the magnitude of the angle of twist in terms of T, L, G, and r. Assume elastic action and a slight taper.

FIG. 5-7

SOLUTION

Note that Eq. 5–2 was developed on the assumption that plane cross sections remain plane and all longitudinal elements have the same length. Neither of these assumptions is strictly valid for the tapered shaft, but if the taper is slight, the error involved is negligible; hence, Eq. 5–2 will be used as follows:

$$d\theta = \frac{\gamma}{\rho}\, dx$$

also

$$\gamma = \frac{\tau}{G}$$

and

$$\tau = \frac{T\rho}{J} = \frac{2T\rho}{\pi\rho^4} = \frac{2T}{\pi\rho^3}$$

Therefore,

$$d\theta = \frac{\tau}{G\rho}\, dx = \frac{2T}{G\pi\rho^4}\, dx$$

The radius can be expressed as a function of x; thus,

$$\rho = r + \frac{2r - r}{L}\, x = \frac{r}{L}\, (L + x)$$

which, substituted into the expression for $d\theta$, gives

$$d\theta = \frac{2TL^4}{G\pi r^4(L + x)^4}\, dx$$

Integration gives

$$\theta = \frac{2TL^4}{G\pi r^4} \int_0^L \frac{dx}{(L+x)^4} = -\frac{2TL^4}{3G\pi r^4}\left(\frac{1}{8L^3} - \frac{1}{L^3}\right)$$

$$\theta = \frac{7TL}{12G\pi r^4} \qquad\qquad \text{Ans.}$$

An alternate setup is to place the origin of coordinates at a distance L to the left of 0 in Fig. 5–7 (point 0'). The function for ρ then becomes

$$\rho = \frac{r}{L} x$$

and

$$\theta = \frac{2TL^4}{G\pi r^4} \int_L^{2L} \frac{dx}{x^4} = \frac{7TL}{12G\pi r^4}$$

PROBLEMS

> *Note: Assume linearly elastic action for all of the following problems.*

5-1*   A solid circular steel shaft 2 in. in diameter is subjected to a pure torque of 20,000 in.-lb. Determine the maximum shearing stress in the shaft and the magnitude of the angle of twist in a 6-ft length. The modulus of rigidity (shear modulus) for the steel is 12,000,000 psi.

5-2*   A hollow steel shaft has an outside diameter of 100 mm and an inside diameter of 75 mm. Determine the maximum shearing stress in the shaft and the magnitude of the angle of twist in a 2-m length when the shaft is subjected to a pure torque of 7.5 kN · m. The modulus of rigidity (shear modulus) for the steel is 80 GPa.

5-3*   Determine the maximum allowable torque to which a solid circular steel ($G = 12,000$ ksi) bar 9 ft long and 3 in. in diameter can be subjected when it is specified that the shearing stress must not exceed 12,000 psi and the angle of twist must not exceed 0.075 rad.

5-4*   Specifications for a solid circular aluminum alloy ($G = 28$ GPa) rod 1.5 m long require that it shall be adequate to resist a torque of 3.0 kN · m without twisting more than 15° or exceeding a shearing stress of 100 MPa. What minimum diameter is required?

5-5* A torque of 50,000 in.-lb is supplied to the 3-in. diameter factory drive shaft of Fig. P5–5 by a belt that drives pulley A. A torque of 30,000 in.-lb is taken off by pulley B and the remainder by pulley C. Shafts AB and BC are 4 ft and 3 ft long, respectively. Both shafts are made of steel ($G = 12,000$ ksi). Determine

(a) The maximum shearing stress in each of the shafts.

(b) The magnitudes of the angles of twist of pulleys B and C with respect to pulley A.

5-6 A torque of 8.5 kN · m is supplied to the 75-mm diameter factory drive shaft of Fig. P5–6 by a belt that drives pulley A. A torque of 5.6 kN · m is taken off by pulley B and the remainder by pulley C. Shafts AB and AC are 1.6 m and 1.2 m long, respectively. Both shafts are made of steel ($G = 80$ GPa). Determine

(a) The maximum shearing stress in each of the shafts.

(b) The magnitudes of the angles of twist of pulleys B and C with respect to pulley A.

5-7 A torque of 90,000 in.-lb is supplied to the driving gear B of Fig. P5–7 by a motor. Gear A takes off 30,000 in.-lb of torque and the remainder is taken off by gear C. Both shafts are made of steel ($G = 12,000$ ksi). Determine

(a) The maximum shearing stress in each of the shafts.

(b) The angle of twist of gear C with respect to gear A.

5-8* A torque of 12.0 kN · m is supplied to the driving gear B of Fig. P5–8 by a motor. Gear A takes off 4.0 kN · m of torque, and the remainder is taken off by gear C. For an allowable shearing stress of 80 MPa, determine

(a) The minimum permissible diameters for the two shafts.

(b) The angle of twist of gear A with respect to gear C if both shafts are made of steel ($G = 80$ GPa) and have diameters of 75 mm.

5-9* A solid circular steel ($G = 12,000$ ksi) shaft with diameters as shown in Fig. P5–9 is subjected to a torque T. The allowable shearing stress is

FIG. P5-7

FIG. P5-8

10,000 psi and the maximum allowable angle of twist in the 7-ft length is 0.05 rad. Determine the maximum allowable value of T.

5-10   Solve Prob. 5-9 with an additional torque of 3000 ft-lb applied at the right end of the 3-in. portion with the same sense as that of T.

5-11   A solid circular steel ($G = 80$ GPa) shaft is fastened securely to a solid circular bronze ($G = 40$ GPa) shaft as indicated in Fig. P5-11. The allowable shearing stress is 70 MPa and the maximum allowable angle of twist in the 3.5-m length is 0.04 rad. Determine the maximum allowable value of T.

5-12   Solve Prob. 5-11 with an additional torque of $3T$ applied at the right end of the bronze portion with the same sense as that of T.

5-13*   The 4-in. diameter shaft shown in Fig. P5-13 is composed of brass ($G = 5000$ ksi) and steel ($G = 12,000$ ksi) sections rigidly connected. Determine the maximum allowable torque, applied as shown, if the maximum shearing stresses in the brass and the steel are not to exceed 7500 psi and 10,000 psi, respectively, and the distance AC, through which the end of the 10-in. pointer AB moves, is not to exceed 0.60 in.

FIG. P5-9

FIG. P5-11

FIG. P5–13   FIG. P5–14

5-14   A stepped steel ($G = 80$ GPa) shaft has the dimensions and is subjected to the torques shown in Fig. P5–14. Determine

(a)   The maximum shearing stress on a section 3 m from the left end.

(b)   The rotation of a section 2 m from the left end with respect to its no-load position.

(c)   The rotation of the section at the right end of the shaft with respect to its no-load position.

5-15   The maximum shearing stress in the stepped steel ($G = 12,000$ ksi) shaft shown in Fig. P5–15 is 20,000 psi. Determine

(a)   The magnitude of the torque T_3.

(b)   The rotation of a section at B with respect to its no-load position.

(c)   The rotation of a section at D with respect to its no-load position.

FIG. P5–15   FIG. P5–16

FIG. P5–17

FIG. P5–18

5-16*   The steel ($G = 80$ GPa) shaft of Fig. P5–16 is in equilibrium under the torques applied as shown. The maximum shearing stress in the 160-mm section is 120 MPa, and the rotation of end C with respect to the section at A is 0.018 rad counterclockwise looking left. Determine the torques T_1 and T_2.

5-17*   The hollow circular steel ($G = 12,000$ ksi) shaft of Fig. P5–17 is in equilibrium under the torques indicated. Determine

(a)   The maximum shearing stress in the shaft.

(b)   The rotation of a section at D with respect to a section at B.

5-18   The hollow circular steel ($G = 80$ GPa) shaft of Fig. P5–18 is in equilibrium under the torques indicated. Determine

(a)   The minimum permissible outside diameter d if the maximum shearing stress in the shaft is not to exceed 100 MPa.

(b)   The rotation of a section at D with respect to a section at A for a shaft with an outside diameter of 120 mm.

FIG. P5–19

FIG. P5–20

5-19 A steel (G = 12,000 ksi) shaft is loaded and supported as shown in Fig. P5–19. Determine

(a) The maximum shearing stress in the shaft.

(b) The angle of twist of the right end with respect to its no-load position.

(c) The rotation of a section 3 ft from the left end with respect to its no-load position.

5-20* The 100-mm diameter shaft shown in Fig. P5–20 is composed of aluminum alloy (G = 28 GPa) and steel (G = 80 GPa) sections rigidly connected. Determine

(a) The maximum shearing stress in the shaft.

(b) The rotation of a section 3 m from the left end with respect to its no-load position.

(c) The rotation of a section at the right end with respect to its no-load position.

5-21 The solid circular tapered shaft of Fig. P5–21 is subjected to a constant torque T. Determine the angle of twist in terms of T, L, G, r, and m.

5-22* The hollow tapered shaft of Fig. P5–22 has a constant wall thickness t. Determine the angle of twist for a constant torque T in terms of T, L, G, t, and r. Note that when t is small, the approximate expression for the polar moment of inertia ($J = \rho^2 A$) may be used.

5-23* The hollow circular tapered shaft of Fig. P5–23 is subjected to a constant torque T. Determine the angle of twist in terms of T, L, G, and r.

5-24 The tapered circular shaft of Fig. P5–24 has an axial hole of constant diameter throughout its length. Determine the angle of twist due to a constant torque T in terms of T, L, G, R, and r.

FIG. P5–21

FIG. P5–22

FIG. P5–23

FIG. P5–24

5-25* The solid cylindrical shaft of Fig. P5–25 is subjected to a uniformly distributed torque of q in.-lb per in. of length. Determine, in terms of q, L, G, and c, the rotation of the left end under the applied torque.

5-26 The solid cylindrical shaft of Fig. P5–26 is subjected to a distributed torque that varies linearly from zero at the left to q N · m per meter of length at the right end. Determine, in terms of q, L, G, and c, the rotation of the left end under the applied torque.

5-5 STRESSES ON OBLIQUE PLANES

As stated in Art. 5–3, the torsion formula can be used to evaluate the shearing stress on a transverse plane. It is necessary to ascertain if the transverse plane is a plane of maximum shearing stress and if there are other significant stresses induced by torsion. For this study, the stresses at point A in the shaft of Fig. 5–8a will be analyzed. Figure 5–8b shows a differential cube taken from the shaft at A and the stresses acting on transverse and longitudinal planes. The stress τ_{xy} may be determined by means of the torsion formula, and $\tau_{yx} = \tau_{xy}$ (see Art. 1–4). If the equations of equilibrium are applied to the free-body diagram of Fig. 5–8c, the following results are obtained:

$$\circlearrowleft+ \qquad \Sigma F_t = 0$$

$$\tau_{nt}\,dA - \tau_{xy}(dA\cos\alpha)\cos\alpha + \tau_{yx}(dA\sin\alpha)\sin\alpha = 0$$

FIG. P5–25

FIG. P5–26

FIG. 5–8

(a)

(b)

(c)

from which

$$\tau_{nt} = \tau_{xy}(\cos^2\alpha - \sin^2\alpha) = \tau_{xy}\cos 2\alpha$$

and

$$(+\nearrow) \quad \Sigma F_n = 0$$

$$\sigma_n dA - \tau_{xy}(dA\cos\alpha)\sin\alpha - \tau_{yx}(dA\sin\alpha)\cos\alpha = 0$$

from which

$$\sigma_n = 2\tau_{xy}(\sin\alpha)\cos\alpha = \tau_{xy}\sin 2\alpha$$

$$J_{xy} = \frac{Tr}{J}$$

$$= \frac{2T}{\pi r^3} = \frac{16T}{\pi D^3}$$

These results are shown in the graph of Fig. 5–9, from which it is apparent that the maximum shearing stress occurs on transverse and longitudinal (diametral) planes. The graph also shows that the maximum normal stresses occur on planes oriented at 45° with the axis of the bar and perpendicular to the surface of the bar. On one of these planes ($\alpha = 45°$ for Fig. 5–8) the stress is tension, and on the other ($\alpha = 135°$) the stress is compression. Furthermore, *all these stresses have the same magnitude*; hence, the torsion

FIG. 5-9

formula will give the magnitude of the maximum normal and shearing stresses at a point in a circular shaft subjected to pure torsion.

Any of the stresses discussed in the preceding paragraph may be significant in a particular problem. Compare, for example, the failures shown in Fig. 5–10. In Fig. 5–10a, the steel truck rear axle split longitudinally. One would also expect this type of failure to occur in a shaft of wood with the grain running longitudinally. In Fig. 5–10b, the compressive stress caused the thin-walled aluminum alloy tube to buckle along one 45° plane, while the tensile stress caused tearing on the other 45° plane. Buckling of thin-walled tubes (and other shapes) subjected to torsional loading is a matter of paramount concern to the designer. In Fig. 5–10c, the tensile stresses caused the gray cast iron to fail in tension—typical of any brittle material subjected to torsion. In Fig. 5–10d, the low-carbon steel failed in shear on a plane that is almost transverse—a typical failure for ductile material. The reason the fracture in Fig. 5–10d did not occur on a transverse plane is that under the large plastic twisting deformation before rupture (note the spiral lines indicating elements originally parallel to the axis of the bar), longitudinal elements were subjected to axial tensile loading because the grips of the testing machine would not permit the bar to shorten as the elements were twisted into spirals. This axial tensile stress (not shown in Fig. 5–8) changes the plane of maximum shearing stress from a transverse to an oblique plane (resulting in a warped surface of rupture).[2]

For problems involving stresses on a given oblique plane, the methods of Arts. 1–6, 1–7, and 1–8 are all available, and there is no particular advantage for one method over the others. None of these methods is required for

---

[2] The tensile stress is not entirely due to the grips because the plastic deformation of the outer elements of the bar is considerably greater than that of the inner elements. This results in a spiral tensile stress in the outer elements and a similar compressive stress in the inner elements.

FIG. 5–10

(a)

(b)

(c)

(d)

the principal stresses because they are equal in magnitude to the maximum shearing stress as given by the torsion formula.

PROBLEMS

Note: Assume linearly elastic action for all of the following problems.

5-27* An aluminum alloy tube is to be used to transmit a torque in a control mechanism. The tube has an outside diameter of 1.25 in. and a wall thickness of 0.065 in. Because of the tendency of thin sections to buckle, the maximum compressive stress is specified as 8000 psi. Determine the maximum allowable torque that may be applied.

5-28* Determine the maximum torque that can be resisted by a hollow circular shaft having an inside diameter of 60 mm and an outside diameter of 100 mm without exceeding a normal stress of 70 MPa T or a shearing stress of 75 MPa.

5-29 When the two torques are applied to the steel ($G = 12,000$ ksi) shaft of Fig. P5–29, point A moves 0.172 in. in the direction indicated by torque T. Determine

(a) The torque T.

(b) The maximum tensile stress in the shaft.

5-30 The hollow circular steel ($G = 80$ GPa) shaft shown in Fig. P5–30 has an outside diameter of 120 mm and an inside diameter of 60 mm. Determine

(a) The maximum compressive stress in the shaft.

(b) The distance point B moves as the torque is applied.

(c) The maximum compressive stress in the shaft after the inside diameter is increased to 100 mm.

5-31* A cylindrical tube is fabricated by butt-welding a 1/4-in. steel plate along a spiral seam as shown in Fig. P5–31. If the maximum compressive stress in the tube must be limited to 12,000 psi, determine

(a) The maximum torque T that can be applied to the tube.

(b) The factor of safety with respect to failure by fracture for the weld if the ultimate strengths of the weld metal are 36 ksi in shear and 60 ksi in tension.

FIG. P5–29

FIG. P5–30

FIG. P5–31

FIG. P5–32

5-32   When the two torques are applied to the steel ($G = 80$ GPa) shaft of Fig. P5–32, point A moves 5 mm in the direction indicated by torque T_1. A strain gage bonded to the surface of the 50-mm shaft at an angle of 45° with the axis of the shaft indicates a strain of 1000μ. Determine

(a)   The magnitudes of the two torques.

(b)   The maximum tensile stress in the 50-mm shaft.

(c)   The rotation of the right end of the 100-mm shaft.

5-6   POWER TRANSMISSION

Almost anyone will recognize that one of the most common uses for the circular shaft is the transmission of power; therefore, no discussion of torsion would be adequate without including this topic. Power is the time rate of doing work, and the basic relationship for work done by a constant torque is $W_k = T\phi$ where W_k is work and ϕ is the angular displacement of the shaft in radians. The time derivative of this expression gives

$$\frac{dW_k}{dt} = T\frac{d\phi}{dt} = T\omega$$

where dW_k/dt is power in foot-pounds per minute (or similar units), T is a constant torque in foot-pounds, and ω is the angular velocity (of the shaft) in radians per minute. All units, of course, may be changed to any other consistent set of units. Since the angular velocity is usually given in revolutions per minute (rpm), the conversion of revolutions to radians will often be necessary. Also, in the English system, power is usually given in units of horsepower, and the relation 1 hp = 33,000 ft-lb per min will be found useful. In the metric (SI) system, power is given in watts (N · m/sec). An example of shaft design follows.

EXAMPLE 5–3

A Diesel engine for a small commercial boat is to operate at 200 rpm and deliver 800 hp through a gearbox with a ratio of 1 to 4 (increase) to the propeller shaft. Both the shaft from engine to gearbox and the propeller shaft are to be solid and made of heat-treated alloy steel. Determine the minimum permissible diameters for the two shafts if the allowable stress is 20 ksi and the angle of twist in a 10-ft length of the propeller shaft is not to exceed 4°. Neglect power loss in the gearbox and assume (incorrectly because of thrust stresses) that the propeller shaft is subjected to pure torsion.

SOLUTION

The first step is the determination of the torques to which the shafts are to be subjected. By means of the expression, power = $T\omega$, the torques are obtained as follows:

$$800(33,000) = T_1(200)(2\pi)$$

from which

$$T_1 = 66,000/\pi \text{ ft-lb}$$

which is the torque at the crank shaft of the engine. Because the propeller shaft speed is four times that of the crankshaft and power loss in the gearbox is to be neglected, the torque on the propeller shaft is one-fourth that on the crankshaft and is equal to $66,000/(4\pi)$ ft-lb. The torsion formula can be used to determine the shaft sizes necessary to satisfy the stress specification. For the main shaft,

$$\frac{J}{c} = \frac{T}{\tau} = \frac{66,000(12)}{\pi(20)(10^3)} = \frac{(\pi/2)c_1^4}{c_1}$$

$$c_1^3 = 8.02 \quad \text{and} \quad c_1 = 2.002 \text{ in.}$$

or the shaft from engine to gearbox should be

<u>4.00 in. in diameter</u>   Ans.

The torque on the propeller shaft is one-fourth that on the main shaft, and this is the only change in the expression for c_1^3; therefore

$$c_2^3 = 8.02/4 \quad \text{and} \quad c_2 = 1.261 \text{ in.}$$

The size of propeller shaft necessary to satisfy the distortion specification will be determined as follows:

$$\gamma = \frac{c\theta}{L} = \frac{\tau}{G} = \frac{Tc}{JG}$$

from which

$$\frac{4(\pi/180)}{10(12)} = \frac{66,000(12)/(4\pi)}{(\pi c_3^4/2)(12)(10^6)}$$

$$c_3^4 = 5.75 \quad \text{and} \quad c_3 = 1.548 > 1.261$$

Therefore, the propeller shaft must be

<u>3.10 in. in diameter</u>   Ans.

PROBLEMS

> *Note: In the following problems, the specified allowable stresses are all below the shearing proportional limits of the materials.*

5-33*   A solid circular steel ($G = 12,000$ ksi) shaft is 3 in. in diameter and 4 ft long. The maximum permissible angle of twist is 0.024 rad. The working shearing stress is 10,000 psi. Determine the maximum horsepower that this shaft can deliver when rotating at 225 rpm.

5-34*   A 3-m long hollow steel ($G = 80$ GPa) shaft has an outside diameter of 100 mm and an inside diameter of 60 mm. The maximum shearing stress in the shaft is 80 MPa and the angular velocity is 200 rpm. Determine

   (a)   The power being transmitted by the shaft.

   (b)   The magnitude of the angle of twist in the shaft.

5-35*   A solid circular steel ($G = 12,000$ ksi) shaft 10 ft long transmits 300 hp at 180 rpm. The allowable shearing stress in the shaft is 9000 psi and the maximum allowable angle of twist is 0.033 rad. Determine the minimum allowable diameter of the shaft.

5-36   A solid circular steel ($G = 80$ GPa) shaft 1.5 m long transmits 200 kW at a speed of 400 rpm. If the allowable shearing stress is 70 MPa and the allowable angle of twist is 0.045 rad, determine

   (a)   The minimum permissible diameter for the shaft.

   (b)   The speed at which this power can be delivered if the stress is not to exceed 50 MPa in a shaft with a diameter of 75 mm.

5-37   A solid circular steel ($G = 12,000$ ksi) shaft has a diameter of 4 in. and is 15 ft long. Determine

   (a)   The torque required to produce an angle of twist of 2°.

(b)   The minimum permissible angular velocity (in rpm) for an allow-able shearing stress of 10,000 psi when the shaft is transmitting 3000 hp.

5-38*   A hollow shaft of aluminum alloy ($G = 28$ GPa) is to transmit 1200 kW at 1800 rpm. The shearing stress is not to exceed 100 MPa and the angle of twist is not to exceed 0.20 rad in a 3-m length. Determine the minimum permissible outside diameter if the inside diameter is to be three-fourths of the outside diameter.

5-39   The motor shown in Fig. P5–39 develops 50 hp at a speed of 360 rpm. Gears A and B deliver 20 hp and 30 hp, respectively, to operating units in a factory. If the maximum shearing stress in the shafts is to be limited to 12,000 psi, determine

(a)   The minimum permissible diameter for each of the shafts.

(b)   The magnitude of the total angle of twist between the motor and gear B if the shafts are made of steel ($G = 12,000$ psi).

5-40   A motor delivers 120 kW at 1800 rpm to a gear box that reduces the speed to 200 rpm to drive a crusher. If the maximum shearing stress in the shafts is not to exceed 70 MPa, determine

(a)   The minimum permissible diameter for each of the shafts.

(b)   The magnitude of the total angle of twist between the motor and the crusher if the shafts are made of steel ($G = 80$ GPa) and each shaft is 2-m long.

5-7   TORSIONAL SHEARING STRESS—INELASTIC ACTION

One of the limitations of the torsion formula (Eq. 5–5) is that stresses must be within the proportional limit of the material. For many design problems, this imposes too severe a restriction; therefore, the analysis must be extended to include plastic action in the material. In Section 5–2, the assumption was made that a plane section remains plane and a diameter remains

FIG. P5–39

straight for circular sections under pure torsion. This assumption led to the development of Eq. 5–3

$$\gamma = \frac{\gamma_c}{c}\, \rho \tag{5-3}$$

which states that the shearing strain at any point on a transverse cross section of a circular shaft is proportional to the distance of the point from the axis of the shaft. Equation 5–3 is valid for elastic or inelastic action provided the strains are not too large ($\tan \gamma \approx \gamma$). Once the shearing strain distribution is known, the law of variation of the shearing stress τ (meaning $\tau_{z\theta}$) with radius ρ on the transverse plane can be determined from a shear stress-strain diagram for the material. As an aid to the thought sequence, it is suggested that sketches be drawn of γ vs. ρ, then τ vs. γ, and finally τ vs. ρ, as illustrated for two typical examples in Fig. 5–11. The resisting torque (or other unknown if the torque is known) can then be evaluated by substituting the expression for τ in terms of ρ into Eq. 5–1.

$$T = T_r = \int_{\text{area}} \rho\tau\, dA \tag{5-1}$$

The procedure is illustrated in the following examples.

EXAMPLE 5–4
A solid circular steel shaft 4 in. in diameter is subjected to a pure torque of 7π ft-kips. Assume the steel is elastoplastic, having a yield point in shear of 18 ksi and a modulus of rigidity of 12,000 ksi. Determine the maximum shearing stress in the shaft and the magnitude of the angle of twist in a 10-ft length.

SOLUTION
Assume first that the maximum stress is less than the proportional limit. In this case, if the material is homogeneous and isotropic, the expression $\tau = G\gamma$ applies over the entire cross section (see Fig. 5–11d), and the γ vs. ρ diagram would appear like the γ vs. ρ diagram (Fig. 5–11a), with γ_c replaced by τ_c, making the equation of the curve $\tau = (\gamma_c/c)\rho$. Substituting this expression for τ into Eq. (5–1) yields

$$7\pi(12) = \int_0^2 \int_0^{2\pi} \rho\left(\frac{\tau_c}{c}\,\rho\right)\cdot(\rho\, d\rho\, d\theta) = \int_0^2 \rho\left(\frac{\tau_c}{2}\,\rho\right)(2\pi\rho\, d\rho) = \pi\tau_c\left(\frac{2^4}{4}\right)$$

from which τ_c is 21 ksi, which is greater than 18 ksi. Therefore, the assumption is not valid, and the stress distribution curve should be as in Fig. 5–11e with the maximum shearing stress

$$\tau_{\max} = \underline{18 \text{ ksi}} \qquad\qquad \text{Ans.}$$

FIG. 5–11

Observe that the first integration above replaces the original dA with an annular area of width $d\rho$. In the future dA will be written in this form.

To obtain the angle of twist, Eq. (5–2) will be rewritten thus

$$\theta = \frac{\gamma L}{\rho} = \frac{\gamma_y L}{r} \qquad (a)$$

where the shearing strain at the yield point γ_y equals $\tau_y/G = 18/12{,}000 = 1.5(10)^{-3}$ rad and r is the radius of that part of the cross section that is deforming elastically (Fig. 5–11e). The value of r will be found by means of Eq. (5–1). Thus,

$$T_r = \int_0^r \rho \left(\frac{18}{r} \rho\right) (2\pi\rho \, d\rho) + \int_r^2 \rho(18)(2\pi\rho \, d\rho)$$

$$7\pi(12) = 2\pi \left(\frac{18}{r}\right) \left(\frac{r^4}{4}\right) + 2\pi(18)(1/3)(2^3 - r^3)$$

from which $r^3 = 4$ and $r = 1.587$ in. Now, from Eq. a, the magnitude of the angle of twist is

$$\theta = \frac{1.5(10)^{-3}(10)(12)}{1.587} = \underline{0.1134 \text{ rad}} \qquad \text{Ans.}$$

EXAMPLE 5-5

A straight shaft with a hollow circular cross section with outside and inside diameters of 3 and 2 in., respectively, is made of a magnesium alloy having the shear stress-strain diagram of Fig. 5–12. Determine the torque required to twist a 5-ft length of the shaft through 0.240 rad.

SOLUTION

From Eq. 5–2 the shear strains at the outside and inside surfaces are obtained. Thus,

$$\gamma_{1.5} = \frac{\rho\theta}{L} = \frac{(3/2)(0.240)}{5(12)} = 6(10)^{-3} \text{ rad}$$

$$\gamma_{1.0} = (2/3)\gamma_{1.5} = 4(10)^{-3} \text{ rad}$$

Since both the minimum and maximum shear strains lie within the right-hand portion of the curve of Fig. 5–12, the shear stress-strain function given for this curve applies for the entire cross section. Thus, the γ vs. ρ function is

$$\gamma = (\gamma_c/c)\rho = \frac{6}{(3/2)}\rho = 4\rho$$

where γ is in 10^{-3} rad. The τ vs. ρ function then becomes

$$\tau = 3.70\gamma^{0.44} = 3.70(4)^{0.44}\rho^{0.44} = 3.70(1.840)\rho^{0.44}$$

Substituting this expression into Eq. 5–1 yields

$$T_r = \int_{1.0}^{1.5} \rho[3.70(1.84)\rho^{0.44}](2\pi\rho \, d\rho) = \left| \frac{2\pi(3.70)(1.84)\rho^{3.44}}{3.44} \right|_{1.0}^{1.5}$$

FIG. 5–12

from which

$$T = T_r = \underline{37.7 \text{ in.-kips}} \qquad \text{Ans.}$$

PROBLEMS

5-41*   A solid circular shaft is made of elastoplastic steel ($G = 12,000$ ksi) that has a shearing yield point of 24 ksi. Determine

- (a)   The shearing strain at the surface of the shaft when the shearing stress at one-half of the radius from the center of the shaft reaches the yield point.
- (b)   The ratio of the torque carried by the shaft of part (a) to the torque carried by an identical shaft when the shearing stress at the outer surface reaches the yield point.

5-42*   A hollow shaft has an inside diameter of 50 mm and an outside diameter of 100 mm. The shaft is made of an elastoplastic steel ($G = 80$ GPa) that has a shearing yield point of 140 MPa. Determine

- (a)   The maximum torque that the shaft can transmit if the allowable shearing stress is 125 MPa.
- (b)   The percentage increase in the torque of part (a) if the shearing stress at the inside surface is allowed to reach the yield point.

5-43   A 4-in. diameter solid circular shaft is made of an elastoplastic steel ($G = 12,000$ ksi) that has a shearing yield point of 24 ksi. Determine

- (a)   The applied torque when the plastic zone is 1/2-in. deep.
- (b)   The applied torque when the plastic zone is 1-in. deep.

5-44*   An 80-mm diameter solid circular shaft is made of an elastoplastic steel ($G = 80$ GPa) that has a shearing yield point of 120 MPa. Determine the magnitude of the torque required to produce inelastic action throughout the entire cross section. Note: There will always be some elastic action at the center of the shaft; however, this effect is so small that it can be neglected without producing significant error.

5-45   A 4-in. diameter solid circular shaft is twisted through an angle of 5° in a 5-ft length. The shaft is made of an elastoplastic steel that has a yield point in shear of 18 ksi and a modulus of rigidity of 12,000 ksi. Determine

- (a)   The maximum shearing strain in the shaft.
- (b)   The magnitude of the applied torque.
- (c)   The shearing stress at a point on a transverse plane 1 in. from the center of the shaft.

5-46   A hollow shaft with an outside diameter of 100 mm and an inside diameter of 50 mm is twisted through an angle of 6° in a 2-m length. The shaft is made of an elastoplastic steel that has a yield point in shear of 140 MPa and a modulus of rigidity of 80 Gpa. Determine

(a)   The maximum shearing strain in the shaft.

(b)   The magnitude of the applied torque.

(c)   The shearing stress at a point on a transverse plane at the inside surface of the shaft.

5-47*   A hollow shaft has an inside diameter of 2 in., an outside diameter of 3 in., and is 3 ft long. The shaft is made of an aluminum alloy that has a shearing stress-strain diagram that can be approximated by the two straight lines shown in Fig. P5–47. Determine

(a)   The torque required to develop a shearing stress of 30 ksi at a point on a transverse plane at the inner surface.

(b)   The angle of twist when the torque of part (a) is applied.

5-48   A 50-mm diameter solid circular shaft is 1 m long. The shaft is made of an aluminum alloy that has a shearing stress-strain diagram that can be approximated by the two straight lines shown in Fig. P5–48. Determine

(a)   The torque required to develop a maximum shearing stress of 230 MPa in the shaft.   *6.8 × 10³ Nm*

(b)   The angle of twist when the torque of part (a) is applied.   *= .5rad*

5-49*   A hollow shaft with an inside diameter of 2 in. and an outside diameter of 4 in. is made of a material with the shearing stress-strain diagram shown in Fig. P5–49. If the diagram is approximated by the two straight lines OA and AB for stresses less than 36 ksi, determine

(a)   The torque that will be developed in the shaft when the maximum shearing stress is 32 ksi.

FIG. P5–47

FIG. P5–48

FIG. P5–49

(b)   The magnitude of the angle of twist per foot of length of the shaft when the maximum shearing stress is 32 ksi.

5-50   A 3-in. diameter solid circular shaft is made of the material shown in Fig. P5–49. If the diagram is approximated by the two straight lines OA and AB for stresses less than 36 ksi, determine

(a)   The torque that will be developed in the shaft when the maximum shearing stress is 32 ksi.

(b)   The angle of twist per foot of length when the maximum shearing stress is 32 ksi.

5-51*   A hollow shaft with an inside diameter of 60 mm and an outside diameter of 80 mm is made of a magnesium alloy that has the shearing stress-strain diagram shown in Fig. P5–51. Determine the torque required to twist a 2-m length of the shaft through 0.300 rad.

5-52   An 80-mm diameter solid circular shaft is made of a magnesium alloy that has the shearing stress-strain diagram shown in Fig. P5–51. Determine the torque required to twist a 2-m length of the shaft through 0.300 rad.

5-8 MODULUS OF RUPTURE

Certain routine design problems involving circular torsion members must be designed against failure by fracture (or buckling in the case of tubing) rather than slip, and the method of Art. 5–7 for inelastic action may be too involved for the type of problem considered. In such cases, the torsion formula may possibly be used with modification.

Consider a circular shaft of a material for which the stress-radius distribution under the torque necessary to produce fracture is as shown by line OB in Fig. 5–13. A linear stress distribution, represented by line OC, may be assumed such that $\int_A \tau \, dA(\rho)$ for this distribution will have the same value as that for the stress distribution of line OB. Naturally, if the distribution of OC is assumed, the torsion formula applies thus

$$T_r = \frac{RJ}{c}$$

where R is a fictitious stress known as the *modulus of rupture in torsion*. The magnitude of R is obtained by loading a member to fracture (for a brittle material) or buckling (for thin-walled tubing), substituting the magnitude of

FIG. 5–13

the torque at fracture (buckling) and J/c for the specimen into the torsion formula and solving for R. The magnitude of the modulus of rupture is dependent on the kind of material and, in general, on some size parameter (such as the ratio of diameter to wall thickness for tubing); hence, the method is useful only if applicable values of the modulus are available.

5-9   STATICALLY INDETERMINATE MEMBERS

In all problems discussed in the preceding articles, the equations of equilibrium sufficed to determine the resisting torque at any section. Occasionally, torsionally loaded members are so constructed and loaded that the number of independent equilibrium equations is less than the number of unknowns, and the member is statically indeterminate. When this occurs, it becomes necessary to write the requisite number of distortion equations so that the total number of equations agrees with the number of unknowns to be determined.

The distortion equations will involve the angles of twist, and a simplified twist diagram will often be of assistance in obtaining the correct equations. Where there is a possibility of inelastic action but such action is not specifically indicated, the recommended procedure is to assume elastic action and verify that the resulting stresses are or are not below the proportional limit of the material. The following examples will serve to illustrate the application of the principles.

EXAMPLE 5–6

The circular shaft AC in Fig. 5–14a is fixed to rigid walls at A and C. The section AB is solid annealed bronze ($G = 6500$ ksi) and the hollow section BC is made of aluminum alloy 2024-T4 ($G = 4000$ ksi). There is no stress in the shaft before the 20,000-ft-lb torque T is applied. Determine the maximum shearing stress in each shaft after T is applied.

SOLUTION

Figure 5–14b is a free-body diagram of the shaft. The torques T_A and T_C, produced by the supports, are unknown, and the summation of moments about the axis of the shaft gives

$$T_A + T_C = 20,000(12) \qquad (a)$$

This is the only independent equation of equilibrium; thus, the problem is statically indeterminate. A second equation can be obtained from the deformation of the shaft, since each section must undergo the same angle of twist; thus,

FIG. 5–14

(a)

(b)

$$\theta_{AB} = \theta_{BC} \qquad\qquad (b)$$

Equations *a* and *b* can both be expressed in terms of the shearing stresses in the two sections. When the torsion formula is applied to Eq. *a*, it becomes

$$\frac{\tau_{AB}\pi(2^4)/2}{2} + \frac{\tau_{BC}\pi(2^4 - 1^4)/2}{2} = 240{,}000 \qquad (c)$$

Since the angle of twist is

$$\theta = \frac{\gamma L}{c} = \frac{\tau L}{Gc}$$

Eq. *b* becomes

$$\frac{\tau_{AB}(6.5)12}{6.5(10^6)2} = \frac{\tau_{BC}(5)12}{4(10^6)2}$$

from which

$$\tau_{AB} = 5\tau_{BC}/4 \qquad\qquad (d)$$

Equations *c* and *d* can be solved simultaneously to give

$$\tau_{AB} = \underline{10{,}900 \text{ psi}} \qquad\qquad \text{Ans.}$$

$$\tau_{BC} = \underline{8{,}730 \text{ psi}} \qquad\qquad \text{Ans.}$$

Note: These maximum stresses, which occur at the surfaces of the shafts, do not exceed the proportional limits of the materials; therefore, only elastic behavior is involved and the results are valid.

EXAMPLE 5–7
In Exam. 5–6 change the length of segment AB to 6 ft and the material to elastoplastic steel with a proportional limit of 18 ksi and modulus of rigidity of 12,000 ksi. Change the applied torque to 30 ft-kips. Make no other changes.

SOLUTION
First assume elastic action; then Eqs. a and c become

$$4\pi\tau_{AB} + \frac{15\pi}{4}\tau_{BC} = 360 \tag{e}$$

and Eq. b becomes

$$\frac{\tau_{AB}(6)(12)}{12(10^3)(2)} = \frac{\tau_{BC}(5)(12)}{4(10^3)(2)} \tag{f}$$

from which

$$\tau_{AB} = 5\tau_{BC}/2 \tag{g}$$

Solving Eqs. e and g gives

$$\tau_{AB} = 20.9 \text{ ksi} > 18 \text{ ksi}$$

Therefore, the steel undergoes some inelastic action, and since for the elastic solution the corresponding aluminum stress is 8.36 ksi, which is much less than the elastic strength of 28 ksi (see Appendix A), it would be reasonable to assume that the aluminum will undergo elastic action only. If r is assumed to be the radius of the elastic-plastic boundary in the steel segment, the equilibrium equation, Eq. e, becomes

$$\int_0^r \frac{18}{r}\rho(2\pi\rho \, d\rho)\rho + 18 \int_r^2 (2\pi\rho \, d\rho)\rho + \frac{15\pi}{4}\tau_{BC} = 360 \tag{h}$$

Also, since the maximum elastic shear strain in the steel is

$$\gamma = \frac{18}{12,000} = \frac{r\theta_{AB}}{6(12)}$$

the angle of twist in segment AB is

$$\theta_{AB} = \frac{108(10^{-3})}{r} \tag{i}$$

Equation *f* now becomes

$$\frac{108(10^{-3})}{r} = \frac{\tau_{BC}(5)(12)}{4(10^3)(2)}$$

from which

$$\tau_{BC} = \frac{72}{5r} \tag{j}$$

Now, substituting Eq. *j* into Eq. *h* and integrating gives

$$9\pi r^3 + 12\pi(8 - r^3) + \frac{15\pi}{4}\left(\frac{72}{5r}\right) = 360$$

which after collecting terms and simplifying becomes

$$r^4 + 6.2r = 18$$

A trial-and-error solution of this equation gives a value of 1.665 in. for *r*, which when substituted into Eq. *j* gives

$$\tau_{BC} = \underline{8.65 \text{ ksi}} \qquad \text{Ans.}$$

which is well below the elastic strength of the aluminum. Also,

$$\tau_{AB} = \underline{18.0 \text{ ksi}} \qquad \text{Ans.}$$

EXAMPLE 5–8

Segment *CD* of Fig. 5–15a is a solid shaft of cold-rolled bronze ($G = 45$ GPa), and segment *EF* is a hollow shaft of aluminum alloy 2024-T4 ($G = 28$ GPa) with a core of low-carbon steel ($G = 80$ GPa). The ends *C* and *F* are to be considered fixed to rigid walls and the steel core of *EF* is connected to the flange of *E* so that the aluminum and steel act together as a unit. The two flanges *D* and *E* are bolted together, and the bolt clearance permits flange *D* to rotate through 0.03 rad before *EF* will carry any of the load. Determine the maximum shearing stress in any shaft material when the torque of 54 kN · m is applied.

SOLUTION

Figure 5–15b is a free-body diagram of the assembly. It is necessary to place two unknown torques T_A and T_S on the right end because the torques and

FIG. 5–15

(a)

(b)

(c)

stresses will need to be related by the torsion formula, which is limited to cross sections of homogeneous material. Summing moments with respect to the axis of the shaft gives the following equation:

$$54(10^3) = T_B + T_A + T_S \tag{k}$$

This is the only independent equilibrium equation for this problem and there are three unknown torques; hence, two distortion equations need to be written. A three-dimensional distortion diagram is not easy to draw, and the cross sectional drawing of Fig. 5–15c is adequate. The angles indicated represent the twist in segment CD at D and in segment EF at E (the coupling is considered rigid). The fact that segment EF is nonhomogeneous does not invalidate the assumption of the plane section remaining plane and the diameter remaining straight; the strains are still proportional to the distance from the shaft axis. However, the stresses are not proportional to the radii throughout the entire cross section because G is not single-valued.

From Fig. 5–15c, the distortion equations are

$$\theta_B = \theta_A + 0.03 \tag{l}$$

and

$$\theta_A = \theta_S \tag{m}$$

Eqs. k, l, and m can be written in terms of the same three unknowns (torque, angle, or stress) and solved simultaneously. Since the maximum stress is required, Eqs. l and m will be written in terms of the maximum stress in each material; thus,

$$\theta = \frac{\gamma L}{c} = \frac{\tau L}{Gc}$$

$$\frac{\tau_B(2)}{45(10^9)(0.06)} = \frac{\tau_A(1.4)}{28(10^9)(0.06)} + 0.03 \tag{l}$$

from which

$$\tau_B = (9/8)\tau_A + 40.5(10^6) \qquad (\text{therefore } \tau_B > \tau_A) \tag{n}$$

and

$$\frac{\tau_A(1.4)}{28(10^9)(0.06)} = \frac{\tau_S(1.4)}{80(10^9)(0.03)} \tag{m}$$

from which

$$\tau_S = (10/7)\tau_A = [10/7][(8\tau_B/9) - 36(10^6)] \tag{o}$$

Writing Eq. k in terms of maximum stresses in each material and using the torsion formula gives

$$54(10^3) = \frac{\tau_B(\pi/2)(0.06^4)}{0.06} + \frac{\tau_A(\pi/2)(0.06^4 - 0.03^4)}{0.06} + \frac{\tau_S(\pi/2)(0.03^4)}{0.03}$$

or

$$4(10^9)/\pi = 8\tau_B + 7.5\tau_A + \tau_S \tag{p}$$

Solving Eqs. n, o, and p simultaneously gives the results

$$\tau_B = 100.1 \text{ MPa}$$

$$\tau_A = 52.9 \text{ MPa}$$

and

$$\tau_S = 75.6 \text{ MPa}$$

Assume for the steel and the bronze that the elastic strengths in shear are approximately six-tenths the elastic strengths in tension. Recall also that the proportional limits may be somewhat lower than the elastic strengths (which may be yield strengths). The above stresses indicate that the elastic solution is valid, and the maximum shearing stress is at the surface of the bronze. Thus,

$$\tau_{max} = \underline{100.1 \text{ MPa}} \qquad\qquad \text{Ans}$$

PROBLEMS

> *Note: In the following problems, the specified allowable stresses are all below the shearing proportional limits of the materials.*

 5-53*   For the steel ($G = 12,000$ ksi) shaft of Fig. P5–53, determine

(a)   The maximum shearing stress in the shaft.

(b)   The angle of rotation of the section where the given torque is applied.

FIG. P5–53

FIG. P5–54   FIG. P5–55

5-54* Two 80-mm diameter solid circular steel ($G = 80$ GPa) and bronze ($G = 45$ GPa) shafts are rigidly connected and supported as shown in Fig. P5–54. A torque T is applied at the junction of the two shafts as indicated. The allowable shearing stresses are 140 MPa for the steel and 50 MPa for the bronze. Determine

    (a)   The maximum torque T that can be applied.

    (b)   The angle of rotation of a section 1.25 m from the left support when the torque of part (a) is applied.

5-55* The steel ($G = 12,000$ ksi) shaft of Fig. P5–55 is rigidly secured at the ends. The right 10 ft of the shaft is hollow, having an inside diameter of 2 in. Determine

    (a)   The maximum shearing stress in the shaft.

    (b)   The angle of rotation of the section where the given torque is applied.

5-56* A disk and two circular shafts are connected and supported between rigid walls as shown in Fig. P5–56. Shaft A is made of brass ($G = 39$ GPa) and has a diameter of 100 mm and a length of 500 mm. Shaft B is made of Monel ($G = 65$ GPa) and has a diameter of 75 mm and a length of 675 mm. If a torque of 20 kN · m is applied to the disk, determine

    (a)   The maximum shearing stress in each of the shafts.

    (b)   The angle of rotation of the disk.

5-57 A hollow circular aluminum alloy ($G = 4000$ ksi) cylinder has a steel ($G = 12,000$ ksi) core as shown in Fig. P5–57. The steel and aluminum parts are securely connected at the ends. If the working stresses in the steel and aluminum must be limited to 12 ksi and 10 ksi respectively, determine

    (a)   The maximum torque that can be applied to the right end of the composite shaft.

    (b)   The rotation of the right end of the composite shaft when the torque of part (a) is applied.

FIG. P5–56

FIG. P5–57

5-58*   A composite shaft consists of a hollow bronze ($G = 40$ GPa) cylinder with a 150-mm outside diameter and a 100-mm inside diameter over a 50-mm diameter solid steel ($G = 80$ GPa) core. The two members are rigidly connected to a bar at the right end and to the wall at the left end as shown in Fig. P5–58. When the couple shown is applied to the shaft, determine

(a)   The rotation of bar AB.

(b)   The maximum shearing stress in each of the members.

5-59   The aluminum alloy ($G = 4000$ ksi) shaft of Fig. P5–59 is rigidly fixed to the wall at C, but the flange at A allows the left end of the shaft to rotate 0.012 rad before the bolts provide rigid support. Determine the maximum torque that can be applied at section B if the shearing stress is not to exceed 12 ksi in either section of the shaft.

FIG. P5–58

FIG. P5–59

5-60* A steel ($G = 80$ GPa) shaft 2.5 m long extends through and is attached to a hollow bronze ($G = 40$ GPa) shaft 1.5 m long, as shown in Fig. P5–60. Both shafts are fixed at the wall. When the two couples shown are applied to the shaft, determine

   (a) The maximum shearing stress in each of the materials.

   (b) The rotation of the right end of the shaft.

5-61 The 4-in. diameter shaft of Fig. P5–61 is composed of brass ($G = 6000$ ksi) and steel ($G = 12{,}000$ ksi) segments. Determine the maximum permissible magnitude for the torque T, applied at C, if the allowable shearing stresses are 5 ksi for the brass and 12 ksi for the steel.

5-62 The torsional assembly of Fig. P5–62 consists of an aluminum alloy ($G = 28$ GPa) segment AB securely connected to a steel ($G = 80$ GPa) segment BCD by means of a flange coupling with four bolts. The diameters of both segments are 80 mm, the cross-sectional area of each bolt is 130 mm²,

FIG. P5–61

FIG. P5–62

and the bolts are located 65 mm from the center of the shaft. If the cross shearing stress in the bolts is limited to 55 MPa, determine

    (a)   The maximum torque T that can be applied at section C.

    (b)   The maximum shearing stress in the steel.

5-63*   The inner surface of the aluminum alloy (G = 4000 ksi) sleeve A and the outer surface of the steel (G = 12,000 ksi) shaft B of Fig. P5–63 are smooth. Both the sleeve and the shaft are rigidly fixed to the wall at D. The 0.500-in diameter pin C fills a hole drilled completely through a diameter of the sleeve and shaft. The shearing deformation in the pin and the bearing deformation between the pin and shaft can be neglected. Determine the maximum torque T that can be applied to the right end of the steel shaft B without exceeding an average shearing stress of 3200 psi on the cross-sectional area of the pin at the interface between the shaft and the sleeve.

FIG. P5–63

FIG. P5–64   FIG. P5–65

5-64   The circular shaft of Fig. P5–64 consists of a steel ($G = 80$ GPa) segment ABC securely connected to a bronze ($G = 40$ GPa) segment CD. Ends A and D of the shaft are fastened securely to rigid supports. Determine the maximum shearing stresses in the bronze and in the steel when T has a magnitude of 22 kN · m.

5-65   The shaft of Fig. P5–65 consists of a 6-ft hollow steel ($G = 12{,}000$ ksi) section AB and a 4-ft solid aluminum alloy ($G = 4000$ ksi) section CD. The torque T of 34 ft-kips is applied initially only to the steel section AB. Section CD is then connected and the torque T is released. When the torque is released, the connection slips 0.012 rad before the aluminum section takes any load. Determine the final maximum shearing stress in the aluminum alloy.

5-66*   A torque T of 8 kN · m is applied to the steel ($G = 80$ GPa) shaft of Fig. P5–66 without the brass ($G = 40$ GPa) shell. The brass shell is then slipped into place and attached to the steel. After the original torque is released, determine the maximum shearing stresses in the brass and in the steel.

FIG. P5–66

FIG. P5–67

(a)

(b)

> *Note: In the following problems, the action is elastic unless (1) it is specifically stated to be inelastic, or (2) data are provided that will permit a judgment to be made on whether the action is elastic or plastic or both.*

5-67* The composite shaft shown in Fig. P5–67*a* consists of a hollow steel cylinder with a 6-in. outside diameter and a 4-in. inside diameter over a 2-in. diameter solid cold-rolled brass core. The two members are rigidly connected to the bar *AB* at the right end and to the wall at the left end. The shearing stress-strain diagram for the steel is shown in Fig. P5–67*b*. The brass has a modulus of rigidity *G* of 5000 ksi and a proportional limit in shear of 30 ksi. Determine the torque required to rotate bar *AB* through an angle of 0.24 rad.

5-68* Two 80-mm diameter solid circular steel and bronze shafts are rigidly connected and supported as shown in Fig. P5–68*a*. The shearing stress-

FIG. P5–68   *(a)*   *(b)* strain diagram for the steel is shown in Fig. P5–68b. The bronze has a modulus of rigidity of 45 GPa and a proportional limit in shear of 84 MPa. Determine the torque required to produce a maximum shearing stress of 40 MPa in the bronze.

5-69   The steel shaft of Fig. P5–69 is rigidly secured at the ends. The left 2 ft of the shaft is solid and the right 3 ft of the shaft is hollow with an inside diameter of 2 in. The steel has a modulus of rigidity G of 12,000 ksi and a yield point in shear of 18 ksi (see Fig. P5–67b). Determine the minimum torque, applied as shown, that will produce completely plastic action in the hollow segment.

5-70*   The 160-mm diameter steel shaft of Fig. P5–70 has a 100-mm diameter bronze core inserted in 3 m of the right end and securely bonded to the steel. The steel has a modulus of rigidity G of 80 GPa and a yield point in shear of 120 MPa (see Fig. P5–68b). The bronze has a modulus of rigidity G of 40 GPa and a proportional limit in shear of 240 MPa. Determine

(a)   The maximum shearing stress in each of the materials.

(b)   The rotation of the free end of the shaft.

FIG. P5–71

5-71 The circular shaft of Fig. P5–71 consists of a steel segment *ABC* securely connected to a bronze segment *CD*. Ends *A* and *D* of the shaft are fastened securely to rigid supports. The steel has a modulus of rigidity *G* of 12,000 ksi and a yield point in shear of 18 ksi (see Fig. P5–67*b*). The bronze has a modulus of rigidity *G* of 6000 ksi and a proportional limit in shear of 35 ksi. When the two equal torques *T* are applied, the section at *B* rotates 0.072 rad in the direction of *T*. Determine

(a) The maximum shearing stresses in the steel and in the bronze.

(b) The magnitude of the applied torque *T*.

5-72 The 100-mm diameter shaft of Fig. P5–72 is composed of a brass segment and a steel segment. The steel has a modulus of rigidity *G* of 80 GPa and a yield point in shear of 120 MPa (see Fig. P5–68*b*). The brass has a modulus of rigidity *G* of 40 GPa and a proportional limit in shear of 60 MPa. Determine the magnitude of the torque, applied as shown, that is required to produce a maximum shearing stress in the brass of 50 MPa.

FIG. P5–72

5-10 NONCIRCULAR SECTIONS

Prior to 1820, when A. Duleau published experimental results to the contrary, it was thought that the shearing stresses in any torsionally loaded member were proportional to the distance from its axis. Duleau proved experimentally that this is not true for rectangular cross sections. An examination of Fig. 5–16 will verify Duleau's conclusion. If the stresses in the rectangular bar were proportional to the distance from its axis, the maximum stress would occur at the corners. However, if there was a stress of any magnitude at the corner, as indicated in Fig. 5–16a, it could be resolved into the components indicated in Fig. 5–16b; if these components existed, the two components shown dashed would also exist. These last components cannot exist, since the surfaces on which they are shown are free boundaries. Therefore, the shearing stresses at the corners of the rectangular bar must be zero.

The first correct analysis of the torsion of a prismatic bar of noncircular cross section was published by Saint Venant in 1855; however, the scope of this analysis is beyond the elementary discussions of this book.[3] The results of Saint Venant's analysis indicate that, in general, *except for members with circular cross sections,* every section will warp (not remain plane) when the bar is twisted.

FIG. 5–16

[3] A complete discussion of this theory is presented in various books, such as *Mathematical Theory of Elasticity,* I. S. Sokolnikoff, Second Edition, McGraw-Hill, New York, 1956, pp. 109–134.

For the case of the rectangular bar shown in Fig. 5–4*d*, the distortion of the small squares is maximum at the midpoint of a side of the cross section and disappears at the corners. Since this distortion is a measure of shearing strain, Hooke's law requires that the shearing stress be maximum at the midpoint of a side of the cross section and zero at the corners. Equations for the maximum shearing stress and angle of twist for a rectangular section obtained from Saint-Venant's theory are

$$\tau_{max} = \frac{T}{\alpha a^2 b}$$

$$\theta = \frac{TL}{\beta a^3 bG}$$

where *a* and *b* are the lengths of the short and long sides of the rectangle, respectively. The numerical factors α and β can be obtained from Fig. 5–17.

Since the mathematical analysis of torsional stresses in noncircular solid sections is relatively difficult, other means such as the *membrane analogy* are frequently used to identify regions of large shearing stress. The membrane analogy is derived from the observation that the differential equation of the deflection of a membrane has the same form as the theoretical equation (from the theory of elasticity) for the shearing stresses in any uniform rod subjected to twisting moments.

When a thin membrane (such as a soap film) is stretched over a plane opening having the shape of the cross section of the bar, a slight pressure difference *p* applied to it will cause the membrane to distend. It can be shown that the volume between the membrane and the plane of the opening is proportional to the torque required to produce a given angle of twist per

FIG. 5–17

unit length of the shaft; and the slope of the membrane at any point is proportional to the shearing stress in the cross section at that point, the direction of the stress being normal to the direction in which the slope is measured. This last observation makes the method very useful in estimating stress distributions, since for many cross sections the shape of the distended membrane may be visualized; hence, points of maximum and minimum shearing stress may be located. For example, Fig. 5–18 indicates several sections through a membrane for a rectangular cross section, and these sections indicate that the shearing stress is zero (where the slope of the membrane must be zero) at the corners and at the center and is a maximum at points a and a. Thus, one observes that for a rectangular section, the maximum shearing stress is in the boundary material closest to the axis of the shaft rather than in the most distant material. In like manner, one can visualize the location of maximum torsional stresses in shaft keyways and how the addition of fillets in the reentrant corners of the keyway reduces the magnitude of the stress. The discussion of the techniques involved in the evaluation of stresses by means of the membrane analogy will be left to others.[4]

5-11 SHEAR FLOW

Although the elementary torsion theory presented in Art. 5–1 is limited to circular sections, one class of noncircular sections that can be readily analyzed by elementary methods is the thin-walled section such as the one illustrated in Fig. 5–19a that represents a noncircular section with a wall of variable thickness (t varies).

FIG. 5–18

[4] *Handbook of Experimental Stress Analysis,* M. Hetenyi, Wiley, New York, 1950.

FIG. 5–19

(a)   (b)

A useful concept associated with the analysis of thin-walled sections is *shear flow q,* defined as internal shearing force per unit of length of the thin section. Typical units for q are pounds per inch or newtons per meter. In terms of stress, q equals $\tau(1)(t)$, where τ is the average shearing stress across the thickness t. It will be demonstrated that the shear flow on a cross section is constant even though the thickness of the section wall varies. Figure 5–19*b* shows a block cut from the member of Fig. 5–19*a* between A and B, and as the member is subjected to pure torsion, the shear forces $V_1 \cdots V_4$ alone (no normal forces) are necessary for equilibrium. Summing forces in the x direction gives

$$V_1 = V_3$$

or

$$q_1\,dx = q_3\,dx$$

from which

$$q_1 = q_3$$

and, as $q = \tau(t)$,

$$\tau_1 t_A = \tau_3 t_B \tag{a}$$

The shearing stresses at point A on the longitudinal and transverse planes have the same magnitude; likewise, the shearing stresses at point B have the same magnitude on the two orthogonal planes; hence, Eq. *a* may be written

$$\tau_A t_A = \tau_B t_B$$

or

$$q_A = q_B$$

which was to be proved.

$dF = q\,ds$

Median line

dA

ρ

O

FIG. 5–20

To develop the expression for the resulting torque on the section, consider the force dF acting through the center of a differential length of perimeter ds, as shown in Fig. 5–20. The resisting torque T_r is the resultant of the forces dF; that is,

$$T_r = \int (dF)\rho = \int (q\,ds)\rho = q\int \rho\,ds$$

This integral may be difficult to evaluate by formal calculus; however, the quantity $\rho\,ds$ is twice the area of the triangle shown shaded in Fig. 5–20, which makes the integral equal to twice the area A *enclosed by the median line*. The resulting expression is

$$T_r = q(2A)$$

or, in terms of stress,

$$\tau = T/(2At) \qquad (5\text{–}7)$$

where τ is the *average* shearing stress across the thickness t (and tangent to the perimeter) and is reasonably accurate when t is relatively small. For example, in a round tube with a diameter to wall thickness ratio of 20, the stress as given by Eq. 5–7 is 5 percent less than that given by the torsion formula. It must be emphasized that Eq. 5–7 applies only to "closed" sections—that is, sections with a continuous periphery. If the member were slotted longitudinally (see, for example, Fig. 5–21), the resistance to torsion would be diminished considerably from that for the closed section.

It is sometimes necessary to evaluate the twist of a thin-walled member subjected to torsion. Since the ratio γ/ρ for a section such as that of Fig. 5–20

FIG. 5–21

is not a constant as it is for a circular section, Eq. 5-6 ($\theta = L\gamma/\rho$) is not easily adapted to such a section. Instead, the principle of strain energy will be found most useful in the development of an expression for the angle of twist.

If it is assumed that energy is conserved, the work done by the torque in twisting a member through an angle θ is equal to the strain energy stored in the member, and if the torque-twist relation is linear (which limits stresses to values within the proportional limit), the expression becomes

$$W_k = U = \frac{T\theta}{2} = \int_{\text{Vol.}} u \, dv \qquad (b)$$

where u is the strain energy per unit volume of the material in the member. All elements of the member are subjected to pure shear (see Fig. 5-19); therefore, the expression for u is $u = \tau^2/(2G)$ (see Art. 3-4) and Eq. b becomes

$$U = \int_{\text{Vol.}} \frac{\tau^2}{2G} \, dv \qquad (c)$$

When τ in Eq. c is replaced by Eq. 5-7 and dv by $t \, dx \, ds$ (see Figs. 5-19 and 5-20), the expression becomes

$$U = \int_{\text{Vol.}} \frac{T^2}{8A^2tG} \, dx \, ds \qquad (d)$$

In general, t is a function of s, but for many practical problems, the integral may be replaced by a summation of n elements, each having a constant thickness t_i and a length L; thus,

$$U = \sum_{i=1}^{n} \frac{T^2L}{8A^2G} \left(\frac{\Delta s_i}{t_i} \right) \qquad (e)$$

Substituting Eq. e into Eq. b yields

$$\theta = \frac{TL}{4A^2G} \sum_{i=1}^{n} \frac{\Delta s_i}{t_i} \qquad (5\text{-}8a)$$

or, in terms of the shear flow q,

$$\theta = \frac{qL}{2AG} \sum_{i=1}^{n} \frac{\Delta s_i}{t_i} \qquad (5\text{-}8b)$$

where $\Delta s_i/t_i$ is the ratio of plate width (measured along the periphery of the section) to thickness. Note that L and A have been treated as constants; hence, the shape and size of the cross section must be maintained constant (possibly by diaphragms, bulkheads, or cross bracing), and all the elements must have the same length. The following example will illustrate the use of Eqs. 5–7 and 5–8.

EXAMPLE 5–9
A rectangular box section of aluminum alloy ($G = 28$ GPa) has outside dimensions 100×50 mm. The plate thickness is 2 mm for the 50-mm sides and 3 mm for the 100-mm sides. Determine

(a) The maximum torque that can be applied to the section if the allowable stress is 95 MPa.
(b) The angle of twist in a 900-mm length of the member when the torque of part (a) is applied.

SOLUTION
The maximum stress will occur in the thinnest plate; therefore,

$$q = 95(10^6)(0.002) = 190(10^3) \text{ N/m}$$

(a) The torque is

$$T = 2qA$$
$$= 2(190)(10^3)(0.1 - 0.002)(0.05 - 0.003)$$
$$= \underline{1750 \text{ N} \cdot \text{m}}$$

Ans.

(b) The angle of twist is

$$\theta = \frac{qL}{2AG} \sum_{i=1}^{n} \frac{\Delta s_i}{t_i}$$
$$= \frac{190(10^3)(0.900)}{2(0.098)(0.047)(28)(10^9)} \left[\frac{0.098(2)}{0.003} + \frac{0.047(2)}{0.002} \right]$$
$$= \underline{0.0745 \text{ rad}}$$

Ans.

PROBLEMS

Note: In all the following problems, assume that buckling of the shell is prevented and that the shape of the section is maintained by ribs, diaphragms, and stiffeners.

5-73* A 24-in. wide by 0.125-in. thick steel (G = 12,000 ksi) sheet is to be formed into a hollow section by bending through 360° and welding (butt-weld) the long edges together. The shape may be (1) circular, (2) an equilateral triangle, or (3) square. Assume a median length of 24 in. (no stretching of the sheet due to bending) and square corners for noncircular sections. The allowable shearing stress is 12 ksi. For each of the shapes listed, determine

    (a)   The maximum torque that can be carried by the section.

    (b)   The angle of twist in a 100-in. length.

5-74* A 500-mm wide by 3-mm thick steel (G = 80 GPa) sheet is to be formed into a hollow section by bending through 360° and welding (butt-weld) the long edges together. The shape may be (1) circular, (2) square, or (3) a 150 by 100-mm rectangle. Assume a median length of 500 mm (no stretching of the sheet due to bending) and square corners for noncircular shapes. The allowable shearing stress is 70 MPa. For each of the shapes listed, determine

    (a)   The maximum torque that can be carried by the section.

    (b)   The angle of twist in a 2-m length.

5-75* A torque of 150 in.-kips is applied to the hollow, thin-walled, aluminum (G = 4000 ksi) section shown in Fig. P5–75. If the section has a uniform thickness of 1/8 in., determine

    (a)   The shearing stress developed in the section.

    (b)   The angle of twist per foot of length.

5-76 A torque of 2.5 kN · m is applied to the hollow, thin-walled, aluminum alloy (G = 28 GPa) section shown in Fig. P5–76. If the section has a uniform thickness of 3 mm, determine

    (a)   The shearing stress developed in the section.

    (b)   The angle of twist per meter of length.

5-77 A hollow, thin-walled, aluminum alloy (G = 4000 ksi) torsional member, with the cross section shown in Fig. P5–77, is being designed to carry a torque of 100 in.-kips. If the shearing stress is not to exceed 12 ksi and the angle of twist is not to exceed 1° per foot of length, determine the minimum permissible thickness.

FIG. P5–75   3″ | 8″ | 3″

FIG. P5–76   100 mm

FIG. P5–77

FIG. P5–78

10″

150 mm   25 mm

5-78* A hollow, thin-walled, aluminum alloy ($G = 28$ GPa) torsional member, with the cross section shown in Fig. P5–78, is being designed to carry a torque of 2.5 kN · m. If the shearing stress is not to exceed 70 MPa and the angle of twist is not to exceed 2° per meter of length, determine the minimum permissible thickness.

5-79 A cross section of the leading edge of an airplane wing is shown in Fig. P5–79. The length along the curve is 26 in. and the enclosed area is 82 sq. in. Sheet thicknesses are shown on the diagram and the material is 2024-T4 aluminum alloy ($G = 4000$ ksi). For a torque of 80 in.-kips, determine the magnitude of

    (a) The maximum shearing stress.

    (b) The angle of twist per foot of length.

5-80* A cross section of an airplane fuselage made of 2024-T4 aluminum alloy ($G = 28$ GPa) is shown in Fig. P5–80. The sheet thicknesses are 1.7 mm from A to B and C to D, 1.3 mm from B to C, and 1.0 mm from D to A. The fuselage is subjected to a torque of 170 kN · m. Determine

    (a) The maximum shearing stress.

    (b) The angle of twist per meter of length of fuselage.

5-81 If an airplane fuselage with the cross section shown in Fig. P5–80 is subjected to a torque of 200 kN · m and the allowable shearing stress is 70 MPa, what is the minimum thickness of sheet (constant for the entire periph-

FIG. P5–79

FIG. P5–80

ery) needed to resist the torque? For the given data, what is the magnitude of the angle of twist per meter of length of the fuselage if the material is 2024-T4 aluminum alloy ($G = 28$ GPa)?

5-82* A cross section of an airplane fuselage made of 2024-T4 aluminum alloy ($G = 4000$ ksi) is shown in Fig. P5–82. The sheet thicknesses are 0.050 in. from A to B and C to D, 0.040 in. from B to C, and 0.032 in. from D to A. For an allowable shearing stress of 12 ksi, determine

(a) The maximum permissible torque.

(b) The angle of twist per foot of length.

5-83 If an airplane fuselage with the cross section shown in Fig. P5–82 is subjected to a torque of 1000 in.-kips and the allowable shearing stress is 8 ksi, what is the minimum thickness of sheet (constant for the entire periphery) needed to resist the torque? For the given data, what is the magnitude of the angle of twist per foot of length of the fuselage if the material is 2024-T4 aluminum alloy ($G = 4000$ ksi)?

5-84 A torque box from an airplane wing is shown in Fig. P5–84. Curves AB and CD are 1005 mm long. The mean depth of the box is 350 mm. The box is made of an aluminum alloy ($G = 28$ GPa). If the angle of twist must not exceed 0.50° per meter of length of the box, determine

(a) The maximum permissible torque.

(b) The maximum shearing stress resulting from this torque.

FIG. P5–82

FIG. P5–84

SOLVE FOR T_{max}

$$\theta = \frac{TL}{4A_m^2 t} \sum_{i=1}^{n} \frac{\Delta s_i}{t_i}$$

$$\tau = \frac{T}{2A_m t}$$

USE TRAPAZOID $\ln\left[\frac{.325 + .25}{2}\right]$

$$\sum_{i=1}^{n} \frac{\Delta s_i}{t_i} - \text{ARC LENTH OR SIDE LENGTH}$$
$$t_i - \text{THICKNESS}$$

$$4 - \frac{1.005}{.0012} + \frac{1.005}{.001} + \frac{.325}{.0012} + \frac{.25}{.001}$$

$$\frac{\theta}{L} = \frac{.008727 \, rad}{m}$$

$$\tau_{max} = \frac{T_{max}}{2A_t}$$

CHAPTER 6

FLEXURAL LOADING: STRESSES

6-1 INTRODUCTION

The beam, or flexural member, is frequently encountered in structures and machines, and its elementary stress analysis constitutes one of the more interesting facets of mechanics of materials. A beam is a member subjected to loads applied transverse to the long dimension, causing the member to bend. For example, Fig. 6–1 is a photograph of an aluminum alloy I-beam, *AB*, simply supported in a testing machine and loaded at the one-third points, and Fig. 6–2 depicts the shape (exaggerated) of the beam when loaded.

Before proceeding with a discussion of stress analysis for flexural members, it may be well to classify some of the various types of beams and loadings encountered in practice. Beams are frequently classified on the basis of the supports or reactions. A beam supported by pins, rollers, or smooth surfaces at the ends is called a *simple beam*. A simple support will develop a reaction normal to the beam but will not produce a couple. If either or both ends of a beam project beyond the supports, it is called a

FIG. 6–1 Setup for measuring longitudinal strains in a beam.

simple beam with overhang. A beam with more than two simple supports is a *continuous beam*. Figures 6–3a, b, and c shows, respectively, a simple beam, a beam with overhang, and a continuous beam. A *cantilever beam* is one in which one end is built into a wall or other support so that the built-in end cannot move transversely or rotate. The built-in end is said to be *fixed* if no rotation occurs and *restrained* if a limited amount of rotation occurs. The supports shown in Figs. 6–3d and e represent fixed ends unless otherwise stated. The beams in Figs. 6–3d, e, and f are, in order, a cantilever beam, a beam fixed (or restrained) at the left end and simply supported near the other end (which has an overhang), and a beam fixed (or restrained) at both ends.

Cantilever beams and simple beams have only two reactions (two forces or one force and a couple), and these reactions can be obtained from a free-body diagram of the beam by applying the equations of equilibrium. Such beams are said to be *statically determinate* since the reactions can be obtained from the equations of equilibrium. Continuous and other beams, with only transverse loads, with more than two reaction components are called *statically indeterminate* since there are not enough equations of equi-

FIG. 6–2

FIG. 6–3

librium to determine the reactions. Statically indeterminate beams are discussed in Chapter 8.

The beams shown in Fig. 6–3 are all subjected to uniformly distributed loads and to concentrated loads, and, although shown as horizontal, they may have any orientation. Distributed loads will be shown on the side of the beam on which they are acting; that is, if drawn on the bottom of the beam, the load is pushing upward, and if drawn on the right side of a vertical beam, the load is pushing to the left. The deflection curves shown beneath the beams are greatly exaggerated to assist in visualizing the shape of the loaded beam.

A free-body diagram of the portion of the beam of Fig. 6–3a between the left end and plane a–a is shown in Fig. 6–4a. A study of this diagram reveals that a transverse force V_r and a couple M_r at the cut section and a force R (a reaction) at the left support are needed to maintain equilibrium.

FIG. 6–4

(a)

(b)

xy plane is a plane of symmetry for the beam and loading

(c)

(d)

The force V_r is the resultant of the shearing stresses acting on the cut section (on plane *a–a*) and is called the resisting shear. The couple M_r is the resultant of the normal stresses acting on the cut section (on plane *a–a*) and is called the resisting moment. The magnitudes and senses of V_r and M_r can be obtained from the equations of equilibrium $\Sigma F_y = 0$ and $\Sigma M_0 = 0$ where 0 is any axis perpendicular to the *xy* plane. The reaction R must be evaluated from a free body of the entire beam.

The normal and shearing stresses σ and τ on plane *a–a* are related to the resisting moment M_r and the shear V_r by equations of the form

$$V_r = -\int_{\text{Area}} \tau \, dA \qquad (a)$$

$$M_r = -\int_{\text{Area}} y \, \sigma \, dA \qquad (b)$$

The resisting shear and moment (V_r and M_r), as shown on Fig. 6–4*a*, will be defined as positive quantities later in Section 6–6. Previously, in Section 1–5, the normal and shearing stresses (σ and τ), as shown on Fig. 6–4*c*, were defined as positive stresses. The minus signs in Eqs. *a* and *b* are required to bring these two definitions into agreement.

Ideally, a positive shearing stress should result from a positive shear and a positive normal stress (a tensile stress) should result from a combination of a positive moment and a positive moment arm *y*. Both of these conditions can be satisfied by choosing the *y* direction to be positive downward as shown in Fig. 6–4*d*. Therefore, in this book, *the y direction will be positive downward for the computation of stresses in beam type members.* In all other discussions and applications (beam deflections, column problems, and the like), the positive direction for *y* will be upward. When the downward positive direction for *y* is used, Eqs. *a* and *b* become

$$V_r = \int_{\text{Area}} \tau \, dA$$
$$M_r = \int_{\text{Area}} y \, \sigma \, dA$$

$$(6–1)$$

It is obvious from Eqs. 6–1 that the laws of variation of the normal and shearing stresses must be known before the integrals can be evaluated. For the present, the shearing stresses will be ignored while the normal stresses are studied.

The first recorded hypothesis on the normal stress distribution is that of Galileo Galilei (1564–1642) who, possibly observing that beams of stone when subjected to bending loads broke somewhat as indicated in Fig. 6–5, concluded that the material at point *A* acted as a fulcrum and that the resultant of the tensile stresses was at the center of the cross section (as for axial loading).[1] This theory would lead to the stress distribution diagram of Fig. 6–6*a* in which F_T is the resultant of the tensile stresses. About fifty years after Galileo's observations, E. Mariotte (1620–1684), a French physicist, still retaining the concept of the fulcrum at the compression surface of the beam, but reasoning that the extensions of the longitudinal elements of the beam (fibers) would be proportional to the distance from the fulcrum, suggested that the tensile stress distribution was as indicated in Fig. 6–6*b*. Mariotte later rejected the concept of the fulcrum and observed that part of the beam on the compression side was subjected to compressive stress having a triangular distribution; however, his expression for the ultimate load was still based on his original concept and was accepted by famous scientists such as Jacob Bernoulli, Leonard Euler, and others until C. A. Coulomb (1736–1806), a French military engineer, in a paper published in 1773 dis-

---

[1] *Two New Sciences,* Galileo Galilei, 1638. Actually, Galileo's published work contains no record of test results; therefore, the manner in which he arrived at this conclusion is mere speculation.

FIG. 6–5 Failure of plaster-of-paris beam subjected to flexure.

carded the fulcrum concept and proposed the distribution of Fig. 6–6c in which both the tensile and compressive stresses have the same linear distribution.[2]

6-2 FLEXURAL STRAINS

The correctness of the Coulomb theory can be demonstrated as follows. A segment of the beam of Fig. 6–4 between the planes *a–a* and *b–b* is shown in Fig. 6–7 with the distortion greatly exaggerated. The assumption is made that *a plane section before bending remains plane after bending.* For this to be strictly true, it is necessary that the beam be bent only with couples (no

[2] Sixty years before Coulomb's paper, Parent (1666–1716), a French mathematician, proposed the correct triangular stress distribution. However, his work remained in obscurity, and the majority of engineers continued to use formulas based on Mariotte's theory. For a more complete discussion of the history of the flexure problem, the reader is referred to the *History of Strength of Materials,* S. P. Timoshenko, McGraw-Hill, New York, 1953, from which most of the foregoing discussion is taken.

FIG. 6–6

shear on transverse planes); the beam must be so proportioned that it will not buckle and the loads applied so that no twisting occurs (this last limitation will be satisfied if the loads are applied in a plane of symmetry—a sufficient though not a necessary condition). One observes in Fig. 6–7 that at some distance c above the bottom of the beam, the longitudinal elements (sometimes referred to as fibers) undergo no change in length. The curved surface formed by these elements is referred to as the *neutral surface,* and the intersection of this surface with any cross section is called the *neutral axis* of the section. All elements (fibers) on one side of the neutral surface are compressed, and those on the opposite side are elongated. The assumption is made that *all longitudinal elements have the same length*. This assumption imposes the restriction that the beam be initially straight and of constant cross section; however, in practice, considerable deviation from these last restrictions is often tolerated. If Fig. 6–7 is referred to,

$$\frac{\delta_y}{y} = \frac{\delta_c}{c}$$

from which

$$\delta_y = \frac{\delta_c}{c} y$$

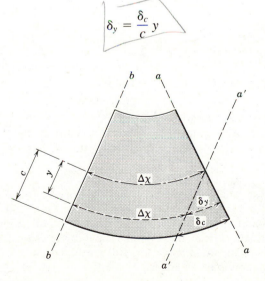

FIG. 6–7

Since all elements had the same initial length Δx, the strain of any element can be determined by dividing the deformation by the length of the element; the strain becomes

$$\varepsilon = \frac{\varepsilon_c}{c} y \qquad (6-2)$$

where ε is the strain of a longitudinal element at a distance y from the neutral surface, and ε_c is the strain of a longitudinal element at the outer surface of the beam. On the basis of the general notation presented in Art. 1–5, both ε and ε_c should be designated ε_x. However, for simplicity, the subscript x is deleted on ε_x and σ_x throughout the chapter, except when the general notation is necessary. The preceding equation indicates that the strain of any fiber is directly proportional to the distance of the fiber from the neutral surface. This variation can be demonstrated experimentally by means of strain gages attached to a beam, as shown in Fig. 6–1. The strains, as measured by gages on two different sections, are plotted against the vertical position of the gages on the beam in Fig. 6–8. Curve 1 represents strains on a section at the center of the beam where pure bending occurs (no transverse shear), and curve 2 shows strains at a section near one end of the beam where both fiber stresses and transverse shearing stresses exist. These curves are both straight lines within the limits of the accuracy of the measuring equipment.[3]

Note that Eq. 6–2 is valid for elastic or inelastic action so long as the beam does not twist or buckle and the transverse shearing stresses are relatively small. Problems in this book will be assumed to satisfy these restrictions.

6-3   FLEXURAL STRESSES

With the acceptance of the premise that the longitudinal strains ε_x are proportional to the distance from the neutral surface, the law of variation of the normal stress σ_x on the transverse plane can be determined by referring to a tensile-compressive stress-strain diagram and making the appropriate substitutions into Eq. 6–2 to obtain the stress distribution diagram σ vs. y. For most real materials, the tension and compression stress-strain diagrams are

[3] A more exact analysis using principles developed in the theory of elasticity indicates that curve 2 should be curved slightly. *Note*: Other experiments indicate that a plane section of an initially curved beam will also remain plane after bending and that the deformations will still be proportional to the distance of the fiber from the neutral surface. The strain, however, will not be proportional to this distance since each deformation must be divided by a different original length.

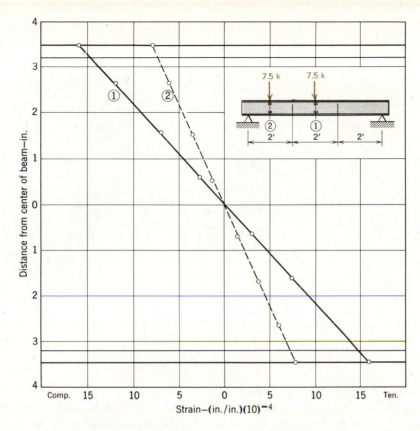

FIG. 6–8

identical in the elastic range; although the diagrams may differ somewhat in the inelastic range, the differences can be neglected for most real problems. For beam problems in this book, the compressive stress-strain diagram will be assumed to be identical to the tensile diagram. As an aid to the thought sequence, it is suggested that sketches be drawn of ε vs. y, then σ vs. ε, and finally σ vs. y, as illustrated for some typical examples in Fig. 6–9. Note that the y axis is taken positive downward so that a combination of positive moment (defined in Art. 6–6) and positive y gives a positive, or tensile, stress. *The downward positive direction for y is used only for computation of stresses. In all other discussions and applications, y is positive upward.* Since no load is applied in the z direction, σ_z is zero throughout the beam. Because of loads applied in the y direction, σ_y is not zero throughout the beam; however, for the elementary solution considered here, σ_y is small enough to neglect (see Art. 6–14). Also, σ_x is uniformly distributed in the z direction because the plane cross section remains plane.

Since the stress-strain diagrams of Fig. 6–9 are for specific materials, the stress distribution diagrams are for the same materials throughout the

FIG. 6–9

(a) (b) (c)

(d) (e) (f)

beam. This imposes the limitation that *the beam be of a homogeneous material*. Figures 6–9c and e are self explanatory; however, Fig. 6–9f might require some explanation. This diagram indicates complete plastic action throughout the section for an ideal elastoplastic material such as that of Fig. 6–9d. Actually, there must be some elastic strain in the material near the neutral surface; however, the contribution of the stresses near the neutral surface to the resisting moment is so small that the assumption of plastic action as indicated in Fig. 6–9f is valid.

With the law of variation of flexural stress known, Fig. 6–4 can now be redrawn as in Fig. 6–10. The vectors F_C and F_T are the resultants of the compressive and tensile flexural stresses, and, because the sum of the forces in the x direction must be zero, F_C is equal to F_T; hence, they form a couple of magnitude M_r.

6-4 FLEXURAL STRESSES—LINEARLY ELASTIC ACTION

The resisting moment M_r of Fig. 6–11, which equals the applied moment M, is given by Eq. 6–1 as

FIG. 6–10

$$M_r = \int_A y \, \sigma \, dA = \int_A y \, dF \qquad (a)$$

and since y is measured from the neutral surface, it is first necessary to locate this surface by means of the equilibrium equation $\Sigma F_x = 0$, which gives

$$\int_A \sigma \, dA = 0 \qquad (b)$$

where σ is some function of y, as indicated in Figs. 6–9c, e, or f.

For the special case of elastic action (which in this book always means *linearly* elastic action) over the entire cross section, the stress distribution diagram would be an extension of the elastic part of Figs. 6–9c or e to the top

FIG. 6–11

and bottom of the beam. The relation between σ and y can be obtained from generalized Hooke's law

$$\sigma_x = \frac{E}{(1 + \nu)(1 - 2\nu)} [(1 - \nu)\varepsilon_x + \nu(\varepsilon_y + \varepsilon_z)] \qquad (3\text{--}8)$$

which, since $\varepsilon_y = \varepsilon_z = -\nu\varepsilon_x$, reduces to $\sigma_x = E\varepsilon_x$: thus,

$$\sigma = E \left(\frac{\varepsilon_c}{c}\right) y = \frac{\sigma_c}{c} y \qquad (c)$$

When Eq. c is substituted into Eq. b, the result is

$$\frac{\sigma_c}{c} \int_A y \, dA = \frac{\sigma_c}{c} A\bar{y} = 0$$

where $\bar{y}$ is the distance from the neutral axis to the centroidal axis (c–c) of the cross section. Since σ_c/c is not zero, $\bar{y}$ must be zero, indicating that the *neutral axis coincides with the centroidal axis (perpendicular to the plane of bending) of the cross section for flexural loading and elastic action.* Instances in which the neutral axis (the line of zero longitudinal strain) does not coincide with the centroidal axis include (1) a combined axial and flexural load (see Chap. 9), (2) a beam with cross section unsymmetrical with respect to the neutral axis and subjected to inelastic action, and (3) a curved beam.

The use of the concepts established thus far will now be demonstrated.

> *Note: Unless otherwise stated, as a simplifying assumption in all beam problems in this book, the weight of the beam itself is to be neglected.*

Obviously this would not, as a rule, be done in professional practice. When recording normal stresses as answers to problems, it is necessary to indicate whether the stress is tension or compression.

EXAMPLE 6–1

The beam of Fig. 6–3a has a rectangular cross section 4 in. wide by 12 in. deep and a span of 12 ft. It is subjected to a uniformly distributed load of 100 lb/ft and concentrated loads of 600 lb and 1200 lb at 2 and 7 ft, respectively, from the left support. Assuming elastic action, determine the maximum fiber stress at the center of the span by computing the magnitudes and positions of the vectors F_T and F_C in terms of σ_{max} and substituting into the moment equilibrium equation.

SOLUTION

The reactions are evaluated from a free-body diagram of the entire beam and moment equations with respect to each of the two reactions. By this method, the reactions are evaluated independently, and summation of forces may be used to check the results. The left reaction R_L is 1600 lb, and the right reaction R_R is 1400 lb, both upward. Figure 6–12a is a free-body diagram of the left half of the beam. The equation $\Sigma M_O = 0$ gives

$$\text{(\!+)} \qquad M_r + 600(4) + 100(6)(3) - 1600(6) = 0$$

from which $M_r = 5400$ ft-lb or $5400(12)$ in.-lb as shown. The moment M_r is the resultant of the flexural stresses indicated in Figs. 6–12b and c in which

(a)

(b)   (c)   FIG. 6–12

F_C and F_T are the resultants of the compressive and tensile stresses, respectively. The resultants F_C and F_T (which act through the centroids of the wedge-shaped stress distribution diagram of Fig. 6–12c) are located through the centroids of the triangular stress distribution diagrams of Fig. 6–12b. The magnitude of the couple M_r equals $F_C(8)$ or $F_T(8)$, and $F_C = (\frac{1}{2})(\sigma_c)(6)(4) = 12(\sigma_c)$; therefore,

$$M_r = 8(12)\sigma_c = 5400(12) \text{ in.-lb}$$

from which

$$\sigma_{max} = \sigma_c = \underline{675 \text{ psi } C \text{ at top and } T \text{ at bottom}} \qquad \text{Ans.}$$

EXAMPLE 6–2

Solve Exam. 6–1 if the beam is composed of two 2×10-in. timbers[4] securely fastened together to form a symmetrical T-section with the flange on top. Use the method of Exam. 6–1 and check the result by integration of Eq. a.

SOLUTION

The horizontal centroidal axis of the cross section is first determined to be 4 in. down from the top surface (see any text on statics). The fiber stress distribution diagrams are shown in Fig. 6–13. For convenience, the compression diagram is subdivided into three parts composed of two triangular distribution (wedge-shaped) diagrams and one uniform distribution diagram. This is done because the locations of the centroids of these diagrams are known. The values of the force components and their locations are shown in the figure. It will be noted that each component F_1, F_2, F_3, being the average stress multiplied by the area on which the stress acts is equal to the volume of the corresponding part of the three-dimensional diagram of Fig. 6–13b. The moment of the component forces with respect to O is

$$M_r = F_1 \left(\frac{10}{3}\right) + F_2 \left(\frac{9}{3}\right) + F_3 \left(\frac{4}{3}\right) + F_4 \left(\frac{16}{3}\right)$$

which from the preceding example is equal to 5400(12); therefore,

$$\left(\frac{5}{2}\right) \sigma_c \left(\frac{10}{3}\right) + 5\sigma_c \left(\frac{9}{3}\right) + \left(\frac{1}{2}\right) \sigma_c \left(\frac{4}{3}\right) + 8\sigma_c \left(\frac{16}{3}\right) = 5400(12)$$

[4] Structural timbers specified as rough sawn will be approximately full size. Cross-sectional dimensions of dressed timbers will be from $\frac{3}{8}$ to $\frac{1}{2}$ in. smaller than nominal. For convenience, all timbers in this book will be considered full size.

FIG. 6–13

$$F_1 = \tfrac{1}{2}(\sigma_c/4)(10)(2) = \tfrac{5}{2}\sigma_c$$
$$F_2 = (\sigma_c/4)(10)(2) = 5\sigma_c$$
$$F_3 = \tfrac{1}{2}(\sigma_c/4)(2)(2) = \tfrac{1}{2}\sigma_c$$

$$F_4 = \tfrac{1}{2}(\sigma_c)(8)(2) = 8\sigma_c$$

(a)

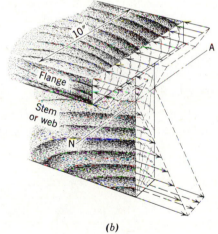

(b)

from which

$$\sigma_{\max} = \sigma_c = \underline{972 \text{ psi } T \text{ at bottom}} \qquad \text{Ans.}$$

A more direct solution can be obtained by integration of Eq. *a*; thus,

$$M_r = \int_A \sigma \, dA \, y = \int_{-4}^{-2} \left(\frac{\sigma_c}{8} y\right)(10 \, dy)(y) + \int_{-2}^{8} \left(\frac{\sigma_c}{8} y\right)(2 \, dy)(y)$$

$$= \left[\frac{5\sigma_c y^3}{12}\right]_{-4}^{-2} + \left[\frac{\sigma_c y^3}{12}\right]_{-2}^{8} = 5400(12)$$

from which

$$\sigma_c = \sigma_{\max} = \underline{972 \text{ psi } T \text{ at bottom}} \qquad \text{Ans.}$$

PROBLEMS

Note: In the following problems, all beams are loaded in a plane of symmetry parallel to the largest cross-sectional dimension except as noted otherwise. Also, assume elastic action.

6-1*   The maximum fiber stress at a certain section in a rectangular beam 4 in. wide by 8 in. deep is 1200 psi. Determine the resisting moment M_r in the beam at that section by the free-body diagram method of Example 6–1.

6-2*   The maximum fiber stress at a certain section in a rectangular beam 100 mm wide by 200 mm deep is 12 MPa. Use the free-body diagram method of Example 6–1 to determine

(a)   The resisting moment M_r in the beam at that section.

(b)   The percentage decrease in this moment if the dotted central portion of the beam (see Fig. P6–2) is removed.

6-3*   A beam, loaded in the vertical plane of symmetry, has the cross section shown in Fig. P6–3. If the fiber stress at point A of a certain section is 2000 psi T, determine

FIG. P6–2

FIG. P6–3

(a)   The maximum fiber stress on this section of the beam.

(b)   The resisting moment M_r in the beam at that section (use the free-body diagram method of Example 6–1).

6-4*   A timber beam is made of three 50×100-mm members fastened together to make an I-beam 100 mm wide by 200 mm deep. Use the free-body diagram method to determine the maximum fiber stress on a section where the bending moment is 8 kN · m.

6-5   A timber beam is made of three 2 by 6-in. planks fastened together to make an I-beam 6 in. wide by 10 in. deep. The allowable fiber stress is 1500 psi. Use the free-body diagram method of Example 6–1 to determine the maximum bending moment to which the beam may be subjected.

6-6   The beam of Fig. P6–6 is made of a material that has a yield point of 250 MPa. Use the free-body diagram method of Example 6–1 to determine the maximum moment that the beam can support if yielding must be avoided.

6-7*   A beam is loaded so that the maximum fiber stress on the cross-section shown in Fig. P6–7 is 8000 psi C. Use the free-body diagram method of Example 6–1 to determine the magnitude of the flexural force carried by the flange.

6-8   An I-beam is fabricated by welding two 400×50-mm flange plates to a 600×25-mm web plate. The beam is loaded in the plane of symmetry parallel to the web. Use the free-body diagram method of Example 6–1 to determine the percentage of the bending moment carried by the flanges.

6-9   Use the free-body diagram method of Example 6–1 to determine the

FIG. P6–6

FIG. P6–7

percentage of the bending moment carried by the flanges of a W36 × 160 beam (see Appendix C).

6-10* Use the free-body diagram method of Example 6–1 to determine the percentage of the bending moment carried by the flanges of a W914 × 342 beam (see Appendix C).

6-5 FLEXURAL STRESSES—THE ELASTIC FLEXURE FORMULA

As the sophistication of structural and machine design increases, problems involving inelastic action are receiving increased attention. However, the design of beams on the basis of linearly elastic action is still a very important area of engineering activity; for such design, the fundamental approach outlined in Art. 6–4 becomes too time-consuming. Therefore, a formula will be developed that has general applicability within the limitations noted in Arts. 6–2 and 6–3.

Substitution of Eq. *c* into Eq. *a* of Art. 6–4 yields the following:

$$M_r = \int_A \sigma \, dA \, y = \frac{\sigma_c}{c} \int_A y^2 \, dA = \frac{\sigma}{y} \int_A y^2 \, dA$$

The integral is the familiar second moment of the cross-sectional area, commonly called "moment of inertia," (see Appendix B), *with respect to the centroidal axis*. When this integral is replaced by the symbol I, the elastic flexure formula is obtained as

$$\sigma = \frac{My}{I} \tag{6–3}$$

where σ is the normal stress (σ_x in general notation) at a distance y from the neutral surface and on a transverse plane, and M is the resisting moment of the section.

At any section of the beam, the fiber stress will be maximum at the surface farthest from the neutral axis, and Eq. 6–3 becomes

$$\sigma_{max} = \frac{Mc}{I} = \frac{M}{S} \tag{6–4}$$

where $S = I/c$ is called the section modulus of the beam. Although the section modulus can be readily calculated for a given section, values of the modulus are often included in tables to simplify calculations. Observe that for a given area, S becomes larger as the shape is altered to concentrate more of the area as far as possible from the neutral axis. Commercial rolled shapes such as I- and H-beams and the various built-up sections are intended to optimize the area-section modulus relation.

   Thus far, the discussion of flexural behavior has been limited to structural members with symmetric cross sections that are loaded in a plane of symmetry. Many other shapes are subjected to flexural loadings, and methods are needed to determine stress distributions in these nonsymmetric shapes. The flexure formula (Eq. 6–3) provides a means for relating the resisting moment M_r at a section of a beam to the normal stress at a point on the transverse cross section. Further insight into the applicability of the flexure formula to nonsymmetric sections can be gained by considering the requirements for equilibrium when the applied moment M does not have a component about the y axis of the cross section. Thus, from Eq. c of Art. 6–4 and the equilibrium equation $\Sigma M_y = 0$,

$$\int_A z\,\sigma\,dA = \int_A z\,\frac{\sigma_c}{c}\,y\,dA = \frac{\sigma_c}{c}\int_A z\,y\,dA = \frac{\sigma_c}{c}\,I_{yz} = 0 \qquad (a)$$

The quantity I_{yz} is commonly known as the "product of inertia" of the cross-sectional area with respect to the centroidal y and z axes. Obviously, Eq. a can be satisfied only if $I_{yz} = 0$. For symmetric cross sections, $I_{yz} = 0$ when the y and z axes coincide with the axes of symmetry. For nonsymmetric cross sections, $I_{yz} = 0$ when the y and z axes are centroidal principal axes (see Appendix B) for the cross section. Thus, the flexure formula is valid for any cross section, provided y is measured along the principal direction, I is a principal moment of inertia, and M is a moment about a principal axis. Further discussion of flexural stresses in nonsymmetric sections is provided in Art. 6–12.

EXAMPLE 6–3
Solve Exam. 6–2 by means of the flexure formula.

SOLUTION
The second moment of the area can be obtained by integration or by composite areas (see Appendix B) and is 1600/3 in.⁴. The maximum fiber stress is

$$\sigma_{\max} = \frac{Mc}{I} = \frac{5400(12)(8)}{1600/3} = \underline{972 \text{ psi } T \text{ at the bottom}} \qquad \text{Ans.}$$

EXAMPLE 6–4
An S152 × 19 steel beam (see Appendix C) is loaded and supported as shown in Fig. 6–14. On a section 2 m to the right of B, determine

   (a)   The fiber stress at a point 25 mm below the top of the beam.
   (b)   The maximum fiber stress on the section.

FIG. 6–14

SOLUTION

On a section 2 m to the right of B, the resisting moment M_r is

$$M_r = -8 + 3(2)(3) + 3(2) - 4(2)(1) = 8 \text{ kN} \cdot \text{m}$$

For an S152 × 19 section (see Appendix C), $I = 9.20(10^6) \text{ mm}^4 = 9.20(10^{-6}) \text{ m}^4$, $S = 121(10^3) \text{ mm}^3 = 121(10^{-6}) \text{ m}^3$, and the depth of the beam is 152.4 mm = 0.1524 m.

(a) At a point 25 mm below the top of the beam, $y = -51.2 \text{ mm} = -0.0512$ m; therefore, with all quantities expressed in newtons and meters, the stress is

$$\sigma = \frac{My}{I} = \frac{8(10^3)(-0.0512)}{9.20(10^{-6})} = -44.5(10^6) \text{ N/m}^2 = \underline{44.5 \text{ MPa } C} \quad \text{Ans.}$$

(b) The maximum fiber stress is

$$\sigma_{max} = \frac{M}{S} = \frac{8(10^3)}{121(10^{-6})} = 66.1(10^6) \text{ N/m}^2 = \underline{66.1 \text{ MPa } T \text{ on the bottom}}$$
$$\underline{C \text{ on the top}}$$

Ans.

PROBLEMS

> *Note: The action is linearly elastic in all of the following problems. In problems involving rolled shapes (see Appendix C), the beam is so oriented that the maximum section modulus applies.*

6-11* Use the flexure formula to solve Prob. 6-1.

6-12* Use the flexure formula to solve Prob. 6-2*a*.

FIG. P6–17

(a)

6-13* Use the flexure formula to solve Prob. 6-3*b*.

6-14* Use the flexure formula to solve Prob. 6-4.

6-15   Use the flexure formula to solve Prob. 6-5.

6-16   Use the flexure formula to solve Prob. 6-6.

6-17* An S8 × 23 steel beam (see Appendix C) is loaded and supported as shown in Fig. P6–17. On a section 9 ft from the left end, determine

    (a)   The fiber stress at a point 1/2 in. below the top of the beam.

    (b)   The maximum fiber stress on the section.

6-18   An S178 × 30 steel beam (see Appendix C) is loaded and supported as shown in Fig. P6–18. On a section 2.7 m from the left end, determine

    (a)   The fiber stress at a point 12 mm above the bottom of the beam.

    (b)   The maximum fiber stress on the section.

6-19   An S7 × 20 steel beam (see Appendix C) is loaded and supported as shown in Fig. P6–19. On a section 7 ft to the right of *B*, determine

    (a)   The fiber stress at a point 3/4 in. above the bottom of the beam.

    (b)   The maximum fiber stress on the section.

6-20* An S203 × 27 steel beam (see Appendix C) is loaded and supported as shown in Fig. P6–20. On a section 3 m from the left end, determine

FIG. P6–18

FIG. P6–19

FIG. P6–20

FIG. P6–21

(a)   The fiber stress at a point 15 mm below the top of the beam.

(b)   The maximum fiber stress on the section.

6-21*   A timber beam having the cross section shown in Fig. P6–21 is loaded and supported as shown in Fig. P6–17. On a section 4 ft from the right end, determine

(a)   The fiber stress at point A of the cross section.

(b)   The fiber stress at point B of the cross section.

6-22   A timber beam having the cross section shown in Fig. P6–22 is loaded and supported as shown in Fig. P6–18. On a section 1.2 m from the left end, determine

(a)   The fiber stress at point A of the cross section.

(b)   The fiber stress at point B of the cross section.

6-23   A timber beam having the cross section shown in Fig. P6–23 is loaded and supported as shown in Fig. P6–19. On a section 1 ft from the left end, determine

(a)   The fiber stress at point A of the cross section.

(b)   The fiber stress at point B of the cross section.

6-24*   A timber beam having the cross section shown in Fig. P6–24 is

FIG. P6–22

FIG. P6–23

FIG. P6–24   FIG. P6–25

loaded and supported as shown in Fig. P6–20. On a section 0.6 m from the left end, determine

    (a)   The fiber stress at point A of the cross section.

    (b)   The fiber stress at point B of the cross section.

6-25   A beam is loaded and supported as shown in Fig. P6–25. If the allowable fiber stresses on the section at the load P are 4000 psi T and 6000 psi C, determine the maximum allowable load P.

6-26*   A beam is loaded and supported as shown in Fig. P6–26. If the allowable fiber stresses on a section at the middle of the span are 15 MPa T and 10 MPa C, determine the maximum permissible value for the load w.

6-27   Two C10 × 15.3 steel channels (see Appendix C) are placed back to back to form a 10-in. deep beam. The beam is 10 ft long and is simply supported at its ends. The beam carries a uniformly distributed load of 1000. lb per ft over its entire length and a concentrated load P at the center of the span. If the allowable fiber stress on the section at the center of the span is 16,000 psi, determine the maximum permissible value of the concentrated load P.

6-28   A timber beam 150 mm wide by 300 mm deep and 3 m long is simply supported at its ends and carries a uniformly distributed load w (N per m)

                       (a)

                       (b)

FIG. P6–26

over the central one-half of the span. If the maximum allowable fiber stress on a section at midspan is 10 MPa, determine the maximum permissible value for the load w.

6-6   SHEAR AND MOMENT IN BEAMS

The procedures for determining flexural stresses outlined in Arts. 6–4 and 6–5 are quite adequate if one wishes to compute the stresses at any specified section in a beam. However, if it is necessary to determine the maximum fiber (flexural) stress in a given beam subjected to loading producing a bending moment that varies with position along the beam, it is desirable to have a convenient method for determining the maximum moment.

In Art. 6–9 a relation will be developed between the resisting shear (V_r of Fig. 6–4) and the longitudinal and transverse shearing stresses at the section. Since the maximum transverse shearing stress will occur at the section at which V_r is maximum, a convenient method of determining such sections is likewise desirable.

When the equilibrium equation $\Sigma F_y = 0$ is applied to the free-body diagram of Fig. 6–4, the result can be written as

$$R - wx - P = V_r$$

or

$$V = V_r$$

where V is defined as the resultant of the external transverse forces acting on the part of a beam to either side of a section and is called the transverse shear or just the shear at the section.

As seen from the definitions of V and V_r, the shear is equal in magnitude and opposite in sense to V_r. Since these shear forces are always equal in magnitude they are frequently treated as though they were identical. For simplicity the symbol V will be used henceforth to represent both the transverse shear and the resisting shear. The sign convention selected will have sufficient generality to apply to both quantities.

The resultant of the fiber stresses on any transverse section has been shown to be a couple (if only transverse loads are considered) and has been designated as M_r. When the equilibrium equation $\Sigma M_O = 0$ (where O is any axis parallel to the neutral axis of the section) is applied to the free-body diagram of Fig. 6–4, the result can be written as

$$Rx - wx^2/2 - P(x - h) = M_r$$

or

$$M = M_r$$

FIG. 6–15

+ Shear   – Shear   + Moment   – Moment

(a)   (b)

where *M is defined as the algebraic sum of the moments of the external forces, acting on the part of the beam to either side of the section, with respect to an axis in the section and is called the bending moment or just the moment at the section.*

As seen from the definitions of M and M_r, the bending moment is equal in magnitude and opposite in sense to M_r. Since these moments are always equal in magnitude, they are frequently treated as though they were identical. For simplicity the symbol M will be used henceforth to represent both the bending moment and the resisting moment. The sign convention selected will have sufficient generality to apply to both quantities.

The bending moment and the transverse shear are not normally shown on a free-body diagram. The usual procedure is to show each external force individually as indicated in Fig. 6–4. The variation of V and M along the beam can be shown conveniently by means of equations or shear and moment diagrams.

A sign convention is necessary for the correct interpretation of results obtained from equations or diagrams for shear and moment. The following convention will give consistent results regardless of whether one proceeds from left to right or from right to left. By definition, the shear at a section is positive when the portion of the beam to the left of the section (for a horizontal beam) tends to move upward with respect to the portion to the right of the section as shown in Fig. 6–15a. Also by definition, the bending moment in a horizontal beam is positive at sections for which the top of the beam is in compression and the bottom is in tension as shown in Fig. 6–15b. Observe that the signs of the terms in the preceding equations for V and M agree with this convention.

Since M and V vary with x, they are functions of x, and equations for M and V can be obtained from free-body diagrams of portions of the beam. The procedure is illustrated in Exam. 6–5.

EXAMPLE 6–5

A beam is loaded and supported as shown in Fig. 6–16a. Write equations for the shear V and bending moment M for any section of the beam in the interval *BC*.

FIG. 6–16

SOLUTION

A free-body diagram, or load diagram, for the beam is shown in Fig. 6–16b. The reactions shown on the load diagram are determined from the equations of equilibrium. A free-body diagram of a portion of the beam from the left end to any section between B and C is shown in Fig. 6–16c. Note that the resisting shear V and the resisting moment M are shown as positive values. From the definition of V or from the equilibrium equation $\Sigma F_y = 0$,

$$V = 1400 - 400(x - 2) = \underline{2200 - 400x} \qquad 2 < x < 6 \qquad \text{Ans.}$$

From the definition of M, or from the equilibrium equation $\Sigma M_O = 0$,

$$M = 1400x - 400(x - 2)\left(\frac{x - 2}{2}\right) = \underline{-200x^2 + 2200x - 800} \qquad 2 < x < 6$$

$$\text{Ans.}$$

The equations for V and M in the other intervals can be determined in a similar manner.

PROBLEMS

6-29* A beam is loaded and supported as shown in Fig. P6–29. Using the coordinate axes shown, write equations for the shear and bending moment for any section of the beam in the interval $2 < x < 8$.

6-30* A beam is loaded and supported as shown in Fig. P6–30. Using the coordinate axes shown, write equations for the shear and bending moment for any section of the beam in the interval $1 < x < 4$.

6-31* A beam is loaded and supported as shown in Fig. P6–31. Using the coordinate axes shown, write equations for the shear and bending moment for any section of the beam in the interval $15 < x < 20$.

6-32 A beam is loaded and supported as shown in Fig. P6–32. Using the coordinate axes shown, write equations for the shear and bending moment for any section of the beam in the interval from B to C.

6-33 A beam is loaded and supported as shown in Fig. P6–33. Using the coordinate axes shown, write equations for the shear and bending moment for any section of the beam between the supports.

6-34* A beam is loaded and supported as shown in Fig. P6–34. Using the coordinate axes shown, write equations for the shear and bending moment for any section of the beam

    (a)   In the interval $0 < x < 2$.

    (b)   In the interval $2 < x < 4$.

6-35 A beam is loaded and supported as shown in Fig. P6–35. Using the coordinate axes shown, write equations for the shear and bending moment for any section of the beam

    (a)   In the interval $0 < x < L$.

    (b)   In the interval $L < x < 2L$.

    (c)   Use the results of parts (a) and (b) to determine the magnitudes and locations of the maximum shear and the maximum bending moment in the beam.

FIG. P6–29   FIG. P6–30

FIG. P6–31

FIG. P6–32

FIG. P6–33

FIG. P6–34

FIG. P6–35

FIG. P6–36

FIG. P6–37

FIG. P6–38

6-36* A beam is loaded and supported as shown in Fig. P6–36. Using the coordinate axes shown, write equations for the shear and bending moment for any section of the beam

$V = w(L+x)$

$M = \frac{w}{2}(x^2 + 2Lx - 2L^2)$

(a) In the interval $0 < x < L$.

(b) In the interval $L < x < 2L$.

(c) Use the results of parts (a) and (b) to determine the magnitudes and locations of the maximum shear and the maximum bending moment in the beam.

6-37 A beam is loaded and supported as shown in Fig. P6–37. Using the coordinate axes shown, write equations for the shear and bending moment for any section of the beam

(a) In the interval $0 < x < L$.

(b) In the interval $L < x < 3L$.

6-38 A beam is loaded and supported as shown in Fig. P6–38. Using the coordinate axes shown, write equations for the shear and bending moment for any section of the beam

(a) In the interval $0 < x < L$.

(b) In the interval $L < x < 2L$.

6-39 A beam is loaded and supported as shown in Fig. P6–39. Using the coordinate axes shown, write equations for the shear and bending moment for any section of the beam

(a) In the interval from C to D.

(b) In the interval from E to F.

6-40 A beam is loaded and supported as shown in Fig. P6–40. Using the coordinate axes shown, write equations for the shear and bending moment for any section of the beam

(a) In the interval from B to C.

(b) In the interval from C to D.

FIG. P6–39

FIG. P6–40

FIG. P6–41

6-41*   A barge weighing 1000 lb per ft of length (not to be neglected) is to be loaded as shown in Fig. P6–41.

(a)   Determine the length of barge required if the barge must float level (that is, the water pressure must be uniform on the bottom of the barge).

(b)   Write equations for the shear and bending moment for any section of the barge in the interval BC. Use the coordinate axes shown.

6-7   LOAD, SHEAR, AND MOMENT RELATIONSHIPS

The mathematical relationships between loads and shears, and between shears and moments, can be used to facilitate construction of shear and moment diagrams. The relationships can be developed from a free-body diagram of an elemental length of a beam as shown in Fig. 6–17, in which the upward direction is considered positive for the applied load w and the shears and moments are shown as positive according to the sign convention established in Art. 6–6. The element must be in equilibrium, and the equation $\Sigma F_y = 0$ gives

$$(\uparrow+) \qquad \Sigma F_y = V + w\,dx - (V + dV) = 0$$

from which

$$dV = w\,dx \qquad \text{or} \qquad \frac{dV}{dx} = w \tag{6–5}$$

The preceding equation indicates that at any section in the beam *the slope of the shear diagram is equal to the intensity of loading.* When w is known as a function of x, the equation can be integrated between definite limits as follows:

$$\int_{V_1}^{V_2} dV = \int_{x_1}^{x_2} w\,dx = V_2 - V_1 \tag{6–6}$$

That is, *the change in shear between sections at x_1 and x_2 is equal to the area under the load diagram between the two sections.*

FIG. 6–17

The equation $\Sigma M_O = 0$, applied to Fig. 6–17, gives

$$\circlearrowleft + \qquad \Sigma M_O = M + V\,dx + w\,dx(dx/2) - (M + dM) = 0$$

from which

$$dM = V\,dx + w(dx)^2/2$$

This expression reduces to

$$dM = V\,dx \qquad \text{or} \qquad V = \frac{dM}{dx} \qquad (6\text{--}7)$$

when second-order differentials are considered as being negligible compared with first-order differentials. The preceding equation indicates that at any section in the beam *the slope of the moment diagram is equal to the shear.* The equation can be integrated between definite limits to give

$$\int_{M_1}^{M_2} dM = \int_{x_1}^{x_2} V\,dx = M_2 - M_1 \qquad (6\text{--}8)$$

Thus, the change in moment between sections at x_1 and x_2 is equal to the area under the shear diagram between the two sections.

Note that the equations in this article were derived with the x axis positive to the right, the applied loads positive upward, and the shear and moment signs as indicated in Fig. 6–15. If one or more of these assumptions are changed, the algebraic signs in the equation may need to be altered.

The relations just developed will be useful in drawing shear and moment diagrams and in computing values of shear and moment at various sections along a beam, as illustrated in the next article.

6-8 SHEAR AND MOMENT DIAGRAMS

As stated in Art. 6–6, shear and moment diagrams provide a convenient method of obtaining maximum values of shear and moment and other information, to be discussed later. *A shear diagram is a graph in which abscissas represent distances along the beam and ordinates represent the transverse shear at the corresponding sections. A moment diagram is a graph in which abscissas represent distances along the beam and ordinates represent the bending moment at the corresponding sections.*

Shear and moment diagrams can be drawn by calculating values of shear and moment at various sections along the beam and plotting enough points to obtain a smooth curve. Such a procedure is rather time-consuming, and, although it may be desirable for graphic solutions of certain structural problems, other more rapid methods will be developed.

A convenient arrangement for constructing shear and moment diagrams is to draw a free-body diagram of the entire beam and construct shear and moment diagrams directly below. Two methods of procedure are presented in this article.

The first method consists of writing algebraic equations for the shear V and the moment M and constructing curves from the equations. This method has the disadvantage that unless the load is uniformly distributed or varies according to a known equation along the entire beam, no single elementary expression can be written for the shear V or the moment M, which applies to the entire length of the beam. Instead, it is necessary to divide the beam into intervals bounded by the abrupt changes in the loading. An origin should be selected (different origins may be used for different intervals); positive directions should be shown for the coordinate axes, and the limits of the abscissa (usually x) should be indicated for each interval.

Complete shear and moment diagrams should indicate values of shear and moment at each section where the load changes abruptly and at sections where they are maximum or minimum (negative maximum values). Sections where the shear and moment are zero should also be located.

The second method consists of drawing the shear diagram from the load diagram and the moment diagram from the shear diagram by means of the mathematical relationships developed in Art. 6–7. The latter method, though it may not produce a precise curve, is less time-consuming than the first and it does provide the information usually required.

When all loads and reactions are known, the shear and moment at the ends of the beam can be determined by inspection. Both shear and moment are zero at the free end of a beam unless a force or a couple or both is applied there; in this case, the shear is the same as the force and the moment the same as the couple. At a simply supported or pinned end, the shear must equal the end reaction and the moment must be zero. At a built-in or fixed-end, the reactions are the shear and moment values.

Once a starting point for the shear diagram is established, the diagram can be sketched by using the definition of shear and the fact that the slope of the shear diagram can be obtained from the load diagram. When positive directions are chosen as upward and to the right, a positive distributed load, one acting upward, will result in a positive slope on the shear diagram, and a negative load will give a negative slope. A concentrated force will produce an abrupt change in shear. The *change* in shear between any two sections is given by the area under the load diagram between the same two sections. The change of shear at a concentrated force is equal to the concentrated force.

The moment diagram is drawn from the shear diagram in the same manner. The slope at any point on the moment diagram is given by the shear at the corresponding point on the shear diagram, a positive shear representing a positive slope and a negative shear representing a negative slope, *when upward and to the right are positive*. The *change* in moment between any two sections is given by the area under the shear diagram between the corresponding sections. A couple applied to a beam will cause the moment to change abruptly by an amount equal to the moment of the couple. Examples 6–6 and 6–7 illustrate these two methods for construction of shear and moment diagrams.

EXAMPLE 6–6

A beam is loaded and supported as shown in Fig. 6–18a.

- (a) Write equations for the shear and bending moment for any section of the beam in the interval BC.
- (b) Draw complete shear and moment diagrams for the beam.

SOLUTION

A free-body diagram, or load diagram, for the beam is shown in Fig. 6–18b. The reactions at C are computed from the equations of equilibrium. It is not necessary to compute the reactions on a cantilever beam in order to write equations for the shear and bending moment or to draw shear and moment diagrams from the load diagram, but the reactions provide a convenient check.

- (a) A free-body diagram of a portion of the beam from the left end to any section between B and C is shown in Fig. 6–18c. Note that the resisting shear V and the resisting moment M are shown as positive values. From the definition of V or from the equilibrium equation $\Sigma F_y = 0$,

$$V = -500 - 200(x - 4) = \underline{300 - 200x} \qquad 4 < x < 10 \qquad \text{Ans.}$$

From the definition of M, or from the equilibrium equation $\Sigma M_O = 0$,

FIG. 6–18

(a)

(b)

(c)

(d)

(e)

$$M = -500x - 200(x - 4)(x - 4)/2$$

$$= -100x^2 + 300x - 1600 \qquad 4 < x < 10 \qquad \text{Ans.}$$

(b) The equations for V and M in the interval AB can be written in the same manner as in part (a) after which the two equations for V can be plotted in the appropriate intervals to give the shear diagram in Fig. 6–18d. Likewise, the two equations for M can be plotted in the appropriate intervals to give the moment diagram in Fig. 6–18e.

  The shear and moment diagrams can also be drawn, working directly from the load diagram without writing the shear and moment equations.

The shear diagram is drawn below the load diagram (Fig. 6–18d). The shear just to the right of the 500-lb load is −500 lb. The slope of the shear diagram is equal to the load, and since no load is applied between A and B, the slope of the diagram is zero. From B to C the load is uniform; therefore, the slope of the shear diagram is constant. The *change* of shear from B to C is equal to the applied load (1200 lb) from B to C, and the shear is −1700 lb at C. This shear is the same as the reaction at C, which provides a check.

The moment diagram is drawn below the shear diagram as Fig. 6–18e. The moment is zero at the free end A. From A to B the shear is constant (−500); therefore, the slope of the moment diagram is constant, and the moment diagram is a straight line. The shear increases (negatively) from −500 lb at B to −1700 lb at C; therefore, the slope of the moment diagram is negative from B to C and increases uniformly in magnitude. The change in moment from A to B is equal to the area under the shear diagram and is 500(4) = 2000 ft-lb. The area under the shear diagram from B to C is $(\frac{1}{2})(500 + 1700)(6) = 6600$ ft-lb, which represents the change in moment from B to C. Thus, the moment at C is −2000 + (−6600) = −8600 ft-lb.

EXAMPLE 6–7

A beam is loaded and supported as shown in Fig. 6–19a.

(a)   Write equations for the shear and bending moment for any section of the beam in the interval CD.

(b)   Draw complete shear and moment diagrams for the beam.

SOLUTION

The reactions shown on the load diagram (Fig. 6–19b) are determined from the equations of equilibrium.

(a)   A free-body diagram of a portion of the beam from the left end to any section between C and D is shown in FIg. 6–19c. From the definition of V, or from the equilibrium equation $\Sigma F_y = 0$,

$$V = 10 - 4x - 8 = \underline{-4x + 2} \qquad 3.5 < x < 5 \qquad \text{Ans.}$$

From the definition of M, or from the equilibrium equation $\Sigma M_O = 0$,

$$M = 10x - 4x(x/2) + 4 - 8(x - 3.5)$$

$$= \underline{32 + 2x - 2x^2} \qquad 3.5 < x < 5 \qquad \text{Ans.}$$

FIG. 6–19

(b) The equations for V and M in the other intervals can be written and the shear and moment diagrams obtained by plotting these equations. In this example, the second method is used and the shear diagram is drawn directly from the load diagram.

The shear just to the right of A is 10 kN and decreases at the rate of 4 kN each meter to -4 kN at C. The concentrated downward load at C causes the shear to change suddenly from -4 kN to -12 kN. The shear changes uniformly to -18 kN just to the left of D. The reaction at D changes the shear by 26 kN (from -18 kN to $+8$ kN). The shear decreases uniformly to zero at E. Note that the distributed load is uniform over the entire beam; consequently, the slope of the shear diagram is constant. Points of zero shear, such as F in Fig. 6–19d, are located from the geometry of the shear diagram. For example, the slope of the left portion of the shear diagram is 4 kN/m. Therefore,

$$\text{slope} = 4 = 10/x_1$$

and

$$x_1 = 2.50 \text{ m}$$

The moment is zero at A, and the slope of the moment diagram (equal to the shear) is 10 kN · m/m. From A to C the shear and, hence, the slope of the moment diagram decreases uniformly to zero at F and to -4 at C. The abrupt change of shear at C indicates a sudden change of slope of the moment diagram; thus, the two parts of the moment diagram at C are not tangent. The slope of the moment diagram changes from -12 at C to -18 just to the left of D. From D to E the slope changes from $+8$ to zero.

The change of moment from A to F is equal to the area under the shear diagram from A to F and is

$$\Delta M = 10(2.5)/2 = 12.5 \text{ kN} \cdot \text{m}$$

which is the moment at F since the moment at A is zero. From F to B, $\Delta M = -0.5$ kN · m and $M_B = +12$ kN · m. The 4-kN · m couple (point moment) is applied at B. Since the moment just to the left of B is positive and the couple contributes an additional positive moment to sections to the right of B, the moment changes abruptly from $+12$ kN · m to $+16$ kN · m. Moments at C, D, and E can be determined from the shear diagram areas that give the changes in moment from section to section. If the moment at E is not equal to zero, it indicates that an error has occurred.

The point of inflection G, where the moment is zero, can be determined by setting the expression for the moment from part (a) equal to zero and solving for x. The result is

$$x_2 = 0.5 \pm \sqrt{16.25} = 4.531 = \underline{4.53 \text{ m}} \qquad \text{Ans.}$$

Note in the example that maximum and minimum moments may occur at sections where the shear curve passes through zero. In general, the shear curve may pass through zero at a number of points along the beam, and each such crossing indicates a point of *possible* maximum moment (in engineering, the moment with the largest absolute value is the maximum moment). It should be emphasized that the shear curve does not indicate the presence of abrupt discontinuities in the moment curve; hence, the maximum moment may occur where a couple is applied to the beam, not where the shear passes through zero. All possibilities should be examined to determine the maximum moment.

Sections where the bending moment is zero, called points of inflection or contraflexure, can be located by equating the expression for M to zero. The fiber stress is zero at such sections, and if a beam must be spliced, the splice should be located at or near a point of inflection if there is one.

PROBLEMS

6-42* Draw complete shear and moment diagrams for the beam of Fig. P6–29.

6-43* Draw complete shear and moment diagrams for the beam of Fig. P6–18.

6-44* Draw complete shear and moment diagrams for the beam of Fig. P6–17.

6-45* Draw complete shear and moment diagrams for the beam of Fig. P6–34.

6-46 Draw complete shear and moment diagrams for the beam of Fig. P6–33.

FIG. P6–49

FIG. P6–50

FIG. P6–51   FIG. P6–52

6-47 Draw complete shear and moment diagrams for the beam of Fig. P6–20.

6-48* Draw complete shear and moment diagrams for the beam of Fig. P6–31.

6-49* Draw complete shear and moment diagrams for the beam of Fig. P6–49 and write the moment equation for the interval BC of the beam with the origin of the coordinate axes at the left end of the beam.

6-50 Draw complete shear and moment diagrams for the beam of Fig. P6–50 and write the moment equation for the interval CD of the beam with the origin of the coordinate axes at the left end of the beam.

6-51 Draw complete shear and moment diagrams for the beam of Fig. P6–51 and write moment equations for (1) the interval BC and (2) the interval EF of the beam with the origin of the coordinate axes at the left support.

6-52* Draw complete shear and moment diagrams for the beam of Fig. P6–52 and write moment equations for (1) the interval AB and (2) the interval BC of the beam with the origin of the coordinate axes at the left end of the beam.

6-53 Draw complete shear and moment diagrams for the beam of Fig. P6–53.

FIG. P6–53

FIG. P6–54

6-54 Draw complete shear and moment diagrams for the beam of Fig. P6–54.

6-55* Draw complete shear and moment diagrams for the beam of Fig. P6–55.

6-56* Draw complete shear and moment diagrams for the beam of Fig. P6–56.

6-57 Draw complete shear and moment diagrams for the beam of Fig. P6–40.

6-58 Draw complete shear and moment diagrams for the beam of Fig. P6–39.

6-59* The shear and moment at end *A* of the beam segment (in equilibrium) of Fig. P6–59 are +3 kN and −2 kN · m, respectively, and the shear and moment at end *D* are unknown. Draw complete shear and moment diagrams for the beam segment and write the moment equation for the interval *CD* of the beam with the origin of the coordinate axes at the left end of the beam segment.

FIG. P6–55

FIG. P6–56   FIG. P6–59

6-60 The beam of Fig. P6–60 has a pin-connected joint at C. The moment and shear at A are $+6000$ ft-lb and $+1200$ lb, respectively. Draw complete shear and moment diagrams for the length $ABCD$ of the beam and write the moment equation for the interval BCD of the beam with the origin of the coordinate axes at support B.

6-61 Select the lightest steel wide-flange or American standard beam (see Appendix C) that may be used for the beam of Fig. P6–61 if the working fiber stress must be limited to 125 MPa.

6-62 Select the lightest steel wide-flange or American standard beam (see Appendix C) that may be used for the beam of Fig. P6–62 if the working fiber stress must be limited to 18,000 psi.

6-63* The beam shown in Fig. P6–63a has the cross section shown in Fig. P6–63b. Determine

    (a) The maximum tensile fiber stress in the beam.

    (b) The maximum compressive fiber stress in the beam.

6-64* The beam shown in Fig. P6–64a has the cross section shown in Fig. P6–64b. Determine

    (a) The maximum tensile fiber stress in the beam.

    (b) The maximum compressive fiber stress in the beam.

FIG. P6–60   FIG. P6–61

FIG. P6–62

FIG. P6–63

FIG. P6–64

FIG. P6–71

FIG. P6–72

6-65   A W10 × 30 steel beam (see Appendix C) is loaded and supported as shown in Fig. P6–35. The total length of the beam is 10 ft. If the maximum allowable fiber stress is 15,000 psi, determine the maximum permissible value for the distributed load w.

6-66*   An S254 × 52 steel beam (see Appendix C) is loaded and supported as shown in Fig. P6–36. The total length of the beam is 4 m. If the maximum allowable fiber stress is 100 MPa, determine the maximum permissible value for the distributed load w.

6-67   A W12 × 50 steel beam (see Appendix C) is loaded and supported as shown in Fig. P6–37. The total length of the beam is 15 ft. If the maximum allowable fiber stress is 18,000 psi, determine the maximum permissible value for the distributed load w.

6-68   An S152 × 26 steel beam (see Appendix C) is loaded and supported as shown in Fig. P6–38. The total length of the beam is 3 m. If the maximum allowable fiber stress is 125 MPa, determine the maximum permissible value for the distributed load w.

6-69*   A timber beam with a rectangular cross section (width equal one-half the depth) is to be loaded and supported as shown in Fig. P6–19. If the maximum allowable fiber stress for the wood is 2000 psi, determine the size timber required for the beam.

6-70*   A timber beam with a rectangular cross section (width equal one-half the depth) is to be loaded and supported as shown in Fig. P6–30. If the maximum allowable fiber stress in the wood is 15 MPa, determine the size timber required for the beam.

6-71   The moment diagram for a beam is shown in Fig. P6–71. Construct the shear and load diagrams for the beam.

6-72   The moment diagram for a beam is shown in Fig. P6–72. Construct the shear and load diagrams for the beam.

6-73   The moment diagram for a beam is shown in Fig. P6–73. Construct the shear and load diagrams for the beam.

6-74   The moment diagram for a beam is shown in Fig. P6–74. Construct the shear and load diagrams for the beam.

6-75   The moment diagram for a beam is shown in Fig. P6–75. Construct the shear and load diagrams for the beam.

FIG. P6–73

FIG. P6–74

FIG. P6–75

6-9 SHEARING STRESSES BY EQUILIBRIUM

In Art. 6–1, the discussion of shearing stresses in beams was bypassed while flexural stresses were studied. This procedure seems to be in keeping with the historical record on the study of beam stresses. From the time of Coulomb's paper, which contained the correct theory of distribution of flexural stresses, approximately seventy years elapsed before the Russian engineer D. J. Jourawski (1821–1891), while designing timber railroad bridges in 1844–1850, developed the elementary shear stress theory used today. In 1856 Saint-Venant developed a rigorous solution for beam shearing stresses, but the elementary solution of Jourawski, being much easier to apply and quite adequate, is the one in general use today by engineers and architects and the one that will be presented here. Note that the method, because it involves the elastic flexure formula, is limited to elastic action. This limitation need be no cause for concern because, for practical engineering problems, the shearing stress evaluation discussed in this and following articles is important only in (1) the evaluation of principal elastic stresses at interior points in certain metallic beams (see Art. 6–11), and (2) the design of timber beams, because of the longitudinal plane of low shear resistance. The allowable stresses for timber are always in the elastic range. In beams of elastoplastic steel designed on the basis of plastic theory, methods for evaluation of shear stresses are available[5]; however, this area is beyond the scope of this book.

If one constructs a beam by stacking flat slabs one on top of another without fastening them together, and then loads this beam in a direction normal to the surface of the slabs, the resulting deformation will appear somewhat like that in Fig. 6–20a. The same type of deformation can be observed by taking a pack of cards and bending them, noting the relative motion of the ends of the cards with respect to each other. The fact that a solid beam does not exhibit this relative movement of longitudinal elements (see Fig. 6–20b, in which the beam is identical to that of Fig. 6–20a except that the layers are glued together) indicates the presence of shearing stresses on longitudinal planes. The evaluation of these shearing stresses will now be studied by means of the free-body diagram and equilibrium approach, as outlined in the following example.

EXAMPLE 6–8

A 12-in. long segment of a beam having the T cross section of Fig. 6–21b is subjected to a moment of +5400 ft-lb at the left end and a constant shear of +900 lb. Determine the average shearing stress on a horizontal plane 4 in. above the bottom of the beam. Assume elastic action.

---

[5] See, for example, *Commentary on Plastic Design in Steel,* A.S.C.E. Manual of Engineering Practice No. 41, 1961.

FIG. 6–20

SOLUTION
A free-body diagram of the segment is shown in Fig. 6–21a, where AB designates the plane on which the shearing stress is to be evaluated. The moment M_2, obtained from a moment equilibrium equation, is +6300 ft-lb. A free-body diagram of the portion of the beam segment below plane AB is shown in Fig. 6–21c. The force V_H is the resultant of the shearing stresses on plane AB. The forces F_1 and F_2 are the resultants of the normal stresses on the ends of the segment and are equal to the average fiber stress (which occurs 6 in. below the neutral surface) on the end area multiplied by the area; that is,

$$F_1 = \frac{M_1 y}{I} A = \frac{5400(12)(6)}{1600/3} (2)(4) = 108(54) \text{ lb}$$

and

$$F_2 = \frac{M_2 y}{I} A = \left(\frac{M_2}{M_1}\right) F_1 = \frac{63}{54}(108)(54) = 108(63) \text{ lb}$$

FIG. 6–21

Summing forces in the longitudinal direction in Fig. 6–21c gives

$$V_H = F_2 - F_1 = 108(63) - 108(54) = 972 \text{ lb}$$

The average shearing stress on plane AB is

$$\tau = \frac{V_H}{A_H} = \frac{972}{12(2)} = \underline{40.5 \text{ psi}} \qquad \text{Ans.}$$

In this example, the shearing stress on plane AB does not vary with the length; therefore, since the shearing stresses on orthogonal planes have the same magnitude, the shearing stresses on the transverse (vertical) planes at A and B are also 40.5 psi in the directions shown.

PROBLEMS

Note: Assume elastic action for the following problems.

6-76*   A 12-in. section of a 6-in. deep by 3-in. wide timber beam is shown in Fig. P6–76a. A free-body diagram of the bottom 2 in. of the beam is shown in Fig. P6–76b. Determine

(a)   Values for the moment M_2 and the forces F_1 and F_2.

(a)

FIG. P6–76   (b)

(a)

(b)

(b) The resultant of the shearing stresses on the area *ABCD*.

(c) The average shearing stress on area *ABCD*.

6-77* A timber T-beam consists of two 40 × 120-mm planks glued together as shown in Fig. P6–77a. A free-body diagram of the flange is shown in Fig. P6–77b. The vertical shear at sections 1–1 and 2–2 is 4000 N. The bending moment is +2720 N · m at section 1–1 and +4080 N · m at section 2–2. Determine

(a) Values for forces F_1 and F_2.

(b) The total resistance offered by the glue on surface *ABCD*.

(c) The average shearing stress at the joint.

6-78 If the beam of Prob. 6-77 consists of one 40 × 160-mm plank and two 40 × 40-mm sticks as shown in Fig. P6–78, determine

(a) Values for forces F_1 and F_2.

(b) The total resistance offered by the glue on surface *ABCD*.

(c) The average shearing stress at the joint.

FIG. P6–78

(a)

(b)

6-79   A timber beam fabricated from 2-in. planks may have either of the cross sections shown in Fig. P6–79. A 6-in. length of the beam for which the shear is constant will be subjected to bending moments of +12 ft-kips and +16 ft-kips at the ends.

(a)   Which method of fabrication will result in the least average shearing stress in the glued joint?

(b)   Compute the average shearing stress in each of the joints.

6-80*   A timber beam is fabricated from two 40 × 120-mm and one 40 × 80-mm pieces of lumber to form the cross section shown in Fig. P6–80. In a

(a)   (b)

FIG. P6–79

$I_{NA} = 56.75(10^6) \text{ mm}^4$

FIG. P6–80

FIG. P6–81   FIG. P6–82   FIG. P6–83

$\frac{1}{4}$ by 2–in. welds

region of constant shear, the two ends of a 400-mm segment of the beam are subjected to negative bending moments of 8 and 6 kN · m, respectively. Determine the average shearing stress in each of the two glued joints between the flanges and the web.

6-81   A W24 × 104 steel beam (see Appendix C) will have 10 × 1/2-in. cover plates welded to the top and bottom flanges as shown in Fig. P6–81. A 20-in. length of the beam for which the shear is constant will be subjected to bending moments of +6600 in.-kips and +3300 in.-kips at the ends. The fillet weld has an allowable load of 2400 lb per in. Determine the number of fillet welds, each 2-in. long, required on each side of the cover plate.

6-82   A box beam will be fabricated by bolting two 15 × 260-mm steel plates to two C305 × 45 steel channels (see Appendix C) as shown in Fig. P6–82. The beam will be simply supported and will carry a concentrated load of 150 kN at the center of a 4-m span. Determine the bolt spacing required if the bolts have a diameter of 15 mm and an allowable shearing stress of 140 MPa.

6-83*   A W21 × 101 steel beam (see Appendix C) is simply supported and carries a concentrated load at the center of a 20-ft span. The concentrated load must be increased to 100 kips which requires that the beam be strengthened. It has been decided that two 3/4 × 16-in. steel plates will be bolted to the flanges as shown in Fig. P6–83. Determine the bolt spacing required if the bolts have a diameter of 3/4 in. and an allowable shearing stress of 20 ksi.

6-10   THE SHEARING STRESS FORMULA

The shearing stress as determined in Exam. 6–8 is reasonably accurate only if the width of the section on plane AB is not large compared to the depth of the beam and the transverse shear in the beam is constant throughout the length of the segment AB. If the transverse shear varies, the length L must be

FIG. 6–22

reduced to a short length Δx. The method outlined in the example, though excellent for emphasizing the concept of horizontal shear, is too laborious for general use; therefore, a formula will be developed that will be generally applicable within limits imposed by the assumptions. For this development refer to Fig. 6–22, which is a free-body diagram similar to that of Fig. 6–21c. The force dF_1 is the normal force acting on a differential area dA and is equal to $\sigma\, dA$. The resultant of these differential forces is F_1 (not shown). Thus, $F_1 = \int \sigma\, dA$ integrated over the shaded area of the cross section, where σ is the fiber stress at a distance y from the neutral surface and is given by the expression $\sigma = My/I$. When the two expressions are combined, the force F_1 becomes

$$F_1 = \frac{M}{I} \int y\, dA = \frac{M}{I} \int_h^c ty\, dy$$

Similarly, the resultant force on the right side of the element is

$$F_2 = \frac{(M + \Delta M)}{I} \int_h^c ty\, dy$$

The summation of forces in the horizontal direction in Fig. 6–22 yields

$$V_H = F_2 - F_1 = \frac{\Delta M}{I} \int_h^c ty\, dy$$

The average shearing stress is V_H divided by the area, from which

$$\tau = \lim_{\Delta x \to 0} \frac{\Delta M}{\Delta x} \left(\frac{1}{It}\right) \int_h^c ty\, dy = \frac{dM}{dx} \left(\frac{1}{It}\right) \int_h^c ty\, dy \qquad (a)$$

and from Eq. 6–7, dM/dx equals V, the shear at the beam section where the stress is to be evaluated. Also, the integral is the first moment of that portion of the cross-sectional area between the transverse line where the stress is to be evaluated and the extreme fiber of the beam. The integral is designated Q, and when V and Q are substituted into Eq. a, the formula for horizontal (or longitudinal) shearing stress becomes

$$\tau = \frac{VQ}{It}$$

(6–9)

Because the flexure formula was used in the derivation, Eq. 6–9 is subject to the same assumptions and limitations. Although the stress given by Eq. 6–9 is associated with a particular point in a beam, it is averaged across the thickness t and hence is accurate only if t is not too great. For a rectangular section having a depth twice the width, the maximum stress as computed by Saint-Venant's more rigorous method is about 3 percent greater than that given by Eq. 6–9. If the beam is square, the error is about 12 percent. If the width is four times the depth, the error is almost 100 percent, from which one must conclude that, if Eq. 6–9 were applied to a point in the flange of the beam of Exam. 6–8, the result would be worthless. Furthermore, if Eq. 6–9 is applied to sections where the sides of the beam are not parallel, such as a triangular section, the average stress is subject to additional error because the transverse variation of stress is greater when the sides are not parallel.

As noted previously, at each point in a beam, the horizontal (longitudinal) and vertical (transverse) shearing stresses have the same magnitude; hence, Eq. 6–9 gives the vertical shearing stress at a point in a beam (averaged across the width). The variation of the transverse shearing stress in a beam will be demonstrated for the T-shaped beam of Exam. 6–8. Note that in Eq. 6–9, V and I are constant for any section, and only Q and t vary for different points in the section. The transverse shearing stress at any point in the stem of the T-section a distance y_O from the neutral axis is, from Fig. 6–23a and Eq. 6–9,

$$\tau = \frac{V}{It_1} \int_{y_O}^{8} yt_1 \, dy = \frac{V}{2I}(8^2 - y_O^2) \qquad (-2 \le y_O \le 8)$$

Note that y_O is shown as positive downward. An expression for the average shearing stress in the flange can be written in a similar manner and is

$$\tau = \frac{V}{2I}(4^2 - y_O^2) \qquad (-4 \le y_O \le -2)$$

These are parabolic equations for the theoretical stress distribution, and the results are shown in Fig. 6–23b. The diagram has a discontinuity at the junction of the flange and stem because the thickness of the section changes

FIG. 6-23

(a)

(b)

abruptly. As pointed out previously, the distribution in the flange is fictitious, since the stress at the bottom of the flange must be zero (a free
surface). Also, because of the abrupt change in section, a stress concentration will exist, and the maximum shearing stress at the junction of flange and
stem will be more than the average stress (see Art. 1–10). From Fig. 6–23b
and Eq. 6–9, one may conclude that in general the maximum[6] longitudinal
and transverse shearing stress occurs at the neutral surface at a section
where the transverse shear is maximum. There may be some exceptions
such as a beam with a cross section in the form of a triangle or a Greek cross
with Q/t at the neutral surface less than the value some distance from the
neutral surface. The following example illustrates the use of the shearing
stress formula.

EXAMPLE 6–9
For the beam segment of Exam. 6–8, determine

(a)   The vertical shearing stress 4 in. from the bottom.
(b)   The maximum vertical shearing stress.

SOLUTION

(a)   The moment of inertia taken from Fig. 6–21b is 1600/3 in.[4]. With
      reference to Fig. 6–24 the first moment Q is the shaded area
      multiplied by the distance from the centroid of the area to the
      neutral axis, which is 4(2)(6) in.[3]

[6]   In this book the term *maximum*, as applied to a longitudinal and transverse
shearing stress, will mean the average stress across the thickness t at a point where
such average has the maximum value.

FIG. 6–24

Then,

$$\tau = \frac{VQ}{It} = \frac{900(4)(2)(6)}{(1600/3)(2)} = \underline{40.5 \text{ psi}} \qquad \text{Ans.}$$

(b) The maximum stress will occur at the neutral surface; it is

$$\tau_{max} = \frac{900(8)(2)(4)}{(1600/3)(2)} = \underline{54.0 \text{ psi}} \qquad \text{Ans.}$$

Note that, in the computation of the value for Q, it is immaterial whether one takes the area above or below a transverse line such as $a–a$ in Fig. 6–24. For example, in part (b), Q for the area above the neutral axis is $Q_{flange} + Q_{web}$ equals $10(2)(2 + 1) + 2(2)(1)$ equals $60 + 4$, which is the same as that used in part (b) above.

PROBLEMS

Note: Assume elastic action for the following problems.

6-84* The transverse shear V at a certain section of a timber beam is 6000 lb. If the beam has the cross section shown in Fig. P6–84, determine

(a) The vertical shearing stress 3 in. below the top of the beam.

(b) The magnitude and location of the maximum vertical shearing stress on the cross section.

FIG. P6–84

FIG. P6–85

6-85*   A timber beam is simply supported and carries a concentrated load P of 36 kN at the center of a 4-m span. If the beam has the cross section shown in Fig. P6–85, determine

(a)   The vertical shearing stress at a point 1 m from the left end and 80 mm above the bottom of the beam.

(b)   The magnitude and location of the maximum horizontal shearing stress in the beam.

6-86   A timber beam is simply supported and carries a uniformly distributed load w of 360 lb per ft over its entire 18-ft span. If the beam has the cross section shown in Fig. P6–86, determine

(a)   The vertical shearing stress at a point 2 ft from the right end and 3 in. below the top surface of the beam.

(b)   The magnitude and location of the maximum horizontal shearing stress in the beam.

(c)   The magnitude and location of the maximum tensile fiber stress in the beam.

6-87   A timber beam 4-m long is simply supported at its ends and carries a

FIG. P6–86

FIG. P6–87

uniformly distributed load w of 8 kN per m over its entire length. If the beam has the cross section shown in Fig. P6–87, determine

   (a)   The fiber stress and the vertical shearing stress at a point 500 mm from the left end and 40 mm below the neutral surface.

   (b)   The magnitude and location of the maximum horizontal shearing stress in the beam.

   (c)   The magnitude and location of the maximum horizontal shearing stress in the glued joints between the web and the flanges of the beam.

6-88* The segment of a timber beam shown in Fig. P6–88a has the cross section shown in Fig. P6–88b. At section b–b, the shear is −2800 lb, and the bending moment is +6400 ft-lb. Determine

   (a)   The fiber stress and the vertical shearing stress at point A of the cross section at section a–a.

   (b)   The magnitude and location of the maximum horizontal shearing stress in the beam segment.

   (c)   The magnitude and location of the maximum compressive stress in the beam segment.

6-89 The segment of a timber beam shown in Fig. P6–89a has the cross section shown in Fig. P6–89b. At section a–a, the shear is +8 kN, and the bending moment is −6 kN · m. Determine

(a)

(b)

FIG. P6–89

FIG. P6–90

(a)

(b)

(a)   The fiber stress and the vertical shearing stress 40 mm below the top of the beam at section *b–b*.

(b)   The magnitude and location of the maximum horizontal shearing stress in the glued joint between the flange and the stem in the beam segment.

(c)   The magnitude and location of the maximum tensile fiber stress in the beam segment.

6-90   The beam shown in Fig. P6–90*a* is composed of two 1 × 6-in. and two 1 × 3-in. hard maple boards that are glued together as shown in Fig. P6–90*b*. Determine the magnitude and location of

(a)   The maximum tensile fiber stress in the beam.

(b)   The maximum horizontal shearing stress in the beam.

6-91*   The beam shown in Fig. P6–91*a* is composed of three pieces of timber that are glued together as shown in Fig. P6–91*b*. If the maximum horizontal shearing stresses are not to exceed 700 kPa in the glued joints or 800 kPa in the wood and if the maximum tensile and compressive fiber stresses in the beam are not to exceed 12 MPa, determine the maximum permissible value for *P*.

6-92*   The cross section of the timber beam shown in Fig. P6–92 is a rectangle 6-in. wide by 12-in. deep. The allowable fiber stress is 950 psi and the

FIG. P6–91   (a)   (b)

FIG. P6–92   FIG. P6–93

allowable vertical shearing stress is 75 psi. Determine the maximum allowable value for the distributed load w.

6-93   The cross section of the simple beam shown in Fig. P6–93 is a T-section formed by glueing two 40 × 120-mm boards together. If the maximum horizontal shearing stresses are not to exceed 800 kPa in the glued joint or 900 kPa in the wood, and if the maximum tensile and compressive fiber stresses in the beam are not to exceed 13.5 MPa, determine the maximum permissible value for the distributed load w.

6-94   The timber beam shown in Fig. P6–94a is fabricated by glueing together two 1 × 5-in. and two 1 × 4-in. boards as shown in Fig. P6–94b. If the maximum horizontal shearing stresses are not to exceed 200 psi in the glued joints or 240 psi in the wood, and if the maximum tensile and compressive fiber stresses in the beam are not to exceed 2400 psi, determine the maximum permissible magnitude for the distributed load w.

6-95   A WT305 × 70 structural tee (see Appendix C) is loaded and supported as a beam (with the flange on top) as shown in Fig. P6–95. Determine

   (a)   The maximum tensile and compressive fiber stresses in the beam.

   (b)   The maximum vertical shearing stress in the beam.

   (c)   The vertical shearing stress at a point in the stem just below the flange on the cross section where the maximum vertical shearing stress occurs.

FIG. P6–94

FIG. P6–95

10 kN

10 kN

15 kN/m

1 m   1 m   2 m   1 m

FIG. P6–96

1750 lb

6'   4'   4'   4'

100 lb/ft

R_1

R_2

6-96* A WT8 × 20 structural tee (see Appendix C) is loaded and supported as a beam (with the flange on top) as shown in Fig. P6–96. Determine

(a) The maximum tensile and compressive fiber stresses in the beam.

(b) The maximum vertical shearing stress in the beam.

(c) The vertical shearing stress at a point in the stem just below the flange on the cross section where the maximum vertical shearing stress occurs.

6-97* A W457 × 113 steel beam (see Appendix C) is loaded and supported as shown in Fig. P6–97. Determine

(a) The maximum tensile fiber stress in the beam.

(b) The maximum vertical shearing stress in the beam.

(c) The vertical shearing stress at a point in the web just below the flange on the cross section where the maximum vertical shearing stress occurs.

6-98 A WT12 × 52 structural tee (see Appendix C) is loaded and supported as a beam (with the flange on the bottom) as shown in Fig. P6–98. Determine

(a) The maximum tensile and compressive fiber stresses in the beam.

(b) The maximum vertical shearing stress in the beam.

4 kN

2 m

30 kN/m   20 kN/m   60 kN

R_L 4 m   4 m R_R 2 m

FIG. P6–97

(c) The vertical shearing stress at a point in the stem just above the flange on a cross section where the maximum vertical shearing stress occurs.

6-11 PRINCIPAL STRESS FOR ELASTIC FLEXURE

Methods for finding the fiber stress at any point in a beam were presented in Arts. 6–4 and 6–5. Procedures for locating the critical sections of a beam (maximum M and V) were developed in Arts. 6–7 and 6–8. Methods for determining the transverse and longitudinal shearing stresses at any point in a beam were presented in Arts. 6–9 and 6–10. However, the discussion of stresses in beams is incomplete without careful consideration of the principal and maximum shearing stresses at carefully selected points on the sections of maximum shear and bending moment.

The fiber stress is maximum at the top or bottom edge of a section, and the transverse and longitudinal shearing stresses are zero at these points. Consequently, the maximum tensile and compressive fiber stresses on a given section are also the principal stresses at these points, and the corresponding maximum shearing stress is equal to one-half the fiber stress $[\tau_{max} = (\sigma_p - 0)/2]$. The longitudinal and transverse shearing stresses are normally maximum for a given section at the neutral axis, where the fiber stress is zero. Thus, the evidence presented so far is limited to the extremes in a cross section. One might well wonder what the stress situation might be between these points. Unfortunately, the magnitude of the principal stresses throughout a cross section cannot be expressed for all sections as a simple function of position. However, in order to provide some insight into the nature of the problem, two typical sections will be discussed in the following paragraphs.

For a cantilever beam of rectangular cross section subjected to a concentrated load, the theoretical principal stress variation is indicated for two sections in Fig. 6–25. One observes from this figure that, at a distance from the load of one-fourth the depth of the beam, the maximum normal stress does not occur at the surface. However, at a distance of one-half the depth, the maximum fiber stress is the maximum normal stress. For either of these

FIG. 6–25

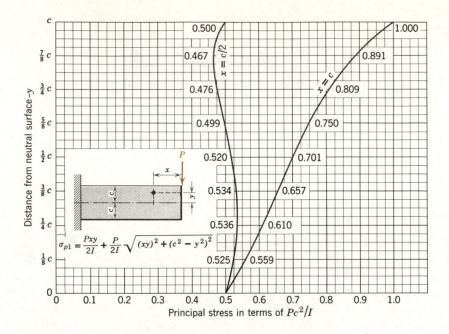

sections, Saint-Venant's principle (see Art. 1–10) indicates that the flexure (and transverse shearing stress) formula is inapplicable. Since such a small increase in bending moment is required to overcome the effect of the transverse shearing stress, the conclusion may be drawn that, for a rectangular cross section, in regions where the flexure formula applies, the maximum fiber stress is the maximum normal stress. Although for the rectangular cross section the maximum shearing stress will usually be one-half the maximum normal stress (at a surface of the beam), for materials having a longitudinal plane of weakness (for example, the usual timber beam), the longitudinal shearing stress may frequently be the significant stress—hence, the emphasis on this stress.

The other section to be discussed is the deep, wide flange section subjected to a combination of large shear and large bending moment. For this combination, the high fiber and transverse shearing stresses occurring simultaneously at the junction of flange and web sometimes yields a principal stress greater than the fiber stress at the surface of the flange (see Exam. 6–10). In general, at any point in a beam, a combination of large M, V, Q, and y and a small t should suggest a check on the principal stresses at such a point. Otherwise, the maximum flexural (fiber) stress will very likely be the maximum normal stress, and the maximum shearing stress will probably occur at the same point.

FIG. 6–26

A knowledge of the directions of the principal stresses may aid in the prediction of the directions of cracks in a brittle material (concrete, for example) and thus may aid in the design of reinforcement to carry the tensile stresses. Curves drawn with their tangents at each point in the directions of the principal stresses are called *stress trajectories*. Since there are, in general, two nonzero principal stresses at each point (plane stress), there are two stress trajectories passing through each point. These curves will be perpendicular since the principal stresses are orthogonal; one set of curves will represent the maximum stresses, whereas the other represents the minimum stresses. The trajectories for a simply supported rectangular beam carrying a concentrated load at the midpoint are shown in Fig. 6–26, with dashed lines representing the directions of the compressive stresses and solid lines showing tensile stress directions. In the vicinities of the load and reactions there are stress concentrations, and the trajectories become much more complicated. Figure 6–26 neglects all stress concentrations.

In order to determine the principal stresses and the maximum shearing stresses at a particular point in a beam, it is necessary to calculate the fiber stresses and transverse (or longitudinal) shearing stresses at the point. With the stresses on orthogonal planes through the point known, the methods of Art. 1–7 or 1–8 can be used to calculate the maximum stresses at the point. The following example illustrates the procedure and provides an example in which the principal stresses at some interior point are greater than the maximum fiber stresses.

EXAMPLE 6–10

A cantilever beam ($I = 1243(10^6)$ mm^4, $S = 4079(10^3)$ mm^3), supported at the left end, carries a uniformly distributed load of 160 kN/m on a span of 2.5 m. Determine the maximum normal and shearing stresses in the beam.

SOLUTION

For a cantilever beam with a uniformly distributed load, both the maximum

FIG. 6–27

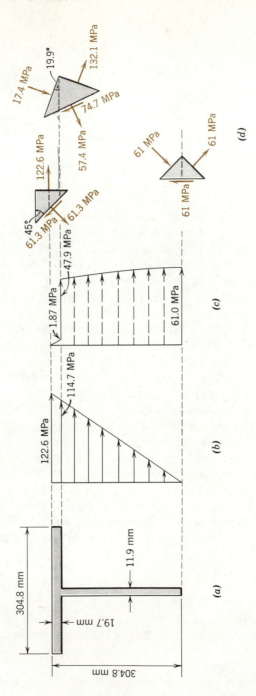

bending moment and the maximum transverse shear occur on the section at the support. In this case they are

$$M = \frac{wL^2}{2} = \frac{160(2.5^2)}{2} = 500 \text{ kN} \cdot \text{m}$$

and

$$V = wL = 160(2.5) = 400 \text{ kN}$$

The upper half of the cross section of the beam at the wall is shown in Fig. 6–27a. The distribution of fiber stress for this half of the section is shown in Fig. 6–27b, and the distribution of the average transverse shearing stress is as shown in Fig. 6–27c. The vertical stresses σ_y due to the pressure of the load on the top of the beam are considered negligible.

Values will be calculated for three points—namely, at the neutral axis, in the web at the junction of the web and the top flange, and at the top surface. Both the fillets and the stress concentrations at the junction of the web and flange will be neglected. At the neutral axis, the fiber stress is zero; however,

$$Q = 304.8(19.7)(295) + 285.1(11.9)(142.6) = 2.255(10^6) \text{ mm}^3$$

and

$$\tau = \frac{VQ}{It} = \frac{400(10^3)(2.255)(10^{-3})}{1243(10^{-6})(0.0119)} = 61.0(10^6) \text{ N/m}^2 = 61.0 \text{ MPa}$$

In the web at the junction with the top flange, the fiber stress is

$$\sigma = \frac{My}{I} = \frac{500(10^3)(0.2851)}{1243(10^{-6})} = 114.7(10^6) \text{ N/m}^2 = 114.7 \text{ MPa } T$$

and since

$$Q = 304.8(19.7)(295) = 1.771(10^6) \text{ mm}^3$$

the transverse shearing stress is

$$\tau = \frac{VQ}{It} = \frac{400(10^3)(1.771)(10^{-3})}{1243(10^{-6})(0.0119)} = 47.9(10^6) \text{ N/m}^2 = 47.9 \text{ MPa}$$

At the top surface, the transverse shearing stress is zero, and the fiber stress is

$$\sigma = \frac{My}{I} = \frac{M}{S} = \frac{500(10^3)}{4079(10^{-6})} = 122.6(10^6) \text{ N/m}^2 = 122.6 \text{ MPa } T$$

The principal stresses and maximum shearing stresses for each of the three selected points are shown in Fig. 6–27d. The calculations, from the equations of Art. 1–7, for the point at the junction of web and flange are

$$\sigma_p = \frac{114.7}{2} \pm \sqrt{\left(\frac{114.7}{2}\right)^2 + 47.9^2}$$

$$= 57.35 \pm 74.72$$

which gives

$$\sigma_{p1} = 132.1 \text{ MPa } T, \qquad \sigma_{p2} = 17.4 \text{ MPa } C, \qquad \text{and} \qquad \tau_{max} = 74.7 \text{ MPa}$$

The angle is not particularly important in this case, but it can be obtained from the equation

$$\tan 2\theta_p = -\frac{47.90}{57.35} = -0.835$$

The result is

$$2\theta_p = -39.9°$$

or

$$\theta_p = 19.9° \,\rotatebox{-30}{$\downarrow$}$$

The other stresses of Fig. 6–27d are obtained by inspection. It should be noted that the maximum tensile stress of 132.1 MPa is 7.75 percent above the maximum tensile fiber stress of 122.6 MPa and that the maximum shearing stress of 74.7 MPa is 22.5 percent above the maximum transverse shearing stress of 61.0 MPa.

PROBLEMS

> *Note: In the following problems neglect both the fillets and the stress concentrations at the junction of the web and flange. The action is elastic in all problems.*

6-99* The steel beam shown in Fig. P6–99a has the cross section shown in Fig. P6–99b. At point A, which is just to the right of the 20-kN load and just

FIG. P6–99

FIG. P6–100

above the flange, determine and show on a sketch the principal stresses and the maximum shearing stress.

6-100* The timber beam shown in Fig. P6–100a has the cross section shown in Fig. P6–100b. At point A, which is 2 ft from the left end of the beam and 2 in. below the top surface of the beam, determine and show on a sketch the principal stresses and the maximum shearing stress.

6-101 The simple beam shown in Fig. P6–101a has the cross section shown in Fig. P6–101b. If the allowable stresses are 75 MPa shear and 120 MPa tension at point A. (just above the flange), determine the maximum allowable load P.

FIG. P6–101

FIG. P6–102   FIG. P6–104

$I_{x-x} = 1073$ in.^{4}

6-102*   The cantilever beam shown in Fig. P6–102a has the cross section shown in Fig. P6–102b. If the allowable stresses are 9000 psi shear and 15,000 psi tension at point A (just below the flange), determine the maximum allowable load P.

6-103   A W610 × 155 cantilever beam (see Appendix C) with a span of 3 m carries a uniformly distributed load of 160 kN/m. Determine the maximum principal and shearing stresses in the beam.

6-104   A W12 × 65 steel beam (see Appendix C) is loaded and supported as shown in Fig. P6–104. Determine the maximum principal and shearing stresses in the beam if the load P is 36 kips and the length L is 16 ft.

6-105*   A W356 × 122 continuous beam (see Appendix C) has an overhanging end 2-m long. A uniformly distributed load of 80 kN/m is being supported by the overhang. Determine the maximum principal and shearing stresses on a section 1.9 m from the free end of the beam.

6-106   A W16 × 100 cantilever beam (see Appendix C) with a span of 8 ft carries a concentrated load of 40 kips at the free end of the beam. Determine the maximum principal and shearing stresses on a section 7.5 ft from the free end of the beam.

6-107   A W356 × 179 steel beam (see Appendix C) is loaded and supported as shown in Fig. P6–104. Determine the maximum principal and shearing stresses in the beam if

   (a)   The load P is 700 kN and the length L is 3 m.

   (b)   The load P is 350 kN and the length L is 6 m.

6-108*   A W14 × 120 steel beam (see Appendix C) is loaded and supported as shown in Fig. P6–104. Determine the maximum principal and shearing stresses in the beam if

   (a)   The load P is 160 kips and the length L is 8 ft.

   (b)   The load P is 80 kips and the length L is 16 ft.

(a)   (b)

6-12 FLEXURAL STRESSES—UNSYMMETRICAL BENDING

The discussion of flexural stresses to this point has been limited to beams with at least one longitudinal plane of symmetry and with the load applied in the plane of symmetry. The work will now be extended to cover pure bending (bent with couples only; no transverse forces) of (1) beams with a plane of symmetry but with the load (couple) applied not in or parallel to the plane of symmetry, or (2) beams with no plane of symmetry. Additional discussion needed for the problem of unsymmetrical beams loaded with transverse forces will be found in Art. 6–15.

Figure 6–28 depicts a beam of unsymmetrical cross section loaded with a couple M in a plane making an angle α with the xy plane, where the origin of coordinates is at the centroid of the cross section. The neutral axis, which passes through the centroid for linearly elastic action (see Art. 6–4), makes an unknown angle β with the z axis. The beam is straight and of uniform cross section, and a plane cross section is assumed to remain plane after bending. Also the following development is restricted to linearly elastic action.

Since the orientation of the neutral axis is not known, the flexural stress distribution function cannot be expressed in terms of one variable, as in Eq. c of Art. 6–4. However, since the plane section remains plane, the stress variation can be written as

$$\sigma = k_1 y + k_2 z \qquad (a)$$

Then the resisting moments with respect to the z and y axes can be written as follows:

$$M_{rz} = \int_A \sigma \, dA \, y = \int_A k_1 y^2 \, dA + \int_A k_2 zy \, dA = k_1 I_z + k_2 I_{yz}$$

$$M_{ry} = \int_A \sigma \, dA \, z = \int_A k_1 yz \, dA + \int_A k_2 z^2 \, dA = k_1 I_{yz} + k_2 I_y$$

(b)

where I_z and I_y are the moments of inertia of the cross-sectional area with respect to the z and y axes, respectively, and I_{yz} is the product of inertia (see Appendix B) with respect to these two axes. Equating the applied moment to the resisting moment and solving for k_1 and k_2 gives

$$M_{rz} = M_z = +M \cos \alpha = k_1 I_z + k_2 I_{yz}$$

$$M_{ry} = M_y = -M \sin \alpha = k_1 I_{yz} + k_2 I_y$$

(c)

and

$$k_1 = \frac{M_z I_y - M_y I_{yz}}{I_y I_z - I_{yz}^2}$$

$$k_2 = \frac{M_y I_z - M_z I_{yz}}{I_y I_z - I_{yz}^2}$$

(d)

Substituting Eqs. d into Eq. a yields the elastic flexure formula for unsymmetrical bending. Thus,

$$\sigma = \left[\frac{M_z I_y - M_y I_{yz}}{I_y I_z - I_{yz}^2} \right] y + \left[\frac{M_y I_z - M_z I_{yz}}{I_y I_z - I_{yz}^2} \right] z$$

(6–10)

or

$$\sigma = \left[\frac{I_z z - I_{yz} y}{I_y I_z - I_{yz}^2} \right] M_y + \left[\frac{I_y y - I_{yz} z}{I_y I_z - I_{yz}^2} \right] M_z$$

(6–11)

Since σ is zero at the neutral surface, the orientation of the neutral axis is found by setting Eq. 6–10 equal to zero. Thus,

$$(M_z I_y - M_y I_{yz}) y + (M_y I_z - M_z I_{yz}) z = 0$$

or

$$y = - \left[\frac{M_y I_z - M_z I_{yz}}{M_z I_y - M_y I_{yz}} \right] z$$

which is the equation of the neutral axis in the yz plane. The slope of the line is dy/dz; therefore, since $dy/dz = \tan \beta$ in Fig. 6–28, the orientation of the neutral axis is given by the expression

$$\tan \beta = -\frac{M_y I_z - M_z I_{yz}}{M_z I_y - M_y I_{yz}} \qquad (6\text{--}12)$$

Equations 6–10, 6–11, and 6–12 have been developed by assuming that *a moment is positive if the z component (M cos α) produces tension in the bottom of the beam*. The applied moment *M* shown in Fig. 6–28*b* is a positive moment. For the orientation of the plane of loading shown in Fig. 6–28*b*, the resisting moments M_{ry} and M_{rz} are both positive. Thus, M_{ry} results from tensile stresses in regions where *z* is positive and from compressive stresses in regions where *z* is negative. Similarly, M_{rz} results from tensile stresses in regions where *y* is positive and from compressive stresses in regions where *y* is negative. This sign convention for moments is consistent with the sign convention used in Article 6–5 where the plane of loading coincided with a plane of symmetry in the beam. Since the positive direction for *y* has been chosen downward, the angles α and β (measured from the *z* axis) are positive in the clockwise direction. Thus, the angles α and β associated with the moment *M* and the neutral axis in Fig. 6–28*b* are negative angles.

In some problems, it will be convenient to select the *y* and *z* axes such that they are principal axes *Y* and *Z*. The product of inertia I_{YZ} is then zero and Eqs. 6–10, 6–11, and 6–12 reduce to

$$\sigma = \frac{M_Z Y}{I_Z} + \frac{M_Y Z}{I_Y} \qquad (6\text{--}13)$$

$$\tan \beta = -\frac{M_Y I_Z}{M_Z I_Y} = \frac{I_Z}{I_Y} \tan \alpha \qquad (6\text{--}14)$$

Equation 6–14 indicates that the neutral axis is not perpendicular to the plane of loading unless (1) the angle α is zero, in which case the plane of loading is (or is parallel to) a principal plane, or (2) the two principal moments of inertia are equal; this reduces to the special kind of symmetry in which all centroidal moments of inertia are equal (square, circle, and the like). The following examples will illustrate the application of Eqs. 6–10 through 6–14.

EXAMPLE 6–11
A beam having the T cross section shown in Fig. 6–29*a* is subjected to a positive couple of 13,600 in.-lb. The couple is applied in plane *a–a* (see Fig. 6–29*a*) that is oriented at an angle α of −36.87°, with respect to the *xy* plane (*x* is the longitudinal axis of the beam).

(a) Determine the maximum flexural stress in the beam.

(b) Determine and show on a sketch the orientation of the neutral axis.

FIG. 6–29

(a)

(b)

(c)

(c)   Check the stress obtained in part (a) by using Eq. 6–3 and the orientation of the neutral axis as determined in part (b).

SOLUTION

(a)   Because of symmetry, the y axis is a principal axis (see Appendix B) of the cross section; therefore, the z axis is the other principal axis. The principal moments of inertia are $I_Z = I_z = 136$ in.^{4} and $I_Y = I_y = 40$ in.^{4}. The point of maximum flexural stress is not apparent by inspection; however, the two possible locations are points A and B, as they are located farthest from the centroid in regions where the two components of stress are additive. Since the angle α is $-36.87°$,

$$M_Y = -M \sin \alpha = -13,600 \sin(-36.87°) = 8,160 \text{ in.-lb}$$

$$M_Z = +M \cos \alpha = +13,000 \cos(-36.87°) = 10,880 \text{ in.-lb}$$

Substituting M_Y and M_Z into Eq. 6–13 yields

$$\sigma_A = \frac{10880(-3)}{136} + \frac{8160(-3)}{40} = -852$$

$$\sigma_B = \frac{10880(5)}{136} + \frac{8160(1)}{40} = +604$$

Therefore, the maximum flexural stress is

$$\underline{852 \text{ psi } C} \qquad\qquad \text{Ans.}$$

(b)   The angle β is obtained from Eq. 6–14; thus,

$$\tan \beta = \frac{I_Z}{I_Y} \tan \alpha = \frac{136}{40} \tan(-36.87°) = -2.550$$

Therefore,

$$\beta = -68.59° = \underline{68.6°} \text{ as shown in Fig. 6–29}b \qquad \text{Ans.}$$

(c) The flexure formula $\sigma = My/I$ applies when M is the moment in the plane perpendicular to the neutral axis, y is the perpendicular distance c_A from the neutral axis to point A, and I is I_{NA}. The distance c_A from geometry is (see shaded triangles in Fig. 6–29b)

$$c_A = -3 \sin(21.4°) - 3 \cos(21.4°) = -3.888 \text{ in.}$$

The moment of inertia about the neutral axis can be found by using Mohr's circle (see Fig. 6–29c); thus,

$$I_{NA} = 88 + 48 \cos(137.2°) = 88 - 35.2 = 52.8 \text{ in.}^4$$

Then, the flexural stress at point A is

$$\sigma_A = \frac{M \cos(\beta - \alpha)c_A}{I_{NA}} = \frac{13600 \cos(-31.72°)(-3.888)}{52.8}$$

$$= -852 = \underline{852 \text{ psi } C} \qquad \text{Ans.}$$

EXAMPLE 6–12

A beam having the cross section shown in Fig. 6–30a is subjected to a positive bending moment of 300 in.-kips in the xy plane (where x is the longitudinal axis of the beam). The moments of inertia with respect to the y and z axes are $I_y = 135$ in.^{4}, $I_z = 240$ in.^{4}, and $I_{yz} = 108$ in.^{4}. Determine

(a) The flexural stress at point B by using the general flexure formula for unsymmetrical bending (Eq. 6–10 or 6–11).

(b) The flexural stress at point B by using the reduced form of the general flexure formula that is valid if principal axes are used (Eq. 6–13).

(c) The orientation of the neutral axis.

(a)   (b)   (c)   FIG. 6–30

SOLUTION

(a) Since the moment is applied in the xy plane, $M_y = 0$ and $M_z = 300$ in.-kips. Equations 6–10 and 6–11 then reduce to

$$\sigma = \left[\frac{I_y y - I_{yz} z}{I_y I_z - I_{yz}^2}\right] M_z$$

Thus, the stress at point B is

$$\sigma_B = \left[\frac{135(-4) - 108(1.5)}{135(240) - (108)^2}\right](300)$$

$$= -10.16 \text{ ksi} = \underline{10.16 \text{ ksi } C} \qquad \text{Ans.}$$

(b) Since the product of inertia I_{yz} is not zero, the y and z axes are not principal axes. The principal moments of inertia and their orientation can be found by using Mohr's circle as shown in Fig. 6–30b. The results are $I_Y = 67.5$ in.^{4}, $I_Z = 307.5$ in.^{4}, and $\alpha = 32.04°$ counterclockwise. The coordinates Y_B and Z_B (see Fig. 6–30c) and the moments M_Y and M_Z are

$$Y_B = -4 \cos(32.04°) + 1.5 \sin(32.04°) = -2.595 \text{ in.}$$

$$Z_B = +4 \sin(32.04°) + 1.5 \cos(32.04°) = +3.394 \text{ in.}$$

$$M_Y = -M \sin \alpha = -300 \sin(32.04°) = -159.2 \text{ in.-kips}$$

$$M_Z = +M \cos \alpha = +300 \cos(32.04°) = +254.3 \text{ in.-kips}$$

The stress at point B obtained from Eq. 6–13 is

$$\sigma = \frac{M_Z Y}{I_Z} + \frac{M_Y Z}{I_Y}$$

$$\sigma_B = \frac{254.3(-2.595)}{307.5} + \frac{-159.2(3.394)}{67.5}$$

$$= -10.15 \text{ ksi} = \underline{10.15 \text{ ksi } C} \qquad \text{Ans.}$$

The small difference in answers for parts (a) and (b) results from round-off errors in the determinations of the quantities used in Eq. 6–13.

(c) The angle β_z measured with respect to the Cartesian z axis is obtained from Eq. 6–12. Thus,

$$\tan \beta_z = \frac{I_{yz}}{I_y} = \frac{108}{135} = 0.800$$

$$\beta_z = \underline{38.66°} \qquad \text{Ans.}$$

The angle β_Z measured with respect to the principal Z axis is obtained from Eq. 6–14. Thus,

$$\tan \beta_Z = \frac{I_Z}{I_Y} \tan \alpha \ = \frac{307.5}{67.5} \tan(32.04°) = 2.851$$

$$\beta_Z = \underline{70.70°}$$

Ans.

PROBLEMS

> *Note: In the following problems, the x axis refers to the longitudinal centroidal axis of the beam.*

6-109*   A beam having the I cross section shown in Fig. P6–109 is subjected to a positive bending moment of 10,000 in.-lb. The moment is applied in plane *a–a*. Determine

(a) The magnitude of the maximum flexural stress.

(b) The orientation of the neutral axis; show its location on a sketch of the cross section.

6-110*   A beam having the T cross section shown in Fig. P6–110 is subjected to a positive bending moment of 20 kN · m. The moment is applied in plane *a–a*. Determine

(a) The magnitude of the maximum flexural stress.

FIG. P6–109

FIG. P6–110

FIG. P6–111

FIG. P6–112

(b)   The orientation of the neutral axis; show its location on a sketch of the cross section.

6-111*   An 8 × 8 × 1-in. angle is used for a beam that supports a positive bending moment of 7500 ft-lb. The moment is applied in the xy plane (see Fig. P6–111). The moments of inertia from a structural steel handbook are $I_y = I_z = 89.0$ in.^{4} and $I_{min} = 36.5$ in.^{4}. Determine

(a)   The flexural stress at point A.

(b)   The maximum flexural stress and its location on the cross section.

(c)   The orientation of the neutral axis; show its location on a sketch of the cross section.

6-112   A beam having the Z cross section shown in Fig. P6–112 is subjected

FIG. P6–113

FIG. P6–114

FIG. P6–115

to a positive bending moment of 8000 N · m. The moment is applied in the xy plane. The moments of inertia from a structural steel handbook are $I_y = 4.02(10^6)$ mm⁴, $I_z = 16.64(10^6)$ mm⁴, and $I_{min} = 1.59(10^6)$ mm⁴. Determine

(a) The flexural stresses at points A and B.

(b) The orientation of the neutral axis; show its location on a sketch of the cross section.

6-113   A beam having the angle cross section shown in Fig. P6–113 is subjected to a negative bending moment of 50 in.-kips. The moment is applied in the xy plane. Determine

(a) The flexural stresses at points A and B.

(b) The orientation of the neutral axis; show its location on a sketch of the cross section.

6-114*   A beam having the triangular cross section shown in Fig. P6–114 is subjected to a negative bending moment of 2 kN · m. The moment is applied in the xy plane. Determine

(a) The flexural stresses at points A and B.

(b) The orientation of the neutral axis; show its location on a sketch of the cross section.

6-115   A beam having the rectangular cross section shown in Fig. P6–115 is subjected to a bending moment M. The moment is applied in plane $a–a$ that is oriented at an angle α with respect to the xy plane.

(a) Develop an expression for the maximum flexural stress in terms of b, h, M, and α.

(b) Determine α in terms of b and h to produce the maximum flexural stress for a given moment.

(c) Use the expression for α obtained in part (b) and determine the orientation of the neutral axis when $h = 2b$.

FIG. P6–116

FIG. P6–117

6-116   A beam having the triangular cross section shown in Fig. P6–116 is subjected to a positive bending moment of 10 in.-kips. The moment is applied in plane a–a. Determine

(a)   The flexural stresses at points A and B.

(b)   The orientation of the neutral axis; show its location on a sketch of the cross section.

6-117*   A beam having the Z cross section shown in Fig. P6–117 is subjected to a positive bending moment of 20 in.-kips. The moment is applied in plane a–a. The moments of inertia from a structural steel handbook are $I_y = 8.83$ in.^{4}, $I_z = 25.40$ in.^{4}, and $I_{min} = 3.08$ in.^{4}. Determine

(a)   The flexural stresses at points A and B.

(b)   The orientation of the neutral axis; show its location on a sketch of the cross section.

6-118*   A beam with a rectangular cross section 50 mm wide by 300 mm deep is designed for a system of loads that produces a moment in the xy plane (the y axis is parallel to the 300-mm side). In construction, the beam is positioned such that the moment acts in a plane inclined 3 degrees with respect to the xy plane. Determine the increase in the maximum flexural stress produced by the construction error.

6-119   Determine the minimum moment M and the orientation α of the plane in which it must be applied required to produce a maximum flexural stress of 15 MPa in a beam with a rectangular cross section 150 mm wide by 300 mm deep.

6-13   FLEXURAL STRESSES—INELASTIC ACTION

A large proportion of structural and machine designs are based on elastic analysis, for which the flexure formula is applicable. However, for some designs, particularly when the weight of the structure is important (aircraft design, for example), the limitation requiring stresses to remain below the proportional limit of the material results in uneconomical or inefficient designs. Therefore, this limitation is sometimes discarded and higher stress levels are tolerated in the design. This section is concerned with the analysis of stresses in instances in which the proportional limit of the material is exceeded or the material does not exhibit a linear stress-strain relationship. In both of these cases, Hooke's law does not apply.

The basic approach to problems involving inelastic action is the same as that outlined in Arts. 6–4 and 6–5 for linearly elastic action. A plane section is assumed to remain plane; therefore, a linear distribution of strain exists. If a stress-strain diagram is available for the material, it can be used with the strain distribution to obtain a stress distribution for the beam. The stress distribution shown in Fig. 6–31a is for a beam of symmetrical cross section made of a material having the stress-strain diagram shown in Fig. 6–31c. If an equation can be written for the stress distribution, the resisting moment can be obtained by using Eq. 6–1. If it is not practical to obtain an equation for the stress distribution, the beam depth can be divided into layers of finite thickness and the stress associated with each layer determined. The method of Art. 6–4 can then be used to evaluate the resisting moment. As a simplifying technique, the stress-strain diagram can be approximated by a series of straight lines, resulting in a simplified stress distribution diagram and a reduction in the amount of work involved.

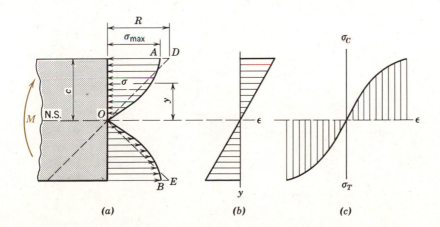

(a)   (b)   (c)   FIG. 6–31

An alternate approach to the problem of inelastic action involves replacing the stress distribution *AOB* of Fig. 6–31*a* by a linear distribution *DOE* (shown dotted) with a maximum stress *R*. The stress *R* (known as the modulus of rupture in bending) is selected to give the same resisting moment as that obtained from the actual stress distribution *AOB*. For the assumed linear distribution *DOE*, the flexure formula is valid; therefore, $R = Mc/I$. The value of R is obtained by loading a beam to rupture and substituting the magnitude of the bending moment at rupture into the flexure formula. The modulus of rupture is useful only for predicting the rupture load. It is also valid only for a particular material and cross section and, in general, may be dependent on some shape parameter (as, for example, the diameter to thickness ratio for tubing).

If the material is elastoplastic (see Art. 3–2), the large strains associated with yielding will permit plastic action to take place as indicated in Fig. 6–32. The stress distribution for fully elastic action is shown in Fig. 6–32*a*, for partially plastic action in Fig. 6–32*b*, and for essentially complete plastic action in Fig. 6–32*c*. There must always be a slight amount of elastic action if the strain at the neutral surface is zero. The plastic action shown in Fig. 6–32*c* is idealized in Fig. 6–32*d*; this is the assumed distribution used for plastic analysis of structural steel beams.

In members subjected to inelastic action, the neutral axis does not necessarily coincide with a centroidal axis as it does when the action is elastic. If the section is symmetrical and the stress-strain diagrams for tension and compression are identical, the neutral axis for plastic action coincides with the centroidal axis. However, if the section is unsymmetrical (for example, a T-section) or if the stress-strain relations for tension and compression differ appreciably (for example, cast iron), the neutral axis shifts away from the fibers that first experience inelastic action, and it is necessary to locate the axis before the resisting moment can be evaluated. The neutral axis is located by using the equation

(a)   (b)   (c)   (d)

FIG. 6–32

$$\Sigma F_x = \int_{Area} \sigma \, dA = 0 \qquad (a)$$

and solving for the location of the axis where σ is zero.

The following examples illustrate the concepts of inelastic analysis as applied to the pure bending of beams that are loaded in a plane of symmetry.

EXAMPLE 6–13

A beam having the cross section of Fig. 6–33*a* is made of magnesium alloy having the approximate stress-strain diagram of Fig. 6–33*b*. Determine the magnitude of the bending moment applied in the vertical plane of symmetry necessary to produce a flexural stress of magnitude 14.4 ksi at point A, which is at the lower surface of the top flange.

SOLUTION

Because of symmetry, the neutral axis coincides with the centroidal z axis, and the strain and stress distribution diagrams are as shown in Figs. 6–33*c* and *d*. Since the specified stress is 14.4 ksi at $y = 2$ in., the corresponding strain from Fig. 6–33*b* is 3.5 (in./in.)$(10)^{-3}$, and the strain at $y = 3$ in. is 5.25 (in./in.)(10^{-3}); therefore, the second stress-strain function of Fig. 6–33*b* is valid for the entire flange. The first function is valid for the web. Substituting

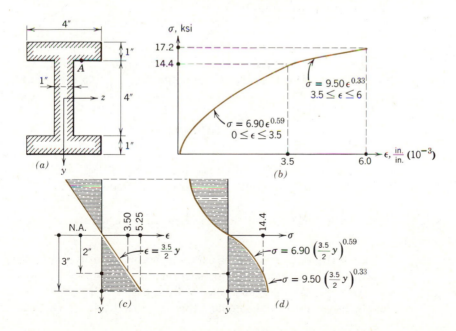

FIG. 6–33

the ε vs. y function from Fig. 6–33c into the σ vs. ε functions gives the σ vs. y functions of Fig. 6–33d. The solution of Eq. a of Article 6.4 follows:

$$M_r = \int_A \sigma \, dA \, y = 2\left\{\int_0^2 [6.90(1.75)^{0.59}y^{0.59}][1 \; dy]y\right.$$

$$+ \left.\int_2^3 [9.50(1.75)^{0.33}y^{0.33}][4 \; dy]y\right\}$$

$$= 2\left[\frac{6.90(1.391)y^{2.59}}{2.59}\right]_0^2 + 2\left[\frac{38(1.203)y^{2.33}}{2.33}\right]_2^3 = 354.8$$

from which

$$M = M_r = \underline{355 \text{ in.-kips}} \qquad \text{Ans.}$$

EXAMPLE 6–14

A beam is to be fabricated from three 6 by 1-in. plates that will be welded together to form a symmetrical I-section. Determine the maximum elastic and plastic bending moments that the beam will sustain if the material is

(a)   Elastoplastic steel with a proportional limit (equal to the yield point) of 40 ksi.
(b)   An aluminum alloy with the stress-strain diagram shown in Fig. 6–34a. Assume the diagram is the same in tension and compression and limit the strain to 0.010.

SOLUTION

(a)   For the elastic moment, the flexure formula $\sigma = Mc/I$ applies. The moment of inertia I about the neutral axis is 166 in.^{4}. Therefore,

$$M_e = \frac{\sigma I}{c} = \frac{40(166)}{4} = \underline{1660 \text{ in.-kips}} \qquad \text{Ans.}$$

For this material, the ideal stress distribution of Fig. 6–32d may be assumed when computing the plastic moment, which is

$$M_p = \left[40(6)(1)\,\frac{7}{2} + 40(3)(1)\,\frac{3}{2}\right][2] = \underline{2040 \text{ in.-kips}} \qquad \text{Ans.}$$

To check the degree of approximation involved in assuming the ideal stress distribution for this section, the maximum strain before strain hardening takes place will be assumed to be 16 times the elastic strain (a reasonable value for structural steel). When

FIG. 6–34

(a)

(b)   (c)   (d)

this strain is reached in the outer fibers, the maximum elastic strain occurs at $(1/16)$ (4) or $1/4$ in. above and below the neutral surface; this means that the actual stress distribution diagram is like that of Fig. 6–32c, the distance from the neutral surface to the point where the stress is constant being $1/4$ in. This distribution makes the moment of 2040 in.-kips computed above too large by the amount

$$\Delta M_r = \frac{40}{2} \frac{1}{4} (1) \frac{1}{3} \frac{1}{4} (2) = 0.833 \text{ in.-kip}$$

or 0.04 percent.

(b)   The elastic moment is given by the flexure formula, where σ is the proportional limit that, from Fig. 6–34a, is 30 ksi. Then,

$$M_e = 30(166/4) = \underline{1245 \text{ in.-kips}} \qquad \text{Ans.}$$

The plastic moment will first be determined with the aid of the stress distribution diagram of Fig. 6–34c that was obtained from Fig. 6–34a by assuming the maximum strain in the beam to be 0.010, and since the strains are proportional to the distance from the neutral surface of the beam, the stress for any point in the cross section may be read from the diagram of Fig. 6–34a. With the assumption that the curve in Fig. 6–34c is a series of chords connecting the points at 1-in. vertical intervals, the following solution is obtained:

$$M_p = 2 \left[6(1)(48) \frac{7}{2} + (50 - 48) \frac{1}{2} (6)(1) \left(3 + \frac{2}{3} \right) \right.$$

$$+ 43.3(1)(1) \frac{5}{2} + (48 - 43.3) \frac{1}{2} \left(2 + \frac{2}{3} \right)$$

$$+ 26.5(1)(1) \frac{3}{2} + (43.3 - 26.5) \frac{1}{2} \left(1 + \frac{2}{3} \right)$$

$$\left. + \frac{26.5}{2} \frac{2}{3} \right]$$

$$= \underline{2414 \text{ in.-kips}} \qquad \text{Ans.}$$

This result will be checked by assuming the curve below the flange to be a parabola. The flange moment will be the same as above and equals 2060 in.-kips; then,

$$M_p = 2060 + \frac{2}{3} (48)(3)(1) \frac{5}{8} (3)(2) = \underline{2420 \text{ in.-kips}} \qquad \text{Ans.}$$

One more technique will be applied to this problem. The stress-strain diagram of Fig. 6–34a will be approximated by the dotted lines shown, and the resulting stress distribution diagram for the beam will be as in Fig. 6–34d. With the use of this diagram, the plastic moment becomes

$$M_p = 2 \left[49(6)(1) \left(\frac{7}{2}\right) + 40\left(\frac{3}{2}\right)(1)\left(\frac{3}{2} + \frac{3}{4}\right) \right.$$

$$+ (49 - 40) \left(\frac{1}{2}\right) \left(\frac{3}{2}\right)(1)\left(\frac{3}{2} + 1\right)$$

$$\left. + 40 \left(\frac{1}{2}\right) \left(\frac{3}{2}\right)(1)\left(\frac{2}{3}\right)\left(\frac{3}{2}\right) \right]$$

$$= \underline{2422 \text{ in.-kips}} \qquad\qquad \text{Ans.}$$

The last three results indicate that any of the three techniques is adequate for this problem.

EXAMPLE 6–15
A beam having the *T* cross section shown in Fig. 6–35 is made of elastoplastic steel (E = 200 GPa) with a proportional limit (equal to the yield point) of 240 MPa. Determine

(a) The bending moment (applied in the vertical plane of symmetry) that will produce a longitudinal strain of −0.0012 at point *B* on the lower face of the flange.

(b) The bending moment required to produce completely plastic action in the beam.

SOLUTION

(a) The strain distribution, stress-strain, and stress distribution dia-

FIG. 6–35

FIG. 6–36

(a)   (b)   (c)

grams are shown in Figs. 6–36a, b, and c. Because of the inelastic action and unsymmetrical section, the location of the neutral axis must first be determined by using Eq. a. Thus,

$$\int_A \sigma \, dA = -240(10^6)(0.025)(0.100) - \frac{240(10^6)}{2}(a)(0.025)$$

$$+ \frac{240(10^6)}{2}(a)(0.025) + 240(10^6)(0.200 - 2a)(0.025) = 0$$

from which

$$a = 0.050 \text{ m} = 50 \text{ mm}$$

Once the neutral axis is located, the easiest way to find the moment is by using the free-body diagram method of Example 6–1. Thus, from Fig. 6–37,

$$M_r = 240(10^6)(0.100)(0.025)(0.0625)$$

$$+ \tfrac{1}{2}(240)(10^6)(0.050)(0.025)(2/3)(0.050)(2)$$

$$+ 240(10^6)(0.100)(0.025)(0.100) = 107.5(10^3) \text{ N} \cdot \text{m}$$

from which

$$M = M_r = \underline{107.5 \text{ kN} \cdot \text{m as shown}} \qquad \text{Ans.}$$

(b)   In this case, the stress distribution diagram would be similar to the diagram shown in Fig. 6–32d. If the distance from the neutral surface to the bottom of the stem is designated as distance a, Eq. a yields

FIG. 6–37

$$\int_A \sigma \, dA = \sigma(0.25)(a) - \sigma(0.025)(0.200 - a) - \sigma(0.100)(0.025) = 0$$

from which $a = 150$ mm; therefore, the neutral surface is 75 mm below the top of the flange. The plastic moment obtained by using the free-body diagram method is

$$M_r = 240(10^6)(0.100)(0.025)(0.0625) + 240(10^6)(0.050)(0.025)(0.025)$$

$$+ \; 240(10^6)(0.150)(0.025)(0.075) = 112.5(10^3) \text{ N} \cdot \text{m}$$

$$M_p = M_r = \underline{112.5 \text{ kN} \cdot \text{m}} \qquad\qquad \text{Ans.}$$

From the preceding discussion and examples, one may observe that, if the load capacity of a beam is based on the plastic moment rather than the elastic moment, a considerable saving in material may be realized (with the same factor of safety). Furthermore, in the design of statically indeterminate beams and frames, the plastic method of analysis is considerably less time-consuming than the elastic analysis. This is not to say that the elastic method of analysis should be discarded as outmoded, for there are many designs that must be based on elastic action, and for some situations, particularly when repeated loading is involved, the application of plastic analysis may be dangerous.

PROBLEMS

Note: In the following problems, all beams are loaded in a plane of symmetry parallel to the largest cross-sectional dimension except as noted otherwise.

6-120*   Determine the plastic moment for a W18 × 60 steel beam having a proportional limit (equal to the yield point) of 36 ksi.

6-121*   Determine the plastic moment for a W610 × 125 steel beam having a proportional limit (equal to the yield point) of 250 MPa.

6-122   Determine the ratio of the plastic moment to the maximum elastic moment for a beam of elastoplastic material with a rectangular cross section.

6-123*   Determine the ratio of the plastic moment to the maximum elastic moment for a beam of elastoplastic material with a circular cross section.

6-124*   Determine the ratio of the plastic moment to the maximum elastic moment for a W36 × 160 steel beam.

6-125   Determine the ratio of the plastic moment to the maximum elastic moment for a box beam fabricated from two 150 × 25-mm and two 100 × 25-mm elastoplastic steel plates to form a section 100 mm wide by 200 mm deep.

6-126*   A beam of elastoplastic material (yield point of 36 ksi) has the cross section shown in Fig. P6–126. Determine

  (a)   The location of the neutral axis when the stress in the outer fibers of the flange reaches the yield point.

  (b)   The moment required to produce the condition of part (a).

  (c)   The ratio of the plastic moment to the maximum elastic moment for this cross section.

6-127   A beam of elastoplastic material (yield point of 240 MPa) has the cross section shown in Fig. P6–127. Determine

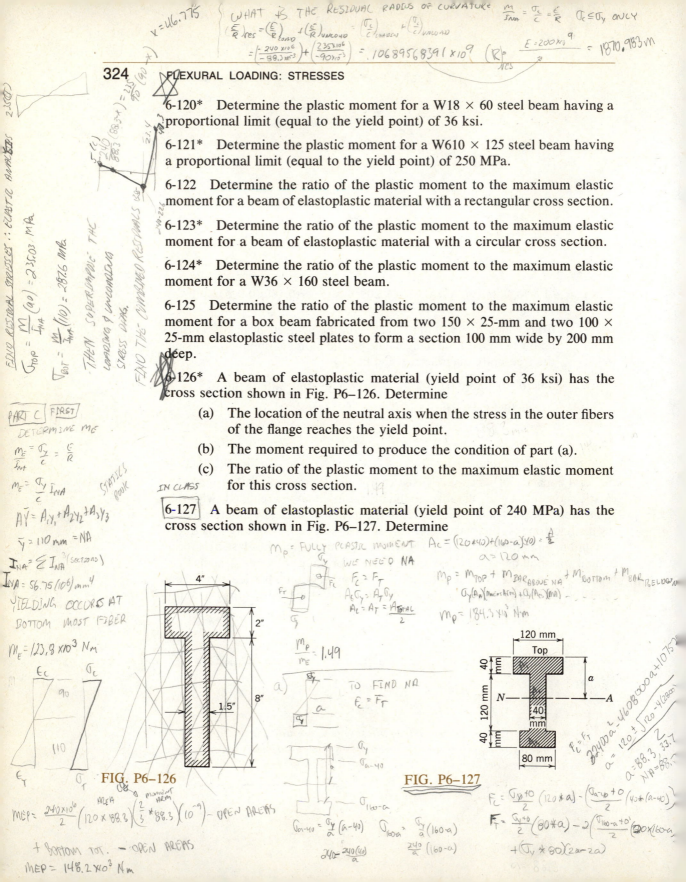

FIG. P6–126

FIG. P6–127

FIG. P6–129

(a)   (b)

(a) The location of the neutral axis when the stress in the outer fibers of the top flange reaches the yield point. *88.3*

(b) The moment required to produce the condition of part (a). *148.2×10³ Nm*

(c) The ratio of the plastic moment to the maximum elastic moment for this cross section. *1.49*

6-128 A beam of rectangular cross section is made of a material for which the stress-strain diagram in tension can be represented by the expression $\sigma = K\varepsilon^{1/2}$. The shape of the diagram is the same in tension and compression. Develop an equation similar to the flexure formula for relating fiber stress and applied moment.

6-129* A beam having the T cross section shown in Fig. P6–129a is made of a magnesium alloy that has the stress-strain diagram shown in Fig. P6–129b. The beam is subjected to a bending moment that produces a maximum flexural stress of 99.3 MPa *T*. When this moment is applied, the neutral surface is located at the junction of the flange and stem. Determine

(a) The dimension c.

(b) The bending moment applied to the beam.

6-14 NEGLECTED EFFECTS OF DISTRIBUTED LOADING

In Arts. 6–4 and 6–5 the flexure formula was developed by assuming that plane sections remain plane and that the beam is subjected to couples only. The formula for transverse shearing stresses was developed in Art. 6–10 for use with the flexure formula when transverse shearing forces are present. If shearing stresses exist, plane sections cannot remain plane, as Fig. 6–38 indicates, because the element b at the left section will undergo the maximum shearing strain, while elements a and c will experience no shearing strain. The result is a warped surface, as indicated. If the segment is in a

FIG. 6–38

region of constant shear, the warping of section *def* will match that of *abc*; consequently, the ends of any longitudinal element will be displaced the same amount, and the normal strains will be proportional to the distance from the neutral surface just as though the section had remained plane. In a region of varying shear (for example, the segment of a beam subjected to distributed loading), the warping of section *def* will not be the same as that of *abc*, and the normal strains will not be proportional to the distance from the neutral surface. For most engineering applications, the effects of cross-sectional warping can be neglected. Observe, for example, that curve 2 in Fig. 6–8, which represents measured longitudinal elastic strains in a region including flexural and transverse shearing stress, is apparently a straight line although the discussion above indicates it should not be. From the mathematical theory of elasticity, the error involved in using the elastic flexure formula to compute the maximum fiber stress in the cantilever beam of Fig. 6–39 is less than 0.2 percent for a span-depth ratio of 6.

Also neglected in previous articles are normal stresses in the transverse direction σ_y due to distributed loading. In order to complete the elementary picture of the state of stress in a beam, an expression for σ_y will now

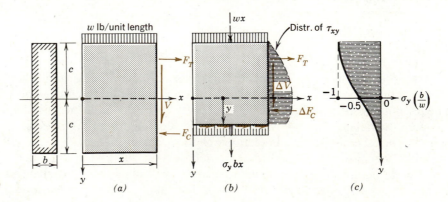

FIG. 6–39

be developed for a rectangular cantilever beam of width b and depth $2c$ carrying a uniformly distributed load w over its entire length. The beam cross section and a free-body diagram of a portion of the free end of the beam are shown in Fig. 6–39a. Passing a horizontal plane through the beam segment results in the free-body diagram of Fig. 6–39b, from which the force equilibrium equation in the y direction gives the following:

$$\sigma_y bx + \Delta V + wx = 0 \qquad (a)$$

where σ_y is uniformly distributed over the top of the beam (in both the x and z directions). Also, the resultant of the shearing stresses is

$$\Delta V = \int_{-c}^{y} \tau_{xy} b \, dy$$

and from Art. 6–10,

$$\tau_{xy} = \frac{V}{Ib} \int_{y}^{c} yb \, dy = \frac{V}{2I} (c^2 - y^2)$$

Therefore,

$$\Delta V = \frac{Vb}{2I} \int_{-c}^{y} (c^2 - y^2) \, dy = \frac{Vb}{2I} \left[c^2 y - \frac{y^3}{3} \right]_{-c}^{y} = \frac{Vb}{6I} (2c^3 + 3c^2 y - y^3)$$

From Fig. 6–39a, $V = -wx$, which when substituted into the expression for ΔV yields

$$\Delta V = -\frac{wxb}{6I} (2c^3 + 3c^2 y - y^3) \qquad (b)$$

Substituting b into a, replacing I with $2bc^3/3$, and solving for σ_y, yields

$$\sigma_y = \frac{w}{4b} \left(-\frac{y^3}{c^3} + \frac{3y}{c} + 2 \right) - \frac{w}{b} \qquad (6\text{--}15)$$

The graph of Eq. 6–15 is given in Fig. 6–39c, which shows that σ_y decreases rapidly from a maximum at the surface where the load is applied to zero at the opposite surface.

It will be of interest to compare the magnitudes of σ_x and σ_y at some point where both have maximum values. For the cantilever beam of Fig. 6–39, the maximum fiber stress σ_x at any section in the beam is at the top surface and is

$$\sigma_x = \frac{M(-c)}{I} = -\frac{3M}{2bc^2}$$

and since $M = -wx^2/2$,

$$\sigma_x = \frac{3wx^2}{4bc^2}$$

If x is written as Kc,

$$\sigma_x = \frac{3K^2}{4}\left(\frac{w}{b}\right) = -\frac{3K^2}{4}\sigma_y$$

If K is taken as 10, which would mean a span-depth ratio of 5 for a cantilever beam (since σ_x is maximum at the fixed end),

$$\sigma_{x\,max} = \frac{3(100)}{4}\sigma_y = -75\sigma_{y\,max}$$

or

$$\sigma_{y\,max} = 1.33 \text{ percent of } \sigma_{x\,max}$$

The above discussion does not complete the picture. For a beam section composed of flanges connected by a thin web, the stress σ_y might have some effect on the maximum stresses. For example, refer to Exam. 6–10 in which the beam has a span depth ratio of 4. The stress σ_y in the web at the bottom of the flange is −13.4 MPa, and when this value is included in the computations for the maximum stresses, they would be changed from 132.1 MPa T and 74.7 MPa to 130.6 MPa T and 80.0 MPa, or −1.11 percent for σ_{max} and +7.10 percent for τ_{max}. A common specification for span-depth ratios (to put a practical limit on deflections) for simple beams and girders is 20 to 25. Such ratios would result in even smaller variations than those noted above; hence, the conclusion may be drawn that for real problems, neglecting σ_y is reasonable.

6-15   SHEARING STRESSES IN THIN-WALLED OPEN SECTIONS—SHEAR CENTER

One of the assumptions made in the development of the elastic flexure formula was that the loads were applied in a plane of symmetry. When this assumption is not satisfied, the beam will, in general, twist about a longitudinal axis. It is possible, however, to place the loads in such a plane that the beam will not twist. When any load is applied in such a plane, the line of action of the load will pass through the *shear center* (also known as the center of flexure or center of twist). *The shear center is defined as the point in the plane of the cross section through which the resultant of the transverse shearing stresses due to flexure (no torsion) will pass for any orientation of transverse loads.* For cross-sectional areas having two axes of symmetry, the shear center coincides with the centroid of the cross-sectional area. For those sections with one axis of symmetry, the shear center is always located on the axis of symmetry. If a beam having a cross section with one axis of symmetry is positioned such that the plane of symmetry is the neutral axis for flexural stresses, the plane of the loads must be perpendicular to the neutral axis but cannot pass through the centroid if bending is to occur without twisting. The channel is a common structural shape with one axis of symmetry. It is normally used with the axis of symmetry as the neutral axis since the section modulus is relatively large in this position.

Thin-walled open sections such as channels, angles, wide-flange sections, and the like, develop significant shearing stresses during bending. Since the shearing stresses can be assumed to be uniform through the thickness, procedures based on longitudinal equilibrium similar to those of Art. 6–10 can be used to establish shearing stress distributions and to locate the shear centers for these sections.

Consider a cantilever beam of arbitrary cross section but constant thickness (see Fig. 6–40a) loaded in a plane parallel to one of its principal planes. The normal stress σ in any longitudinal fiber is then given by the elastic flexure formula as $\sigma = My/I$. The shearing stresses on section a–a can be determined by considering longitudinal equilibrium of the small block B shown in Fig. 6–40a. An enlarged free-body diagram of the block is shown in Fig. 6–40b. Since the moments on sections at the ends of the block are M and $M + \Delta M$, the resultant forces F_1 and F_2 are

FIG. 6–40

$$F_1 = \int_{A_s} \sigma \, dA = \int_{A_s} \frac{My}{I} \, dA = \int_0^s \frac{My}{I} \, (t \, ds)$$

and

$$F_2 = \int_{A_s} \sigma \, dA = \int_{A_s} \frac{(M + \Delta M)y}{I} \, dA = \int_0^s \frac{(M + \Delta M)y}{I} \, (t \, ds)$$

where A_s is the area of the cross section between the free edge of the block and the longitudinal plane b–b at a distance s from the free edge. A summation of forces in the longitudinal direction on the block of Fig. 6–40b yields

$$V_H = F_2 - F_1 = \int_{A_s} \frac{(\Delta M)y}{I} \, (dA) = \frac{\Delta M}{I} \int_0^s y(t \, ds)$$

The average shearing stress on the longitudinal plane b–b is V_H divided by the area. Thus,

$$\tau = \lim_{\Delta x \to 0} \frac{\Delta M}{\Delta x} \left(\frac{1}{It}\right) \int_{A_s} y \, dA = \frac{dM}{dx} \left(\frac{1}{It}\right) \int_{A_s} y \, dA = \frac{dM}{dx} \left(\frac{1}{It}\right) \int_0^s y(t \, ds) \quad (a)$$

where dM/dx equals V, the shear at the beam section where the shearing stresses are to be evaluated.

Equation a is identical in form to Eq. a of Art. 6–10 and the terms have identical meanings. The integral is the first or static moment with respect to the neutral axis of the cross-sectional area to one side of the longitudinal plane b–b. This integral is usually referred to as Q_s. Thus,

$$\tau = \frac{VQ_s}{It} \qquad (6\text{--}16)$$

which is identical in form to Eq. 6–9. The shearing stresses are uniform through the thickness and act tangent to the surface of the beam. This same shearing stress acts on the transverse cross section at a distance s from the free edge of the section. The stresses on the transverse cross section "flow" in a continuous direction, as shown in Fig. 6–40a, and at the neutral axis they have the same sense as the shearing force V. The shearing force per unit length of cross section is frequently referred to as the shear flow q. Shear flow was previously discussed in Art. 5–11.

The procedure for locating a shear center will be illustrated by considering a channel section loaded as a cantilever beam (see Fig. 6–41). Since all fibers in a flange can be considered to be located at a distance $h/2$ from the neutral axis, the shearing stress distribution obtained from Eq. a is

$$\tau = \frac{V}{It} \int_0^s y(t \, ds) = \frac{V}{It} \int_0^s \left(\frac{h}{2}\right) (t \, ds) = \frac{Vhs}{2I}$$

FIG. 6–41

This result shows that the horizontal shearing stress in the flange varies linearly from zero at the outer edge to $Vhb/(2I)$ at the web. The resultant shearing force on the top flange is

$$F_3 = (\tau_{\text{avg}})(\text{area}) = \left(\frac{Vhb}{4I}\right)(bt_1) = \frac{Vhb^2t_1}{4I}$$

as shown in Fig. 6–41. A similar analysis indicates that the force F_4 on the lower flange is equal in magnitude and opposite in direction to F_3. These two forces constitute a clockwise twisting moment that must be balanced by the resultant force in the web and the applied load if the beam is not to twist.

The resultant of the shearing forces in the web is V and must be equal to the applied load P for equilibrium. It is assumed that the vertical shearing force transmitted by the flanges is small enough to neglect (see Art. 6–10). The forces V and P will provide a counterclockwise couple to balance F_3 and F_4 if P is located a distance e from the center of the web as indicated in Fig. 6–41. When the moments of these two couples are equated, the result is

$$Ve = hF_3 = \frac{Vh^2b^2t_1}{4I}$$

from which

$$e = \frac{h^2b^2t_1}{4I}$$

If the load P of Fig. 6–41 was applied in the z direction, the forces corresponding to F_3 and F_4 (entirely different magnitudes) would be equal, and both would be directed to oppose the load P. There would also be two shearing forces in the web (parallel to the web) of equal magnitude and oppositely directed; hence, a moment equation would establish the fact that the shear center lies on the z axis of symmetry. If there is to be no twisting of the beam (buckling is assumed not to occur), any load must pass through the intersection of this axis of symmetry and the line of action of force P of Fig.

6–41. If the load is inclined, it may be resolved into components parallel to these axes, and the stresses may be determined using the principle of superposition.

PROBLEMS

> *In all of the following problems, assume that the loads are applied in a vertical xy plane that passes through the shear center of the cross section.*

6-130*   A channel is 460 mm deep overall and has flanges 100 mm wide by 16 mm thick. The web is 12 mm thick. Assume that all of the section is effective in resisting fiber stresses and that only the web resists vertical shearing stresses.

(a)   Locate the shear center of the section with respect to the center of the web.

(b)   Prepare a sketch showing the distribution of shearing stress throughout the cross section when the vertical shear at the section is 50 kN.

6-131*   An eccentric H-section is made by welding two 1 by 10-in. steel flanges to a 1 by 12-in. steel web as shown in Fig. P6–131. The section is used as a 5-ft cantilever beam that carries a concentrated load of 100 kips at the free end. Assume that all of the section is effective in resisting fiber stresses and that only the web resists vertical shearing stresses.

(a)   Locate the shear center of the section with respect to the center of the web.

(b)   Prepare a sketch showing the distribution of shearing stress throughout the cross section.

6-132   Two 12-mm thick plates are to be welded to the flanges of the channel of Prob. 6-130, as shown in Fig. P6–132. Determine

(a)   The width b of the plates if the shear center of the section must be located at the center of the web.

(b)   The maximum vertical shearing stress at the section when the vertical shear at the section is 50 kN.

6-133   A channel is to be fabricated with a depth of 6 in., a web thickness of

FIG. P6–131

FIG. P6–132

1/4 in., and a flange width of 5 in. Structural considerations require the shear center of the channel to be located 2 in. from the center of the web. Determine the required thickness of the flanges.

6-134*   A cantilever beam with the cross section shown in Fig. P6–134 is subjected to a concentrated load of 4 kN at the free end. The section has a constant thickness of 4 mm.

   (a)   Locate the shear center of the section with respect to the center of the web.

   (b)   Prepare a sketch showing the distribution of shearing stress throughout the cross section. The dimensions shown in Fig. P6–134 are centerline dimensions for the flanges, web, and extensions. Assume that all of the section is effective in resisting fiber stresses, but that the vertical shearing force resisted by the flanges is negligible. The vertical shearing force in the extensions is not negligible.

6-135*   A thin-walled cylindrical tube cut longitudinally to make a semicylinder is used as a cantilever beam. The load acts parallel to the cut section as shown in Fig. P6–135.

   (a)   Locate the shear center of the section with respect to the center of the tube (dimension e).

   (b)   Determine the shearing stress at point A if the radius R of the section is 2 in., the thickness t is 0.10 in., and the load P is 100 lb.

6-136   A thin-walled slotted tube is used as a cantilever beam. The beam is loaded as shown in Fig. P6–136.

FIG. P6–134

FIG. P6–135

(a)   Locate the shear center of the section with respect to the center of the tube (dimension e).

(b)   Determine the shearing stress at point A if the radius R of the section is 50 mm, the thickness t is 2.5 mm, and the load P is 500 N.

6-137   A thin-walled box section (see Fig. P6–137) is used as a cantilever beam. Locate the shear center of the section with respect to the center of the web.

6-138*   All metal in the cross section shown in Fig. P6–138 has a thickness of 1/4 in. The dimensions shown are centerline dimensions for the flanges and the web.

FIG. P6–136

FIG. P6–137

FIG. P6–138

FIG. P6–139

(a) Locate the shear center of the section with respect to the center of the web.

(b) Determine the shearing stress at point O when the vertical shear at the section is 2000 lb.

6-139   All metal in the cross section shown in Fig. P6–139 has a thickness of 2.5 mm. The dimensions shown are centerline dimensions for the flanges and the web.

(a) Locate the shear center of the section with respect to the center of the web.

(b) Determine the shearing stress at point O when the vertical shear at the section is 2.5 kN.

6-140*   Locate the shear center for the cross section shown in Fig. P6–140 and determine the maximum shearing stress produced on the cross section by a vertical shear of 2500 lb.

6-141   Locate the shear center for the cross section shown in Fig. P6–141 and determine the maximum shearing stress produced on the cross section by a vertical shear of 1.5 kN.

FIG. P6–140

FIG. P6–141

FIG. P6–142

FIG. P6–143

6-142   Locate the shear center for the cross section shown in Fig. P6–142 and determine the maximum shearing stress produced on the cross section by a vertical shear of 2000 lb.

6-143   Locate the shear center for the cross section shown in Fig. P6–143 and determine the maximum shearing stress produced on the cross section by a vertical shear of 2.5 kN.

6-16 FLEXURAL STRESSES IN BEAMS OF TWO MATERIALS BY THE TRANSFORMED SECTION METHOD

The method of fiber stress computation covered in Art. 6–4 is sufficiently general to cover symmetrical beams composed of longitudinal elements (layers) of different materials. However, for many real beams of two materials (often referred to as reinforced beams), a method can be developed to allow the use of the elastic flexure formula, thus reducing the computational labor involved. The method is applicable to elastic design only.

The assumption of the plane section remaining plane is still valid, provided the different materials are securely bonded together to provide the necessary resistance to longitudinal shearing stresses. Therefore, the usual linear transverse distribution of longitudinal strains is valid.

The beam of Fig. 6–42, composed of a central portion of material A and two outer layers of material B, will serve as a model for the development of the stress distribution. The section is assumed to be symmetrical with respect to the xy and xz planes, and the moment is applied in the xy plane. As long as neither material is subjected to stresses above the proportional limit, Hooke's law applies and the strain relation

$$\varepsilon_b = \varepsilon_a \frac{b}{a}$$

becomes

FIG. 6–42

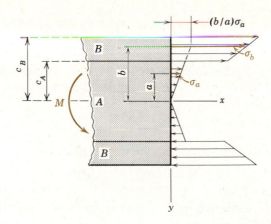

$$\frac{\sigma_b}{E_B} = \frac{\sigma_a}{E_A}\left(\frac{b}{a}\right)$$

or

$$\sigma_b = \sigma_a \left(\frac{E_B}{E_A}\right)\left(\frac{b}{a}\right) \tag{a}$$

From this relation it is evident that, at the junction between the two materials where distances a and b are equal, there is an abrupt change in stress determined by the ratio $n = E_B/E_A$ of the two moduli. If Eq. a is used the normal force on a differential end area of element B is given by the expression

$$dF_B = \sigma_b dA = n\frac{b}{a}\sigma_a dA = \left(\frac{b}{a}\sigma_a\right)(nt)dy \qquad c_A \leqslant y \leqslant c_B$$

where t is the width of the beam at a distance b from the neutral surface. The first factor in parentheses represents the linear stress distribution in a homogeneous material A. The second factor in parentheses may be interpreted as the extended width of the beam from $y = c_A$ to $y = c_B$ if material B were replaced by material A, thus resulting in an equivalent or transformed *cross section* for a beam of homogeneous material. The transformed section is obtained by replacing either material by an equivalent amount of the other material as determined by the ratio n of their elastic moduli. The method is not limited to two materials; however, the use of more than two materials in one beam would be unusual. The method is illustrated by the following example.

EXAMPLE 6–16

A timber beam 4 in. wide by 8 in. deep has an aluminum alloy plate with a net section $\frac{10}{3}$ in. wide by $\frac{1}{2}$ in. deep securely fastened to its bottom face. The working stresses in the timber and aluminum alloy are 1200 and 16,000 psi, respectively. The moduli of elasticity for the timber and aluminum alloy are 1250 and 10,000 ksi, respectively. Determine the maximum allowable resisting moment for the beam.

SOLUTION

The ratio of the moduli of the aluminum and timber is

$$n = 10{,}000/1250 = 8$$

The actual cross section (timber A and aluminum B) and the transformed timber cross section are shown in Figs. 6–43a and b. The neutral axis of the transformed section is located by the principle of moments as

$$\bar{y} = \frac{\Sigma M}{\Sigma A} = \frac{(40/3)(1/4) + 32(9/2)}{(40/3) + 32} = 3.25 \text{ in.}$$

above the bottom of the aluminum plate. The moment of inertia of the cross-sectional area with respect to the neutral axis is

$$I_c = \frac{1}{12}\left(\frac{80}{3}\right)\left(\frac{1}{2}\right)^3 + \frac{40}{3}(3)^2 + \frac{1}{12}(4)(8)^3 + 32\left(\frac{5}{4}\right)^2 = 341 \text{ in.}^4$$

The stress relation

$$\sigma_b = \sigma_a\left(\frac{E_B}{E_A}\right)\left(\frac{b}{a}\right)$$

(a)   (b)

FIG. 6–43

will be used to determine which working stress controls the capacity of the section. From Fig. 6–43 it is seen that the distance to the outermost timber fiber is 5.25 in. and that the distance to the outermost aluminum fiber is 3.25 in. With 1200 psi for σ_a the maximum stress in the aluminum is

$$\sigma_b = 1200(8)(3.25/5.25) = 5940 \text{ psi}$$

which is well below the allowable stress. Therefore, the stress in the timber is the controlling stress. From the flexure formula

$$M = \frac{\sigma I}{c} = \frac{1200(341)}{5.25} = \underline{77,900 \text{ in.-lb}} \qquad \text{Ans.}$$

PROBLEMS

6-144*   A timber beam 4 in. wide by 10 in. deep has a 4-in. wide by 1/2-in. deep steel plate fastened securely to its bottom face. The moduli of elasticity for the timber and steel are 1500 ksi and 30,000 ksi, respectively. Determine the maximum fiber stress in the steel when the maximum fiber stress in the timber is 1500 psi C.

6-145*   A timber beam 120 mm wide by 300 mm deep has a 15-mm deep by 120-mm wide steel plate fastened securely to its top face. The moduli of elasticity for the timber and steel are 10 GPa and 200 GPa, respectively. Determine the maximum fiber stress in the timber when the maximum fiber stress in the steel is 70 MPa T.

6-146*   A 4-in. wide by 6-in. deep timber cantilever beam 7 ft long is reinforced by bolting two 1/2-in. wide by 6-in. deep structural aluminum plates to the sides of the timber beam as shown in Fig. P6–146. The moduli of elasticity of the timber and aluminum are 1000 ksi and 10,000 ksi, respectively. Determine the maximum tensile fiber stress in each of the materials when a static load of 1000 lb is applied to the free end of the beam.

6-147   A composite beam 225 mm wide by 300 mm deep by 5 m long is made by bolting two 100-mm wide by 300-mm deep timber planks to the sides of a 25-mm wide by 300-mm deep steel plate as shown in Fig. P6–147. The moduli of elasticity of the timber and steel are 8 GPa and 200 GPa, respectively. Determine the maximum tensile fiber stress in each of the materials if the composite beam is simply supported and subjected to a uniformly distributed load of 12 kN/m over its entire length.

FIG. P6–146

FIG. P6–147

6-148   A 100-mm wide by 300-mm deep timber beam 5 m long is reinforced with 100-mm wide by 15-mm deep steel plates on the top and bottom faces. The beam is simply supported and carries a uniformly distributed load of 20 kN/m over its entire length. The moduli of elasticity for the timber and steel are 10 GPa and 200 GPa, respectively. Determine the maximum fiber stresses in the timber and in the steel.

6-149   A timber beam 6 in. wide by 12 in. deep has a 6-in. wide by 1/2-in. deep steel plate securely fastened to its bottom face. The beam will be simply supported, will have a span of 16 ft, and will carry a uniformly distributed load over its entire length. The moduli of elasticity for the wood and steel are 1500 ksi and 30,000 ksi, respectively. If the allowable fiber stresses are 1200 psi in the wood and 16,000 psi in the steel, determine the maximum allowable magnitude for the distributed load.

6-150*   A cantilever beam 2 m long carries a concentrated load of 15 kN at the free end. The beam consists of a 75-mm wide by 200-mm deep timber section reinforced with 75-mm wide by 15-mm thick steel plates on the top and bottom faces. The moduli of elasticity for the wood and steel are 10 GPa and 200 GPa, respectively. Determine the maximum fiber stresses in the wood and in the steel.

6-151   A timber beam 10 in. wide by 12 in. deep has a 10-in. wide by 1/2-in. deep steel plate securely fastened to its bottom face. The moduli of elasticity for the wood and steel are 1500 ksi and 30,000 ksi, respectively. If the allowable fiber stresses are 1000 psi in the wood and 10,000 psi in the steel, determine the maximum load P that can be applied at the center of a simply supported beam having a span of 20 ft.

6-17   FLEXURAL STRESSES IN REINFORCED CONCRETE BEAMS

Concrete is widely used in beam construction because it is economical, readily available, fireproof, and exhibits a reasonable compressive strength.

However, concrete has relatively little tensile strength; therefore, concrete beams must be reinforced with another material, usually steel, that can resist the tensile forces.

The transformed section method of Art. 6–16 provides a satisfactory procedure for analyzing reinforced concrete beam problems. The transformed section used for these problems consists of the actual concrete on the compression side of the neutral axis plus the equivalent amount of hypothetical concrete (which is able to develop tensile stresses) on the tension side of the neutral axis required to replace the steel reinforcing rods. The actual concrete on the tension side of the beam is assumed to crack to the neutral surface; therefore, it has no tensile load-carrying ability and is neglected.

The solution for maximum stresses in a given beam or for the maximum bending moment with given allowable stresses consists of three steps.

1. Locate the neutral axis for the transformed section.
2. Determine the moment of inertia of the transformed section with respect to the neutral axis.
3. Use the flexure formula to determine the required stresses or moment.

The procedure is illustrated in the following examples.

EXAMPLE 6–17

A simple reinforced concrete beam carries a uniformly distributed load of 1500 lb per ft on a span of 16 ft. The beam has a rectangular cross section 12 in. wide by 21 in. deep, and 2 sq. in. of steel reinforcing rods are placed with their centers 3 in. from the bottom of the beam. The moduli of elasticity for the concrete and steel are 2500 ksi and 30,000 ksi, respectively. Determine the maximum fiber stress in the concrete and the average fiber stress in the steel.

SOLUTION

The ratio of the moduli of elasticity is

$$n = 30{,}000/2500 = 12$$

which means that 12(2) or 24 sq. in. of hypothetical concrete that resists tension is required for the transformed section. The actual and transformed sections are shown in Figs. 6–44a and b. The principle of moments with respect to the centroidal axis for the transformed section gives

$$12h(h/2) = 24(18 - h)$$

from which

$$h = 6.72 \text{ in.} \quad \text{and} \quad 18 - h = 11.28 \text{ in.}$$

FIG. 6–44

The moment of inertia of the transformed section with respect to the neutral axis is

$$I = 24(11.28)^2 + (12)(6.72)^3/3 = 4270 \text{ in.}^4$$

For a simple beam with a uniformly distributed load, the maximum moment occurs at the center of the span and, from the free-body diagram of Fig. 6–44c, is

$$M = 8w(8) - 8w(4) = +32w \text{ ft-lb}$$

The flexure formula applied to the transformed section gives the maximum stress in the concrete as

$$\sigma = Mh/I = 32(1500)(12)(6.72)/4270 = \underline{907 \text{ psi } C} \qquad \text{Ans.}$$

Since the stresses in the transformed section have a linear distribution, the average stress in the steel is

$$\sigma = \frac{(18 - 6.72)}{6.72}(907)n = \frac{11.28}{6.72}(907)(12) = \underline{18,260 \text{ psi } T} \qquad \text{Ans.}$$

In the design of a reinforced concrete beam to carry a specified moment with a balanced design (a balanced design means that the allowable stresses in the two materials are reached simultaneously), the following four equations can be written in terms of four unknown properties of the cross section and solved simultaneously.

1. The flexure formula for the allowable stress in the concrete.
2. The flexure formula for the allowable stress in the steel.
3. The moment equation for the location of the neutral axis.
4. The equation for the moment of inertia of the cross section with respect to the neutral axis.

The procedure is illustrated in the following example.

EXAMPLE 6–18
Determine the distance from the top of the beam to the center of the steel and the area of steel required for balanced design of a reinforced concrete beam that is to be 10 in. wide and that must resist a bending moment of 600 in.-kips. The working stresses in the concrete and steel are 1000 psi and 18,000 psi, respectively. The moduli of elasticity for the concrete and steel are 2500 ksi and 30,000 ksi, respectively.

SOLUTION
The ratio of the moduli of elasticity is

$$n = 30,000/2500 = 12$$

The actual and transformed sections are shown in Figs. 6–45a and b. The four unknown quantities are d, h, A, and I for the transformed section where A is the area of the steel. The four available equations involving these four unknowns are as follows:

1. The flexure formula for the maximum stress in the concrete is

$$1000 = 600,000h/I$$

2. The flexure formula modified for the average stress in the steel is

$$18,000 = 12(600,000)(d - h)/I$$

3. The principle of moments with respect to the neutral axis gives

$$10h(h/2) = 12A(d - h)$$

4. The moment of inertia with respect to the neutral axis is

$$I = 10h^3/3 + 12A(d - h)^2$$

(a)   (b)   FIG. 6–45

Simultaneous solution of these four equations yields

$$h = \underline{7.44 \text{ in.}} \qquad d = \underline{18.61 \text{ in.}} \qquad \text{and} \qquad A = \underline{2.07 \text{ sq. in.}} \qquad \text{Ans.}$$

The design of a reinforced concrete beam requires a consideration of other factors such as the bond (shearing) stresses between the concrete and reinforcing steel, the diagonal tensile stresses that may be developed, and the amount of concrete that is needed beyond the reinforcing bars. Discussion of such topics can be found in textbooks devoted to reinforced concrete design.

PROBLEMS

6-152*   A simply supported reinforced concrete beam carries a uniformly distributed load of 1100 lb/ft on a span of 16 ft. The beam has a rectangular cross section 10 in. wide by 18 in. deep and 2 sq. in. of steel reinforcing rods are placed 2 in. from the bottom of the beam. The moduli of elasticity of the concrete and steel are 2500 ksi and 30,000 ksi, respectively. Determine the average tensile stress in the steel and the maximum compressive stress in the concrete at the section of maximum moment.

6-153*   A 3.5-m long simply supported reinforced concrete beam has a 200-mm wide by 350-mm deep cross section and 625 mm^2 of steel reinforcing rods are placed 60 mm from the bottom of the beam. The beam supports a uniformly distributed load of 12 kN/m over the entire span. The moduli of elasticity of the concrete and steel are 15 GPa and 200 GPa, respectively. Determine the maximum average tensile stress in the steel and the maximum compressive stress in the concrete at the section of maximum moment.

6-154   A simply supported reinforced concrete beam 10 in. wide with a depth to the center of the steel of 18 in. has a span of 20 ft. The tension reinforcement consists of three 1-in. diameter steel bars. If the working stresses are 1000 psi in the concrete and 18,000 psi in the steel, determine the maximum allowable uniformly distributed load per foot for the beam if the ratio of the moduli of elasticity is 12.

6-155   A 7.5-m long simply supported reinforced concrete beam has a width of 300 mm and a depth to the center of the steel of 450 mm. The tension reinforcement consists of four 25-mm diameter steel bars. If the working stresses are 8.5 MPa in the concrete and 125 MPa in the steel, determine the maximum allowable concentrated load that can be placed on the beam 3 m from the left support. The ratio of the moduli of elasticity is 12.

FIG. P6–158   FIG. P6–159

6-156*   Determine the depth to center of steel and the area of steel required for a reinforced concrete beam of balanced design that is to resist a maximum bending moment of 450 in.-kips. The allowable stresses are 1000 psi in the concrete and 18,000 psi in the steel. The width of the beam is to be 8 in. and the ratio of the moduli of elasticity is 12.

6-157   Determine the depth to center of steel and the area of steel required for a reinforced concrete beam of balanced design that is to resist a maximum bending moment of 70 kN · m. The allowable stresses are 7 MPa in the concrete and 125 MPa in the steel. The width of the beam is to be 250 mm and the ratio of the moduli of elasticity is 12.

6-158*   The effective cross section for a steel reinforced concrete beam is shown in Fig. P6–158. For allowable stresses of 1200 psi in the concrete and 18,000 psi in the steel, determine the maximum permissible bending moment that the section can resist. Use 3000 ksi and 30,000 ksi for the moduli of elasticity of the concrete and steel, respectively.

6-159   The effective cross section for a steel reinforced concrete beam is shown in Fig. P6–159. For allowable stresses of 8 MPa in the concrete and 120 MPa in the steel, determine the maximum permissible bending moment that the section can resist. Use 20 GPa and 200 GPa for the moduli of elasticity of the concrete and steel, respectively.

6-18   FLEXURAL STRESSES IN CURVED BEAMS

One of the assumptions made in the development of the flexural stress theory in Art. 6–2 was that all longitudinal elements of a beam have the same length, thus restricting the theory to initially straight beams of constant cross section. Although considerable deviation from this restriction can be tolerated in real problems, when the initial curvature of the beam becomes significant, the linear variation of strain over the cross section is no longer valid, even though the assumption of the *plane cross section remaining plane is valid*. A theory will now be developed for a beam, subjected to pure bending, having a constant cross section and a constant or slowly varying initial radius of curvature in the plane of bending. The development will be limited to linearly elastic action.

Figure 6–46a is the elevation of part of such a beam. The *xy* plane is the plane of bending and a plane of symmetry. Since a plane section before bending remains plane after bending, the longitudinal deformation of any element will be proportional to the distance of the element from the neutral surface, as indicated in Fig. 6–46a, from which

$$\delta = \frac{\delta_o}{b} y \qquad (a)$$

The unstrained length of any longitudinal element is $\rho\theta$; therefore, Eq. *a* in terms of strain becomes

$$\rho\theta\varepsilon = \frac{r_o\theta\varepsilon_o}{b} y$$

from which the expression for the longitudinal strain distribution becomes

$$\varepsilon = \frac{r_o\varepsilon_o}{b} \frac{y}{\rho} \qquad (b)$$

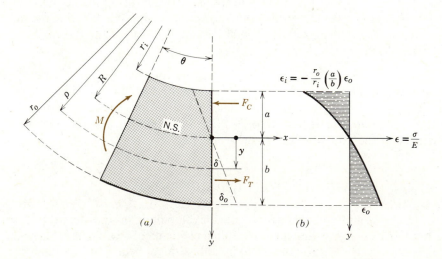

FIG. 6–46

The plot of Eq. b is shown in Fig. 6–46b. Since the action is elastic, Hooke's law applies; therefore, $\varepsilon = \sigma/E$ (for $\sigma_y = \sigma_z = 0$), which when substituted into Eq. b yields

$$\sigma = \frac{r_o\sigma_o}{b}\frac{y}{\rho} \qquad (c)$$

Thus, the curve in Fig. 6–46b indicates the flexural stress distribution as well as the strain distribution.

The location of the neutral surface is obtained from the equation $\Sigma F_x = 0$; thus,

$$\int_A \sigma\, dA = \frac{r_o\sigma_o}{b}\int_A \frac{y}{\rho}\, dA = 0$$

Since $y = \rho - R$, where R is the radius of the neutral surface,

$$\frac{r_o\sigma_o}{b}\int_A \frac{\rho - R}{\rho}\, dA = 0 \qquad (6\text{–}17)$$

In general, Eq. 6–17 should be solved for R for each specific problem; however, the general solution for a *rectangular cross section* of width t is easily obtained as follows:

$$\int_{r_i}^{r_o} \frac{\rho - R}{\rho}\, t\, d\rho = \int_{r_i}^{r_o} t\, d\rho - \int_{r_i}^{r_o} t\,\frac{R\, d\rho}{\rho} = 0$$

Since R is a constant,

$$r_o - r_i - R\ln(r_o/r_i) = 0$$

from which the radius of the neutral surface is

$$R = \frac{r_o - r_i}{\ln(r_o/r_i)} \qquad (d)$$

The resisting moment in terms of flexural stress is obtained from the equilibrium equation $\Sigma M = 0$; thus,

$$M_r = \int_A \sigma\, dA\, y = \frac{r_o\sigma_o}{b}\int_A \frac{y^2 dA}{\rho} = \frac{r_o\sigma_o}{b}\int_A \frac{(\rho - R)^2}{\rho}\, dA \qquad (e)$$

The value of R for a given problem, obtained from Eq. 6–17, can be substituted in Eq. e and a solution for M_r thus obtained. However, it will, in general, be found more convenient to write Eq. e in the following form:

$$M_r = \frac{r_o \sigma_o}{b} \left[\int_A (\rho - R)dA - R \int_A \frac{\rho - R}{\rho} \, dA \right]$$

From Eq. 6–17, the second integral in the brackets is zero, and when $\rho - R$ is replaced by y in the first integral, the resisting moment is given by the expression

$$M_r = \frac{r_o \sigma_o}{b} \int_A y \, dA = \frac{r_o \sigma_o}{b} A\bar{y} \qquad (f)$$

where $\bar{y}$ is the y coordinate of the centroid of the cross-sectional area A measured from the neutral axis N. A. Since r_o, b, σ_o, and A are all positive for a positive moment (*defined as a moment producing tension at the boundary farthest from the center of curvature*), $\bar{y}$ must always be positive, indicating that the neutral axis of the cross section is always displaced from the centroid toward the center of curvature. Replacing $(r_o\sigma_o)/b$ in Eq. f by $(\rho\sigma)/y$ from Eq. c and solving for σ gives

$$\sigma = \frac{My}{\rho A\bar{y}} = \frac{My}{(R + y)A\bar{y}} \qquad (6\text{--}18)$$

which is the expression for the elastic flexural stress at any point in an initially curved beam.

    The preceding development is for pure bending and neglects radial compressive stresses that occur within the material. These compressive stresses are usually very small. If the beam is loaded with forces (instead of couples), additional stresses will occur on the radial planes. Because the action is elastic, the principle of superposition applies and the additional normal stresses can be added to the flexural stresses obtained from Eq. 6–18. Such problems are covered in Art. 9–4.

    The application of Eqs. 6–17 and 6–18 is illustrated in the following examples.

FIG. 6–47

EXAMPLE 6–19

The segment of a curved beam (see Fig. 6–47) of high-strength steel (proportional limit = 95 ksi) having a trapezoidal cross section is subjected to a moment M of −100 in.-kips.

(a) Determine the flexural stresses at the top and bottom surfaces.
(b) Sketch the flexural stress distribution in the beam.
(c) Determine the percentages of error if the flexure formula for a straight beam (Eq. 6–3) were used for part (a).

SOLUTION

The first step will be to locate the neutral surface by setting the integral of Eq. 6–17 equal to zero; thus,

$$\int_A \frac{\rho - R}{\rho} \, dA = \int_{r_i}^{r_o} \frac{\rho - R}{\rho} (2z) d\rho = 0$$

and from the cross section in Fig. 6–47,

$$2z = 2 - \frac{2 - 1}{3} (\rho - r_i) = 2 - \frac{1}{3} (\rho - 3) = 3 - \frac{\rho}{3}$$

which when substituted into the integral above yields

$$\int_3^6 \frac{\rho - R}{\rho} \left(3 - \frac{\rho}{3}\right) d\rho = 0$$

Integrating and substituting limits gives

$$3(6 - 3) - 3R \ln(6/3) - (36 - 9)/6 + (R/3)(6 - 3) = 0$$

from which

$$R = 4.1688 \text{ in.}$$

In order to apply Eq. 6–18, it is necessary to locate the centroid of the cross section, which is found to be 4/3 in. below the top edge. Therefore,

$$\bar{y} = 1.3333 - (4.1688 - 3) = 0.1645 \text{ in.}$$

(a) The required stresses from Eq. 6–18 are

$$\sigma_1 = \frac{-100(6 - 4.1688)}{6(4.5)(0.1645)} = \underline{41.2 \text{ ksi } C \text{ at bottom}}$$

$$\sigma_2 = \frac{-100(3 - 4.1688)}{3(4.5)(0.1645)} = \underline{52.6 \text{ ksi } T \text{ at top}} \qquad \text{Ans.}$$

FIG. 6–48

These stresses are well below the proportional limit of the material.

(b) Plotting the two stresses from part (a) and zero stress at the neutral surface will indicate that the curve must be shaped as in Fig. 6–48.

(c) The cross-sectional moment of inertia with respect to the centroidal axis parallel to the neutral axis is found to be 13/4 in.^{4} and Eq. 6–3 yields

$$\sigma_1 = \frac{My}{I} = \frac{-100(5/3)}{(13/4)} = 51.3 \text{ ksi } C \text{ at bottom}$$

$$\sigma_2 = \frac{-100(-4/3)}{(13/4)} = 41.0 \text{ ksi } T \text{ at top}$$

Therefore, the errors are

$$\frac{51.3 - 41.2}{41.2} (100) = \underline{24.5\% \text{ high at bottom}}$$

$$\frac{52.6 - 41.0}{52.6} (100) = \underline{22.0\% \text{ low at top}} \qquad \text{Ans.}$$

EXAMPLE 6–20
The elevation and cross section of a segment of a gray cast-iron (proportional limit = 100 MPa T) punch press frame are shown in Fig. 6–49. Determine the maximum flexural tensile and compressive stresses due to a moment M of -40 kN · m.

SOLUTION
Assume that the neutral axis of the cross section is to the right of the flange and apply Eq. 6–17; thus,

FIG. 6–49

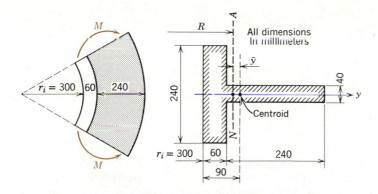

$$\int_A \frac{\rho - R}{\rho}\, dA = \int_{300}^{360} \frac{\rho - R}{\rho}\,(240)dp + \int_{360}^{600} \frac{\rho - R}{\rho}\,(40)d\rho = 0$$

Integrating and substituting limits gives

$$240(360 - 300) - 240R \ln(360/300) + 40(600 - 360) - 40R \ln(600/360) = 0$$

from which

$$R = 373.89 \text{ mm}$$

Therefore, the assumption about the location of the neutral axis is correct. The centroid is 90 mm from the left edge of the section, and the distance from the neutral axis to the centroid is

$$\bar{y} = 90 - 73.89 = 16.11 \text{ mm}$$

From Eq. 6–18 the stresses are

$$\sigma = \frac{My}{\rho A \bar{y}} = \frac{-40(10^3)(0.22611)}{0.600(24000)(10^{-6})(0.01611)}$$

$$= -39.0(10^6) \text{ N/m}^2 = \underline{39.0 \text{ MPa } C \text{ at the outside of stem}} \quad \textbf{Ans.}$$

and

$$\sigma = \frac{-40(10^3)(-0.07389)}{0.300(24000)(10^{-6})(0.01611)} = 25.5(10^6) \text{ N/m}^2$$

$$= \underline{25.5 \text{ MPa } T \text{ at inside of flange}} \qquad\qquad \textbf{Ans.}$$

The low stresses indicate elastic action.

PROBLEMS

> *Note: The action is elastic in the following problems, and moments are applied in a plane of symmetry.*

6-160   A curved rectangular beam with a width t, a depth d (in the radial direction), and an inside radius r_i of $10d$ is subjected to a moment M in the plane of curvature. Show that the error in computing the maximum flexural stress by Eq. 6–3 (the flexure formula) instead of Eq. 6–18 (the curved beam formula) is approximately 3.7 percent low.

6-161*   A curved beam having an inside radius r_i of 500 mm has a rectangular cross section 80 mm wide by 160 mm deep (in the radial direction). The beam is subjected to a moment M of +40 kN · m in the plane of curvature. Determine the flexural stresses at the inside and outside curved surfaces of the beam.

6-162*   A curved beam having an inside radius of 6 in. has the cross section shown in Fig. P6–162. The beam is subjected to a moment M in the plane of curvature. Determine the dimension b needed to make the flexural stress at the inside curved surface equal in magnitude to the flexural stress at the outside curved surface of the beam.

6-163   Determine the maximum tensile and compressive flexural stresses in the curved beam of Fig. P6–163 due to a moment M of +25 kN · m.

6-164*   Determine the maximum tensile and compressive flexural stresses in the curved beam of Fig. P6–164 due to a moment M of −35 ft-kips.

6-165   Solve Example 6–20 with the depth of the stem of the T cross section changed from 240 mm to 180 mm and the moment changed from −40 kN · m to +45 kN · m. All other given data remain the same except the 90-mm distance to the centroidal axis, which now must be determined for the new section.

6-166*   Determine the maximum tensile and compressive flexural stresses in the curved beam of Fig. P6–166 due to a moment M of −500 in.-kips.

6-167*   The curved beam of Fig. P6–167 is subjected to a positive bending moment M in the plane of curvature. Determine the maximum permissible magnitude for M if the flexural stress is not to exceed 35 MPa T or 140 MPa C.

FIG. P6–162

All dimensions in millimeters

Cross section

FIG. P6–163

FIG. P6–164

FIG. P6–166

6-168 The cross section of a segment of a crane hook is shown in Fig. P6–168. The curvature and loading are in the xy plane. Determine the flexural stresses at the inside curved surface and at the outside curved surface due to a moment M of -80 in.-kips.

FIG. P6–167

FIG. P6–168

CHAPTER 7

FLEXURAL LOADING: DEFLECTIONS

7-1 INTRODUCTION

The important relations between applied load and stress (fiber and shear) developed in a beam were presented in Chapter 6. However, a beam design is frequently not complete until the amount of deflection has been determined for the specified load. Failure to control beam deflections within proper limits in building construction is frequently reflected by the development of cracks in plastered walls and ceilings. Beams in many machines must deflect just the right amount for gears or other parts to make proper contact. In innumerable instances the requirements for a beam involve a given load-carrying capacity with a specified maximum deflection.

The deflection of a beam depends on the stiffness of the material and the dimensions of the beam as well as on the applied loads and supports. Four common methods for calculating beam deflections due to fiber stresses are presented here: (1) the double integration method, (2) the area-moment method, (3) the superposition method, and (4) an energy method. The material is presented so that methods 1 and 2 can be studied independently of each other. Deflections due to shearing stresses are also discussed.

FIG. 7–1

7-2   THE RADIUS OF CURVATURE

When a straight beam is loaded and the action is elastic, *the longitudinal centroidal axis of the beam becomes a curve defined as the elastic curve.* In regions of constant bending moment, the elastic curve is an arc of a circle of radius ρ as indicated in Fig. 7–1, in which the portion AB of a beam is bent only with couples. Therefore, the plane sections A and B remain plane (see Art. 6–2), and the deformation (elongation and compression) of the fibers is proportional to the distance from the neutral surface, which is unchanged in length. From Fig. 7–1,

$$\theta = \frac{L}{\rho} = \frac{L + \delta}{\rho + c}$$

from which

$$\frac{c}{\rho} = \frac{\delta}{L} = \varepsilon_x = \frac{\sigma_x}{E} = \frac{Mc}{EI} \qquad (7\text{–}1)$$

Therefore,

$$\frac{1}{\rho} = \frac{M}{EI} \qquad (7\text{–}2)$$

which relates the radius of curvature of the neutral surface of the beam to the bending moment M, the stiffness of the material E, and the moment of inertia of the cross-sectional area I.

PROBLEMS

7-1*   For the W6 × 16 steel ($E = 30{,}000$ ksi) beam shown in Fig. P7–1, determine

(a)   The radius of curvature at any point between the supports.

FIG. P7–1   **FIG. P7–2**

(b)   The deflection at a section midway between the supports.

(c)   The maximum fiber stress in the beam.

7-2*   For the WT203 × 74 steel (E = 200 GPa) beam shown in Fig. P7–2, determine

(a)   The radius of curvature at any point between the supports.

(b)   The deflection at a section midway between the supports.

(c)   The maximum fiber stress in the beam.

7-3   The cantilever beam shown in Fig. P7–3a is fabricated from two 1 by 3-in. steel (E = 30,000 ksi) bars as shown in Fig. P7–3b. Determine

(a)   The maximum fiber stress in the beam.

(b)   The radius of curvature of the beam.

(c)   The deflection at the right end of the beam.

7-4   The cantilever beam shown in Fig. P7–4a is fabricated from three 30 by 120-mm aluminum (E = 70 GPa) bars as shown in Fig. P7–4b. Determine .

(a)   The maximum fiber stress in the beam.

(b)   The radius of curvature of the beam.

(c)   The deflection at the left end of the beam.

7-5*   The boards for a concrete form are to be bent to a circular curve of 10-ft radius. What maximum thickness can be used if the stress is not to exceed 2000 psi? The wood is saturated, and Young's modulus may be taken as 800,000 psi.

(a)   (b)   **FIG. P7–3**

FIG. P7–4

(a)   *(b)*

7-6* A high-strength steel (E = 200 GPa) tube having an outside diameter of 80 mm and a wall thickness of 3 mm is bent into a circular curve of 24-m radius. Determine the maximum fiber stress in the tube.

7-7 A 1-in. thick board is simply supported at the ends of a span of 10 ft. Loads are applied such that the elastic curve between supports is an arc of a circle. The deflection at the midpoint of the span is 4 in. Determine the maximum fiber stress in the wood. Use 800,000 psi for the modulus of elasticity of the wood.

7-8 Determine the minimum diameter required for a bandsaw pulley if the maximum fiber stress due to bending in a 0.75-mm thick steel (E = 200 GPa) bandsaw blade is not to exceed 200 MPa.

7-3 THE DIFFERENTIAL EQUATION OF THE ELASTIC CURVE

The expression for the curvature of the elastic curve derived in Art. 7–2 is useful only when the bending moment is constant for the interval of the beam involved. For most beams the bending moment is a function of the position along the beam and a more general expression is required.

The curvature from the calculus (see any standard calculus textbook) is

$$\frac{1}{\rho} = \frac{d^2y/dx^2}{[1 + (dy/dx)^2]^{3/2}}$$

For actual beams the slope dy/dx is very small, and its square can be neglected in comparison to unity. With this approximation,

$$\frac{1}{\rho} = \frac{d^2y}{dx^2}$$

and Eq. 7–2 becomes

$$EI\frac{d^2y}{dx^2} = M_x \tag{7–3}$$

which is the differential equation for the elastic curve of a beam where the

FIG. 7–2

subscript x on M is used as a reminder that M is a function of x.

The differential equation of the elastic curve can also be obtained from the geometry of the bent beam as shown in Fig. 7–2, where it is evident that $dy/dx = \tan \theta = \theta$ (approximately) for small angles and that $d^2y/dx^2 = d\theta/dx$. Again from Fig. 7–2,

$$d\theta = \frac{dL}{\rho} = \frac{dx}{\rho}$$

(approximately) for small angles. Therefore,

$$\frac{d^2y}{dx^2} = \frac{d\theta}{dx} = \frac{1}{\rho} = \frac{M_x}{EI}$$

or

$$EI \frac{d^2y}{dx^2} = M_x \qquad (7\text{–}3)$$

The sign convention for bending moments established in Art. 6–6 will be used for Eq. 7–3. Both E and I are always positive; therefore, the signs of the bending moment and the second derivative must be consistent. With the coordinate axes selected as shown in Fig. 7–3, the slope changes from positive to negative in the interval from A to B; therefore, the second derivative is negative, which agrees with the sign convention of Art. 6–6. For the interval BC, both d^2y/dx^2 and M_x are seen to be positive.

FIG. 7–3

Careful study of Fig. 7–3 reveals that the signs of the bending moment and the second derivative are also consistent when the origin is selected at the right with x positive to the left and y positive upward. Unfortunately, the signs are inconsistent when y is positive downward. *In this book, for horizontal beams, y will always be selected positive upward for beam deflection problems.*

Before proceeding with the solution of Eq. 7–3, it is desirable to correlate the successive derivatives of the deflection y of the elastic curve with the physical quantities that they represent in beam action. They are

$$\text{deflection} = y$$

$$\text{slope} = \frac{dy}{dx}$$

$$\text{moment} = EI \frac{d^2y}{dx^2} \text{ (from Eq. 7–3)}$$

$$\text{shear} = \frac{dM}{dx} \text{ (from Eq. 6–7)} = EI \frac{d^3y}{dx^3} \text{ (for } EI \text{ constant)}$$

$$\text{load} = \frac{dV}{dx} \text{ (from Eq. 6–5)} = EI \frac{d^4y}{dx^4} \text{ (for } EI \text{ constant)}$$

where the signs are as given in Arts. 6–6 and 6–7.

In Art. 6–8 a method based on these differential relations was presented for starting from the load diagram and drawing first the shear diagram and then the moment diagram. This method can readily be extended to the construction of the slope diagram from the relation

$$M = EI \frac{d\theta}{dx}$$

from which

$$\int_{\theta_A}^{\theta_B} d\theta = \int_{x_A}^{x_B} \frac{M}{EI} \, dx$$

This relation shows that except for a factor EI, the area under the moment diagram between any two points along the beam gives the change in slope between the same two points. Likewise, the area under the slope diagram between two points along a beam gives the change in deflection between these points. These relations have been used to construct the complete series of diagrams shown in Fig. 7–4 for a simply supported beam with a concentrated load at the center of the span. The geometry of the beam was used to locate the points of zero slope and deflection, required as starting

FIG. 7–4

points for the construction. More commonly used methods for calculating beam deflections will be developed in succeeding articles.

Before proceeding with specific methods for calculating beam deflections, it is advisable to consider the assumptions used in the development of the basic relation Eq. 7–3. All of the limitations that apply to the flexure formula apply to the calculation of deflections because the flexure formula was used in the derivation of Eq. 7–3. It is further assumed that

1.   The square of the slope of the beam is negligible compared to unity.
2.   The beam deflection due to shearing stresses is negligible (a plane section is assumed to remain plane). See Art. 7–12 for further information.
3.   The values of E and I remain constant for any interval along the beam. In case either of them varies and can be expressed as a function of the distance x along the beam, a solution of Eq. 7–3 that takes this variation into account may be possible.

7-4   DEFLECTIONS BY INTEGRATION

Whenever the assumptions of the previous paragraph are essentially correct and the bending moment can be readily expressed as an integrable function of x, Eq. 7–3 can be solved for the deflection y of the elastic curve of a beam at any point x along the beam. The constants of integration can be evaluated from the applicable boundary or matching conditions.

A boundary condition is defined as a known set of values for x and y, or x and dy/dx, at a specific location along the beam. One boundary condition (one set of values) can be used to determine one and only one constant of integration.

Many beams are subjected to abrupt changes in loading along the beam, such as concentrated loads, reactions, or even distinct changes in the amount of a uniformly distributed load. Since the expressions for the bending moment on the left and right of any abrupt change in load are different functions of x, it is impossible to write a single equation for the bending moment in terms of ordinary algebraic functions that is valid for the entire length of the beam. This can be resolved by writing separate bending moment equations for each interval of the beam. Although the intervals are bounded by abrupt changes in load, the beam is continuous at such locations and, consequently, the slope and the deflection at the junction of adjacent intervals must match. *A matching condition is defined as the equality of slope or deflection, as determined at the junction of two intervals from the elastic curve equations for both intervals.* One matching condition (for example, at x equals $L/3$, y from the left equation equals y from the right equation) can be used to determine one and only one constant of integration.

The procedure for obtaining beam deflections when matching conditions are required is lengthy and tedious. A method is presented in the next article in which *singularity functions* are used to write a single equation for the bending moment that is valid for the entire length of the beam; this eliminates the need for matching conditions and, accordingly, reduces the labor involved.

Calculating the deflection of a beam by the double integration method involves four definite steps, and the following sequence for these steps is strongly recommended.

1. Select the interval or intervals of the beam to be used; next, place a set of coordinate axes on the beam *with the origin at one end of an interval* and then indicate the range of values of x in each interval. For example, two adjacent intervals might be

$$0 \le x \le L \quad \text{and} \quad L \le x \le 3L$$

2. List the available boundary and matching (where two or more adjacent intervals arc used) conditions for each interval selected. Remember that two conditions are required to evaluate the two constants of integration for each interval used.

FIG. 7–5

(a)   (b)

(c)   (d)

3. Express the bending moment as a function of x for each interval selected and equate it to $EI\,d^2y/dx^2$.

4. Solve the differential equation or equations from item 3 and evaluate all constants of integration. Check the resulting equations for dimensional homogeneity. Calculate the deflection at specific points when required.

A roller or pin at any point in a beam (see Figs. 7–5a and b) represents a simple support at which the beam cannot deflect (unless otherwise stated in the problem) but can rotate. At a fixed end, as represented by Figs. 7–5c and d, the beam can neither deflect nor rotate unless otherwise stated.

The following examples illustrate the use of the double integration method for calculating beam deflections.

EXAMPLE 7–1

For the beam loaded and supported as shown in Fig. 7–6a, locate the point of maximum deflection between the supports.

FIG. 7–6

SOLUTION

From a free-body diagram of the beam and the equations of equilibrium, the left reaction is found to be $7wL/12$ upward. As indicated in Fig. 7–6b, the origin of coordinates is selected at the left support, and the interval to be used is $0 \le x \le L$.

The two required boundary conditions are $y = 0$ when $x = 0$ and $y = 0$ when $x = L$. From the free-body diagram of a portion of the beam as shown in Fig. 7–6b, the expression for the bending moment at any point in the selected interval is seen to be

$$EI\,\frac{d^2y}{dx^2} = M_x = \frac{7wL}{12}\,x - \frac{wL^2}{12} - wx\left(\frac{x}{2}\right)$$

Successive integration gives

$$EI\,\frac{dy}{dx} = \frac{7wL}{24}\,x^2 - \frac{wL^2}{12}\,x - \frac{w}{6}\,x^3 + C_1$$

and

$$EIy = \frac{7wL}{72}\,x^3 - \frac{wL^2}{24}\,x^2 - \frac{w}{24}\,x^4 + C_1 x + C_2$$

Substitution of the boundary condition $y = 0$ when $x = 0$ yields $C_2 = 0$. Substitution of the remaining boundary condition $y = 0$ when $x = L$ gives $C_1 = -wL^3/72$, and therefore the elastic curve equation is

$$EIy = \frac{7wL}{72}\,x^3 - \frac{wL^2}{24}\,x^2 - \frac{w}{24}\,x^4 - \frac{wL^3}{72}\,x$$

The maximum deflection occurs where the slope dy/dx is zero or

$$0 = \frac{7wL}{24}\,x^2 - \frac{wL^2}{12}\,x - \frac{w}{6}\,x^3 - \frac{wL^3}{72}$$

The solution of this cubic equation in x gives the point of maximum deflection at

$$x = \underline{0.541\ L \text{ to the right of the left support}} \qquad \textbf{Ans.}$$

Although not requested here, the amount of the maximum deflection can readily be obtained by substituting $0.541L$ for x in the elastic curve equation. The result is

$$y = \frac{-7.88wL^4}{10^3 EI} = \frac{7.88wL^4}{10^3 EI} \text{ downward}$$

EXAMPLE 7–2

For the beam loaded and supported as shown in Fig. 7–7a, determine the deflection of the right end.

SOLUTION

From the free-body diagram of Fig. 7–7b the equations of equilibrium give the shear at the wall as $wL/3$ upward and the moment at the wall as $5wL^2/18$ counterclockwise. With the sign conventions established in Art. 6–6, these results are

$$V_W = + \frac{wL}{3} \quad \text{and} \quad M_W = - \frac{5wL^2}{18}$$

In Fig. 7–7b there are two intervals to be considered, namely, the loaded and the unloaded portions of the beam. The loaded portion of the beam must be used because it contains the point where the deflection is required. However, a quick check reveals the absence of boundary conditions in this interval. It therefore becomes necessary to use both intervals as well as matching and boundary conditions; hence, the origin of coordinates is selected, as shown. For clarity, the two intervals are written side by side with available boundary conditions, matching conditions, and equations listed under the appropriate interval. The intervals are

FIG. 7–7

$$0 \leq x \leq 2L/3 \qquad\qquad 2L/3 \leq x \leq L$$

The available boundary conditions are

$$\frac{dy}{dx} = 0 \quad \text{when} \quad x = 0$$

and

$$y = 0 \quad \text{when} \quad x = 0$$

The available matching conditions are

$\dfrac{dy}{dx}$ from the left equation $= \dfrac{dy}{dx}$ from the right equation

$$\text{when} \quad x = 2L/3$$

and

y from the left equation $= y$ from the right equation

$$\text{when} \quad x = 2L/3$$

Four conditions (two boundary and two matching) are sufficient for the evaluation of the four constants of integration (two in each of the two elastic curve differential equations); therefore, the problem can be solved in this manner.

From the free-body diagram of an unloaded portion of the beam in Fig. 7–7c, the expression for the bending moment at any point in the left interval is seen to be

$$EI \frac{d^2y}{dx^2} = M_x = \frac{wL}{3} x - \frac{5wL^2}{18}$$

From the free-body diagram of Fig. 7–7d where the beam is cut in the loaded interval, the bending moment at any point in the right interval is found to be

$$EI \frac{d^2y}{dx^2} = M_x = \frac{wL}{3} x - \frac{5wL^2}{18}$$
$$- w \left(x - \frac{2L}{3} \right) \left(\frac{x - 2L/3}{2} \right)$$

Integration of each differential equation gives

$$EI\frac{dy}{dx} = \frac{wL}{6}x^2 - \frac{5wL^2}{18}x + C_1 \qquad\qquad EI\frac{dy}{dx} = \frac{wL}{6}x^2 - \frac{5wL^2}{18}x$$

$$- \frac{w}{6}\left(x - \frac{2L}{3}\right)^3 + C_3$$

Substitution of the boundary condition

$$\frac{dy}{dx} = 0 \quad \text{when } x = 0$$

gives

$$C_1 = 0$$

Substitution of the matching condition

$$\frac{dy}{dx} \text{ from the left equation} \qquad = \frac{dy}{dx} \text{ from the right equation}$$

$$\text{when} \quad x = \frac{2L}{3}$$

gives

$$C_1 = C_3 = 0$$

Integration of the resulting differential equations gives

$$EIy = \frac{wL}{18}x^3 - \frac{5wL^2}{36}x^2 + C_2 \qquad\qquad EIy = \frac{wL}{18}x^3 - \frac{5wL^2}{36}x^2$$

$$- \frac{w}{24}\left(x - \frac{2L}{3}\right)^4 + C_4$$

Use of the remaining boundary condition yields $C_2 = 0$. Use of the final matching condition yields $C_2 = C_4 = 0$.

The deflection of the right end of the beam can now be obtained from the elastic curve equation for the right interval by replacing x by its value L. The result is

FIG. P7–9

FIG. P7–9

FIG. P7–10

$$EIy = \frac{wL}{18} L^3 - \frac{5wL^2}{36} L^2 - \frac{w}{24}\left(L - \frac{2L}{3}\right)^4$$

from which

$$y = -\frac{163}{1944}\frac{wL^4}{EI} = \frac{163}{1944}\frac{wL^4}{EI} \quad \text{downward} \qquad\qquad \textbf{Ans.}$$

PROBLEMS

7-9*   A beam is loaded and supported as shown in Fig. P7–9.

   (a)   Derive the equation of the elastic curve in terms of $E, I, w,$ and L. Use the designated axes.

   (b)   Determine the deflection at the left end of the beam.

7-10*   A beam is loaded and supported as shown in Fig. P7–10.

   (a)   Derive the equation of the elastic curve in terms of $E, I, w,$ and L for the interval of the beam between the supports. Use the designated axes.

   (b)   Determine the deflection midway between the supports.

FIG. P7–11

FIG. P7–12

FIG. P7–13   FIG. P7–14

7-11* A beam is loaded and supported as shown in Fig. P7–11.

(a) Derive the equation of the elastic curve in terms of E, I, w, and L for the interval of the beam between the supports. Use the designated axes.

(b) Determine the deflection midway between the supports.

7-12* A beam is loaded and supported as shown in Fig. P7–12.

(a) Derive the equation of the elastic curve in terms of E, I, w, and L for the interval BC of the beam. Use the designated axes.

(b) Determine the deflection at the midpoint of interval BC.

7-13 A beam is loaded and supported as shown in Fig. P7–13.

(a) Derive the equation of the elastic curve in terms of E, I, w, and L for the interval of the beam between the supports. Use the designated axes.

(b) Determine the maximum deflection between the supports.

7-14 A beam is loaded and supported as shown in Fig. P7–14. The bending moment at the second support is $-wL^2/12$ and the vertical shear just to the right of the second support is $+5wL/12$. Determine

(a) The equation of the elastic curve in terms of E, I, w, and L for the portion of the beam between the second and third supports. Use the designated axes.

(b) The maximum deflection between the second and third supports.

7-15* A beam is loaded and supported as shown in Fig. P7–15. Determine

(a) The equation of the elastic curve in terms of E, I, w, and L. Use the designated axes.

(b) The deflection of the free end of the beam.

FIG. P7–15

$$x = L$$
$$y = 0$$
$$y' = 0$$

FIG. P7–16

$$x = 0 \qquad x = 0$$
$$y = 0 \qquad y' = 0$$

7-16* A beam is loaded and supported as shown in Fig. P7–16. Determine

    (a)   The equation of the elastic curve in terms of E, I, w, and L. Use the designated axes.

    (b)   The deflection of the free end of the beam.

7-17 Determine the maximum deflection, in terms of E, I, w, and L, for the beam shown in Fig. P7–17.

7-18 Determine the deflection, in terms of E, I, w, and L, at the free end of the cantilever beam shown in Fig. P7–18.

7-19* A timber beam 6 in. wide by 12 in. deep is loaded and supported as shown in Fig. P7–19. The modulus of elasticity of the timber is 1200 ksi. A pointer is attached to the right end of the beam. Determine

    (a)   The deflection of the right end of the pointer.

    (b)   The maximum deflection of the beam.

7-20* A 100 × 300-mm timber having a modulus of elasticity of 8 GPa is loaded and supported as shown in Fig. P7-20. Determine

    (a)   The deflection at the 7-kN load.

    (b)   The deflection at the free end of the beam.

FIG. P7–17

FIG. P7–18

$$x = 0$$
$$y = 0$$
$$y' = 0$$

FIG. P7–19   FIG. P7–20

7-21   A cantilever beam of length L ft is fixed at the left end and carries a uniformly distributed load of w lb per ft. The right end is subjected to a moment of $+3wL^2/8$. Determine the maximum deflection in the beam if $I = 6$ in.^{4}, $E = 30(10^6)$ psi, $L = 10$ ft, and $w = 100$ lb/ft.

7-22   A W356 × 45 steel beam 12 m long is simply supported at its ends and carries a concentrated load P at midspan. Determine the value of P that will produce deflections of 20 mm at points 3 m from the supports. The web of the cross section is vertical.

7-23*   Determine the maximum deflection, in terms of M, L, E, and I, for the beam shown in Fig. P7–23.

7-24   A beam is loaded and supported as shown in Fig. P7–24. Determine in terms of P, L, E, and I

(a)   The deflection at the center of the span.

(b)   The maximum deflection in the beam.

7-25   A beam is loaded and supported as shown in Fig. P7–25. Determine the maximum deflection in the beam in terms of P, L, E, and I.

7-26   A beam is loaded and supported as shown in Fig. P7–26. Determine in terms of P, L, E, and I

FIG. P7–23

FIG. P7–24

FIG. P7–25

FIG. P7–26

(a)   The maximum deflection between the supports.

(b)   The deflection at the right end of the beam.

7-27*   A beam is loaded and supported as shown in Fig. P7–27. Determine the deflection at the center of the beam in terms of w, L, E, and I.

7-28*   Determine the deflection at the left end of the beam shown in Fig. P7–28 when $E = 70$ GPa and $I = 42(10^6)$ mm^4.

7-29   A beam is loaded and supported as shown in Fig. P7–29. Determine

(a)   The equation of the elastic curve, in terms of E, I, w, and a, for the loaded portion of the beam. Use the designated axes.

(b)   The maximum deflection when $I = 100$ in.^{4}, $E = 12(10^6)$ psi, $w = 1250$ lb/ft, and $a = 3$ ft.

FIG. P7–27

FIG. P7–28

FIG. P7–29

FIG. P7–30

7-30 A beam is loaded and supported as shown in Fig. P7–30. Determine

  (a) The equation of the elastic curve, in terms of E, I, w, and L, for the loaded portion of the beam. Use the designated axes.

  (b) The maximum deflection when $I = 50(10^6)$ mm^4, $E = 80$ GPa, $w = 20$ kN/m, and $L = 1$ m.

7-5   SINGULARITY FUNCTIONS FOR BEAM DEFLECTIONS

The double integration method of Art. 7–4 becomes extremely tedious and time-consuming when several intervals and several sets of matching conditions are required. The labor involved in solving problems of this type can be diminished by making use of *singularity functions* following the method originally developed in 1862 by the German mathematician A. Clebsch (1833–1872).[1]

Singularity functions are closely related to the unit step function used by the British physicist O. Heaviside (1850–1925) to analyze the transient response of electrical circuits. Singularity functions will be used here for writing one bending moment equation that applies in all intervals along a beam, thus eliminating the need for matching conditions.

A singularity function of x is written as $\langle x - x_0 \rangle^n$, where n is any *integer* (positive or negative) including zero, and x_0 is a constant equal to the value of x at the initial boundary of a specific interval along a beam. Selected properties of singularity functions required for beam-deflection problems are listed here for emphasis and ready reference.

$$\langle x - x_0 \rangle^n = \begin{cases} (x - x_0)^n & \text{when } n > 0 \text{ and } x \geq x_0 \\ 0 & \text{when } n > 0 \text{ and } x < x_0 \end{cases}$$

$$\langle x - x_0 \rangle^0 = \begin{cases} 1 & \text{when } x \geq x_0 \\ 0 & \text{when } x < x_0 \end{cases}$$

$$\int \langle x - x_0 \rangle^n \, dx = \frac{1}{n + 1} \langle x - x_0 \rangle^{n+1} + C \qquad \text{when } n \geq 0$$

$$\frac{d}{dx} \langle x - x_0 \rangle^n = n \langle x - x_0 \rangle^{n-1} \qquad \text{when } n \geq 1$$

Several examples of singularity functions are shown in Fig. 7–8.

[1] For a rather complete history on the Clebsch method and the numerous extensions thereof, see "Clebsch's Method for Beam Deflections," Walter D. Pilkey, *Journal of Engineering Education,* Jan. 1964, p. 170.

FIG. 7–8

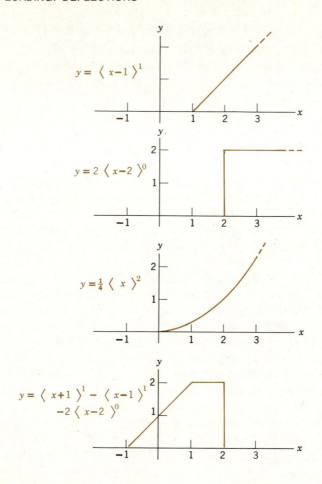

By making use of these properties of singularity functions, one is able to write a single equation for the bending moment for a beam and obtain the correct value of the moment in any interval along the beam. To amplify this statement, consider the beam of Fig. 7–9. By the methods of Arts. 6–8 and 7–4, the moment equations at the four designated sections are:

FIG. 7–9

$$M_1 = R_L x \qquad\qquad\qquad\qquad\qquad\qquad 0 < x < x_1$$

$$M_2 = R_L x - P(x - x_1) \qquad\qquad\qquad\qquad x_1 < x < x_2$$

$$M_3 = R_L x - P(x - x_1) + M_A \qquad\qquad\qquad x_2 < x < x_3$$

$$M_4 = R_L x - P(x - x_1) + M_A - (w/2)(x - x_3)^2 \qquad x_3 < x < L$$

These four moment equations can be combined into a single equation by means of singularity functions to give

$$M_x = R_L x - P\langle x - x_1 \rangle^1 + M_A \langle x - x_2 \rangle^0 - \frac{w}{2} \langle x - x_3 \rangle^2 \qquad 0 \le x \le L$$

Distributed beam loadings that are sectionally continuous (the distributed load cannot be represented by a single function of x for all values of x) require special consideration. Moment equations for such loads can be readily obtained by superposition, as illustrated in the following example. Further applications of singularity functions to the determination of beam deflections are given in Exams. 7–4 and 7–5.

EXAMPLE 7–3

Use singularity functions to write a single equation for the bending moment at any section of the beam of Fig. 7–10a.

SOLUTION

The loading on the beam of Fig. 7–10a can be considered a combination of the loadings shown in Fig. 7–10b, c, d, and e, where downward-acting loads are shown on top of the beam and upward-acting loads on the bottom. The moment of the linearly varying load can be obtained from the geometry of the load diagram. The moment equation is

$$M_x = R_L x - \frac{w_1}{6(x_2 - x_1)} \langle x - x_1 \rangle^3 + \frac{w_1}{6(x_2 - x_1)} \langle x - x_2 \rangle^3$$

$$+ \frac{w_1 - w_2}{2} \langle x - x_2 \rangle^2 + \frac{w_2}{2} \langle x - x_3 \rangle^2 \qquad\qquad \text{Ans.}$$

EXAMPLE 7–4

Solve Exam. 7–2 (Art. 7–4) by use of singularity functions.

SOLUTION

The boundary conditions, in the interval $0 \le x \le L$, are $dy/dx = 0$ when $x = 0$ and $y = 0$ when $x = 0$ with the axes as shown in Fig. 7–7b. From the free-body diagram of Fig. 7–7d, the expression for the bending moment is

FIG. 7–10

FIG. 7–11

$$EI \frac{d^2y}{dx^2} = -\frac{5wL^2}{18} + \frac{wLx}{3} - \frac{w}{2} \left\langle x - \frac{2L}{3} \right\rangle^2$$

The first integration gives

$$EI \frac{dy}{dx} = -\frac{5wL^2x}{18} + \frac{wLx^2}{6} - \frac{w}{6} \left\langle x - \frac{2L}{3} \right\rangle^3 + C_1$$

Substituting the boundary condition $dy/dx = 0$ when $x = 0$ and noting that for $x < 2L/3$ the term in the brackets is zero gives $C_1 = 0$. Integrating again gives

$$EIy = -\frac{5wL^2}{36} x^2 + \frac{wL}{18} x^3 - \frac{w}{24} \left\langle x - \frac{2L}{3} \right\rangle^4 + C_2$$

The boundary condition $y = 0$ when $x = 0$ gives $C_2 = 0$. The deflection at the right end is obtained by substituting $x = L$ in the above elastic curve equation. The result is

$$y = -\frac{163}{1944} \frac{wL^4}{EI} = \frac{163}{1944} \frac{wL^4}{EI} \quad \text{downward} \qquad \text{Ans.}$$

The result agrees with that of Exam. 7–2.

EXAMPLE 7–5
Determine the deflection at the left end of the beam of Fig. 7–11a.

SOLUTION
The equations of equilibrium, applied to the free-body diagram of Fig. 7–11b, give the reaction R_L as $3wL/8$ upward. By the use of singularity functions and the superposed loads of Fig. 7–11c, the bending-moment equation becomes

$$EI \frac{d^2y}{dx^2} = -\frac{w}{2} (x + L)^2 + \frac{w}{2} \left\langle x + \frac{L}{2} \right\rangle^2 + \frac{3wL}{8} \langle x \rangle^1 + \frac{wL^2}{2} \left\langle x - \frac{L}{2} \right\rangle^0$$

The boundary conditions are $y = 0$ when $x = 0$, and $y = 0$ when $x = L$. Two integrations of the moment equation give

$$EI \frac{dy}{dx} = -\frac{w}{6} (x + L)^3 + \frac{w}{6} \left\langle x + \frac{L}{2} \right\rangle^3 + \frac{3wL}{16} \langle x \rangle^2 + \frac{wL^2}{2} \left\langle x - \frac{L}{2} \right\rangle^1 + C_1$$

and

$$EIy = -\frac{w}{24} (x + L)^4 + \frac{w}{24} \left\langle x + \frac{L}{2} \right\rangle^4 + \frac{wL}{16} \langle x \rangle^3$$

$$+ \frac{wL^2}{4} \left\langle x - \frac{L}{2} \right\rangle^2 + C_1 x + C_2$$

The first boundary condition, $y = 0$ when $x = 0$, gives

$$0 = -\frac{wL^4}{24} + \frac{wL^4}{384} + 0 + 0 + 0 + C_2$$

from which

$$C_2 = +\frac{5}{128} wL^4$$

The second boundary condition, $y = 0$ when $x = L$, gives

$$0 = -\frac{16wL^4}{24} + \frac{81wL^4}{384} + \frac{wL^4}{16} + \frac{wL^4}{16} + C_1 L + \frac{5wL^4}{128}$$

from which

$$C_1 = +\frac{7}{24} wL^3$$

The deflection at $x = -L$ is found, by substitution, to be

$$y = -\frac{97}{384} \frac{wL^4}{EI} = \frac{97}{384} \frac{wL^4}{EI} \quad \text{downward} \qquad \text{Ans.}$$

PROBLEMS

7-31*   Use singularity functions to determine the deflection, in terms of P, L, E, and I, at the right end of the cantilever beam shown in Fig. P7–31.

7-32*   Use singularity functions to determine the deflection, in terms of w, L, E, and I, at the right end of the cantilever beam shown in Fig. P7–32.

7-33*   Use singularity functions to determine the deflection, in terms of P, L, E, and I, at the center of the simply supported beam shown in Fig. P7–33.

FIG. P7–31

FIG. P7–32

FIG. P7–33   FIG. P7–34

7-34*   Use singularity functions to determine the deflection, in terms of w, L, E, and I, at the left end of the distributed load of the simply supported beam shown in Fig. P7–34.

7-35   Use singularity functions to determine the deflection, in terms of w, L, E, and I, at the right end of the cantilever beam shown in Fig. P7–35.

7-36   A beam is loaded and supported as shown in Fig. P7–36. Use singularity functions to determine, in terms of w, L, E, and I

(a)   The deflection at the center of the beam.

(b)   The maximum deflection in the beam.

7-37*   Use singularity functions to solve Prob. 7-24.

7-38   Use singularity functions to solve Prob. 7-25.

7-39   Use singularity functions to solve Prob. 7-26.

7-40*   Use singularity functions to solve Prob. 7-27.

7-41*   A cantilever beam is loaded and supported as shown in Fig. P7–41. Use singularity functions to determine, in terms of w, L, E, and I

(a)   The deflection at the left end of the beam.

(b)   The deflection at the center of the distributed load.

7-42*   A beam is loaded and supported as shown in Fig. P7–42. Use singularity functions to determine, in terms of w, L, E, and I

(a)   The deflection at the right end of the beam.

(b)   The maximum deflection between the supports.

FIG. P7–35

FIG. P7–36

FIG. P7–41

FIG. P7–42

7-43   A cantilever beam is loaded and supported as shown in Fig. P7–43. Use singularity functions to determine, in terms of w, L, E, and I

(a)   The deflection at the point of application of the couple.

(b)   The deflection at the right end of the beam.

7-44   A beam is loaded and supported as shown in Fig. P7–44. Use singularity functions to determine the maximum deflection in the beam in terms of w, L, E, and I.

7-6   THE DEVELOPMENT OF THE AREA-MOMENT METHOD

The area-moment method of determining the deflection at any specified point along a beam is a semigraphic method utilizing the relations between successive derivatives of the deflection y and the moment diagram. For problems involving several changes in loading, the area-moment method is usually much faster than the double-integration method; consequently, it is widely used in practice. The method is based on two propositions or theorems derived as follows.

From Fig. 7–2 it is seen that for any arc length dL along the elastic curve of a loaded beam

$$d\theta = dL/\rho$$

FIG. P7–43

FIG. P7–44

and from Eq. 7–2 this becomes

$$d\theta = \frac{M}{EI}\,dL$$

For actual beams the curvature is so small that dL can be replaced by the distance dx along the unloaded beam, and the relation becomes

$$d\theta = \frac{M}{EI}\,dx$$

Reference to Fig. 7–12 (which consists of a loaded beam, its moment diagram, and a sketch, greatly exaggerated, of the elastic curve of the deflected beam) shows that $d\theta$ is also the angle between tangents to the elastic curve at two points a distance dL apart. When any two points A and B along the elastic curve are selected as shown in Fig. 7–12c and the relation for $d\theta$ is integrated between these limits, the result is

$$\int_{\theta_A}^{\theta_B} d\theta = \theta_B - \theta_A = \int_{x_A}^{x_B} \frac{M}{EI}\,dx$$

The integral on the left, designated θ_{AB}, is the angle between tangents to the elastic curve at A and B. The theorem can be written as

$$\theta_{AB} = \int_{x_A}^{x_B} \frac{M\,dx}{EI} \tag{7–4}$$

FIG. 7–12

From Fig. 7–12b, the integral is clearly equal to $1/(EI)$ times the area under the moment diagram when E and I are constants (as they are in many beams). In case either E or I vary along the beam, it is advisable to construct an $M/(EI)$ diagram instead of a moment diagram. The first area-moment theorem can be expressed in words as follows: *The angle between the tangents to the elastic curve of a loaded beam at any two points A and B along the elastic curve equals the area under the M/(EI) diagram of the beam between ordinates at A and B.*

The correct interpretation of numerical results frequently depends on a reliable sign convention, and the following convention is recommended. Areas under positive bending moments are considered positive. A positive area means that the angle is positive or counterclockwise when measured from the tangent to the point on the left to the other tangent, as shown in Fig. 7–13.

In order to obtain the second area-moment theorem, let B' in Fig. 7–12c be the intersection of the tangent to the elastic curve at B with a vertical from point A on the elastic curve. Since the deflections of a beam are assumed to be small, it can be seen from Fig. 7–12c that

$$dt = x\, d\theta = \frac{M}{EI} x\, dx$$

Integration gives

$$\int_0^{t_{A/B}} dt = t_{A/B} = \int_{x_A}^{x_B} \frac{M}{EI} x\, dx \tag{7–5}$$

A study of Fig. 7–12c shows that the left integral is the vertical (actually perpendicular to the beam) distance $t_{A/B}$ of any point A on the elastic curve from a tangent drawn to any other point B on the elastic curve. This distance is frequently called the *tangential deviation* to distinguish it from the beam deflection y. The integral on the right, as seen from Fig. 7–12b, is the first moment, with respect to an axis through A parallel to the y axis, of the area under the $M/(EI)$ diagram between ordinates at A and B.

The second area-moment theorem (Eq. 7–5) can be stated in words as

FIG. 7–13

follows: *The vertical distance (tangential deviation) of any point A on the elastic curve of a beam from a tangent drawn at any other point B on the elastic curve equals the first moment, with respect to an axis at A, of the area under the M/(EI) diagram between ordinates at A and B.*

A sign convention is required for those instances in which the shape of the elastic curve is not immediately evident from the given loads. The following sign convention is recommended. The first moment of the area under a positive $M/(EI)$ diagram is considered positive, and from the second area-moment theorem a positive first moment gives a positive tangential deviation. A positive tangential deviation means that point A on the beam (the moment center) is above the tangent drawn from the other point B. Examples of the sign conventions for both the first and second area-moment theorems are shown in Fig. 7–13.

The effectiveness of the area-moment method for determining beam deflections depends, to a considerable extent, on the ease of calculating areas and first moments of areas under the $M/(EI)$ diagrams. Naturally, these values can be obtained by integration of the general expressions from Eqs. 7–4 and 7–5. In general, however, this procedure offers little or no advantage over the double integration method of Arts. 7–3 and 7–4. The method of drawing a separate moment diagram for each load and reaction on a beam, as explained in the next section, greatly simplifies the finding of areas and moments of areas because in this way nearly all areas are triangular, rectangular, or parabolic.

7-7   MOMENT DIAGRAMS BY PARTS

In Chapter 6 a composite or complete moment diagram was drawn that included the contribution or effect of all loads and reactions acting on a beam. In the application of the area-moment method for obtaining deflections, it is sometimes easier and more practical to draw a series of diagrams showing the moment due to each load or reaction on a separate sketch. For this purpose, each moment diagram is drawn from the load to an arbitrarily selected section called the reference section. It is preferable to select the reference section at one end of the beam or at one end of a uniformly distributed load. With a reference section at one end, start with the load or reaction at the opposite end of the beam and draw the bending moment diagram immediately below the beam as if no other loads or reactions were acting on the beam except at the reference section. Proceed toward the right (or left), drawing the moment diagram for each load and reaction below the previous one. The composite moment diagram can be obtained, when necessary, by superposition (addition) of the component diagrams. The actual or total moment on the beam at any section is the algebraic sum of the moments

FIG. 7–14

from the separate diagrams, except at an interior reference section where the sum of the moments on either side of the reference section gives the total moment. The following example illustrates the procedure for drawing moment diagrams by parts.

EXAMPLE 7–6
For the beam in Fig. 7–14a, construct the moment diagram by parts.

SOLUTION
From a free-body diagram of the beam, the equations of equilibrium give the left reaction as 2500 lb upward and the right reaction as 3500 lb upward. The reference section can be selected anywhere along the beam, and each different reference section yields a different series of component moment diagrams. Three different solutions of the example are given to emphasize the importance of a careful selection of the reference section.

(a)   The reference section is selected at the right end of the beam. The bending moment due to the 2500-lb left reaction is positive and equal to 2500 times the distance from the left end, as shown in Fig. 7–14b. The bending moment due to the 2400-lb concentrated load is negative and equal to 2400 times the distance from the load, as shown in Fig. 7–14c. The bending moment due to the distributed load is negative and equal to 300 times the square of the distance from C (the left edge of the uniform load), as shown in Fig. 7–14d. Since the diagrams were drawn from left to right, the right reaction makes no contribution to the moment. It is in the reference section.

(b) The reference section is selected at C, the left edge of the distributed load. The bending moment due to the 3500-lb right reaction is positive, varies directly as the distance from the right end, and is shown in Fig. 7–15b. The bending moment due to the distributed load is negative, equal to 300 times the square of the distance from the right end, as shown in Fig. 7–15c. Likewise, the moments due to the left reaction and the concentrated load are shown in Figs. 7–15d and e. The total bending moment at the reference section C is the sum of the individual moments just to the left or right of the section—namely, $+10{,}200$ ft-lb.

(c) The reference section is selected at the left end. The bending moment due to the 3500-lb right reaction is shown in Fig. 7–16b. The bending moment due to the distributed load is shown in Fig. 7-16c, and the area is subdivided into a triangle, a rectangle, and a

FIG. 7–16

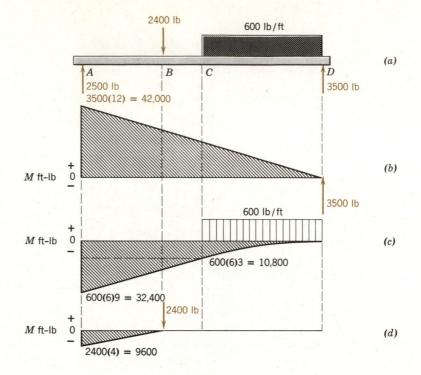

parabolic area. This shows the disadvantage of selecting the reference section away from one end of a distributed load. The bending moment due to the concentrated load is shown in Fig. 7–16d.

A study of the three solutions reveals that the amount of numerical work involved in finding areas and first moments is greatly altered by the choice of reference section to be used. In solution (a) there are three component areas, whereas solution (b) involves four component areas, and solution (c) is the least desirable with five component areas, three of which are due to the distributed load.

PROBLEMS

7-45 Construct the moment diagram by parts for the beam of Fig. P7–10. Use a reference section at the right support.

7-46 Construct the moment diagram by parts for the beam of Fig. P7–11. Use a reference section at the right support.

7-47   Construct the moment diagram by parts for the beam of Fig. P7–12. Use a reference section at the right support.

7-48   Construct the moment diagram by parts for the beam of Fig. P7–29. Use a reference section at the right end of the distributed load.

7-49   Construct the moment diagram by parts for the beam of Fig. P7–43. Use a reference section at the left end of the beam.

7-50   Construct the moment diagram by parts for the beam of Fig. P7–44. Use a reference section at the right support.

7-51   Construct the moment diagram by parts for the beam of Fig. P7–51. Use a reference section at the right end of the beam.

7-52   Construct the moment diagram by parts for the beam of Fig. P7–52. Use a reference section at the right support.

7-53   Construct the moment diagram by parts for the beam of Fig. P7–53. Use a reference section at the right support.

7-54   Construct the moment diagram by parts for the beam of Fig. P7–54. Use a reference section at the left end of the distributed load.

7-55   Construct the moment diagram by parts for the beam of Fig. P7–55. Use a reference section at the right support.

7-56   Construct the moment diagram by parts for the beam of Fig. P7–56. Use a reference section at the right support.

FIG. P7–52

FIG. P7–53

FIG. P7–54

FIG. P7–55

FIG. P7–56

(a)

7-8 DEFLECTIONS BY AREA MOMENTS

The deflection of a specific point along the elastic curve of a beam can normally be determined by one or more applications of the second area-moment theorem. The amount of time and effort required for a given problem depends to a considerable extent on the judgment used in selecting the moment center for the second theorem and in drawing the moment diagram by parts. The ability to construct the correct moment diagram and sketch the elastic curve is essential to the application of the method.

A few general suggestions are presented here to facilitate the application of the method.

1. It is frequently easier to use a composite moment diagram instead of a moment diagram by parts for those cases in which the beam carries only one load.
2. It will be found that selection of the reference section of a cantilever beam at the fixed end for construction of the moment diagram by parts usually gives the minimum number of areas.
3. For overhanging beams it is advisable to avoid taking a tangent to the elastic curve at the overhanging end.
4. It is sometimes advantageous to solve the problem with symbols and substitute numerical values in the last step.

The list of steps given below for the solution of deflection problems by the area-moment method may prove helpful. There is no point in memorizing these steps as they are not a substitute for an understanding of the principles involved. They are:

1. Sketch the loaded beam, the moment [or $M/(EI)$] diagrams (either by parts or composite diagram, depending on the complexity of the problem), and the elastic curve (if the shape is not known, it may be assumed).
2. Visualize which tangent lines may be most helpful and draw such lines on the elastic curve. When a point of zero slope is known, either from symmetry or from supports, a tangent drawn at this point will frequently be useful.
3. By application of the second area-moment theorem, determine the tangential deviation at the point where the beam deflection is desired and at any other points required.
4. From geometry, determine the perpendicular distance from the unloaded beam to the tangent line at the point where the beam deflection is desired, and, using the results of step 3, solve for the required deflection.

FIG. 7–17

(a)

(b)

(c)

For essential data on properties of curves, see Appendix B. Examples 7–7 and 7–8 illustrate the method.

EXAMPLE 7–7
Determine the deflection of the free end of the cantilever beam of Fig. 7–17a in terms of w, L, E, and I.

SOLUTION
Since E and I are constant, a bending moment diagram is used instead of an $M/(EI)$ diagram. The moment diagram is shown in Fig. 7–17b, and the area under it has been divided into rectangular, triangular, and parabolic parts for ease in calculating areas and moments. The elastic curve is shown in Fig. 7–17c with the deflections greatly exaggerated. Points A and B are selected at the ends of the beam because the beam has a horizontal tangent at B and the deflection at A is required. The vertical distance to A from the tangent at B ($t_{A/B}$ in Fig. 7–17c) equals the deflection of the free end of the beam y_A. For this reason, the second area-moment theorem can be used directly to obtain the required deflection. The area of each of the three portions of the area under the moment diagram is shown in Fig. 7–17b along with the distance from the centroid of each part to the moment axis at the free end A. The second area-moment theorem gives

$$EIt_{A/B} = -\frac{wL^3}{6}\left(\frac{5L}{4}\right) - \frac{wL^3}{2}(2L) - \frac{wL^3}{2}\left(\frac{13L}{6}\right)$$

FIG. 7–18

from which

$$y_A = t_{A/B} = -\frac{55wL^4}{24EI} = \frac{55wL^4}{24EI} \quad \text{downward} \qquad \text{Ans.}$$

EXAMPLE 7–8

Determine the deflection of the left end of the beam loaded and supported as shown in Fig. 7–18a, in terms of P, L, E, and I.

SOLUTION

The equations of equilibrium yield the left reaction equal to $4P/3$ upward and the right reaction equal to $P/3$ downward. Since E and I are constant, a moment diagram (instead of an $M/(EI)$ diagram) is drawn, and, with only one applied load, a composite moment diagram as shown in Fig. 7–18b is used.

The elastic curve with a tangent drawn at C is shown in Fig. 7–18c. The tangent line could have been drawn at B for an equally satisfactory solution. The second area-moment theorem gives the tangential deviation at B as

$$EIt_{B/C} = -\frac{3PL^2}{2}(L) = -\frac{3PL^3}{2}$$

where the negative sign indicates that the moment center B is below the tangent drawn from C. Another application of the second area-moment theorem gives the tangential deviation at A as

$$EIt_{A/C} = -\frac{3PL^2}{2}(2L) - \frac{PL^2}{2}\left(\frac{2L}{3}\right) = -\frac{10PL^3}{3}$$

where the negative sign indicates that A is below the tangent from C. Finally,

$$EIy_A = EI\left(t_{A/C} - \frac{4}{3}t_{B/C}\right) = -\frac{10PL^3}{3} - \frac{4}{3}\left(-\frac{3PL^3}{2}\right) = -\frac{4}{3}PL^3$$

and

$$y_A = \underline{\frac{4}{3}\frac{PL^3}{EI}}\quad\text{downward}\qquad\qquad\text{Ans.}$$

PROBLEMS

7-57* Use area moments to determine the deflection at the point of application of the load $3P$ for the beam of Fig. P7–57 in terms of P, L, E, and I.

7-58* Use area moments to determine the deflection at the right end of the beam of Fig. P7–58 in terms of P, L, E, and I.

7-59* Use area moments to determine the deflection at the left end of the beam of Fig. P7–59 in terms of w, L, E, and I.

7-60* Use area moments to determine the deflection at the right end of the distributed load w of the beam of Fig. P7–60 in terms of w, L, E, and I.

7-61 Use area moments to determine the deflection at the right end of the beam of Fig. P7–35 in terms of w, L, E, and I.

7-62 Use area moments to determine the deflection at the right end of the beam of Fig. P7–43 in terms of w, L, E, and I.

7-63* Use area moments to determine the deflection at the right end of the beam of Fig. P7–54 in terms of w, L, E, and I.

7-64 A beam is loaded and supported as shown in Fig. P7–64. Use area

FIG. P7–57   FIG. P7–58

FIG. P7–59

FIG. P7–60

moments to determine the deflection in terms of P, L, E, and I

(a) At the point of application of the load P.

(b) At a point midway between the supports.

7-65 A beam is loaded and supported as shown in Fig. P7–65. Use area moments to determine the deflection in terms of P, L, E, and I

(a) At the point of application of the load P.

(b) At a point midway between the supports.

7-66 A beam is loaded and supported as shown in Fig. P7–66. Use area moments to determine the deflection in terms of P, L, E, and I

(a) At the point of application of the load $2P$.

(b) At a point midway between the supports.

7-67* A beam is loaded and supported as shown in Fig. P7–67. Use area moments to determine the deflection in terms of P, L, E, and I

(a) At the left end of the beam.

(b) At a point midway between the supports.

7-68 Use area moments to determine the deflection at the right end of the beam of Fig. P7–12 in terms of w, L, E, and I.

7-69 Use area moments to determine the deflection at the left end of the beam of Fig. P7–13 in terms of w, L, E, and I.

7-70* Use area moments to determine the deflection at the center of the beam of Fig. P7–34 in terms of w, L, E, and I.

FIG. P7–64

FIG. P7–65

FIG. P7–66

FIG. P7–67

7-71*   Use area moments to determine the deflection at the center of the beam of Fig. P7–44 in terms of w, L, E, and I.

7-72   Use area moments to determine the deflection at the center of the beam of Fig. P7–53 in terms of w, L, E, and I.

7-73*   A beam is loaded and supported as shown in Fig. P7–73. Determine the midspan deflection if $w = 1875$ lb/ft, $L = 12$ ft, $E = 30(10^6)$ psi, and $I = 120$ in.^{4}. Use the area moment method.

7-74*   A beam is loaded and supported as shown in Fig. P7–74. Determine the deflection at the left end of the beam if $w = 45$ kN/m, $L = 1.25$ m, $E = 70$ GPa, and $I = 42(10^6)$ mm^4. Use the area moment method.

7-75   A beam is loaded and supported as shown in Fig. P7–75. Use area moments to determine the deflection in terms of w, L, E, and I

    (a)   At the left end of the beam.

    (b)   At a point midway between the supports.

7-76   A beam is loaded and supported as shown in Fig. P7–76. Use area moments to determine the deflection at a point midway between the supports in terms of w, L, E, and I.

7-77*   A beam is loaded and supported as shown in Fig. P7–29. Determine the deflection at a point midway between the supports if $w = 2000$ lb/ft, $a = 4$ ft, $I = 120$ in.^{4}, and $E = 30(10^6)$ psi. Use the area moment method.

7-78   A beam is loaded and supported as shown in Fig. P7–30. Determine

FIG. P7–73

FIG. P7–74

FIG. P7–75

FIG. P7–76

the deflection at the right end of the beam if $w = 50$ kN/m, $L = 1.5$ m, $I = 75(10^6)$ mm^4, and $E = 200$ GPa. Use the area moment method.

7-9   MAXIMUM DEFLECTIONS BY AREA MOMENTS

In most cantilever beams and in symmetrically loaded simply supported beams, the point of maximum deflection can be determined by inspection. In such instances the procedure outlined in Art. 7–8 is usually adequate for finding the maximum deflection. When the location of the point of maximum deflection is unknown, it is necessary to use both the first and second area-moment theorems for determining the maximum deflection.

Figure 7–19 shows a typical situation in which the maximum deflection, at point B on the elastic curve, is required. The deflection y_B, or the tangential deviation $t_{A/B}$, is wanted, but the distance x along the beam is unknown and must be determined before the deflection can be obtained. Once $t_{C/A}$ is determined from the second area-moment theorem, $t_{C/A}/L$ gives $\tan \theta_A$ that is approximately equal to θ_A. The angle θ_A can also be determined, in terms of x, from the first area-moment theorem because the tangent to the elastic curve at B is horizontal and $\theta_A = \theta_{AB}$. When these two expressions for θ_A are equated, the distance x is found. Finally, $t_{A/B}$ is determined from the second

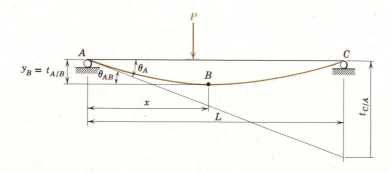

FIG. 7–19

area-moment theorem. The following example illustrates the procedure for a specific case.

EXAMPLE 7–9

For the beam in Fig. 7–20a, determine the maximum deflection in terms of w, L, E, and I.

SOLUTION

The equations of equilibrium give the left reaction as $wL/2$ upward and the right reaction as wL upward. The moment diagram by parts is shown in Figs. 7–20b, c, and d. The elastic curve, very much exaggerated, of the loaded beam is shown in Fig. 7–20e with tangents drawn at A and B. The second area-moment theorem, with the moment axis at C, gives the vertical distance to C from the tangent at A as

$$EIt_{C/A} = L\left(\frac{9wL^3}{4}\right) - \frac{2L}{3}(wL^3) - \frac{L}{4}\left(\frac{wL^3}{6}\right) = \frac{37}{24}wL^4$$

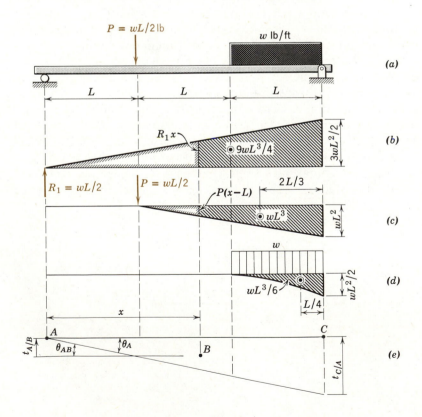

FIG. 7–20

Since the deflections are small,

$$\theta_A = \tan \theta_A = \frac{t_{C/A}}{3L} = \frac{37wL^3}{72EI}$$

Because the tangent at B is horizontal, the first area-moment theorem applied to the segment AB gives

$$\theta_A = \theta_{AB} = \frac{1}{EI}\left[\frac{wL}{2} x \left(\frac{x}{2}\right) - \frac{wL}{2}(x - L)\frac{x - L}{2}\right] = \frac{wL}{4EI}(2Lx - L^2)$$

Equating the two expressions for θ_A gives

$$\frac{37wL^3}{72EI} = \frac{wL}{4EI}(2Lx - L^2)$$

from which

$$x = 55L/36$$

Application of the second theorem to the segment AB, with the moment axis at A, gives the vertical distance to A from the horizontal tangent at B as

$$|y_B| = t_{A/B} = \frac{1}{EI}\left\{\left(\frac{2x}{3}\right)\left(\frac{wLx}{2}\right)\left(\frac{x}{2}\right) - \left[L + \frac{2}{3}(x - L)\right]\frac{wL}{2}\frac{(x - L)^2}{2}\right\}$$

$$= \frac{wL^4}{2EI}\left[\frac{1}{3}\left(\frac{55}{36}\right)^3 - \frac{1}{2}\left(\frac{19}{36}\right)^2 - \frac{1}{3}\left(\frac{19}{36}\right)^3\right]$$

where the positive result means that A is above the tangent line. Therefore,

$$y_B = 0.500\frac{wL^4}{EI} \quad \text{downward} \qquad\qquad \text{Ans.}$$

PROBLEMS

7-79* A beam is loaded and supported as shown in Fig. P7–79. Use area moments to determine the maximum deflection between the supports in terms of P, L, E, and I.

7-80* A beam is loaded and supported as shown in Fig. P7–80. Use area moments to determine the maximum deflection between the supports in terms of P, L, E, and I.

FIG. P7–79

FIG. P7–80

7-81*   A beam is loaded and supported as shown in Fig. P7–10. Use area moments to determine the maximum deflection between the supports in terms of w, L, E, and I.

7-82*   A beam is loaded and supported as shown in Fig. P7–11. Use area moments to determine the maximum deflection between the supports in terms of w, L, E, and I.

7-83   A beam is loaded and supported as shown in Fig. P7–12. Use area moments to determine the maximum deflection between the supports in terms of w, L, E, and I.

7-84   A beam is loaded and supported as shown in Fig. P7–23. Use area moments to determine the maximum deflection between the supports in terms of M, L, E, and I.

7-85*   An S12 × 50 steel beam ($E = 30,000$ ksi) is loaded and supported as shown in Fig. P7–85. Use area moments to determine

(a)   The deflection at the right end.

(b)   The maximum deflection between the supports.

7-86*   A beam is loaded and supported as shown in Fig. P7–86. If $w = 1500$ N/m, $L = 4$ m, $E = 10$ GPa, and $I = 50(10^6)$ mm⁴, use area moments to determine

(a)   The deflection midway between the supports.

(b)   The maximum deflection in the beam.

7-87   A beam is loaded and supported as shown in Fig. P7–34. Use area

FIG. P7–85   FIG. P7–86

moments to determine the maximum deflection between the supports in terms of w, L, E, and I.

7-88 A beam is loaded and supported as shown in Fig. P7–53. Use area moments to determine the maximum deflection between the supports in terms of w, L, E, and I.

7-89* A beam is loaded and supported as shown in Fig. P7–67. Use area moments to determine the maximum deflection between the supports in terms of P, L, E, and I.

7-90 A beam is loaded and supported as shown in Fig. P7–76. Use area moments to determine the maximum deflection between the supports in terms of w, L, E, and I.

7-10 DEFLECTIONS BY SUPERPOSITION

As explained in Art. 3–5, the concept or method of superposition consists of finding the resultant effect of several loads acting on a member simultaneously as the sum of the contributions from each of the loads applied individually. The results for the separate loads are frequently available from previous work or readily determined by a previous method. In such instances the superposition method becomes a powerful concept or tool for finding stresses, deflections, and the like. The method is applicable in all cases in which a linear relation exists between the stresses or deflections and the applied loads.

To show that beam deflections can be accurately determined by superposition, consider the cantilever beam of Fig. 7–21 with loads Q, w, and P. In order to determine the deflection at any point of this beam by the double integration method, it is necessary to express the bending moment in terms of the applied loads. For each interval along the beam, the value of M is the algebraic sum of the moments due to the separate loads. After two successive integrations, the solution for the deflection at any point will still be the algebraic sum of the contributions from each applied load. Furthermore, for any given value of x, the relation between applied load and resulting deflection will be linear. It is evident, therefore, that the deflection of a beam is the sum of the deflections produced by the individual loads. Once the deflections

FIG. 7–21

produced by a few typical individual loads have been determined by one of the methods already presented, the superposition method provides a means of rapidly solving a wide variety of more complicated problems by various combinations of known results. As more data become available, a wider range of problems can be solved by superposition.

The data in Appendix D are provided for use in mastering the superposition method. No attempt is made to give a large number of results because such data are readily available in various handbooks. The data given and the illustrative examples are for the purpose of making the concept and methods clear.

EXAMPLE 7–10
For the beam in Fig. 7–22a, determine the maximum deflection when E is $12(10^6)$ psi and I is 81 in.[4].

SOLUTION
From symmetry and $\Sigma F_y = 0$, the reactions are each 4500 lb upward. Because of the symmetrical loading, the slope of the beam is zero at the center of the span, and the right (or left) half of the beam can be considered a cantilever beam with two loads. As shown in Fig. 7–22b, the cantilever with two loads can be replaced by two beams (designated 1 and 2), each carrying one of the two loads.

The elastic curve (exaggerated) for part 1, as shown in Fig. 7–22c,

FIG. 7–22

gives the deflection at the right end as $y_{4.5} + y_3$, where $y_{4.5}$ is the deflection at the end of the uniformly distributed load and y_3 is the additional deflection of the unloaded 3 ft. From case 2 of Appendix D,

$$y_{4.5} = -\frac{wL^4}{8EI} = -\frac{1000(4.5^4)(12^3)}{8(12)(10^6)(81)} = -\frac{9^3}{8(10^3)} = -0.0911 \text{ in.}$$

Note that in determining deflection y, w lb/ft times L ft gives the applied load in pounds. The remaining L^3 must be expressed in inches if E and I are in inches and the result is to be in inches.

Similarly,

$$\theta_{4.5} = -\frac{wL^3}{6EI} = -\frac{1000(4.5^3)(12^2)}{6(12)(10^6)(81)} = -\frac{9}{4(10^3)} = -0.00225 \text{ rad}$$

from which

$$y_{3.0} = 3(12)(-0.00225) = -0.0810 \text{ in.}$$

Consequently, the total deflection of the right end is 0.1721 in. downward.

The elastic curve (exaggerated) for part 2 is shown in Fig. 7–22d. From case 1 of Appendix D,

$$y_{7.5} = +\frac{PL^3}{3EI} = +\frac{4500(7.5^3)(12^3)}{3(12)(10^6)(81)} = 1.125 \text{ in. upward}$$

The algebraic sum of the deflections for parts 1 and 2 is +0.953 in., which means that the right end of the beam is 0.953 in. above the center. Obviously, the right end does not move, and the maximum deflection is at the center and is

$$y_{\text{max}} = \underline{0.953 \text{ in. downward}} \qquad \text{Ans.}$$

EXAMPLE 7–11
Determine the deflection at midspan for the steel beam ($E = 30,000$ ksi and $I = 180$ in.^{4}) of Fig. 7–23a.

SOLUTION
The given loading is equivalent to the two loads, parts 1 and 2, shown in Fig. 7–23b. For part 1, the deflection at the center of the span is given as case 7 of Appendix D. It is

$$(y_C)_1 = -\frac{5wL^4}{384EI} = -\frac{5(600)(16^4)(12^3)}{384(30)(10^6)(180)} = -0.1638 \text{ in.}$$

FIG. 7–23

(a)

(b)

For part 2, the deflection at the center is given as a part of case 5 of Appendix D. It is

$$(y_C)_2 = -\frac{Pb(3L^2 - 4b^2)}{48EI} = -\frac{2000(6)[3(16^2) - 4(6^2)](12^3)}{48(30)(10^6)(180)} = -0.499 \text{ in.}$$

The algebraic sum of the deflections for midspan is

$$y_C = -0.1638 + (-0.0499) = -0.2137 \text{ in.} = \underline{0.214 \text{ in. downward}} \qquad \text{Ans.}$$

PROBLEMS

> *Note: Use the method of superposition to determine deflections in the following problems.*

7-91*   Determine the deflection at the free end of the cantilever beam of Fig. P7–91 when $w = 600$ lb/ft, $L = 10$ ft, $E = 12(10^6)$ psi, and $I = 14.4$ in.^{4}.

7-92*   Determine the deflection at a point midway between the supports for the beam of Fig. P7–92 when $w = 9$ kN/m, $L = 4$ m, $E = 200$ GPa, and $I = 35(10^6)$ mm^4.

FIG. P7–91

FIG. P7–92

FIG. P7–93   FIG. P7–105

7-93*   Determine the deflection at the free end of the cantilever beam of Fig. P7–93 when $w = 800$ lb/ft, $L = 6$ ft, $E = 30(10^6)$ psi, and $I = 100$ in.^{4}.

7-94   Determine the deflection at the right end of the beam of Fig. P7–10 in terms of w, L, E, and I.

7-95*   Determine the deflection at the left end of the beam of Fig. P7–10 in terms of w, L, E, and I.

7-96   Determine the deflection at the right end of the beam of Fig. P7–11 in terms of w, L, E, and I.

7-97*   Determine the deflection at a point midway between the supports of the beam of Fig. P7–11 in terms of w, L, E, and I.

7-98   Determine the deflection at a point midway between the supports of the beam of Fig. P7–12 in terms of w, L, E, and I.

7-99*   Determine the deflection at the free end of the cantilever beam of Fig. P7–16 in terms of w, L, E, and I.

7-100   Determine the deflection at the right end of the beam of Fig. P7–26 in terms of P, L, E, and I.

7-101   Determine the deflection at the free end of the cantilever beam of Fig. P7–30 in terms of w, L, E, and I.

7-102*   Determine the deflection at the right end of the beam of Fig. P7–35 in terms of w, L, E, and I.

7-103   Determine the deflection at the right end of the beam of Fig. P7–43 in terms of w, L, E, and I.

7-104*   Determine the deflection at the point of application of the load $3P$ on the beam of Fig. P7–57 in terms of P, L, E, and I.

7-105   Determine the deflection at the right end of the beam of Fig. P7–105 in terms of w, L, E, and I.

7-106   Determine the deflection at the left end of the beam of Fig. P7–105 in terms of w, L, E, and I.

7-11   DEFLECTIONS BY ENERGY METHODS—CASTIGLIANO'S THEOREM

Strain energy techniques are frequently used to analyze the deflections of beams and structures. Of the many available methods, the application of Castigliano's theorem, to be developed here, is one of the most widely used. It was presented in 1873 by the Italian engineer Alberto Castigliano (1847–1884). Although the theorem will be derived by considering the strain energy stored in beams, it is applicable to any structure for which the force-deformation relations are linear.

If the beam shown in Fig. 7–24 is slowly and simultaneously loaded by the two forces P_1 and P_2 with resulting deflections y_1 and y_2, the strain energy U of the beam is, by Clapeyron's theorem (see Art. 3–4), equal to the work done by the forces. Therefore,

$$U = \frac{1}{2} P_1 y_1 + \frac{1}{2} P_2 y_2 \qquad (a)$$

Let the force P_1 be increased by a small amount ΔP_1, and Δy_1 and Δy_2 be the changes in deflection due to this incremental load. Since the forces P_1 and P_2 are already present, the increase in the strain energy is

$$\Delta U = \frac{1}{2} \Delta P_1 \Delta y_1 + P_1 \Delta y_1 + P_2 \Delta y_2 \qquad (b)$$

If the order of loading was reversed so that the incremental force, ΔP_1, was applied first, followed by P_1 and P_2, the resulting strain energy would be

$$U + \Delta U = \frac{1}{2} \Delta P_1 \Delta y_1 + \Delta P_1 y_1 + \frac{1}{2} P_1 y_1 + \frac{1}{2} P_2 y_2 \qquad (c)$$

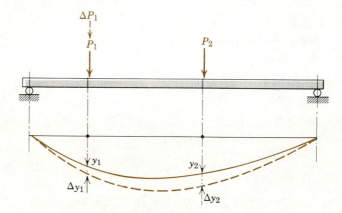

FIG. 7–24

The resulting strain energy must be independent of the order of loading; hence, by combining Eqs. a, b, and c, one obtains

$$\Delta P_1 y_1 = P_1 \Delta y_1 + P_2 \Delta y_2 \qquad (d)$$

Equations b and d can be combined to give

$$\frac{\Delta U}{\Delta P_1} = y_1 + \frac{1}{2} \Delta y_1$$

or upon taking the limit as ΔP_1 approaches zero[2]

$$\frac{\partial U}{\partial P_1} = y_1 \qquad (e)$$

For the general case in which there are many loads involved, Eq. e is written as

$$\frac{\partial U}{\partial P_i} = y_i \qquad (7\text{–}6a)$$

The following is a statement of Castigliano's theorem. *If the strain energy of a linearly elastic structure is expressed in terms of the system of external loads, the partial derivative of the strain energy with respect to a concentrated external load is the deflection of the structure at the point of application and in the direction of that load.* By a similar development, Castigliano's theorem can also be shown to be valid for applied moments and the resulting rotations (or changes in slope) of the structure, Thus,

$$\frac{\partial U}{\partial M_i} = \theta_i \qquad (7\text{–}6b)$$

If the deflection is required either at a point where there is no unique point load or in a direction not aligned with the applied load, a dummy load is introduced at the desired point acting in the proper direction. The deflection is obtained by first differentiating the strain energy with respect to the dummy load and then taking the limit as the dummy load approaches zero. Also, for the application of Eq. 7–6b, either a unique point moment or a dummy moment must be applied at point i. The moment will be in the direction of rotation at the point. Note that if the loading consists of a number of point loads, all expressed in terms of a single parameter (e.g., P, $2P$, $3P$, wL, $2wL$) and if the deflection is wanted at one of the applied loads, one must either write the moment equation with this load as a separate

---

[2] The partial derivative is used because the strain energy is a function of both P_1 and P_2.

identifiable term or add a dummy load at the point, so that the partial derivative can be taken with respect to this load only.

It was shown in Art. 3–4 that the strain energy per unit volume for uniaxial stress is $\sigma^2/(2E)$; hence, the total strain energy under uniaxial stress is

$$U = \int_{\text{vol.}} \frac{\sigma^2}{2E} \, dV$$

For a beam of constant (or slowly varying) cross section subjected to pure bending, the principal stresses are parallel to the axis of the beam; therefore, using Eq. 6–3, the strain energy becomes

$$U = \frac{1}{2E} \int_{\text{vol.}} \left(\frac{My}{I}\right)^2 dV \qquad (f)$$

By writing dV as $dA \, dx$ (where x is measured along the axis of the beam), Eq. f becomes

$$U = \frac{1}{2E} \int_0^L \left(\frac{M^2}{I^2} \int_{\text{area}} y^2 \, dA\right) dx = \frac{1}{2E} \int_0^L \frac{M^2}{I} \, dx \qquad (7\text{--}7)$$

Equation 7–7 was developed for a beam loaded in pure bending. However, most real beams will be subjected to transverse loads that induce shearing stresses and, in the case of distributed loading, transverse normal stresses. In Art. 6–14, it was shown that the transverse normal stress is small enough to neglect, and in the following Art. 7–12, it is shown that, except for short, deep beams, the deflection due to shearing stresses is also small enough to neglect. Hence, Eq. 7–7, which neglects strain energy due to these stresses, is applicable to the usual real beams. In applying Eq. 7–6 or Eq. 7–7, it is usually much simpler to apply Leibnitz's rule[3] to differentiate under the integral sign so that

$$y_i = \frac{\partial U}{\partial P_i} = \frac{1}{E} \int_0^L \frac{M}{I} \frac{\partial M}{\partial P_i} \, dx \qquad (7\text{--}8)$$

EXAMPLE 7–12

Determine the deflection at the free end of a cantilever beam of constant cross section and length L, loaded with a force P at the free end and a distributed load that varies linearly from zero at the free end to w at the support.

[3] *Advanced Calculus,* W. Kaplan, Addison-Wesley, Reading, Mass., 1953, p. 219.

SOLUTION
From the free-body diagram of Fig. 7–25, the moment equation is

$$M = - Px - \frac{wx^3}{6L}$$

The partial derivative of M with respect to P is $-x$, and Eq. 7–8 becomes

$$EIy = \int_0^L \left(Px^2 + \frac{wx^4}{6L} \right) dx = + \frac{PL^3}{3} + \frac{wL^3}{30}$$

The positive signs indicate deflection in the direction of the force P; hence,

$$y = \frac{PL^3}{3EI} + \frac{wL^4}{30EI} \quad \text{downward} \qquad\qquad \text{Ans.}$$

The result may be verified by reference to Appendix D.

EXAMPLE 7–13
Determine the deflection at the center of a simply supported beam of constant cross section and span L carrying a uniformly distributed load w lb/ft over its entire length.

SOLUTION
In the free-body diagram of Fig. 7–26, the dashed force P represents a fictitious load applied at the location and in the direction of the desired deflection. The moment equation is

$$M = M_w + M_P = \frac{wLx}{2} - \frac{wx^2}{2} + \frac{Px}{2} - P \left\langle x - \frac{L}{2} \right\rangle^1$$

where the quantity $\langle x - L/2 \rangle^1$ is zero for all $x \le L/2$ (see Art. 7–5). The partial derivative of M with respect to P is

FIG. 7–25   FIG. 7–26

$$\frac{\partial M}{\partial P} = \frac{x}{2} - \left\langle x - \frac{L}{2} \right\rangle^1$$

The dummy force P is equated to zero and the deflection is given by

$$EIy = \int_0^L \left(\frac{wLx}{2} - \frac{wx^2}{2} \right) \left(\frac{x}{2} - \left\langle x - \frac{L}{2} \right\rangle^1 \right) dx$$

which, for ease of integration, can be written as

$$EIy = \frac{w}{4} \int_0^L (Lx^2 - x^3)dx + \frac{w}{4} \int_{L/2}^L (L^2x - 3Lx^2 + 2x^3)dx = \frac{5wL^4}{384EI}$$

Since the deflection is positive in the direction of the force, the deflection is downward and

$$y = \frac{5wL^4}{384EI} \quad \text{downward} \qquad\qquad \text{Ans.}$$

PROBLEMS

> *Note: Use Castigliano's theorem to solve the following problems.*

7-107* A beam is loaded and supported as shown in Fig. P7–107. Determine the deflection at the concentrated load P in terms of P, L, E, and I.

7-108* A beam is loaded and supported as shown in Fig. P7–108. Determine the deflection at the concentrated load P in terms of P, L, E, and I.

7-109* A beam is loaded and supported as shown in Fig. P7–109. Determine the deflection at the left end in terms of w, L, E, and I.

7-110* Determine the deflection at the left end of the cantilever beam shown in Fig. P7–110 in terms of w, L, E, and I.

FIG. P7–107

FIG. P7–108

FIG. P7–109   FIG. P7–110

7-111   For the beam shown in Fig. P7–111, determine the slope at the section where the couple M is applied in terms of M, L, E, and I.

7-112   For the beam shown in Fig. P7–111, determine the deflection at the section where the couple M is applied in terms of M, L, E, and I.

7-113*   For the beam shown in Fig. P7–113, determine the deflection at the concentrated load P in terms of P, L, E, and I.

7-114   For the beam shown in Fig. P7–113, determine the slope at the left end of the beam in terms of P, L, E, and I.

7-115   For the beam shown in Fig. P7–113, determine the deflection at a section midway between the supports in terms of P, L, E, and I.

7-116*   A beam is loaded and supported as shown in Fig. P7–116. Determine the deflection at the left end when $E = 10(10^6)$ psi and $I = 100$ in.[4].

7-117*   A beam is loaded and supported as shown in Fig. P7–117. Determine the deflection at the left end when $E = 70$ GPa and $I = 20(10^6)$ mm[4].

7-118   A beam is loaded and supported as shown in Fig. P7–116. Determine the deflection at a section midway between the supports when $E = 30(10^6)$ psi and $I = 120$ in.[4].

7-119   A beam is loaded and supported as shown in Fig. P7–117. Determine the deflection at a section midway between the supports when $E = 200$ GPa and $I = 30(10^6)$ mm[4].

FIG. P7–111   FIG. P7–113

FIG. P7–116

w = 3200 lb/ft

$L = 4'$ $2L = 8'$ $L = 4'$

FIG. P7–117

10 kN

20 kN·m

1 m   1.5 m   1.5 m

7-120* Determine the deflection at a section midway between the supports in terms of w, L, E, and I if the beam is loaded and supported as shown in Fig. P7–120.

7-121* Determine the deflection at a section midway between the supports in terms of w, L, E, and I when the beam is loaded and supported as shown in Fig. P7–121.

7-122* Determine the vertical component of the deflection due to flexure at point B of the curved bar of Fig. P7–122 in terms of P, R, E, and I. Assume that the radius of curvature is sufficiently large relative to the depth for the elastic flexure formula to apply.

7-123 Determine the horizontal component of the deflection due to flexure at point B of the curved bar of Fig. P7–122 in terms of P, R, E, and I. Assume that the radius of curvature is sufficiently large relative to the depth for the elastic flexure formula to apply.

7-124* Determine the vertical component of the deflection due to flexure at point B of the curved bar of Fig. P7–124 in terms of P, R, E, and I. Assume that the radius of curvature is sufficiently large relative to the depth for the elastic flexure formula to apply.

7-125 Determine the slope at point B of the curved bar of Fig. P7–124 in terms of P, R, E, and I. Assume that the radius of curvature is sufficiently large relative to the depth for the elastic flexure formula to apply.

$\frac{wL^2}{3}$ ft-lb

w lb/ft

L ft

FIG. P7–120

wL N

w N/m

B

A L m L m C

FIG. P7–121

FIG. P7–122

FIG. P7–124

7-12 DEFLECTIONS DUE TO SHEARING STRESS

As mentioned in Art. 7–3, the beam deflections calculated so far neglect the deflection produced by the shearing stresses in the beam. For short, heavily loaded beams this deflection can be significant, and an approximate method for evaluating such deflections will now be developed. The deflection of the neutral surface dy due to shearing stresses in the interval dx along the beam of Fig. 7–27 is

$$dy = \gamma \, dx = \frac{\tau}{G} \, dx = \frac{VQ}{GIt} \, dx$$

from which, since the shear in Fig. 7–27 is negative,

FIG. 7–27

$$\frac{GIt}{Q}\frac{dy}{dx} = -V \tag{7-9}$$

Since the vertical shearing stress varies from top to bottom of a beam, the deflection due to shear is not uniform. This nonuniform deflection due to shear is reflected in the slight warping of the sections of a beam (see Art. 6-14). Equation 7-9 gives values too high because the maximum shearing stress (at the neutral surface) is used and also because the rotation of the differential shear element (see Art. 6-14) is ignored.

In order to obtain an idea of the relative amount of beam deflection due to shearing stress, consider a rectangular cross section for the beam of Fig. 7-27 and use the maximum stress for which

$$\frac{Q}{It} = \frac{3}{2A}$$

where A is the cross-sectional area of the beam. The expression for dy becomes

$$dy = \frac{3wxdx}{2AG}$$

where V is replaced by its value $-wx$. Integration along the entire beam gives the change in y due to shear as

$$\Delta y = \frac{3w}{2AG}\int_0^L x\,dx = \frac{3wL^2}{4AG}$$

which equals the deflection at the left end. For this same beam the magnitude of the deflection at the left end due to flexural stresses is

$$\frac{wL^4}{8EI} = \frac{3wL^4}{2EAd^2}$$

and the magnitude of the total deflection at the left end becomes

$$y = \frac{3wL^4}{2EAd^2} + \frac{3wL^2}{4AG} = \frac{3wL^2}{4AG}\left(\frac{2L^2G}{d^2E} + 1\right) \tag{a}$$

For a material such as steel the ratio G/E is approximately 0.4, and, with the ratio L/d selected as 6 (a very short, deep beam), the deflection due to flexural stress is 28.8 times the deflection due to shear. In contrast, for a W36 × 160 steel beam and the same span-depth ratio, the flexural stress deflection is 9.88 times the approximate shear deflection. For a working fiber stress of 18 ksi, the two deflections are 0.389 and 0.039 in., respectively.

Equation *a* indicates that the ratio of the two deflections increases as the square of L/d, which means that the deflection due to shear is of importance only in the case of very short, deep beams.

7-13 CLOSURE

Four methods for determining the deflection of a beam were presented in this chapter. The integration method consists of solving the differential equation of the elastic curve and evaluating the constants of integration from the boundary and matching conditions. Singularity functions are quite helpful if more than one interval is involved. The area-moment method consists of applying one or both of the area-moment theorems to the $M/(EI)$ diagram and using the geometry of the beam to obtain the required deflections. Note that the integration method can be adapted to the area-moment method by replacing the indefinite integral with a definite integral, whereby, upon integration, one obtains the change in slope and the change in tangential deviation (measured from the tangent drawn to the curve at the origin) between the limits selected. The superposition method consists of subdividing the applied loads and supports into an equivalent set of less involved components, for which solutions are readily available, and obtaining the required deflections as the algebraic sum of the deflections for the component parts. An energy method—Castigliano's theorem—consists of writing an expression for the energy U stored in the beam due to all applied loads and then evaluating the partial derivative of U with respect to a real or dummy load P applied at the point where the deflection is to be obtained.

The selection of the best method for a given problem requires judgment and experience.

CHAPTER 8

STATICALLY INDETERMINATE BEAMS

8-1 INTRODUCTION

As explained in Art. 6–1, a beam, subjected only to transverse loads, with more than two reaction components is statically indeterminate because the equations of equilibrium are not sufficient to determine all the reactions. In such cases the geometry of the deformation of the loaded beam is utilized to obtain the additional relations needed for an evaluation of the reactions (or other unknown forces). For problems involving elastic action, each additional constraint on a beam provides additional information concerning slopes or deflections. Such information, when utilized with appropriate slope or deflection equations, yields expressions that supplement the independent equations of equilibrium.

Any of the methods for determining beam deflections outlined in Chapter 7 can be utilized to provide the additional equations needed to solve for the unknown reactions or loads in a statically indeterminate beam, provided the action is linearly elastic. In this book no attempt is made at extensive coverage of beams subjected to inelastic action. However, one type of analysis, because of its extensive use in modern design of structures, is presented in Art. 8–6. Once the reactions and loads are known, the methods of Chapters 6 and 7 will give the required stresses and deflections.

8-2 THE INTEGRATION METHOD

For statically determinate beams, known slopes and deflections were used to obtain boundary and matching conditions, from which the constants of integration in the elastic curve equation could be evaluated. For statically indeterminate beams, the procedure is identical. However, the moment equations will contain reactions or loads that cannot be evaluated from the available equations of equilibrium, and one additional boundary condition is needed for the evaluation of each such unknown. For example, if a beam is subjected to a force system for which there are two independent equilibrium equations and if there are four unknown reactions or loads on the beam, two boundary or matching conditions are needed in addition to those necessary for the determination of the constants of integration. These extra boundary conditions, when substituted in the appropriate elastic curve equations (slope or deflection), will yield the necessary additional equations. The following example illustrates the method.

EXAMPLE 8–1

For the beam loaded and supported as shown in Fig. 8–1a, determine the reactions in terms of w and L.

FIG. 8–1

SOLUTION

From the free-body diagram of Fig. 8–1b it is seen that there are three unknown reaction components (M, V, and R) and that only two independent equations of equilibrium are available. The additional unknown requires the use of the elastic curve equation, for which one extra boundary condition is required in addition to the two required for the constants of integration. Because three boundary conditions are available in the interval between the supports, only one elastic curve equation need be written. The origin of coordinates is arbitrarily placed at the wall, and, for the interval $0 \leq x \leq L$, the boundary conditions are: when $x = 0$, $dy/dx = 0$; when $x = 0$, $y = 0$; and when $x = L$, $y = 0$. From Fig. 8–1c the bending moment equation is

$$EI \frac{d^2y}{dx^2} = M_x = Vx + M - \frac{wx^2}{2}$$

Integration gives

$$EI \frac{dy}{dx} = \frac{Vx^2}{2} + Mx - \frac{wx^3}{6} + C_1$$

The first boundary condition gives $C_1 = 0$. A second integration yields

$$EIy = \frac{Vx^3}{6} + \frac{Mx^2}{2} - \frac{wx^4}{24} + C_2$$

The second boundary condition gives $C_2 = 0$, and the last boundary condition gives

$$0 = \frac{VL^3}{6} + \frac{ML^2}{2} - \frac{wL^4}{24}$$

which reduces to

$$4VL + 12M = wL^2$$

The equation of equilibrium $\Sigma M_R = 0$ for the free-body diagram of Fig. 8–1b yields

$$VL + M = w\left(\frac{5L}{4}\right)\left(\frac{5L}{8} - \frac{L}{4}\right) = \frac{15wL^2}{32}$$

The simultaneous solution of these equations gives

$$M = -\frac{7wL^2}{64} = \frac{7wL^2}{64} \text{ ft-lb counterclockwise} \qquad \text{Ans.}$$

and

$$V = \frac{37wL}{64} \text{ lb upward} \qquad\qquad \text{Ans.}$$

Finally, the equation $\Sigma F_y = 0$ for Fig. 8–1b gives

$$R + V = 5wL/4$$

from which

$$R = \frac{43wL}{64} \text{ lb as shown} \qquad\qquad \text{Ans.}$$

An alternate solution would be to place the origin of coordinates at the right support and write the moment equation for the interval $0 \leqslant x \leqslant L$. This equation would involve only one unknown, the reaction R; upon integration and evaluation of the constants, the third boundary condition would directly yield the value of R. The two independent equilibrium equations could then be used to evaluate M and V.

PROBLEMS

Note: Use the integration method to solve the following problems.

8-1*   When the couple Q is applied to the left end of the cantilever beam shown in Fig. P8–1, the deflection at the left end is zero. Determine the magnitude of the couple Q in terms of w and L.

8-2*   When the couple Q is applied to the left end of the cantilever beam shown in Fig. P8–1, the deflection at the left end is $wL^4/(24EI)$ downward. Determine the magnitude of the couple Q in terms of w and L.

8-3   When the couple Q is applied to the left end of the cantilever beam shown in Fig. P8–1, the slope at the left end is zero. Determine

(a)   The magnitude of the couple Q in terms of w and L.

(b)   The deflection at the left end in terms of w, L, E, and I.

8-4*   When the load P is applied to the left end of the cantilever beam shown in Fig. P8–4, the slope at point A is zero. Determine

FIG. P8–1

FIG. P8–4

(a) The magnitude of the force P in terms of w and L.

(b) The deflection at A in terms of w, L, E, and I.

8-5 When the load P is applied to the left end of the cantilever beam shown in Fig. P8–4, the deflection at point A is $2wL^4/EI$ downward. Determine

(a) The magnitude of the force P in terms of w and L.

(b) The deflection at the left end of the beam in terms of w, L, E, and I.

8-6* The right end of the beam shown in Fig. P8–6 is fixed. Determine

(a) The reaction at the left end of the beam.

(b) The maximum deflection in terms of w, L, E, and I.

8-7 The right end of the beam shown in Fig. P8–6 is only partially fixed and under action of the load the slope becomes $wL^3/(36EI)$ upward to the right. Determine

(a) The reaction at the left end of the beam.

(b) The maximum deflection in terms of w, L, E, and I.

(c) The maximum bending moment in the beam.

8-8* When the load P is applied to the right end of the beam shown in Fig. P8–8, the slope becomes zero over the right support. Determine

(a) The magnitude of the force P in terms of w and L.

(b) The maximum deflection between the supports in terms of w, L, E, and I.

FIG. P8–6

FIG. P8–8

FIG. P8–9

FIG. P8–10

8-9 In Fig. P8–9, the slope of the beam over the support B is $wL^3/(36EI)$ downward to the right. Determine

(a) The magnitude of the force P in terms of w and L.

(b) All of the reactions in terms of w and L.

8-10* In Fig. P8–10, the slope of the beam over the support at A is $wL^3/(6EI)$ downward to the left. Determine

(a) The magnitude of the moment M in terms of w and L.

(b) All of the reactions in terms of w and L.

8-11* A beam is loaded and supported as shown in Fig. P8–11. Determine

(a) All of the reactions in terms of w and L.

(b) The deflection midway between supports in terms of w, L, E, and I.

8-12 A beam is loaded and supported as shown in Fig. P8–12. Determine

(a) All of the reactions in terms of w and L.

(b) The deflection midway between the two right supports in terms of w, L, E, and I.

8-13 A beam is loaded and supported as shown in Fig. P8–13. Determine

(a) All of the reactions.

(b) The bending moment over the center support.

8-14* A beam is loaded and supported as shown in Fig. P8–14. Determine

FIG. P8–11

FIG. P8–12   FIG. P8–13

(a)   The left reaction.

(b)   The deflection at B if $E = 200$ GPa and $I = 350(10^6)$ mm^4.

8-15*   A beam is loaded and supported as shown in Fig. P8–15. Determine

(a)   All of the reactions.

(b)   The deflection at the 3000-lb load if $E = 30{,}000$ ksi and $I = 26$ in.^{4}.

8-16   A beam is loaded and supported as shown in Fig. P8–16. Determine

(a)   All of the reactions.

(b)   The deflection at C if $E = 200$ GPa and $I = 150(10^6)$ mm^4.

8-17   A beam is loaded and supported as shown in Fig. P8–17. As a result of poor construction, the beam rotates clockwise at the left support when the load is applied until the slope is $wL^3/(30EI)$. The right support does not rotate and both supports remain on the same level. Determine

(a)   All of the reactions in terms of w and L.

(b)   The deflection at a section midway between the supports in terms of w, L, E, and I.

8-18*   A beam is loaded and supported as shown in Fig. P8–18. Determine

(a)   All of the reactions in terms of w and L.

(b)   The bending moment over the center support.

8-19*   A beam is loaded and supported as shown in Fig. P8–19. Determine the reaction at B in terms of w and L.

8-20   A beam is loaded and supported as shown in Fig. P8–20. Determine the maximum deflection in terms of w, L, E, and I.

FIG. P8–14   FIG. P8–15

FIG. P8–16

FIG. P8–17

FIG. P8–18

FIG. P8–19

FIG. P8–20

FIG. P8–23   FIG. P8–24

8-21   A steel (E = 30,000 ksi) beam with a span of 100 in. is fixed at both ends. When a uniformly distributed load of 100 lb/in. is applied to the beam, the right support settles 0.10 in. There is no rotation at either support. The moment of inertia of the cross section with respect to the neutral axis is 50 in.[4]. Determine all of the reactions.

8-22*   A steel (E = 200 GPa) beam 10 m long is simply supported at the ends and at the center. When a uniformly distributed load of 30 kN/m is applied over the entire length, the center support settles 10 mm. The moment of inertia of the cross section with respect to the neutral axis is 250(10[6]) mm[4]. Determine the reaction at the center support.

8-23*   The beam shown in Fig. P8–23 is simply supported at the left end and framed into a column at the right end. When the load is applied, the ends of the beam remain at the same level, but the right end rotates (due to loading the adjacent span) clockwise until the slope of the elastic curve is $wL^3/(24EI)$ downward to the right. Determine the left reaction in terms of w and L.

8-24   A beam is loaded and supported as shown in Fig. P8–24. When the loads are applied, the center support settles an amount equal to $wL^4/(12EI)$. Determine all of the reactions in terms of w, L, E, and I.

8-25   A beam of span L is fixed at the left support and restrained at the right support. When a uniformly distributed load of w lb/ft is applied over the entire length, the right support settles an amount equal to $wL^4/(48EI)$ and rotates (as a result of loading the adjacent span) until the slope of the elastic curve is $wL^3/(60EI)$ downward to the right. Determine the reaction components at the left support in terms of w, L, E, and I.

8-3   THE AREA-MOMENT METHOD

For statically determinate beams, points with known slopes or deflections were used as tangent points and moment centers for the application of the area-moment theorems. For statically indeterminate beams, each additional constraint provides information regarding a slope or deflection, which makes it possible to write an area-moment equation to supplement the equilibrium equations. The following example illustrates the application of the concepts.

EXAMPLE 8–2
For the beam of Fig. 8–2a determine the reactions in terms of w and L.

SOLUTION
There are four unknown reactions (V_A, M_A, V_B, and M_B) on the free-body diagram of Fig. 8–2b, and there are only two independent equations of equilibrium. Therefore, two area-moment relations are necessary to supplement the equations of equilibrium.

Inspection of the elastic curve (very much exaggerated) of Fig. 8–2c reveals that the angle between tangents drawn at A and B is zero. Also, the distance from a tangent drawn at A to point B is zero.

These geometric statements can be expressed mathematically by the area-moment theorems. Thus, since the angle between the tangents at A and B is zero, the area under the moment diagram between A and B must be zero. Also, since the distance from the tangent drawn at A to point B is zero, the moment of the area under the $M/(EI)$ diagram between A and B, with respect

FIG. 8–2

to an axis at B, is zero. (In this particular problem the moment axis could also have been selected at A since the deflection of A from a tangent drawn at B is also zero). These two area-moment relations give

$$EI\theta_{AB} = 0 = \frac{9}{2} V_A L^2 + 3M_A L - \frac{4}{3} wL^3$$

from which

$$27V_A L + 18M_A = 8wL^2$$

and

$$EIt_{B/A} = 0 = \frac{9}{2} V_A L^2(L) + 3M_A L \left(\frac{3L}{2}\right) - \frac{4wL^3}{3} \left(\frac{L}{2}\right)$$

from which

$$27V_A L + 27M_A = 4wL^2$$

The simultaneous solution of these equations yields

$$M_A = -\frac{4wL^2}{9} = \underline{\frac{4wL^2}{9} \text{ N} \cdot \text{m counterclockwise}}$$ Ans.

and

$$V_A = \underline{\frac{16wL}{27} \text{ N upward}}$$ Ans.

With two reactions known, the equations of equilibrium give the other two as

$$M_B = -\frac{2wL^2}{3} = \underline{\frac{2wL^2}{3} \text{ N} \cdot \text{m clockwise}}$$ Ans.

and

$$V_B = -\frac{38wL}{27} = \underline{\frac{38wL}{27} \text{ N upward}}$$ Ans.

PROBLEMS

Note: Use the area-moment method to solve the following problems.

FIG. P8–26

FIG. P8–27

8-26*   A beam is loaded and supported as shown in Fig. P8–26. Determine
  (a)   The left reaction in terms of w and L.
  (b)   The deflection at the left end of the beam in terms of w, L, E, and I.

8-27*   A beam is loaded and supported as shown in Fig. P8–27. Determine
  (a)   The right reaction in terms of w and L.
  (b)   The deflection at the right end of the beam in terms of w, L, E, and I.

8-28*   Use the area-moment method to solve Prob. 8-6.

8-29   Use the area-moment method to solve Prob. 8-20.

8-30   A beam is loaded and supported as shown in Fig. P8–30. Determine
  (a)   The left reaction in terms of w and L.
  (b)   The maximum deflection in terms of w, L, E, and I.

8-31*   A beam is loaded and supported as shown in Fig. P8–31. Determine
  (a)   The left reaction in terms of w and L.
  (b)   The maximum deflection in terms of w, L, E, and I.

8-32*   A beam is loaded and supported as shown in Fig. P8–17. Determine

FIG. P8–30

FIG. P8–31

FIG. P8–34   FIG. P8–35

(a)   The reaction components at the left end in terms of w and L.

(b)   The deflection at the center of the span in terms of w, L, E, and I.

8-33   A beam is loaded and supported as shown in Fig. P8–18. Determine

(a)   All of the reactions in terms of w and L.

(b)   The deflection at the center of the left span in terms of w, L, E, and I.

8-34   A beam is loaded and supported as shown in Fig. P8–34. When the load P and the moment M are applied, the slope of the elastic curve at the left support is $PL^2(36EI)$ downward to the right. Determine

(a)   The value of M in terms of P and L.

(b)   The maximum deflection in terms of P, L, E, and I.

8-35*   A beam is loaded and supported as shown in Fig. P8–35. When the moments M and PL are applied, the reaction at A is $3P/2$ upward. Determine

(a)   The value of M in terms of P and L.

(b)   The reaction at B in terms of P.

8-36*   Use the area-moment method to solve Prob. 8-11.

8-37   Use the area-moment method to solve Prob. 8-12.

8-38   A beam is loaded and supported as shown in Fig. P8–38. Determine

(a)   All of the reactions.

(b)   The deflection at the left end of the beam if $E = 30,000$ ksi and $I = 15$ in.^{4}.

FIG. P8–38   FIG. P8–39

FIG. P8–40

FIG. P8–41

8-39*   A beam is loaded and supported as shown in Fig. P8–39. Determine

(a)   All of the reactions.

(b)   The deflection at the right end of the beam if $E = 200$ GPa and $I = 16(10^6)$ mm^4.

8-40*   A beam is loaded and supported as shown in Fig. P8–40. Determine

(a)   All of the reactions in terms of w and L.

(b)   The deflection at the right end of the beam in terms of w, L, E, and I.

8-41   A beam is loaded and supported as shown in Fig. P8–41. Determine

(a)   All of the reactions in terms of w and L.

(b)   The deflection at point C in terms of w, L, E, and I.

8-42   A beam is simply supported at the ends and carries a uniformly distributed load w lb/ft over its entire length. A third support is added at midspan. This support is $wL^4/(384EI)$ lower than the end supports, where L is the total length of the beam. Determine the amount the end reactions are changed by adding this third support.

8-4   THE SUPERPOSITION METHOD

The concept (discussed in Art. 7–10) that a slope or deflection due to several loads is the algebraic sum of the slopes or deflections due to each of the loads acting individually is frequently used to provide the deformation equations needed to supplement the equilibrium equations in the solution of statically indeterminate beam problems. The general method of attack is outlined in Art. 3–5. To provide the necessary deformation equations, selected restraints are removed and replaced by unknown loads (forces and couples); the deformation diagrams corresponding to individual loads (both known and unknown) are sketched, and the component deflections or slopes are summed to produce the known configuration. The following examples illustrate the use of superposition for this purpose.

EXAMPLE 8–3

A steel beam 20 ft long is simply supported at the ends and at the midpoint. Under a load of 400 lb/ft uniformly distributed over the entire beam, the center support is 0.12 in. above the end supports. Determine the reactions. The moment of inertia of the cross section with respect to the neutral axis is 100 in.[4].

SOLUTION

With three unknown reactions and only two equations of equilibrium available, the beam is statically indeterminate. Replace the center support with an unknown applied load. The resulting simply supported beam is equivalent to two beams with individual loads as shown in Fig. 8–3. The resulting deflection at the midpoint of the beam is the upward deflection y_R due to R_C, minus the downward deflection y_w due to the uniform load; that is,

$$y = y_R - y_w = 0.12 \text{ in.} \tag{a}$$

The deflection y_w can be obtained from Apendix D, case 7, and is

$$y_w = \frac{5wL^4}{384EI} = \frac{5[400(20)][20(12)]^3}{384(30)(10^6)(100)} = 0.48 \text{ in. downward}$$

The deflection y_R can be obtained in terms of R_C from Appendix D, case 6, and is

$$y_R = \frac{R_C L^3}{48EI} = \frac{R_C[20(12)]^3}{48(30)(10^6)(100)} = \frac{96R_C}{10^6} \text{ in. upward}$$

When these values are substituted in Eq. a, the result is

$$\frac{96R_C}{10^6} - 0.48 = 0.12$$

from which

$$R_C = \underline{6250 \text{ lb upward}} \qquad\qquad \text{Ans.}$$

The equilibrium equation $\Sigma F_y = 0$ and symmetry give

$$R_L = R_R = \underline{875 \text{ lb upward}} \qquad\qquad \text{Ans.}$$

Note that in determining y_w, w lb/ft × L ft gives the applied load in pounds. The remaining L^3 must be expressed in inches if E and I are in inches and the result is to be in inches. The arithmetic will frequently be simplified if the

FIG. 8–3

expressions for the deflections are substituted in the deflection equation in symbol form. In this example Eq. *a* becomes

$$\frac{R_C L^3}{48EI} - \frac{5wL^4}{384EI} = 0.12$$

which reduces (when multiplied by 48 EI/L^3) to

$$R_C - \frac{5wL}{8} = \frac{0.12(48EI)}{L^3}$$

or

$$R_C = \frac{5(400)(20)}{8} + \frac{0.12(48)(30)(10^6)(100)}{[20(12)]^3} = 6250 \text{ lb upward}$$

EXAMPLE 8–4
Determine all reactions for the beam in Fig. 8–4a in terms of *P*, *L*, and *a*.

FIG. 8–4   (b)

SOLUTION

There are four unknown reactions (a shear and moment at each end), and only two equations of equilibrium are available; therefore, the beam is statically indeterminate, and two deformation equations are necessary. Replace the constraint at the right end with an unknown force and couple. The resulting cantilever beam is equivalent to three beams with individual loads, as shown in Fig. 8–4b. Note that the unknown shear and moment at the right end are both shown as positive values so that the algebraic sign of the result will be correct. From the geometry of the constrained beam, the resultant slope and the resultant deflection at the right end are both zero. The slope and deflection at the end of each of the three replacement beams can be obtained from the expressions in Appendix D. Thus, the first beam with load P (see case 1, Appendix D) has a constant slope from P to the end of the beam, which is

$$\theta_P = -\frac{Pa^2}{2EI}$$

The deflection y_P at the end is made up of two parts: y_1 for a beam of length a, and y_2 the added deflection of the tangent segment (straight line) from P to the end of the beam. This deflection is

$$y_P = y_1 + y_2 = -\frac{Pa^3}{3EI} + (L - a)\theta_P$$

$$= -\frac{Pa^3}{3EI} + (L - a)\left(-\frac{Pa^2}{2EI}\right) = \frac{Pa^3}{6EI} - \frac{Pa^2L}{2EI}$$

The slope and deflection at the end of the beam due to the shear V_R are (also from case 1, Appendix D)

$$\theta_V = -\frac{V_RL^2}{2EI} \quad \text{and} \quad y_V = -\frac{V_RL^3}{3EI}$$

Finally, the slope and deflection at the right end of the beam due to M_R are (see case 4, Appendix D)

$$\theta_M = \frac{M_RL}{EI} \quad \text{and} \quad y_M = \frac{M_RL^2}{2EI}$$

Since the resultant slope is zero,

$$\theta_P + \theta_V + \theta_M = -\frac{Pa^2}{2EI} - \frac{V_RL^2}{2EI} + \frac{M_RL}{EI} = 0$$

Similarly,

$$y_P + y_V + y_M = \frac{Pa^3}{6EI} - \frac{Pa^2L}{2EI} - \frac{V_R L^3}{3EI} + \frac{M_R L^2}{2EI} = 0$$

The simultaneous solution of these two equations yields

$$M_R = \frac{-Pa^2(L - a)}{L^2} \quad \text{and} \quad V_R = \frac{-Pa^2(3L - 2a)}{L^3} \qquad \text{Ans.}$$

When the equations of equilibrium are applied to a free-body diagram of the beam, the shear and moment at the left end are found to be

$$M_L = \frac{-Pa(L - a)^2}{L^2} \quad \text{and} \quad V_L = \frac{+P(L^3 - 3a^2L + 2a^3)}{L^3} \qquad \text{Ans.}$$

PROBLEMS

> *Note: Use the method of superposition to solve the following problems.*

8-43*   A beam is loaded and supported as shown in Fig. P8–6. Determine the left reaction in terms of w and L.

8-44*   A beam is loaded and supported as shown in Fig. P8–11. Determine the reaction at the center of the beam in terms of w and L.

8-45   A beam is loaded and supported as shown in Fig. P8–17. Determine the reaction components at the right end of the beam in terms of w and L.

8-46*   A beam is loaded and supported as shown in Fig. P8–18. Determine the reaction at the center of the beam in terms of w and L.

8-47*   A beam is loaded and supported as shown in Fig. P8–19. Determine the reaction at the center of the beam in terms of w and L.

8-48   A beam is loaded and supported as shown in Fig. P8–24. Determine the reaction at the right support in terms of w and L.

8-49   A beam is loaded and supported as shown in Fig. P8–30. Determine the reaction at the left support in terms of w and L.

8-50*   A beam is loaded and supported as shown in Fig. P8–31. Determine the reaction at the left support in terms of w and L.

8-51* A 6-in. wide by 10-in. deep timber beam 20 ft long is simply supported at the ends and at the middle of the span. Under a uniformly distributed load of 400 lb/ft, the middle support settles 3/4 in. relative to the end supports. The flexural modulus of elasticity of the timber is 1200 ksi. Determine the reactions at the center and ends of the beam.

8-52 A timber beam 100 mm wide by 250 mm deep is 4 m long. The beam is fixed at the left end, simply supported at the right end, and subjected to a uniformly distributed load of w N/m. If the right support settles an amount equal to $wL^4/(24EI)$ when the load w is applied, determine the maximum total load that can be carried by the beam if the maximum allowable fiber stress is 10 MPa. The flexural modulus of elasticity of the timber is 8.5 GPa.

8-53* The 4-in. wide by 6-in. deep timber beam of Fig. P8–53 is fixed at the left end and supported at the right end by an aluminum alloy tie rod that has a cross-sectional area of 0.15 sq. in. The moduli of elasticity of the wood and aluminum are 1200 ksi and 10,000 ksi, respectively. Determine the tension in the tie rod if it is unstressed before the load is applied to the timber beam.

8-54 Solve Prob. 8-53 if there is a 0.30-in. gap between the bottom of the beam and the bearing plate attached to the tie rod before the load is applied to the beam.

8-55* A steel beam ($E = 200$ GPa) is loaded and supported as shown in Fig. P8–55. Consider the end supports and the support at D (but not the post) to be rigid. The post BD is a 100×150-mm timber ($E = 8$ GPa) that is braced to

FIG. P8–53

FIG. P8–55

FIG. P8–57

FIG. P8–58

prevent buckling. The moment of inertia of the cross section of the beam with respect to the neutral axis is $61(10^6)$ mm^4. Determine the reaction at B due to the 8-kN/m load on the beam.

8-56   Solve Prob. 8-55 if the post BD is replaced by two coil springs in series at B (one spring on top of the other and a rigid support under the springs). The spring moduli are 1.4 MN/m and 2.1 MN/m (that is, one spring will deflect 1 mm under a load of 1.4 kN and the other will deflect 1 mm under a load of 2.1 kN).

8-57   A timber beam is loaded and supported as shown in Fig. P8–57. Consider the supports at A, C, and D to be rigid and the tension in rod BD to be zero before the load is applied to the beam. The moduli of elasticity for the timber and steel are 1500 ksi and 30,000 ksi, respectively.

(a)   Determine the reaction at B.

(b)   Draw complete shear and bending moment diagrams for the beam.

8-58   In Fig. P8–58, beam A is made of brass and beam B is made of steel. The moment of inertia of B is twice that of A, and the modulus of elasticity of the steel is twice that of the brass. The reaction at C is zero before the load w is applied. Determine the reaction at C on beam B in terms of w and L.

FIG. P8–59   FIG. P8–60

8-59*   In Fig. P8–59, the aluminum alloy tie rod passes through a hole in the aluminum alloy cantilever beam and through the coil spring positioned on the end of the tie rod. Before loading, there is a clearance of 2.5 mm between the bottom of the beam and the top of the spring. The cross-sectional area of the tie rod is 100 mm^2, the moment of inertia of the cross section of the beam with respect to the neutral axis is $40(10^6)$ mm^4, the modulus of elasticity of the aluminum alloy is 70 GPa, and the spring modulus is 1000 kN/m. Determine the axial stress in the tie rod when $M = 9$ kN $\cdot$ m and $w = 90$ kN/m.

8-60   The steel beam AB of Fig. P8–60 is fixed at ends A and B and supported at the center by the pin-connected wooden struts CD and CE. The cross-sectional area of each strut is $12(10^3)$ mm^2, the moment of inertia of the cross section of the beam with respect to the neutral axis is $30(10^6)$ mm^4, and

FIG. P8–62

the moduli of elasticity of the wood and steel are 8 GPa and 200 GPa, respectively. Determine the force in each strut due to the distributed load of 6 kN/m. The constants in the expression for maximum deflection of a beam with fixed ends are 1/384 for a uniformly distributed load and 1/192 for a concentrated load at the center.

8-61*   A uniformly distributed load of 500 lb/ft is supported by two beams arranged as shown in Fig. P8–61. Beam *AB* is fixed at the wall and beam *CD* is simply supported. Before the load is applied, the beams are in contact at *B*, but the reaction at *B* is zero. The moment of inertia of the cross section with respect to the neutral axis for each of the beams is 10 in.[4]. The modulus of elasticity of the material is 1200 ksi. Determine the deflection at *B* when the load is applied.

8-62   Two steel beams ($E = 200$ GPa) support a concentrated load of 7 kN as shown in Fig. P8–62. In the unloaded condition, beam *AB* touches but exerts no force on beam *CD*. The moment of inertia of the cross section with respect to the neutral axis is $2.79(10^6)$ mm[4] for beam *AB* and $6.24(10^6)$ mm[4] for beam *CD*. Determine the deflection of beam *AB* at the point of application of the 7-kN force.

8-5   ENERGY METHOD—CASTIGLIANO'S THEOREM

Castigliano's theorem (see Art. 7–11) is an effective supplement to the equations of equilibrium in the solution of statically indeterminate structures. If the deflection at some point in a structure is known (a boundary condition), Eq. 7–6a may be applied at this point and set equal to the known deflection. The resulting equation will contain one or more unknown forces. If the equation contains more than one unknown, additional equations must be written, such as equilibrium equations or the application of Eq. 7–6b. However, there are certain restrictions necessary to ensure that the equations are independent; also, some remarks regarding procedure may be in order. The following outline may be found helpful:

1.   For any system of loads and reactions, as many independent energy equations (Eq. 7–6a or 7–6b) can be written as there are redundant unknowns—defined as unknowns not necessary to maintain equilibrium of the structure. This restriction is equivalent to the restriction of Art. 8–2 requiring one boundary condition for each extra unknown.

2.   If only one independent energy equation can be written and the moment equation contains two unknowns (forces or couples, including dummy loads), one of the unknowns must be expressed

in terms of the other by means of an equilibrium equation. This case is illustrated in the alternate solution of Exam. 8–5 which follows. If two independent energy equations can be written, these can contain two unknowns that can be obtained from a solution of the two energy equations, as illustrated by Exam. 8–7.

3.   If the loading changes along the beam, the moment equation can be written for the entire span using the singularity notation of Art. 7–5 and, after multiplication by $\partial M/\partial P_i$ or $\partial M/\partial M_i$ (also using singularity notation), integration can be performed over the entire span. However, this procedure requires integration by parts of such expressions as $x\langle x - L\rangle^1 dx$. Sometimes it may be more convenient to write an ordinary algebraic moment equation for each interval, multiply by $\partial M/\partial P_i$ or $\partial M/\partial M_i$ (different for each interval), and integrate over each interval; then add the results of all the integrations.

EXAMPLE 8–5

Solve Exam. 8–1 of Art. 8–2 by the application of Castigliano's theorem. The beam is loaded and supported as shown in Fig. 8–5a.

FIG. 8–5

SOLUTION

The free-body diagram of Fig. 8–5b indicates that the problem is statically indeterminate, since there are three unknown reaction components and only two independent equations of equilibrium available.

If the portion of the beam to the right of the reaction R was removed and replaced by an equivalent shearing force and bending moment at the transverse section above R, neither R nor the elastic curve in the interval $0 \leqslant x \leqslant L$ would be changed. Hence, it is necessary to deal with only the strain energy of the beam in the interval $0 \leqslant x \leqslant L$. With the coordinate system placed as shown, the resulting moment equation obtained from the free-body diagram in Fig. 8–5c is

$$M = Rx - \frac{w}{2}\left(x + \frac{L}{4}\right)^2$$

$$= Rx - \frac{wx^2}{2} - \frac{wLx}{4} - \frac{wL^2}{32}$$

From Eq. 7–6a, the deflection at the right support is given by

$$y_O = \frac{\partial U}{\partial R} = \frac{1}{EI}\int_0^L M\,\frac{\partial M}{\partial R}\,dx$$

Since the partial derivative of M with respect to R is x, the expression for the deflection is

$$y_O = \frac{1}{EI}\int_0^L \left(Rx^2 - \frac{wx^3}{2} - \frac{wLx^2}{4} - \frac{wL^2x}{32}\right)dx$$

which upon integration and substitution of limits becomes

$$y_O = \frac{1}{EI}\left(\frac{RL^3}{3} - \frac{wL^4}{8} - \frac{wL^4}{12} - \frac{wL^4}{64}\right)$$

Since the support is unyielding ($y = 0$ when $x = 0$), the above expression can be equated to zero and solved for R, yielding

$$R = \frac{43wL}{64} \text{ lb upward} \qquad\qquad \text{Ans.}$$

The other reaction components (M_A and V_A) are obtained from the equilibrium equations $\Sigma M_A = 0$ and $\Sigma F_y = 0$ and are

$$M_A = \frac{7wL^2}{64} \text{ ft-lb counterclockwise} \qquad\qquad \text{Ans.}$$

and

$$V_A = \frac{37wL}{64} \text{ lb upward} \qquad\qquad \text{Ans.}$$

ALTERNATE SOLUTION

This solution will make use of the free-body diagram of Fig. 8–5d, from which the moment equation is

$$M = M_A + V_A x - \frac{wx^2}{2}$$

Since there are two unknowns in the moment equation and only one independent energy equation can be written, the equilibrium equation $\Sigma M_B = 0$ will be used to obtain a relation between the two unknowns. Thus,

$$M_A + V_A L - \frac{wL^2}{2} + \frac{wL^2}{32} = 0$$

from which

$$M_A = \frac{15wL^2}{32} - V_A L \qquad \text{or} \qquad V_A = \frac{15wL}{32} - \frac{M_A}{L}$$

Eliminating M_A from the moment equation gives

$$M = V_A(x - L) + \frac{15wL^2}{32} - \frac{wx^2}{2}$$

and

$$\frac{\partial M}{\partial V_A} = x - L$$

From Eq. 7–6a, the deflection at the left support is given by

$$y_A = \frac{\partial U}{\partial V_A} = \frac{1}{EI} \int_0^L M \frac{\partial M}{\partial V_A} \, dx$$

$$= \int_0^L \left[V_A(x - L)^2 - \frac{wx^2}{2}(x - L) + \frac{15wL^2}{32}(x - L) \right] dx = 0$$

from which

$$V_A = \frac{37wL}{64} \text{ lb upward} \qquad\qquad \text{Ans.}$$

The other reaction components (M_A and R) can be obtained from equilibrium equations.

EXAMPLE 8–6

Beam BC of Fig. 8–6a is a rolled steel W203 × 22 beam having a cross-sectional moment of inertia of $20(10^6)$ mm^4. The beam is fixed at B and supported at C by an aluminum alloy tie rod having a cross-sectional area of 100 mm^2. Determine the force in the tie rod due to the distributed load of 12 kN/m. Use 200 GPa and 70 GPa for the moduli of elasticity of the steel and aluminum, respectively.

SOLUTION

The free-body diagram of the beam and tie rod is shown in Fig. 8–6b with the origin of coordinates at the top of the tie rod where the boundary condition is

FIG. 8–6

$x = 0$, $y = 0$. The total strain energy in the system is the sum of the energies in the tie rod and the beam; that is,

$$U = \frac{\sigma^2}{2E_a} Al + \int_0^L \frac{M^2}{2E_sI} \, dx = \frac{P^2l}{2AE_a} + \int_0^L \frac{M^2}{2E_sI} \, dx$$

The deflection at D is

$$y_O = \frac{\partial U}{\partial P} = \frac{Pl}{AE_a} + \frac{1}{E_sI} \int_0^L M \frac{\partial M}{\partial P} \, dx = 0$$

The moment in the beam is

$$M = Px - \frac{wx^2}{2}$$

and

$$\frac{\partial M}{\partial P} = x$$

Therefore,

$$y_O = \frac{Pl}{AE_a} + \frac{1}{E_sI} \int_0^L \left(Px^2 - \frac{wx^3}{2} \right) dx = \frac{Pl}{AE_a} + \frac{PL^3}{3E_sI} - \frac{wL^4}{8E_sI}$$

Since the structure is fixed at D, y_O is zero and substitution of the given data yields

$$\frac{P(7.5)}{100(10^{-6})(70)(10^9)} + \frac{P(3)^3}{3(200)(10^9)(20)(10^{-6})} = \frac{12(10^3)(3)^4}{8(200)(10^9)(20)(10^{-6})}$$

from which

$$P = 9145 \text{ N} = \underline{9.15 \text{ kN } T} \qquad\qquad \text{Ans.}$$

EXAMPLE 8–7

Determine the reactions at the left end of the beam of Fig. 8–7a in terms of P, L, and a.

SOLUTION

The free-body diagram of Fig. 8–7b indicates four unknown reaction components, and since there are only two independent equations of equilibrium, two supplementary equations are necessary. These are obtained by applying

FIG. 8–7

(a)

(b)

Eq. 7–6a and 7–6b and solving the resulting simultaneous equations for M_O and V_O. With the origin at the left support, the boundary conditions are $y = 0$ and $\theta = 0$ at $x = 0$. Since the moment equation is

$$M = M_O + V_O x - P\langle x - a \rangle^1$$

the deflection at the left end is given by Eq. 7–6a as

$$y_O = \frac{\partial U}{\partial V_O} = \frac{1}{EI} \int_0^L M \frac{\partial M}{\partial V_O}\, dx$$

and

$$EIy_O = \left[\frac{M_O x^2}{2} + \frac{V_O x^3}{3} - P\left(\frac{x}{2}\langle x - a \rangle^2 - \frac{1}{6}\langle x - a \rangle^3 \right) \right]_0^L$$

where the last term was integrated by parts. Substitution of the limits gives

$$EIy_O = \frac{M_O L^2}{2} + \frac{V_O L^3}{3} - \frac{PL}{2}(L - a)^2 + \frac{P}{6}(L - a)^3$$

The application of the boundary condition $y = 0$ at $x = 0$ gives

$$\frac{M_O L^2}{2} + \frac{V_O L^3}{3} - \frac{PL^3}{3} + \frac{PL^2 a}{2} - \frac{Pa^3}{6} = 0 \qquad (a)$$

When Eq. 7–6b is applied, the rotation at the left end is given by the expression

$$\theta_O = \frac{\partial U}{\partial M_O} = \frac{1}{EI} \int_0^L M \frac{\partial M}{\partial M_O}\, dx$$

and

$$EI\theta_O = \int_0^L (M_O + V_O x - P\langle x - a \rangle^1) dx$$

$$= \left[M_O x + \frac{V_O x^2}{2} - \frac{P}{2} \langle x - a \rangle^2 \right]_0^L$$

which reduces to

$$M_O L + \frac{V_O L^2}{2} - \frac{P L^2}{2} + PaL - \frac{Pa^2}{2} = 0 \qquad (b)$$

Equations a and b are solved to obtain the following results:

$$V_O = P - \frac{3Pa^2}{L^2} + \frac{2Pa^3}{L^3} \qquad \text{Ans.}$$

and

$$M_O = -Pa + \frac{2Pa^2}{L} - \frac{Pa^3}{L^2} \qquad \text{Ans.}$$

PROBLEMS

Note: Use Castigliano's theorem to solve the following problems.

8-63*   A beam is loaded and supported as shown in Fig. P8–1. When the couple Q is applied, the slope at the left end of the beam is zero. Determine

(a)   The magnitude of the couple Q in terms of w and L.

(b)   The deflection at the left end of the beam in terms of w, L, E, and I.

8-64*   A beam is loaded and supported as shown in Fig. P8–10. When the moment M_O is applied, the slope of the deflected beam over the support at A is $wL^3/(60EI)$ upward to the left. Determine

(a)   The magnitude of the moment M_O in terms of w and L.

(b)   The deflection at the left end of the beam in terms of w, L, E, and I.

FIG. P8–67

FIG. P8–68

8-65*   A beam is loaded and supported as shown in Fig. P8–11. Determine
   (a)   The reaction at the center support in terms of w and L.
   (b)   The deflection at a section midway between supports in terms of w, L, E, and I.

8-66   A beam is loaded and supported as shown in Fig. P8–12. Determine
   (a)   The reaction at support B in terms of w and L.
   (b)   The deflection at a section midway between supports B and C in terms of w, L, E, and I.

8-67   A beam is loaded and supported as shown in Fig. P8–67. Determine
   (a)   The reaction at A in terms of w and L.
   (b)   The deflection at B in terms of w, L, E, and I.

8-68*   A beam is loaded and supported as shown in Fig. P8–68. Determine
   (a)   The components (shear and moment) of the reaction at support A in terms of w and L.
   (b)   The deflection at C in terms of w, L, E, and I.

8-69   A beam is loaded and supported as shown in Fig. P8–17. Determine
   (a)   The components (shear and moment) of the reaction at support C in terms of w and L.
   (b)   The deflection at B in terms of w, L, E, and I.

8-70*   A beam is loaded and supported as shown in Fig. P8–18. Determine the reactions at supports A and B in terms of w and L.

8-71*   A beam is loaded and supported as shown in Fig. P8–19. Determine the reaction at support B in terms of w and L.

8-72   A beam is loaded and supported as shown in Fig. P8–24. Determine
   (a)   The reaction at support B in terms of w and L.
   (b)   The deflection at D in terms of w, L, E, and I.

8-73   Use Castigliano's theorem to solve Prob. 8-17.

8-74*   Use Castigliano's theorem to solve Prob. 8-23.

8-75   Use Castigliano's theorem to solve Prob. 8-24.

8-76*   Use Castigliano's theorem to solve Prob. 8-53.

8-77   The steel (E = 200 GPa) beam of Fig. P8–77 is fixed at the left end and, when loaded, is supported by a helical spring B at the right end. Determine the reaction of the spring B on the beam if the following numerical data apply: I = 20(10^6) mm^4, L = 5 m, w = 3 kN/m, y_0 = 5 mm (when the beam is unloaded), spring modulus = 1200 kN/m.

8-78*   The curved beam of Fig. P8–78 has a constant radius and constant E and I. Assume that the ratio R/c is large enough for the elastic flexure formula to apply and neglect axial stress. Determine the reaction at B on the beam in terms of P.

8-79   The frame ABC of Fig. P8–79 is a continuous member of constant E and I. Assume the corner at B remains a right angle and neglect the energy due to the axial stress in leg BC. Determine

(a)   The reaction at A on member AB in terms of w and L.

(b)   The horizontal deflection of end A in terms of w, L, E, and I.

8-80   A frame and loading are identical to those of Prob. 8-79 except that the end A is pinned as shown in Fig. P8–80. Neglect the energy due to axial

FIG. P8–78

FIG. P8–79

FIG. P8–80

stresses in *AB* and *BC*. Determine the horizontal and vertical components of the pin reaction at *A* on member *AB*.

8-6   PLASTIC ANALYSIS OF STATICALLY INDETERMINATE BEAMS

Because of increasing emphasis on structural analysis based on plastic action, a brief introduction to plastic analyses of statically indeterminate beams is presented in this article. The analysis will be limited to ideal elasto-plastic materials as defined in Art. 3–2. For practical purposes, ordinary structural steel will be satisfactory, and for this material the following assumptions are reasonable and may be verified (within practical limits) by experiment:

1.   As the load on the beam increases, the fiber stress will increase proportionately until the yield point for the material is attained at the section of maximum moment, after which yielding at this section will progress over the entire section (see Art. 6–3 and Fig. 6–9*f*). The resisting moment will increase with the load until the stress in each fiber has reached the yield point, after which the moment will remain constant. In other words, a hinge exerting a constant moment forms at the section, producing a discontinuity in the slope of the elastic curve. The moment at the hinge is termed the *plastic moment M_p*.
2.   As the loading is increased, a second hinge will form and additional hinges will form until the beam is reduced to what has been termed a *mechanism*; that is, a system of members, connected at their ends by pins (each pin exerting a constant moment), which will continue to deflect with no additional load.

The load producing a mechanism is considered the *ultimate load*, although strain hardening of the material might necessitate the application of a larger load to produce complete collapse. The hinge, which actually extends over a finite length of beam, is assumed to form at a discrete point in the beam.

One procedure for analyzing statically indeterminate beams consists of arbitrarily assigning the plastic moment value to the redundant moments; this procedure makes the beam statically determinate. Additional hinges are introduced at points of high moment until a mechanism develops. Finally, a check of the moments should show that nowhere does the moment exceed the magnitude of the plastic moment M_p. This is not the only method used in plastic analysis,[1] but for the simpler cases it is probably the easiest to apply. The method is illustrated by the following example.

EXAMPLE 8–8

The continuous beam of Fig. 8–8a is a steel W36 × 160 section. Consider the steel to be elastoplastic with a yield point of 36 ksi and determine the ultimate load w (which for span BC includes the weight of the beam) for the beam.

SOLUTION

The free-body diagram of Fig. 8–8b indicates three unknown reactions, and for this force system there are only two independent equations of equilibrium; hence, there is one redundant quantity. If one moment, shear, or reaction is determined, the beam becomes statically determinate. However, the ultimate load is not attained until a mechanism forms; therefore, it becomes necessary to assume more than one possible hinge location. To assist in estimating the location of possible hinges, the shape of the moment diagram by parts and the composite moment diagram are shown in Fig. 8–8c, where the solid line indicates the composite diagram and the dashed lines represent the diagrams for two simple beams with a moment at B. This diagram indicates that hinges may possibly form at B, D, and E. Possible deformations are indicated (greatly exaggerated) in Fig. 8–8d, the top curve indicating the shape of the elastic curve when the action is entirely elastic. As the magnitude of the loading is increased beyond that necessary to produce the yield point stress in the outer fibers, a hinge will form at the section where the flexural stress is the highest. The first hinge is assumed to form at B, and under this condition the beam would assume the shape indicated by the middle curve of Fig. 8–8d. Since a mechanism has not yet formed, the loading is increased until a second hinge forms. This hinge is assumed to form at D, and from the lower curve of Fig. 8–8d it is apparent that a mechanism has developed since, with hinges at B and D and with end C supported on rollers, deflection can continue in the span BC without an increase in the load. The second hinge may have formed at E instead of at D; hence, the moment at E must be evaluated as a check on the hinge assumptions.

The magnitude of the load necessary to produce hinges at B and D will first be determined, after which the moment at E will be checked. It should

---

[1] For additional material see *Plastic Design of Steel Frames,* L. S. Beedle, Wiley, New York, 1958.

FIG. 8–8

be noted that when the plastic moment M_p is used in either an equilibrium or a moment equation, it must have the correct sense or sign. With reference to the free-body diagram of Fig. 8–8e, which represents the right portion of span BC, the equation for the moment in span BC is

$$M_x = R_C x - wx^2/2 \qquad (a)$$

When x is 80 ft,

$$M_B = -M_p = R_C(80) - w(80)^2/2$$

from which

$$R_C = 40w - M_p/80 \qquad (b)$$

The negative sign on M_B is obtained from Fig. 8–8c. The section of maximum positive moment in span BC is at D where the shear is zero and is located from the shear equation

$$V_x = -R_C + wx$$

When x is equal to d,

$$V_x = 0 = -R_C + wd$$

from which

$$d = R_C/w \qquad (c)$$

From Eq. a the moment at D (the plastic moment) is

$$M_D = +M_p = R_C d - wd^2/2$$

Since R_C is equal to wd from Eq. c, the plastic moment becomes

$$M_p = wd^2 - \frac{wd^2}{2} = \frac{w}{2}\left(40 - \frac{M_p}{80w}\right)^2 \qquad (d)$$

where the plus sign on M_D is obtained from Fig. 8–8c and R_C is obtained from Eq. b. Solving Eq. d for w gives

$$w = \frac{5.8284\,M_p}{3200} \quad \text{or} \quad \frac{0.1716M_p}{3200}$$

When these expressions are substituted into Eq. b, only the larger value yields a positive result for d; therefore, this expression will give the correct magnitude for w unless the bending moment at E exceeds M_p. The plastic moment for the W36 $\times$ 160 section is

$$\overbrace{\phantom{[\sigma_{yp}(12)(1.02)(18-0.51)]}}^{\text{one flange}} \qquad \overbrace{\phantom{\sigma_{yp}(0.650)(18-1.02)^2(1/2)]}}^{\text{one-half web}}$$

$$M_p = [\sigma_{yp}(12)(1.02)(18 - 0.51) + \sigma_{yp}(0.650)(18 - 1.02)^2(1/2)](2) = 616\sigma_{yp}$$

$$= 616(36)/12 = 1848 \text{ ft-kips}$$

Substituting this value for M_p in the expression for w yields

$$w = 5.828(616)(3)/3200 = 3.366 \text{ kips/ft}$$

To check the bending moment at E, the reaction at A will first be obtained by summing moments with respect to B on the free-body diagram of Fig. 8–8f. The equation

$$60R_A - 40w(36) - 0.16(60)(30) + M_p = 0$$

gives

$$R_A = 24(3.366) + 4.80 - 616(3)/60 = 54.8 \text{ kips}$$

The bending moment at E is

$$M_E = 54.8(24) - 0.16(24)(12) = 1269 \text{ ft-kips}$$

which is less than 1848; therefore, a hinge does not form at E, the moment in the beam is nowhere greater than M_p, and the ultimate load is

$$w = \underline{3.37 \text{ kips/ft}} \qquad \text{Ans.}$$

Some economy in the design could be effected by reducing the size of the beam in the left span to such a section that a hinge might form at E; that is, a beam section could be selected for which the plastic moment is at least 1269 ft-kips. If a table of plastic section moduli was available,[2] the section could be determined from the expression $M_p = Z \sigma_{yp}$, where Z is the plastic section modulus.[3] If such a table is not available, the following method may be used. It has been shown[4] that the ratio Z/S (where S is the elastic section modulus) for wide flange and American standard sections ranges from 1.10 to 1.23, with 1.14 as the average for all wide flange sections and 1.18 as the average for American standard sections. Using the average value of 1.14 gives

$$S = \frac{Z}{1.14} = \frac{M_E}{1.14\sigma_{yp}} = \frac{1269(12)}{1.14(36)} = 371 \text{ in.}^3$$

From Appendix C, a W33 × 130 section with a section module of 406 appears to be a likely choice, and it will be checked as follows. When the cross-sectional dimensions given in Appendix C are used, the plastic moment is

$$\overbrace{[11.51(0.855)(16.12)}^{\text{one flange}} + \overbrace{15.69(0.580)(15.69/2)]}^{\text{one-half web}}(2)(36/12) = 1379 \text{ ft-kips}$$

---

[2] See AISC *Manual of Steel Construction* and BCSA-CONSTRADO *Structural Steelwork Handbook* for standard metric sections.

[3] The plastic modulus is the arithmetic sum of the absolute values of the first moments, with respect to the neutral axis, of the cross-sectional areas on each side of the neutral axis.

[4] See Fig. 7.1 of *Plastic Design of Steel Frames*, L. S. Beedle, Wiley, New York, 1958.

which is more than adequate. Therefore, one could splice the W33 × 130 beam to the W36 × 160 beam at some section between B and E. Theoretically, the splice could be made where the bending moment in span AB has a value of −1379 ft-kips (which is approximately 5.3 ft left of B). Practically, to avoid splicing at a section of large moment, the joint might be located some distance to the left, the distance depending on the judgment of the designer or a pertinent specification. The resulting beam will develop hinges at B and D and will just begin to yield at E under a load of 3.37 kips/ft.

The maximum value for w, based on elastic analysis and with the assumption that the I_{AB} for the full length of 60 ft is $(7/10)I_{BC}$, is found to be 2.76 kips/ft; thus, the plastic analysis for this problem gives a loading approximately 22 percent higher than that obtained from elastic analysis. Part of this increase is due to the plastic moment being greater (about 14 percent) than the elastic moment, and part is due to the increase in load necessary to cause formation of the second plastic hinge. These loads will be considered failure loads that, when divided by an appropriate factor of safety,[5] will yield working loads.

The discussion presented in this article is very limited in scope and hence does no more than present some of the fundamental concepts of plastic design involving an ideal elastoplastic material. A complete analysis would involve the rotation capacity of the beam sections used, to ascertain if the first hinge to form will sustain sufficient plastic deformation to permit the formation of successive hinges. Furthermore, shearing stresses or local buckling may be quite significant for some designs. These and other design problems are beyond the scope of this book.

PROBLEMS

Note: The following problems are to be solved using plastic analysis, and for simplification the weight of the beam is to be neglected. All beams are made of structural steel having a yield point and proportional limit of 36 ksi or 250 MPa. For wide flange beam design, use Z = 1.14 S.

[5] Here, the factor of safety is defined as the ratio of the ultimate (failure) load to the computed (or specified) applied load. This definition, which differs from that of Art. 3–6, will be used in all plastic analysis problems in this book.

8-81*   A beam is loaded and supported as shown in Fig. P8–81. Determine the ultimate load w in terms of M_p and L.

8-82*   A beam is loaded and supported as shown in Fig. P8–82. Determine the ultimate load P in terms of M_p and L.

8-83   A beam is loaded and supported as shown in Fig. P8–83. Determine the ultimate load P in terms of M_p and L.

8-84   A beam is loaded and supported as shown in Fig. P8–84. Determine the ultimate load w in terms of M_p and L.

8-85*   A two-span continuous beam with simple supports is subjected to a uniformly distributed load of 1000 lb/ft on both spans. The span lengths are 32 ft and 40 ft, respectively. Determine the lightest wide-flange beam required if a factor of safety of 2 is specified and if the beam is to have a constant section throughout its length.

8-86*   A wide-flange section having a plastic section modulus $Z = 3667(10^3)$ mm^3 is used as a two-span, simply supported, continuous beam with spans of 8 m and 14 m, respectively. The beam carries uniformly distributed loads of $2w$ kN/m on the 8-m span and w kN/m on the 14-m span. Determine

   (a)   The ultimate load w for the beam.

   (b)   The lightest wide-flange beam required for the 8-m span.

   (c)   The theoretical splice point.

8-87   The beam shown in Fig. P8–81 is a wide-flange section that has a plastic section modulus $Z = 57.6$ in.^{3} and a length $L = 24$ ft. Determine the magnitude of the ultimate load w.

8-88*   The beam shown in Fig. P8–81 has a length $L = 20$ ft and is carrying a load $w = 2500$ lb/ft. Determine the lightest wide-flange section required if a factor of safety of 2 is specified.

8-89   The beam shown in Fig. P8–84 is a wide-flange section that has a plastic section modulus $Z = 900(10^3)$ mm^3 and a length $L = 8$ m. Determine the magnitude of the ultimate load w.

8-90*   The beam shown in Fig. P8–84 has a length $L = 6$ m and is to carry a load $w = 40$ kN/m. Determine the lightest wide-flange section required if a factor of safety of 2 is specified.

8-91   A beam is loaded and supported as shown in Fig. P8–91. Determine the ultimate load w in terms of M_p and L.

8-92   A beam is loaded and supported as shown in Fig. P8–92. Determine the ultimate load w in terms of M_p and L.

FIG. P8–81

FIG. P8–82

FIG. P8–83

FIG. P8–84

FIG. P8–91

FIG. P8–92

FIG. P8–93

8-93   The continuous beam shown in Fig. P8–93 is a wide-flange section that has a plastic section modulus $Z = 38$ in.³. Determine

(a)   The ultimate load w.

(b)   The lightest wide-flange section required for the portion of the beam from near C to E.

(c)   The theoretical splice point.

Hint: Assume three plastic hinges in span BC.

8-7   CLOSURE

In this chapter, four methods of elastic analysis and one method of plastic analysis of statically indeterminate beams were presented. When the analysis is to be based on elastic action, the choice of method will be governed by the particular problem and the analyst's familiarity with certain of the four methods presented. As a possible aid to selection of the method, the following outline is submitted.

1.   The integration method is the fundamental method and may be used to solve any problem although its application may sometimes be laborious. This method requires that, in addition to the boundary conditions necessary for the determination of constants of integration, there must be additional boundary (or matching) conditions equal in number to the difference between the number of unknowns and the number of independent equations of equilibrium.

2.   The area-moment method provides a neat and somewhat less laborious technique to supplement or replace the integration approach. The method requires the application of an area-moment theorem for each unknown in excess of the number of independent equations of equilibrium. The area-moment equations to be written will be dictated by the geometry of the deflected beam. If the expressions for the areas and centroids of the moment diagrams are not known, the areas and their first moments may be obtained by integration. The method is applicable to any problem and is readily applied to a beam with variable moment of inertia by drawing an M/I diagram and subdividing it into finite areas.

3. For many problems the superposition method provides the most rapid solution. The application of this method depends on the availability of a summary (or a memorization) of slope and deflection expressions for various spans and loadings and, except for the simpler cases, depends on the ingenuity of the analyst.

4. The energy method is preferred by many and is frequently much shorter than other methods. The method presented in this book requires one or more applications of Castigliano's theorem (Eq. 7–6a or 7–6b) at points where the deflections or slopes are known. The use of singularity functions to provide a single moment equation for the entire span is frequently advantageous.

CHAPTER 9

COMBINED STATIC LOADING

9-1 INTRODUCTION

The stresses and strains produced by three fundamental types of loads (centric, torsional, and flexural) have been analyzed in the preceding chapters. Many machine and structural elements are subjected to a combination of any two or all three of these types of loads, and a procedure for calculating the stresses on a given plane at a point is required. One method is to replace the given force system by an equivalent system of forces and couples, each of which is selected in such a manner that the stresses produced by it can be calculated by methods already developed. The composite effect can be obtained by the principle of superposition, as explained in Art. 3–5, provided that the combined stresses do not exceed the proportional limit. Various combinations of loads that can be analyzed in this manner are discussed in the following articles.

9-2 COMBINED AXIAL AND TORSIONAL LOADS

There are numerous situations in which a shaft or other machine member is subjected to both an axial and a torsional load. Examples are the drill rod on a well or other drilling machine and the propeller shaft on a ship. Since radial and circumferential normal stresses are zero, the combination of axial and torsional loads results in a state-of-plane stress at any point in the body. The axial stress is the same for all interior and exterior points, while the torsional stress is maximum at the outer surface; therefore, the investigation for critical stresses would normally be made for points on the surface of the shaft.

The following example illustrates the procedure for combined torsional and axial loads.

EXAMPLE 9–1

A hollow circular shaft of outer diameter 4 in. and inner diameter 2 in. is loaded as shown in Fig. 9–1a.

(a) Determine the principal stresses and the maximum shearing stress at the point (or points) where the stress situation is most severe.

(b) On a sketch show the approximate directions of these stresses.

SOLUTION

The axial stress is constant throughout the shaft and is equal to

$$\sigma = \frac{P}{A} = \frac{24{,}000\pi}{\pi(2^2 - 1^2)} = 8000 \text{ psi } C$$

The torsional load is highest to the right of the torque T_1, and the torsional stress is maximum on the outer surface of the shaft; therefore, the magnitude of the maximum torsional shearing stress is

(a)

(b)

(c)

(d)

FIG. 9–1

$$\tau = \frac{Tc}{J} = \frac{22,500\pi(2)}{\pi(2^4 - 1^4)/2} = 6000 \text{ psi}$$

These stresses are shown acting on mutually perpendicular planes through a point on the surface of the right portion of the shaft in Fig. 9–1b. Mohr's circle for the stresses at any point on the surface of the right portion of the shaft is shown in Fig. 9–1c.

(a) The principal stresses OA and OB from Mohr's circle are

$$\sigma_{p1} = OA = -4.00 + 2\sqrt{13} = \underline{3.21 \text{ ksi } T} \qquad \text{Ans.}$$

and

$$\sigma_{p2} = OB = -4.00 - 2\sqrt{13} = -11.21 = \underline{11.21 \text{ ksi } C} \qquad \text{Ans.}$$

The maximum shearing stress CD or CE from Mohr's circle is

$$\tau_{max} = CD = 2\sqrt{13} = \underline{7.21 \text{ ksi}} \qquad \text{Ans.}$$

(b) From the Mohr's circle,

$$\tan 2\theta_p = 6/4 = 1.5$$

which gives

$$2\theta_p = 56.30° \text{ }\rangle$$

and

$$\theta_p = 28.15° \text{ }\rangle$$

The stresses are shown in their proper directions in Fig. 9–1d.

PROBLEMS

> Note: Assume linearly elastic action for all of the following problems

9-1* A 4-in. diameter shaft is subjected to a torque of 4000π in.-lb and an axial tension of $6000\ \pi$ lb as shown in Fig. P9–1.

(a) Determine the principal stresses and the maximum shearing stress at point A on the surface of the shaft.

FIG. P9–1

4000π in–lb

6000π lb

FIG. P9–2

400 mm

300 mm

$T = 300$ kN·m

$P = 800$ kN

(b) Show the stresses of part (a) and their approximate directions on a sketch.

9-2*   A hollow shaft with an outside diameter of 400 mm and an inside diameter of 300 mm is subjected to a torque of 300 kN · m and an axial tension of 800 kN as shown in Fig. P9–2.

(a) Determine the principal stresses and the maximum shearing stress at a point on the outside surface of the shaft.

(b) Show the stresses of part (a) and their approximate directions on a sketch.

9-3   A 6-in. diameter steel shaft is subjected to the loads shown in Fig. P9–3.

(a) Determine the principal stresses and the maximum shearing stress at point A on the surface of the shaft.

8000π in–lb

3′

1′

6″

6′

$27,000\pi$ lb

$19,000\pi$ in–lb

150 mm

$T = 35$ kN·m

P

FIG. P9–3   FIG. P9–4

FIG. P9–5   FIG. P9–6

(b)   Show the stresses of part (a) and their approximate directions on a sketch.

9-4*   A steel shaft is loaded and supported as shown in Fig. P9–4. Determine the maximum allowable value for the axial load P if the maximum shearing stress in the shaft is not to exceed 70 MPa and the maximum compressive stress in the shaft is not to exceed 125 MPa.

9-5*   A steel shaft is loaded and supported as shown in Fig. P9–5. The allowable shearing stress is 8 ksi and the allowable tensile stress is 12 ksi. With T_2 equal to P in.-kips and T_1 equal to $P/2$ in.-kips, determine the maximum allowable axial load P.

9-6   A steel shaft with the left portion solid and the right portion hollow is loaded as shown in Fig. P9–6. The allowable shearing stress is 80 MPa and the allowable normal stress is 140 MPa. Determine the maximum allowable axial load P.

9-3   COMBINED CROSS SHEAR AND TORSIONAL LOADS

A common example of a cross shear load combined with a torsional load occurs in a closely coiled helical spring. A spring formed by winding a rod around a cylinder in such a manner that the rod forms a helix is called a helical spring. A helical spring is said to be closely coiled when the coils are so close to each other that any one coil can be assumed to lie in a plane essentially perpendicular to the axis of the spring. The discussion in this article will be limited to a spring made of a circular rod of radius r with a helix of mean radius R as shown in Fig. 9–2.

For a closely coiled spring, a cross section of the spring rod perpendicular to the axis of the rod can be assumed to be in the same plane as the axis

FIG. 9–2

(a)

$$\tau_1 = \frac{P}{A}$$

(b)

$$\tau_2 = \frac{PRr}{J}$$

(c)

of the helix (a vertical plane for the cross section shown in Figs. 9–2a through c). The tensile load P applied to the spring is held in equilibrium by the resisting shear V (equal to P) and the resisting torque T (equal to PR) as shown in Fig. 9–2a. The cross (frequently called the direct) shearing stress due to the centric load ($V = P$) is

$$\tau_1 = \frac{P}{A} = \frac{P}{\pi r^2}$$

and is usually considered to be uniformly distributed over the cross section of the rod as shown in Fig. 9–2b. The torsional shearing stress due to the torque ($T = PR$) is

$$\tau_2 = \frac{Tc}{J} = \frac{PRr}{\pi r^4/2}$$

and is maximum at the outer surface of the rod and has the same direction as the direct shearing stress at the left edge of the cross section as shown in Fig. 9–2c. Therefore, the critical shearing stress occurs at the inside surface of the coil and is the sum of the direct and torsional shearing stresses; that is,

$$\tau_{max} = \frac{P}{A} + \frac{Tr}{J} \qquad (9-1)$$

Since the mean radius of the helix, R, is usually large compared to the radius of the rod, r, the direct shearing stress is normally small compared to the torsional shearing stress.

Equation 9–1 must be considered a statement of a basic concept (the addition of stresses acting on a common plane) rather than an adequate equation for stress determination in any helical spring, for the following reasons: (1) The cross shearing stress is not uniformly distributed but is a maximum on the transverse diameter of the rod. (2) The torsion formula, which was derived for a straight bar, gives only an approximate value for the torsional stress because the longitudinal elements of the rod in a helical spring have different lengths, those on the inside being shorter than those on the outside of the coil. Since γ is equal to $r\theta/L$, and τ_2 is equal to $G\gamma$, it is observed that the torsional stress will be highest on the inside of the helix where L has the least value.

A more exact analysis (supported by experimental evidence) of helical springs[1] leads to the Wahl formula

$$\tau_{max} = \frac{2PR}{\pi r^3} \left(\frac{4m - 1}{4m - 4} + \frac{0.615}{m} \right) \qquad (9-2)$$

where m is R/r. For a ratio of R/r of 8, this expression gives a maximum stress about 10 percent greater than that from Eq. 9–1.

The deflection of a closely coiled helical spring results primarily from the torsional shearing stress ($\tau_2 = Tr/J$) in the circular rod. The deflection resulting from the direct or cross shearing stress ($\tau_1 = V/A$) is very small; therefore, it can be neglected without introducing significant error into the deflection expression. Since the spring deflection δ results primarily from torsional effects, the increment $d\delta$ contributed by a small increment of length dL of the rod (which twists through an angle $d\theta$) is given by Eq. 5–6b as

$$d\delta = R \, d\theta = R \frac{T \, dL}{G J}$$

[1] *Mechanical Springs*, A. M. Wahl, Penton Publishing Co., Cleveland, Ohio, 1944.

Integrating over the total length of the rod yields

$$\delta = R\theta = \frac{RTL}{GJ} \tag{9-3a}$$

or since $T = PR$ and $L = 2\pi Rn$, where n is the number of effective coils in the spring,

$$\delta = \frac{R(PR)(2\pi Rn)}{G(\pi/2)(r^4)} = \frac{4PR^3n}{Gr^4} \tag{9-3b}$$

The relationship between the deflection of a spring and the force applied (so long as this relationship is linear) is conveniently expressed in terms of the modulus of the spring. The modulus k of a spring is the force required to stretch or compress the spring a unit distance and in equation form is

$$k = \frac{Gr^4}{4R^3n} = \frac{P}{\delta} \tag{9-4}$$

PROBLEMS

9-7* A helical spring with an inside diameter of 3 in. is made from a 1-in. diameter steel rod. If the spring is subjected to a load of 2000 lb, determine the maximum shearing stress in the spring

    (a) By using Eq. 9–1.

    (b) By using Eq. 9–2.

9-8* A helical spring with a mean diameter of 75 mm is made from a 15-mm diameter alloy steel rod. If the maximum shearing stress must be limited to 350 MPa, determine the maximum load that can be carried by the spring

    (a) By using Eq. 9–1.

    (b) By using Eq. 9–2.

9-9 A helical spring is made from a 1-in. diameter brass ($G = 5000$ ksi) rod. The spring has an inside diameter of 4 in., has 10 turns, and carries a load of 600 lb. Determine

    (a) The maximum shearing stress in the rod. Use Eq. 9–1.

    (b) The deflection of the spring.

9-10*   A helical spring with a mean diameter of 250 mm is made from a 25-mm diameter alloy steel ($G = 80$ GPa) rod. The spring has 8 turns and is subjected to a load of 5 kN. Determine

(a)   The maximum shearing stress in the rod. Use Eq. 9–2.

(b)   The deflection of the spring.

9-11   A helical spring is made by winding 2.5 ft of 1/2-in. diameter steel ($G = 12,000$ ksi) rod around a 3.5-in. diameter cylinder. Determine the maximum shearing stress produced by a load that stretches the spring 0.3 in. Compare results from Eqs. 9–1 and 9–2.

9-12   Design a helical spring made of 15-mm diameter steel ($G = 80$ GPa) rod that will deflect at least 60 mm and have a maximum shearing stress of not more than 275 MPa when supporting a 5-kN load. Use Eq. 9–1.

9-13*   One helical spring is made by winding 3 ft of 1/2-in. diameter steel ($G = 12,000$ ksi) rod around a 2.5-in. diameter cylinder. Another helical spring is made from a 5/8-in. diameter bronze ($G = 6500$ ksi) rod. The second spring has an inside diameter of 3 in. and 10 turns. Both springs are closely coiled and they are attached together in series. Determine the modulus of the two-spring system.

9-4   COMBINED CENTRIC AND FLEXURAL LOADS

A common example of a centric load combined with a bending load is a structural or machine part with a force acting parallel to but not along the axis of the member; for example, the vertical portion AB of the loading hook of Fig. 9–3. Such an eccentric load can be replaced by a centric load (a force acting along the centroidal axis of the part AB) and one or more couples. The stresses resulting from the axial load were discussed in Chapter 4. Procedures developed in Chapter 6 can be applied to evaluate the stresses produced by the couples.

The normal stress at any point on a transverse plane through AB is the algebraic sum of the normal stresses due to the centric load and that due to the bending load (the couples). Since the transverse shear is zero, this stress will be a principal stress. The procedure is illustrated in the following example.

EXAMPLE 9–2

The cast-iron frame of a small press is shaped as shown in Fig. 9–4a. Figure 9–4b represents the cross section a–a, and the moment of inertia of the area

FIG. 9–3 Five-ton Littell Coil Loading Hook. (*Courtesy. Steel Founders Society of America*)

is 1000/3 in.[4] with respect to the centroidal axis c–c. For a load P of 16 kips, and assuming linearly elastic action, determine

(a) The normal stress distribution on section a–a.
(b) The principal stresses for the critical points on section a–a.

SOLUTION

(a) When the free-body diagram of the top portion of the frame was drawn, two equal opposite collinear forces P through the centroid of the section a–a were added as shown in Fig. 9–4c. The force P_1 is a centric load equal to the resisting force T on section a–a. The other two forces P and P_2 constitute an 18 P-in.-kip couple equal to the resisting moment M on section a–a.

   The axial stress uniformly distributed over section a–a, as shown in Fig. 9–4d, is

FIG. 9–4

(a)   *(b)*   *(f)*

(c)   *(d)*   *(e)*

$$\sigma_1 = \frac{P}{A} = \frac{16,000}{40} = 400 \text{ psi } T$$

The maximum tensile fiber stress occurs at the left edge of the section a–a and is

$$\sigma_2 = \frac{Mc}{I} = \frac{16,000(18)(3)}{1000/3} = 2592 \text{ psi } T$$

The maximum compressive fiber stress occurs at the right edge of the section a–a and is

$$\sigma_3 = \frac{Mc}{I} = \frac{16,000(18)(7)}{1000/3} = 6048 \text{ psi } C$$

The distribution of fiber stresses is shown in Fig. 9–4e.

When the principle of superposition is applied, the total normal stress at any point on section a–a is the algebraic sum of the component stresses at the point. When the stress distributions of Figs. 9–4d and e are added, the resultant stress distribution for section a–a is as shown in Fig. 9–4f, which is the answer for part (a).

(b)   With no shearing stresses on section a–a, the normal stresses are principal stresses, and the critical points, as observed from the stress distribution of Fig. 9–4f, are the left and right edges of the section. The principal stresses for the right edge are

$$\sigma_p = \underline{5650 \text{ psi } C} \quad \text{and} \quad \underline{0} \qquad \text{Ans.}$$

The principal stresses for the left edge are

$$\sigma_p = \underline{2990 \text{ psi } T} \quad \text{and} \quad \underline{0} \qquad \text{Ans.}$$

Although not requested in this example, the location of the neutral axis (line of zero stress) can be determined as the place where the compressive fiber stress is 400 psi because this stress will just balance the axial tensile stress of 400 psi. The distance on Fig. 9–4f can be obtained by writing

$$400 = \frac{16,000(18)x}{1000/3}$$

from which

$$x = 0.463 \text{ in.}$$

to the right of the centroidal axis of the cross section.

EXAMPLE 9–3

A gray cast-iron compression member is subjected to a vertical load P, as shown in Fig. 9–5a. The allowable normal stresses on plane ABCD are 140 MPa C and 35 MPa T. Determine the maximum allowable load P.

SOLUTION

The specified stresses indicate elastic action. When two vertical forces P (labeled P_1 and P_2 for clarity) of opposite sense are introduced at the center of the top section, EFGH, the eccentric force P is replaced by a centric force P_2 and a couple PP_1, as shown in Fig. 9–5b.

The stresses produced by the centric load are easily calculated; however, the couple does not act in a plane of symmetry of the member, and the flexural stresses produced are not readily computed for this orientation. This difficulty can be eliminated by replacing the resultant couple PP_1 by its components in or parallel to planes of symmetry of the compression member. The magnitudes of the components are 60P N · mm = 0.060P N · m

(a)

(b)

(c)

(d)

(e)

$\sigma_1 = 55.56P$ Pa

$\sigma_2 = 166.67P$ Pa

$\sigma_3 = 166.67P$ Pa

0.060 *P* N·m

0.075 *P* N·m

and $75P$ N · mm $= 0.075P$ N · m. The sense of each of these couples is shown in Figs. 9–5*d* and *e*.

The centric load P_2 produces a uniformly distributed compressive stress on plane *ABCD*. The magnitude of this stress is

$$\sigma_1 = \frac{P}{A} = \frac{P}{(0.150)(0.120)} = 55.56P \text{ N/m}^2 = 55.56P \text{ Pa}$$

The $0.060P$ component of the couple PP_1 produces a compressive stress along edge *CD* of plane *ABCD* and an equal tensile stress along edge *AB*. These stresses are

$$\sigma_2 = \frac{Mc}{I} = \frac{0.060P(0.060)}{(0.150)(0.120)^3/12} = 166.67P \text{ N/m}^2 = 166.67P \text{ Pa}$$

The $0.075P$ component of the couple PP_1 produces a compressive stress along edge BC of plane $ABCD$ and an equal tensile stress along edge AD. These stresses are

$$\sigma_3 = \frac{Mc}{I} = \frac{0.075P(0.075)}{(0.120)(0.150)^3/12} = 166.67P \text{ N/m}^2 = 166.67P \text{ Pa}$$

The component stresses are shown in the three stress distribution diagrams of Figs. 9–5c, d, and e. From the diagrams, one observes that the maximum compressive stress occurs at C and is

$$\sigma_C = 55.56P + 166.67P + 166.67P = 388.9P \text{ Pa}$$

and that the maximum tensile stress occurs at A and is

$$\sigma_T = 166.67P + 166.67P - 55.56P = 277.8P \text{ Pa}$$

When the allowable stresses are substituted in the above expressions, the values of P become

$$P_C = \frac{140(10^6)}{388.9} = 360(10^3) \text{ N} = 360 \text{ kN}$$

and

$$P_T = \frac{35(10^6)}{277.8} = 126(10^3) \text{ N} = 126 \text{ kN}$$

Since P_T is less than P_C, the maximum allowable load is

$$P = \underline{126 \text{ kN}} \qquad\qquad \text{Ans.}$$

FIG. 9–6

Another example involving combined centric and flexural loads is a beam with a force or a component of a load acting perpendicular to the axis of the beam. In this case, the transverse shearing stress in the beam is not zero everywhere and may be an important factor in some instances.

This type of combined load is complicated by having the deflection of the beam and the amount of bending moment dependent on each other. As a beam deflects, the centroid of a specific cross section of the beam moves relative to the line of action of any longitudinal force. This movement alters the moment or couple applied to the beam due to the eccentricity of the longitudinal force, as further explained in Exam. 9–4. For this reason a complete solution suggests the use of successive approximations.[2] Fortunately, the change in stress due to the deflection of the beam can normally be neglected as shown by the following example.

EXAMPLE 9–4

A 5-ft long timber cantilever beam with the cross section of Fig. 9–6*a* is loaded elastically, as shown in Fig. 9–6*b*. The modulus of elasticity of the timber is 1200 ksi. Determine

(a) The normal stress distribution on a vertical plane through the beam at the wall.

(b) The principal stresses and maximum shearing stress at a point 2.50 in. above the centroidal axis on a cross section at the wall.

SOLUTION

The 2000-lb load can be replaced by a 1200-lb vertical force, a 1600-lb horizontal force through the centroid of the cross section of the beam at the wall, and a couple with 1600-lb horizontal forces, as shown in Fig. 9–6*c*.

(a) The axial compressive stress produced by the 1600-lb horizontal force on the section at the wall is

$$\sigma_1 = \frac{P}{A} = \frac{1600}{60 - 12} = 33.3 \text{ psi } C$$

as shown in Fig. 9–6*c*.

If the deflection at the right end of the beam is neglected, the bending moment at the support is $+1200(60) + 1600(5)$, which is equal to $+80,000$ in.-lb. If the deflection of the beam is appreciable, the couple at the free end will be increased since, for the load to be centric, the 1600-lb force must pass through the centroid of

---

[2] It is recognized that the differential equation expressing the moment as a function of both x and y has a solution (see Art. 10–8). However, for many practical problems, the method of successive approximations is more efficient.

the resisting section (which is at the support for this problem). The deflection of the right end due to the 1200-lb load as given by case 1 of Appendix D is

$$y_1 = \frac{PL^3}{3EI} = \frac{1200(60^3)}{3(12)(10^5)[6(10^3) - 2(6^3)]/12} = \frac{9}{58} \text{ in. upward}$$

The deflection at the right end due to the 8000-in.-lb couple as given by case 4 of Appendix D is

$$y_2 = \frac{ML^2}{2EI} = \frac{1600(5)(60^2)}{2(12)(10^5)(464)} = \frac{1.5}{58} \text{ in. upward}$$

The sum of these deflections is 0.1810 in. upward; with this deflection included, the moment of the couple becomes 1600(5.18) or 8288, thus making the total moment 80,290 in.-lb or 0.36 percent higher, which is so slight an increase that it will be neglected. Hence, the maximum flexural stresses at the wall are

$$\sigma_2 = \frac{Mc}{I} = \frac{80,000(5)}{464} = 862 \text{ psi } T \text{ and } C$$

The distributions of the component stresses are shown in Fig. 9–6c. Also shown is the distribution of the combined stresses, which is the answer to part (a).

(b) The point specified is halfway from the centroidal axis to the outermost fiber; therefore, the normal stress on the transverse plane is

$$\sigma = \frac{862}{2} + 33 = 464 \text{ psi } C$$

The horizontal and vertical shearing stresses at the specified point are

$$\tau = \frac{VQ}{It} = \frac{1200[12(4) + 2(2.75)]}{464(4)} = 34.6 \text{ psi}$$

From Eqs. 1–7 and 1–9, the principal stresses and maximum shearing stress at the specified point are

$$\sigma_p = -232 \mp \sqrt{232^2 + 34.6^2}$$

$$= -232 \mp 235$$

$$= \underline{467 \text{ psi } C} \quad \text{and} \quad \underline{3 \text{ psi } T} \qquad \text{Ans.}$$

and

$$\tau_{max} = \underline{235\ psi}$$ Ans.

PROBLEMS

Note: In the following problems, the effect of deflection on the eccentricity is to be neglected and elastic action is to be assumed, unless otherwise noted.

9-14* Determine the distribution of normal stress on section *AB* of the structural member shown in Fig. P9–14.

9-15* Determine the vertical normal stresses at points *A*, *B*, *C*, and *D* of the rectangular post shown in Fig. P9–15.

9-16 The straight portion *AB* of the cast steel machine part shown in Fig. P9–16 has a hollow rectangular cross section with outside dimensions 12 by

FIG. P9–14

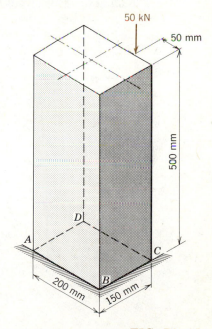

FIG. P9–15

FIG. P9–16

FIG. P9–17

10 in. and walls 2 in. thick. Determine the maximum normal stress on a transverse plane within the portion *AB*.

9-17   The cross section of the straight vertical portion of the Littel coil-loading hook illustrated in Fig. 9–3 is shown in Fig. P9–17. The horizontal distance from the line of action of the applied load to the inside face *AB* of the cross section is 550 mm. Determine the maximum principal stresses on section *ABCD* for a 50-kN load.

Section *B–B*

FIG. P9–18

FIG. P9–19

(a)   (b)

9-18* A short post is loaded by a vertical force P of 4800 lb and a horizontal force H of 1000 lb as shown in Fig. P9–18. Determine the vertical normal stresses at corners A, B, C, and D of the post.

9-19* Determine the maximum normal stress on a transverse plane in the straight portion of the structure shown in Fig. P9–19 and clearly indicate where it occurs. The member is braced in the plane perpendicular to the plane of symmetry.

9-20 The cantilever beam shown in Fig. P9–20a has the cross section shown in Fig. P9–20b. If the allowable stresses are 10,000 psi shear and 16,000 psi tension and compression at point A (just below the flange), determine the maximum allowable load P.

9-21* The simple beam shown in Fig. P9–21a has the cross section shown in Fig. P9–21b. If the allowable stresses are 70 MPa shear and 110 MPa tension and compression at point A (just above the flange and to the right of the load), determine the maximum allowable load P.

9-22* The estimated maximum total force P to be exerted on the C-clamp shown in Fig. P9–22 is 500 lb. If the normal stress on section A–A is not to exceed 19,000 psi, determine the minimum allowable value for the dimension h of the cross section.

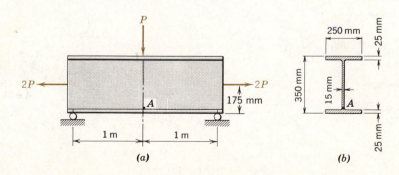

(a)   (b)

FIG. P9–21

FIG. P9–22

Section A–A

FIG. P9–23

9-23 The cross section of the aluminum beam ($E = 70$ GPa) shown in Fig. P9–23 is a rectangle 100 mm wide by 150 mm deep. Determine the maximum normal stress on a vertical section at the wall

(a) By neglecting the deflection of the member.

(b) By including the approximate increase in stress due to the deflection of the horizontal portion of the member.

9-24 The member shown in Fig. P9–24 has a solid circular cross section of radius r. Determine the minimum permissible radius of the member if the maximum tensile stress on section a–a is not to exceed 20 ksi.

9-25* A beam is loaded and supported as shown in Fig. P9–25. Determine the maximum normal stress on plane a–a.

9-26 The machine element shown in Fig. P9–26 is loaded in a plane of symmetry. Determine and show on a sketch the principal and maximum

FIG. P9–24

FIG. P9–25

FIG. P9–26

FIG. P9–27

shearing stresses at point *A* that is in the web just above the junction between the flange and the web.

9-27* The beam shown in Fig. P9–27 has a 3 by 8-in. rectangular cross section and is loaded in a plane of symmetry. Determine and show on a sketch the principal and maximum shearing stresses at point *A*.

9-28 The beam shown in Fig. P9–28 has a triangular cross section and is loaded in a plane of symmetry. Determine

FIG. P9–28

FIG. P9–29

FIG. P9–30

(a)   The maximum compressive stress on plane A–A.

(b)   The longitudinal shearing stress at the centroidal axis.

9-29   Determine and show on a sketch the principal stresses and the maximum shearing stress at point A in the roof beam shown in Fig. P9–29.

9-30*   The frame shown in Fig. P9–30 is constructed of 100 by 100-mm timbers. Determine and show on sketches the principal and maximum shearing stresses at points G and H.

9-31   The cantilever beam shown in Fig. P9–31a has the cross section shown in Fig. P9–31b. If the stresses at point A (in the web) are $\sigma_x = 35$ MPa C and $\tau_{xy} = 7$ MPa, determine

(a)   The magnitudes of loads P and T.

(a)

(b)

FIG. P9–31

Section A-A

(b)   The principal and maximum shearing stresses at point A.

9-32*   A crane hook has the dimensions and cross section shown in Fig. P9–32. The allowable stresses on plane A–A are 12,000 psi T and 16,000 psi C. Determine the capacity P of the hook.

9-33   A split ring has the dimensions and cross section shown in Fig. P9–33. Determine the maximum load P that can be applied to the ring if the normal stresses on plane A–A are not to exceed 80 MPa T or 120 MPa C.

9-5   COMBINED CENTRIC, TORSIONAL, AND FLEXURAL LOADS

In numerous industrial situations, machine members are subjected to all three general types of loads studied so far. When a flexural load is combined with torsional and centric loads, it is frequently difficult to locate the point or points where the most severe stresses occur. The following review or summary statements may be helpful in this regard.

All dimensions in millimeters

Section A-A

The longitudinal and transverse shearing stresses are maximum at the centroidal axis of the section where V is maximum. The flexural stress is maximum at the greatest distance from the centroidal axis on the section where M is maximum. The torsional shearing stress is maximum at the surface of a shaft at the section where T is maximum. With these facts in mind, it is normally possible to locate one or more possible points of high stress. Even so, it may be necessary to calculate the various stresses at more than one point of a member before locating the most severely stressed point. For stresses below the proportional limit of the material, the superposition method can be used to combine stresses on any given plane at any specific point of a loaded member.

After the stresses on a pair of mutually perpendicular planes at a specific point are determined, the methods of Arts. 1–7 and 1–8 can be used to find the principal stresses and the maximum shearing stress at the point.

The following example illustrates the procedure for the solution of elastic combined stress problems.

EXAMPLE 9–5
The 100-mm diameter shaft of Fig. 9–7a is subjected to the loads shown. Determine the maximum permissible value of P for allowable stresses of 120 MPa T and 70 MPa shear.

SOLUTION
When two equal opposite forces P are introduced along line EF as in Fig. 9–7b, it is seen that the shaft is subjected to a couple of $0.6P$ N · m clockwise looking toward the wall and a concentrated load P N acting at E. In other words, the shaft acts as a cantilever beam with a concentrated load P N at the right end, a torsion member subjected to a $0.6P$-N · m torque, and an axially loaded member with a load of $30P$ N.

The couple produces a torsional shearing stress at all surface points of the shaft of

$$\tau_{xt} = \frac{Tc}{J} = \frac{0.6P(0.050)}{(\pi/2)(0.050)^4} = \frac{9600P}{\pi}$$

as shown in Fig. 9–7c.

The axial tensile stress at all points of the shaft is

$$\sigma_{x1} = \frac{P}{A} = \frac{30P}{\pi(0.050)^2} = \frac{12000P}{\pi}$$

as shown in Fig. 9–7d.

The cantilever beam load P produces a maximum tensile fiber stress at point A and an equal compressive stress on the opposite side of the shaft.

FIG. 9–7

The axial stress is tensile, and therefore the compressive flexural stress is not critical. The tensile fiber stress at A is (see Fig. 9–7e)

$$\sigma_{x2} = \frac{Mc}{I} = \frac{0.75P(0.050)}{(\pi/4)(0.050)^4} = \frac{24000P}{\pi}$$

The transverse shearing stress is equal to the longitudinal shearing stress, which is maximum on the xz plane including such surface points as B and C where the torsional shearing stress is also maximum. The maximum shearing stress is (see Fig. 9–7f)

$$\tau_{xy} = \frac{VQ}{It} = \frac{P\left[\dfrac{\pi(0.050)^2}{2}\right]\left[\dfrac{4(0.050)}{3\pi}\right]}{(\pi/4)(0.050)^4(0.100)} = \frac{1600P}{3\pi}$$

The torsional and transverse shearing stresses are in the same direction along the top of the shaft and in opposite directions along the bottom of the shaft. Therefore, any critical situation due to these two stresses will develop along the top of the shaft at some point such as B.

All four stresses are maximum at some point or points in the section of the shaft at the wall since the fiber stress is not maximum at any other section of the shaft. Therefore, a complete analysis for the member can be made on the section at the wall. A complete analysis including interior points of this section is beyond the scope of this book. However, a fairly good solution can be obtained by investigating the combined stresses at two surface points A and B. At A the fiber stress, axial stress, and torsional shearing stress are all maximum, whereas the transverse shearing stress is zero. At B the axial stress, torsional shearing stress, and transverse shearing stress are all maximum, whereas the fiber stress is zero. The fiber stress and transverse shearing stress are never maximum at the same point. Furthermore, the torsional shearing stress is never maximum at an interior point. Therefore, the chance of finding a more severe stress at an interior point is unlikely.

The stresses on orthogonal planes through A and B are as shown in Fig. 9–7g. When these values are substituted in Eq. 1–7 of Art. 1–7 and the resulting expressions compared, it is evident that the combination of stresses of A is more severe than that at B. The maximum stresses at A are

$$\sigma_A = \frac{18000P}{\pi} + \sqrt{\left(\frac{18000P}{\pi}\right)^2 + \left(\frac{9600P}{\pi}\right)^2} = \frac{38400P}{\pi}$$

and

$$\tau_A = \sqrt{\left(\frac{18000P}{\pi}\right)^2 + \left(\frac{9600P}{\pi}\right)^2} = \frac{20400P}{\pi}$$

The allowable values of P from the allowable stresses and these relations are

$$120(10^6) = 38400P_\sigma/\pi$$

or

$$P_\sigma = 9.817(10^3) \text{ N} = 9.82 \text{ kN}$$

and

$$70(10^6) = 20400P_\tau/\pi$$

or

$$P_\tau = 10.780(10^3) \text{ N} = 10.78 \text{ kN}$$

The smaller value controls and the maximum permissible value is

$$P = \underline{9.82 \text{ kN}} \qquad\qquad \text{Ans.}$$

PROBLEMS

> *Note: In the following problems, stress concentrations and the effect of beam deflection on the eccentricity are to be neglected unless otherwise noted. Assume elastic action.*

9-34*   A 2-in. diameter steel rod is loaded and supported as shown in Fig. P9–34. Determine and show on a sketch the principal stresses and the maximum shearing stress at the top surface adjacent to the support.

9-35*   A 50-mm diameter steel rod is loaded and supported as shown in Fig. P9–35. Determine and show on a sketch the principal stresses and the maximum shearing stress at the top surface adjacent to the support.

FIG. P9–34   FIG. P9–35

FIG. P9–36

9-36* A 2-in. diameter shaft is loaded and supported as shown in Fig. P9–36. Determine the principal stresses and the maximum shearing stress at point P on the surface of the shaft.

9-37 A solid shaft 100 mm in diameter is acted on by forces P and Q as shown in Fig. P9–37. Determine the principal stresses and the maximum shearing stress at point A on the surface of the shaft.

9-38 Determine the minimum permissible diameter for the shaft shown in Fig. P9–38 if the shearing and tensile stresses at point A on the surface of the shaft are not to exceed 11,000 psi and 18,000 psi, respectively.

9-39* A solid circular shaft is loaded and supported as shown in Fig. P9–39. If the tensile and shearing stresses at point A on the surface of the shaft are not to exceed 140 MPa and 85 MPa, respectively, determine the maximum value for the load R.

9-40* A steel shaft 4 in. in diameter is supported in flexible bearings at its ends. Two pulleys each 2 ft in diameter are keyed to the shaft. The pulleys

FIG. P9–37

FIG. P9–38   FIG. P9–39

carry belts as shown in Fig. P9–40. Determine and show on a sketch the principal stresses and the maximum shearing stress at point A on the surface of the shaft.

9-41   A steel shaft 120 mm in diameter is supported in flexible bearings at its ends. Two pulleys each 600 mm in diameter are keyed to the shaft. The pulleys carry belts as shown in Fig. P9–41. Determine and show on a sketch the principal stresses and the maximum shearing stress at point A on the surface of the shaft.

9-42   A solid shaft 4 in. in diameter is subjected to a torque of 100 in.-kips and a tensile force of 75 kips applied parallel to the axis of the shaft but with an eccentricity of 1/2 in. Determine and show on a sketch the maximum normal stress and the maximum shearing stress in the shaft.

FIG. P9–40

FIG. P9–41

FIG. P9–44

FIG. P9–45

9-43*   A short hollow shaft, having an outside diameter of 100 mm and an inside diameter of 50 mm, is subjected to a torque of 8.5 kN · m and a compressive force of 150 kN applied parallel to the axis of the shaft but with an eccentricity of 15 mm. Determine and show on a sketch the maximum normal stress and the maximum shearing stress in the shaft.

9-44   A free-body diagram of a thin-walled cylindrical pressure vessel with rigid flat ends is shown in Fig. P9–44. The tank has an inside diameter of 60 in. and a wall thickness of 0.20 in. The tank is subjected to an internal pressure of 100 psi. Additional loads are applied to the cylinder through the rigid end plates as shown in Fig. P9–44. Determine and show on a sketch the principal and maximum shearing stresses acting on element A if it is located on the inside surface of the tank. Neglect the weight of the tank and assume that the walls of the tank are braced to prevent buckling.

9-45*   A thin-walled cylindrical pressure tank is fabricated by butt-welding 15-mm plate with a spiral seam as shown in Fig. P9–45. The pressure in the tank is 1700 kPa. Additional loads are applied to the cylinder through a rigid end plate as shown in Fig. P9–45. Determine

(a)   The normal and shearing stresses on the plane of the weld.

(b)   The principal stresses and the maximum shearing stress at a point on the inside surface of the tank.

Neglect the weight of the tank and assume that the walls of the tank are braced to prevent buckling.

9-46*   A 4-in. diameter solid circular steel shaft is loaded and supported as shown in Fig. P9–46. Determine and show on a sketch the principal stresses and the maximum shearing stress at point A on the surface of the shaft.

FIG. P9–46

FIG. P9–47

9-47 A 100-mm diameter solid circular steel shaft is loaded and supported as shown in Fig. P9–47. Because of discontinuities at points A and B, the normal and shearing stresses at these points must be limited to 90 MPa T and 60 MPa, respectively. Determine the maximum permissible value for the load P.

9-48 A hollow circular shaft has an outside diameter of 12 in. and an inside diameter of 4 in. as shown in Fig. P9–48. The shaft is in equilibrium when the indicated loads and torque are applied. Determine and show on a sketch the principal stresses and the maximum shearing stress at point A on the outside surface of the shaft.

9-6 THEORIES OF FAILURE

A tension test of an axially loaded member is easy to conduct, and the results, for many types of materials, are well known. When such a member

FIG. P9–48

fails, the failure occurs at a specific principal (axial) stress, a definite axial strain, a maximum shearing stress of one-half the axial stress, and a specific amount of strain energy per unit volume of stressed material. Since all of these limits are reached simultaneously for an axial load, it makes no difference which criterion (stress, strain, or energy) is used for predicting failure in another axially loaded member of the same material.

For an element subjected to biaxial or triaxial loading, however, the situation is more complicated because the limits of normal stress, normal strain, shearing stress, and strain energy existing at failure for an axial load are not all reached simultaneously. In other words, the cause of failure, in general, is unknown. In such cases, it becomes important to determine the best criterion for predicting failure, because test results are difficult to obtain and the combinations of loads are endless. Several theories have been proposed for predicting failure of various types of material subjected to many combinations of loads. Unfortunately, none of the theories agree with test data for all types of materials and combinations of loads. Several of the more common theories of failure are presented and briefly explained in the following paragraphs.

Maximum-Normal-Stress Theory[3]

The maximum-normal-stress theory predicts failure of a specimen subjected to any combination of loads when the maximum normal stress at any point reaches the axial failure stress as determined by an axial tensile or compressive test of the same material.

The maximum-normal-stress theory is presented graphically in Fig. 9–8b for an element subjected to biaxial principal stresses in the x and y directions, as shown in Fig. 9–8a. The limiting stress σ_f is the failure stress for this material when loaded axially. Any combination of biaxial principal stresses σ_x and σ_y represented by a point inside the square of Fig. 9–8b is safe according to this theory, whereas any combination of stresses represented by a point outside of the square will cause failure of the element on the basis of this theory.

Maximum-Shearing-Stress Theory[4]

The maximum-shearing-stress theory predicts failure of a specimen subjected to any combination of loads when the maximum shearing stress at any point reaches the failure stress τ_f equal to $\sigma_f/2$, as determined by an axial tensile or compressive test of the same material. For ductile materials the shearing elastic limit, as

---

[3] Often called Rankine's theory after W. J. M. Rankine (1820–1872), an eminent engineering educator in England.

[4] Sometimes called Coulomb's theory because it was originally stated by him in 1773. More frequently called Guest's theory or law because of the work of J. J. Guest in England in 1900.

determined from a torsion test (pure shear), is greater than one-half the tensile elastic limit (with an average value of τ_f about $0.57\sigma_f$). This means that the maximum-shearing-stress theory errs on the conservative side by being based on the limit obtained from an axial test.

The maximum-shearing-stress theory is presented graphically in Fig. 9–8c for an element subjected to biaxial (σ_z is equal to zero) principal

FIG. 9–8

stresses, as shown in Fig. 9–8a. In the first and third quadrants, σ_x and σ_y have the same sign, and the maximum shearing stress is half the numerically larger principal stress σ_x or σ_y, as explained in Art. 1–7. In the second and fourth quadrants, where σ_x and σ_y are of an opposite sign, the maximum shearing stress is half the arithmetical sum of the two principal stresses. In the fourth quadrant, the equation of the boundary, or limit stress, line is

$$\sigma_x - \sigma_y = \sigma_f$$

and in the second quadrant the relation is

$$\sigma_x - \sigma_y = -\sigma_f$$

Maximum-Normal-Strain Theory[5]

The maximum-normal-strain theory predicts failure of a specimen subjected to any combination of loads when the maximum normal strain at any point reaches the failure strain σ_f/E at the proportional limit, as determined by an axial tensile or compressive test of the same material.

The maximum-normal-strain theory is presented graphically in Fig. 9–8d for an element subjected to biaxial principal stresses (not greater than the proportional limit), as shown in Fig. 9–8a. The limiting strain in the positive x direction is

$$\varepsilon_x = \frac{\sigma_f}{E} = \frac{\sigma_x - \nu\sigma_y}{E}$$

from which

$$\sigma_y = \frac{1}{\nu}(\sigma_x - \sigma_f)$$

which is observed to be a straight line through the point $(\sigma_f, 0)$ with a slope of $1/\nu$. The limiting strain in the positive y direction is

$$\varepsilon_y = \frac{\sigma_f}{E} = \frac{\sigma_y - \nu\sigma_x}{E}$$

from which

$$\sigma_y = \nu\sigma_x + \sigma_f$$

which is observed to be a straight line through the point $(0, \sigma_f)$ with a slope equal to ν.

[5] Often called Saint Venant's theory because of the work of Barre de Saint Venant (1797–1886), a great French mathematician and elastician.

Maximum-Strain-Energy Theory   The maximum-strain-energy theory predicts failure of a specimen subjected to any combination of loads when the strain energy per unit volume (see Chapter 3 for the derivation of the relation between σ_f and the elastic strain energy per unit volume) of any portion of the stressed member reaches the failure value of the strain energy per unit volume as determined by an axial tensile or compressive test of the same material. The maximum-strain-energy theory has been largely replaced by the maximum-distortion-energy theory.

Maximum-Distortion-Energy Theory[6]   This theory differs from the maximum-strain-energy theory in that the portion of the strain energy producing volume change is considered ineffective in causing failure by yielding. Supporting evidence comes from experiments showing that homogeneous materials can withstand very high hydrostatic stresses without yielding. The portion of the strain energy producing change of shape of the element is assumed to be completely responsible for the failure of the material by inelastic action.

The strain energy of distortion is most readily computed by determining the total strain energy of the stressed material and subtracting the strain energy corresponding to the volume change. Article 3–4 defined the quantity $\sigma^2/(2E)$ as the strain energy per unit volume for a member subjected to a slowly applied axial load. This expression can also be written as

$$u = \frac{\sigma^2}{2E} = \frac{\sigma\varepsilon}{2}$$

where u is the strain energy per unit volume and σ and ε are the slowly applied axial stress and strain. This equation assumes that the stress does not exceed the proportional limit.

When an elastic element is subjected to triaxial loading, the stresses can be resolved into three principal stresses such as σ_x, σ_y, and σ_z, where x, y, and z are the principal axes. These stresses will be accompanied by three principal strains related to the stresses by Eq. 3–6 (Art. 3–3). If it is assumed that the loads are applied simultaneously and gradually, the stresses and strains will increase in the same manner. The total strain energy per unit volume is the sum of the energies produced by each of the stresses (energy is a scalar quantity and can be added algebraically regardless of the directions of the individual stresses); thus,

[6] Frequently called the Huber-Hencky-von Mises theory, because it was proposed by M. T. Huber of Poland in 1904 and independently by R. von Mises of Germany in 1913. The theory was further developed by H. Hencky and von Mises in Germany and the United States.

$$u = \left(\frac{1}{2}\right)(\sigma_x \varepsilon_x + \sigma_y \varepsilon_y + \sigma_z \varepsilon_z)$$

When the strains are expressed in terms of the stresses, this equation becomes

$$u = [1/(2E)][(\sigma_x^2 + \sigma_y^2 + \sigma_z^2) - 2\nu(\sigma_x\sigma_y + \sigma_y\sigma_z + \sigma_z\sigma_x)]$$

The strain energy can be resolved into two components u_v and u_d, resulting from a volume change and a distortion, respectively, by considering the principal stresses to be made up of two sets of stresses as indicated in Figs. 9–9a, b, and c. The state of stress in Fig. 9–9c will result in distortion only (no volume change) if the sum of the three normal strains is zero. That is,

$$E(\varepsilon_x + \varepsilon_y + \varepsilon_z)_d = [(\sigma_x - p) - \nu(\sigma_y + \sigma_z - 2p)]$$
$$+ [(\sigma_y - p) - \nu(\sigma_z + \sigma_x - 2p)]$$
$$+ [(\sigma_z - p) - \nu(\sigma_x + \sigma_y - 2p)] = 0$$

which reduces to

$$(1 - 2\nu)(\sigma_x + \sigma_y + \sigma_z - 3p) = 0$$

Therefore,

$$p = (\sigma_x + \sigma_y + \sigma_z)/3$$

The three normal strains due to p are, from Eq. 3–6,

$$\varepsilon_v = (1 - 2\nu)p/E$$

and the energy resulting from the hydrostatic stress (the volume change) is

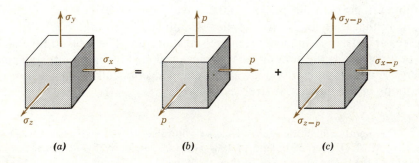

(a)   (b)   (c)

FIG. 9–9

$$u_v = 3\left(\frac{p\varepsilon_v}{2}\right) = \frac{3}{2}\frac{1-2\nu}{E}p^2$$

$$= \frac{1-2\nu}{6E}(\sigma_x + \sigma_y + \sigma_z)^2$$

The energy resulting from the distortion (change of shape) is

$$u_d = u - u_v$$

$$= \frac{1}{6E}[3(\sigma_x^2 + \sigma_y^2 + \sigma_z^2) - 6\nu(\sigma_x\sigma_y + \sigma_y\sigma_z + \sigma_z\sigma_x)$$

$$- (1 - 2\nu)(\sigma_x + \sigma_y + \sigma_z)^2]$$

When the third term in the brackets is expanded, the expression can be rearranged to give

$$u_d = \frac{1+\nu}{6E}[(\sigma_x^2 - 2\sigma_x\sigma_y + \sigma_y^2) + (\sigma_y^2 - 2\sigma_y\sigma_z + \sigma_z^2)$$

$$+ (\sigma_z^2 - 2\sigma_z\sigma_x + \sigma_x^2)]$$

$$= \frac{1+\nu}{6E}[(\sigma_x - \sigma_y)^2 + (\sigma_y - \sigma_z)^2 + (\sigma_z - \sigma_x)^2] \qquad (a)$$

The maximum-distortion-energy theory of failure assumes that inelastic action will occur whenever the energy given by Eq. *a* exceeds the limiting value obtained from a tensile test. For this test, only one of the principal stresses will be nonzero. If this stress is called σ_f, the value of u_d becomes

$$(u_d)_f = \frac{1+\nu}{3E}\sigma_f^2$$

and when this value is substituted in Eq. *a*, it becomes

$$2\sigma_f^2 = (\sigma_x - \sigma_y)^2 + (\sigma_y - \sigma_z)^2 + (\sigma_z - \sigma_x)^2 \qquad (b)$$

for failure by slip.

When a state of plane stress exists, assume σ_z equals zero, Eq. *b* becomes

$$\sigma_x^2 - \sigma_x\sigma_y + \sigma_y^2 = \sigma_f^2$$

This last expression is the equation of an ellipse with its major axis along the line σ_x equals σ_y, as shown in Fig. 9–8e.

An alternate approach to the distortion-energy theory of failure consists of determining the shearing stress on the octahedral planes, defined as any plane whose normal makes equal angles with the three principal axes. Figure 9–10 is a free-body diagram of an element bounded by the three principal planes and the octahedral plane. Since

$$\theta_x = \theta_y = \theta_z$$

$$\cos^2\theta_x + \cos^2\theta_y + \cos^2\theta_z = 3\cos^2\theta_x = 1$$

or

$$\cos\theta_x = \cos\theta_y = \cos\theta_z = 1/\sqrt{3}$$

If the area of the octahedral plane is ΔA, the other three areas are $\Delta A/\sqrt{3}$. The equilibrium equation along the n axis gives

$$\Sigma F_n = \sigma_{\text{oct}}\Delta A - \frac{1}{\sqrt{3}}\left(\sigma_x\frac{\Delta A}{\sqrt{3}}\right) - \frac{1}{\sqrt{3}}\left(\sigma_y\frac{\Delta A}{\sqrt{3}}\right)$$

$$- \frac{1}{\sqrt{3}}\left(\sigma_z\frac{\Delta A}{\sqrt{3}}\right) = 0$$

or

$$\sigma_{\text{oct}} = (\sigma_x + \sigma_y + \sigma_z)/3$$

Note that σ_{oct} is the same as p (the hydrostatic stress) in Fig. 9–9b. The resultant of the normal and shearing forces on the octahedral plane must equal the resultant of the forces on the other three planes; thus,

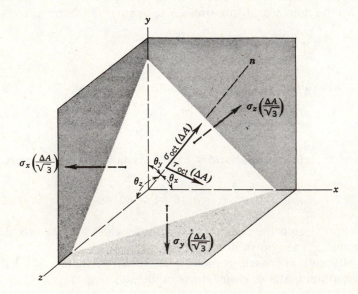

FIG. 9–10

$$(\sigma_{oct}\Delta A)^2 + (\tau_{oct}\Delta A)^2 = \left(\sigma_x \frac{\Delta A}{\sqrt{3}}\right)^2 + \left(\sigma_y \frac{\Delta A}{\sqrt{3}}\right)^2 + \left(\sigma_z \frac{\Delta A}{\sqrt{3}}\right)^2$$

from which

$$(\tau_{oct})^2 = \frac{\sigma_x^2 + \sigma_y^2 + \sigma_z^2}{3} - \left(\frac{\sigma_x + \sigma_y + \sigma_z}{3}\right)^2$$

$$= \frac{1}{9}[3(\sigma_x^2 + \sigma_y^2 + \sigma_z^2) - (\sigma_x^2 + \sigma_y^2 + \sigma_z^2)$$

$$- 2(\sigma_x\sigma_y + \sigma_y\sigma_z + \sigma_z\sigma_x)]$$

$$= \frac{1}{9}[(\sigma_x - \sigma_y)^2 + (\sigma_y - \sigma_z)^2 + (\sigma_z - \sigma_x)^2] \qquad (c)$$

When Eqs. b and c are compared, it is seen that

$$2\sigma_f^2 = 3(\pi_{oct})^2$$

or

$$\tau_{oct} = \frac{\sqrt{2}}{3}\sigma_f = 0.471\sigma_f$$

The octahedral shearing stress is thus related directly to the tensile failure stress and can be used as a criterion of the distortion energy at any point.

The graphic representations of Figs. 9–8b, c, d, and e are superimposed in Fig. 9–8f for convenient comparison of the different theories. The maximum-distortion-energy theory has been found to agree best with available test data for ductile materials, and its use in design practice is increasing. The maximum-normal-strain theory is unsafe for ductile materials. The maximum-normal-stress theory is also unsafe for ductile materials with principal stresses of opposite sign. The maximum-shearing-stress theory is conservative for ductile materials and is used in some design codes. The maximum-normal-stress theory agrees with test data for brittle materials and is generally accepted in design practice for such materials.

The following example illustrates the application of the theories of failure in predicting the load-carrying capacity of a member.

EXAMPLE 9–6

The solid circular shaft of Fig. 9–11 has a proportional limit of 64 ksi and a Poisson's ratio of 0.30. Determine the value of the load R for failure by slip as predicted by each of the four given theories of failure. Assume that point A is the most severely stressed point.

SOLUTION

When two equal opposite forces R are introduced along line EF of Fig. 9–11, it is evident that the bending moment at point A is $36R$ in.-kips, the torsional moment is $24R$ in.-kips, and the axial load is $12R$ kips. The flexural stress at A is

$$\sigma_1 = \frac{Mc}{I} = \frac{36R(2)}{\pi(2^4)/4} = \frac{18R}{\pi} \, T$$

The direct stress at A is

$$\sigma_2 = \frac{P}{A} = \frac{12R}{\pi(2^2)} = \frac{3R}{\pi} \, T$$

and the sum of the two tensile stresses at A becomes

$$\sigma = \sigma_1 + \sigma_2 = \frac{21R}{\pi}$$

The torsional shearing stress at A is

$$\tau = \frac{Tc}{J} = \frac{24R(2)}{\pi(2^4)/2} = \frac{6R}{\pi}$$

The maximum stresses at A are

$$\tau_{\max} = \sqrt{\left(\frac{10.5R}{\pi}\right)^2 + \left(\frac{6R}{\pi}\right)^2} = \frac{12.1R}{\pi}$$

$$\sigma_{p1} = \frac{10.5R}{\pi} + \frac{12.1R}{\pi} = \frac{22.6R}{\pi} \, T$$

FIG. 9–11

and

$$\sigma_{p2} = \frac{10.5R}{\pi} - \frac{12.1R}{\pi} = -\frac{1.6R}{\pi} = \frac{1.6R}{\pi} C$$

According to the maximum-normal-stress theory

$$64 = \sigma_f = \sigma_{p1} = 22.6R/\pi$$

from which

$$R = 64\pi/22.6 = \underline{8.90 \text{ kips}} \qquad\qquad \text{Ans.}$$

According to the maximum-shearing-stress theory

$$64/2 = \tau_f = \tau_{max} = 12.1R/\pi$$

from which

$$R = 32\pi/12.1 = \underline{8.31 \text{ kips}} \qquad\qquad \text{Ans.}$$

The maximum-normal-strain theory gives

$$\frac{\sigma_f}{E} = \frac{\sigma_{p1} - \nu\sigma_{p2}}{E}$$

or

$$64 = \frac{22.6R}{\pi} - 0.3\left(-\frac{1.6R}{\pi}\right)$$

which gives

$$R = 64\pi/23.1 = \underline{8.71 \text{ kips}} \qquad\qquad \text{Ans.}$$

According to the maximum-distortion-energy theory

$$\sigma_f^2 = \sigma_{p1}^2 + \sigma_{p2}^2 - \sigma_{p1}\sigma_{p2}$$

or

$$64^2 = \left(\frac{22.6R}{\pi}\right)^2 + \left(-\frac{1.6R}{\pi}\right)^2 - \frac{22.6R}{\pi}\left(-\frac{1.6R}{\pi}\right)$$

which gives

$$R = 64\pi/23.4 = \underline{8.59 \text{ kips}} \qquad\qquad \text{Ans.}$$

In this example, the maximum-shearing-stress theory is seen to be the most conservative, and the maximum-normal-stress theory gives the least conservative result.

PROBLEMS

9-49* A material with a proportional limit in tension and compression of 60 ksi and a Poisson's ratio of 0.30 is subjected to a biaxial state of stress. The principal stresses are 40 ksi T and 45 ksi C. Determine which, if any, of the theories will predict failure by slip for this state of stress.

9-50* A material with a proportional limit in tension and compression of 360 MPa and a Poisson's ratio of 0.30 is subjected to a biaxial state of stress. The principal stresses are 210 MPa T and 300 MPa C. Determine which, if any, of the theories will predict failure by slip for this state of stress.

9-51 A material with a proportional limit of 36 ksi in tension and compression and a Poisson's ratio of 0.30 is subjected to a biaxial state of stress. The principal stresses are $\sigma_{p1} = 20$ ksi T, $\sigma_{p2} = 12$ ksi T or C, and $\sigma_{p3} = \sigma_z = 0$. Determine the factor of safety with respect to failure by slip according to each of the four theories of failure if

(a) The principal stress σ_{p2} is tension.

(b) The principal stress σ_{p2} is compression.

9-52* A material with a proportional limit of 280 MPa in tension and compression and a Poisson's ratio of 1/3 is subjected to a biaxial state of stress. The principal stresses are $\sigma_{p1} = 175$ MPa T, $\sigma_{p2} = 105$ MPa T or C, and $\sigma_{p3} = \sigma_z = 0$. Determine the factor of safety with respect to failure by slip according to each of the four theories of failure if

(a) The principal stress σ_{p2} is tension.

(b) The principal stress σ_{p2} is compression.

9-53 At a point on the free surface of a steel ($E = 30,000$ ksi and $\nu = 0.30$) machine component the principal stresses are 50 ksi T and 30 ksi C. What minimum proportional limit is required according to each of the four theories of failure if failure by slip is to be avoided?

9-54 The shaft shown in Fig. P9–54 is made of steel having the following properties: proportional limit 250 MPa in tension or compression and 140 MPa in shear; ultimate tensile strength 425 MPa; and Poisson's ratio 0.30. If a factor of safety of 2.5 with respect to failure by slip is specified, determine the minimum permissible diameter D according to

(a) The maximum-normal-stress theory of failure.

(b) The maximum-shearing-stress theory of failure.

(c) The maximum-normal-strain theory of failure.

(d) The maximum-distortion-energy theory of failure.

FIG. P9–54

FIG. P9–55

9-55* The shaft shown in Fig. P9–55 is made of steel having the following properties: proportional limit 40 ksi in tension or compression and 20 ksi in shear; ultimate tensile strength 60 ksi; and Poisson's ratio 0.30. If a factor of safety of 2.0 with respect to failure by slip is specified, determine the maximum permissible value for the axial load P according to

    (a)   The maximum-normal-stress theory of failure.

    (b)   The maximum-shearing-stress theory of failure.

    (c)   The maximum-normal-strain theory of failure.

    (d)   The maximum-distortion-energy theory of failure.

9-56 A steel shaft 100 mm in diameter is supported in flexible bearings at its ends. Two pulleys each 600 mm in diameter are keyed to the shaft. The pulleys carry belts that are loaded as shown in Fig. P9–56. The steel has the

FIG. P9–56

FIG. P9–57

following properties: proportional limit 275 MPa in tension or compression and 160 MPa in shear; ultimate tensile strength 400 MPa; and Poisson's ratio 0.30. If a factor of safety of 2.5 with respect to failure by slip is specified, determine the maximum allowable belt tension P according to

(a)   The maximum-shearing-stress theory of failure.

(b)   The maximum-distortion-energy theory of failure.

9-57*   The shaft shown in Fig. P9–57 is made of steel that has a proportional limit of 64 ksi and a Poisson's ratio of 0.30. If a factor of safety of 2.0 with respect to failure by slip is specified, determine the maximum allowable value for the load R according to

(a)   The maximum-shearing-stress theory of failure.

(b)   The maximum-distortion-energy theory of failure.

9-58   The cast-iron frame shown in Fig. P9–58 is subjected to a load P of 100 kN. The material has a proportional limit of 220 MPa and a Poisson's ratio of 0.20. Determine the factor of safety with respect to failure by slip according to

(a)   The maximum-normal-stress theory of failure.

(b)   The maximum-normal-strain theory of failure.

(c)   The maximum-shearing-stress theory of failure.

9-59   The cast-steel crane hook shown in Fig. P9–59 is subjected to a load P of 20,000 lb. The material has a proportional limit of 60 ksi and a Poisson's ratio of 0.30. Determine the factor of safety with respect to failure by slip according to

(a)   The maximum-normal-stress theory of failure.

FIG. P9–58

FIG. P9–59

(b)   The maximum-normal-strain theory of failure.

(c)   The maximum-shearing-stress theory of failure.

9-60*   Determine the maximum allowable internal pressure to which a closed-end, thick-walled cylinder with an inside diameter of 150 mm and an outside diameter of 250 mm may be subjected if the cylinder is made of an alloy steel with a proportional limit of 550 MPa and a Poisson's ratio of 0.30. A factor of safety of 2.5 with respect to failure by slip according to the maximum-distortion-energy theory of failure is specified.

9-61   A thick-walled hydraulic cylinder with an inside diameter of 6 in. is required to operate under a maximum internal pressure of 7500 psi. The cylinder is to be made of steel with a proportional limit of 40 ksi and a Poisson's ratio of 0.30. Determine the minimum outside diameter required if a factor of safety of 2 with respect to failure by slip according to the maximum-shearing-stress theory of failure is specified. The wall of the cylinder is not required to carry axial load.

9-62   A laminated thick-walled cylinder consists of a sleeve with an inside diameter of 150 mm and an outside diameter of 200 mm shrunk over a tube with an inside diameter of 100 mm. The maximum shrinkage stress developed in the sleeve is 45 MPa T. Determine the maximum permissible internal pressure to which the laminated cylinder may be subjected if a factor of safety of 3 with respect to failure by slip according to the maximum-distortion-energy theory of failure is specified. The cylinder is made of steel with a proportional limit of 275 MPa and a Poisson's ratio of 0.30. The wall of the cylinder is not required to carry axial load.

CHAPTER 10

COLUMNS

10-1 INTRODUCTION

In their simplest form, columns are long, straight, prismatic bars subjected to compressive, axial loads. As long as a column remains straight, it can be analyzed by the methods of Chapter 1; however, if a column begins to deform laterally, the deflection may become large and lead to catastrophic failure. This situation, called *buckling,* can be defined as the sudden large deformation of a structure due to a slight increase of an existing load under which the structure had exhibited little, if any, deformation before the load was increased. A yardstick illustrates this phenomenon. It will support a compressive load of several pounds without discernible lateral deformation, but once the load becomes large enough to cause the yardstick to ''bow out'' a slight amount, any further increase of load produces large lateral deflections.

Buckling of such a column is caused not by failure of the material of which the column is composed but by deterioration of what was a stable equilibrium to an unstable one. The three states of equilibrium can be illus-

503

FIG. 10–1

(a)   (b)   (c)

trated with a ball at rest on a surface, as shown in Figure 10–1. The ball in Fig. 10–1a is in a *stable equilibrium* position at the bottom of the pit because gravity will cause it to return to its equilibrium position if perturbed. The ball in Fig. 10–1b is in a *neutral equilibrium* position on the horizontal plane because it will remain at any new position to which it is displaced, tending neither to return to nor move farther from its original position. The ball in Fig. 10–c, however, is in an *unstable equilibrium* position at the top of a hill because, if it is perturbed, gravity will cause it to move even farther from its original location until it eventually finds a stable equilibrium position at the bottom of another pit.

As the compressive load on a column is gradually increased from zero, the column is at first in a state of stable equilibrium. During this state, if the column is perturbed by inducing small lateral deflections, it will return to its straight configuration when the loads are removed. As the load is increased further, a critical value is reached at which the column is on the verge of experiencing a lateral deflection so, if it is perturbed, it will not return to its straight configuration. The load cannot be increased beyond this value unless the column is restrained laterally; should the lateral restraints be removed, the slightest perturbation will trigger large lateral deflections. For long, slender columns, the *critical buckling load* (the maximum load for which the column is in stable equilibrium) occurs at stress levels much less than the proportional limit for the material. This indicates that this type of buckling is an elastic phenomenon.

10-2 STABILITY OF ELASTIC MECHANICAL SYSTEMS

Some insight into the problem of stability of a column can be gained by considering the stability of a simple elastic mechanical system such as the one shown in Fig. 10–2. The rigid prismatic bar is supported by a smooth pin at the bottom and is loaded by an axial force at the top. A torsional spring at the bottom offers a resisting couple $M_r = k\theta$ to rotation about the pin. The spring is adjusted initially so that the bar will stand in the vertical position. If the bar is displaced from the vertical position (by a momentary lateral force)

FIG. 10–2

(a)   (b)

to the position shown in the free-body diagram (Fig. 10–2*b*), the torsional spring at the bottom will exert a restoring couple that will return the bar to the original (vertical) position, provided the restoring couple exceeds the moment of the force P with respect to the pin. Since the moment of P about the pin is $PL \sin \theta$, this condition for stable equilibrium can be expressed by the inequality

$$k\theta > PL \sin \theta$$

or

$$P < \frac{k}{L} \frac{\theta}{\sin \theta} \qquad (a)$$

By applying L'Hospital's rule to Eq. *a* as $\theta \to 0$, it is clear that the vertical position ($\theta = 0$) is stable as long as $P < k/L$.

   If the load P is increased beyond the value k/L, the vertical position becomes unstable and the bar, if perturbed, will rotate through an angle θ until it finds a new equilibrium position for which the restoring moment M_r balances the moment of the force P about the pin. In this new position

$$k\theta = PL \sin \theta \qquad (b)$$

or

$$PL/k = \theta/\sin \theta \qquad (c)$$

FIG. 10-3

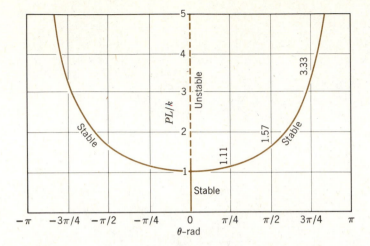

Figure 10-3 shows a plot of the quantity PL/k as a function of θ. For $P < k/L$ (or $PL/k < 1$), the only equilibrium position is $\theta = 0$. For $P > k/L$ (or $PL/k > 1$), there are three possible equilibrium positions. Equation b can be satisfied by equating θ to zero; however, this is already known to be unstable. Thus, the other branches, which are graphs of the solution of Eq. c, are the stable positions for $P > k/L$. There are two branches to the curve because the bar could rotate either right or left (the spring is assumed to be equally effective in either direction). Observe that an increase of only 11 percent in the magnitude of P above k/L will result in a rotation of 45° ($\theta = \pi/4$) before stable equilibrium is established. Displacements of this magnitude are typical of those encountered during elastic buckling. Thus, there is a *critical load* $P = k/L$ in which the nature of the equilibrium of the elastic mechanical system changes markedly. The point ($\theta = 0$, $P = k/L$) where the curve branches is called a bifurcation point.

    If the bar and torsion spring system of Fig. 10-2 is replaced by an elastic bar fixed at the bottom as in Fig. 10-4, the analogy between the action of a column and the mechanical system is apparent. The entire column is now the spring, which resists lateral deflection. The lateral deflection can be initiated by accidental eccentricities due to (1) deviations from absolute straightness of the column, (2) nonuniformity of the stress distribution within the member due to nonhomogeneity of the material or residual stresses, or (3) error in alignment when the load is applied. Only a small increase in load above the critical load will significantly increase the initial deflection—possibly enough to induce inelastic action in the material on the concave side of the deflected column, with the further possibility that such inelastic action could be sufficient to cause collapse (stability failure).

FIG. 10–4

(a)   (b)

In the following article an expression comparable to Eq. *a* is developed for the limiting compressive load that may be applied to a long, slender, elastic, prismatic bar (an axially loaded perfect column).

PROBLEMS

Note: All springs in the following problems can act in either tension or compression.

10-1*   The rigid bar shown in Fig. P10–1 is supported at the top by two springs that hold the bar in a vertical position. Each spring has a modulus of 5 lb/in. Determine the maximum load P for which the bar is in stable equilibrium in the vertical position. Perform the analysis by considering the behavior of the system under small angular rotations from the vertical.

10-2*   The rigid bar shown in Fig. P10–2 is supported at the right end by two springs that hold the bar in a horizontal position. Each spring has a modulus of 750 N/m. Determine the maximum load P for which the bar is in stable equilibrium in the horizontal position. Perform the analysis by considering the behavior of the system under small angular rotations from the horizontal.

10-3*   The system of two rigid bars shown in Fig. P10–3 is restrained at *B* by two springs adjusted so that the system is in equilibrium when bars *AB* and *BC* are aligned. Each of the springs has a modulus of 20 lb/in. Determine the maximum load *P* that can be applied at *C* and have the system remain in stable equilibrium with the bars aligned. Perform the analysis by considering small angular rotations of bar *AB*.

10-4   The system of two rigid bars shown in Fig. P10–4 is restrained at *B* by two springs adjusted so that the system is in equilibrium when bars *AB*

FIG. P10–3

FIG. P10–4

and *BC* are aligned. Each of the springs has a modulus of 20 kN/m. Determine the maximum load *P* that can be applied at *C* and have the system remain in stable equilibrium with the bars aligned. Perform the analysis by considering small angular rotations of bar *AB*.

10-3   BUCKLING OF LONG STRAIGHT COLUMNS

The first solution for the buckling of long slender columns was published in 1757 by the Swiss mathematician Leonhard Euler (1707–1783). Although the results of this article can be used only for long slender columns, the analysis, similar to that used by Euler, is mathematically revealing and helps explain the behavior of columns.

The purpose of this analysis is to determine the minimum axial compressive load for which a column will experience lateral deflections. A straight, slender, pivot-ended column centrically loaded by axial compressive forces *P* at each end is shown in Fig. 10–5*a*. A pivot-ended column is supported such that the bending moment and lateral movement are zero at the ends. In Fig. 10–5*b*, the load *P* has been increased sufficiently to cause a lateral deflection δ at the midpoint of the span. Axes are selected with the origin at the center of the span for convenience. If the column is sectioned at an arbitrary position *x*, a free-body diagram of the portion to the right of the section will appear as shown in Fig. 10–5*c*. The two forces constitute a couple of magnitude $P(\delta - y)$ that must equal the resisting moment M_r; thus, $M_r = P(\delta - y)$. The differential equation for the elastic curve, as given by Eq. 7–3, becomes

$$EI \frac{d^2y}{dx^2} = M_r = P(\delta - y)$$

FIG. 10–5

or

$$\frac{d^2y}{dx^2} + \frac{P}{EI}\, y = \frac{P\delta}{EI} \qquad (a)$$

Equation a is an ordinary second-order linear differential equation with constant coefficients and a constant right-hand side. Established methods for the solution of such equations show it to be of the form

$$y = A \sin px + B \cos px + C \qquad (b)$$

where A, B, C, and p are constants. By differentiating Eq. b, substituting the results into Eq. a, and collecting like terms, the following expression is obtained for evaluating p and C:

$$(-p^2 + P/EI)(A \sin px + B \cos px) + (P/EI)C = (P/EI)\delta$$

from which it follows that

$$p^2 = P/EI \qquad \text{and} \qquad C = \delta$$

The constants A and B can be obtained from the two boundary conditions that the slope and deflection of the elastic curve are zero at the origin; that is, at $x = 0$, $y = 0$ and $dy/dx = 0$. Since the purpose of this analysis is to determine the minimum load P for which lateral deflections occur, it is necessary to have a third boundary condition; namely, at $x = L/2$, $y = \delta > 0$. The midspan deflection δ is required to be nonzero; otherwise, $\delta = 0$, $y \equiv 0$ could be a solution of Eq. a. When the boundary conditions are substituted in Eq. b and its first derivative, they give

$$0 = B + C$$

$$0 = Ap$$

from which

$$A = 0 \qquad \text{and} \qquad B = -C = -\delta$$

The solution to Eq. a thus becomes

$$y = \delta \left(1 - \cos \sqrt{\frac{P}{EI}}\, x \right) \qquad (10\text{--}1)$$

In order to satisfy the condition that at $x = L/2$, $y = \delta$, the cosine term in Eq. 10–1 must vanish. Thus,

$$\cos \sqrt{\frac{P}{EI}}\, \frac{L}{2} = 0$$

This equation is satisfied when the argument of the cosine is an odd multiple of $\pi/2$ or

$$\sqrt{\frac{P}{EI}}\,\frac{L}{2} = \frac{\pi}{2}, \frac{3\pi}{2}, \frac{5\pi}{2}, \cdots$$

Only the first value has physical significance since it determines the minimum value of P for a nontrivial solution. This value for P is called the critical buckling load, is designated by P_{cr}, and has the magnitude

$$P_{cr} = \frac{\pi^2 EI}{L^2} \tag{10--2}$$

The term P_{cr}[1] is usually called the *Euler buckling load* in honor of Leonhard Euler.

The moment of inertia I in Eq. 10–2 refers to the axis about which bending occurs. When I is replaced by Ar^2, where r is the radius of gyration about the axis of bending, Eq. 10–2 becomes

$$\frac{P_{cr}}{A} = \frac{\pi^2 E}{(L/r)^2} \tag{10--3}$$

The quantity L/r is the *slenderness ratio* and is determined for the axis about which bending tends to occur. For a pivot-ended, centrically loaded column with no intermediate bracing to restrain lateral motion, bending occurs about the axis of minimum moment of inertia (minimum radius of gyration).

The Euler buckling load as given by Eq. 10–2 or 10–3 agrees well with experiment only if the slenderness ratio is large, typically in excess of 140 for steel columns. Whereas short compression members can be treated as explained in Chapter 1, many columns lie between these extremes in which neither solution is applicable. These intermediate length columns are analyzed by empirical formulas described in later sections.

EXAMPLE 10–1

A 2.5-m pivot-ended, air dried douglas fir column has a 50 × 100-mm rectangular cross section. Determine

  (a)   The slenderness ratio.
  (b)   The Euler buckling load.
  (c)   The ratio of the axial stress under the action of the buckling load to the elastic strength of the material.
  (d)   The maximum load the column can support with a factor of safety of 2.5.

---

[1] While the analysis predicts the buckling load, it does not determine the corresponding lateral deflection δ. This deflection can assume any nonzero value small enough that the nonlinear factor $[1 + (dy/dx)^2]^{3/2}$ in the curvature expression is approximately unity.

SOLUTION

(a) The length of the column is 2.5 m or 2500 mm. To determine the slenderness ratio, the minimum radius of gyration must be calculated. The moment of inertia of a rectangular cross section is $bh^3/12$ and the area is bh; therefore, the radius of gyration is $\sqrt{I/A}$ or $h/2\sqrt{3}$. The minimum radius of gyration is found by using the centroidal axis parallel to the longer side of the rectangle. Thus, $b = 100$ mm and $h = 50$ mm so that

$$r = h/2\sqrt{3} = 50/2\sqrt{3} = 14.434 \text{ mm}$$

The slenderness ratio is then found to be

$$L/r = 2500/14.434 = \underline{173.2} \qquad \text{Ans.}$$

(b) From Appendix A, the modulus of elasticity for air-dried douglas fir is found to be 13 GPa. Thus, by using Eq. 10–3, the Euler buckling load is found to be

$$P_{cr} = \pi^2 EA/(L/r)^2 = (\pi^2)(13)(10^9)(0.050)(0.100)/(173.2)^2$$

$$= 21.385(10^3) \text{ N} = \underline{21.4 \text{ kN}} \qquad \text{Ans.}$$

(c) The axial stress under the action of the buckling load is

$$\sigma_{cr} = P_{cr}/A = 21.385(10^3)/(0.050)(0.100)$$

$$= 4.277(10^6) \text{ N/m}^2 = 4.277 \text{ MPa}$$

From Appendix A, the elastic strength of air-dried douglas fir is found to be 44 MPa. Thus, the stress ratio is

$$\sigma_{cr}/\sigma_y = 4.277/44 = 0.0972 = \underline{9.72 \text{ percent}} \qquad \text{Ans.}$$

(This demonstrates that buckling can occur at stresses well below the elastic limit of a material for sufficiently slender columns.)

(d) The maximum load that the column can support with a factor of safety of 2.5 must be based on the Euler buckling load. Therefore,

$$P_{max} = P_{cr}/2.5 = 21.385/2.5 = \underline{8.55 \text{ kN}} \qquad \text{Ans.}$$

EXAMPLE 10–2

Two $2 \times 2 \times 1/8$-in. steel angles 9 ft long are to serve as a pivot-ended column. Determine the slenderness ratio and the Euler buckling load if

(a) The two angles are not connected and each acts as an independent member.

(b) The two angles are fastened together as shown in Fig. 10–6 to act as a unit.

FIG. 10–6

SOLUTION

(a) If the angles are not connected and each acts independently, the slenderness ratio is determined by the least radius of gyration of the individual cross sections. From Appendix C, r for the Z–Z axis for this angle is 0.40 in. Thus, the slenderness ratio is

$$L/r = (9)(12)/0.40 = \underline{270} \qquad \text{Ans.}$$

The area for each angle is 0.48 sq. in., making the total area 0.96 sq. in. If the modulus of elasticity for the steel is $30(10^6)$ psi, the buckling load is found to be

$$P_{cr} = \pi^2 EA/(L/r)^2 = \pi^2(30)(10^6)(0.96)/(270)^2 = \underline{3900\ \text{lb}} \qquad \text{Ans.}$$

(b) With the two angles connected as shown in Fig. 10–6, buckling will occur about the X–X axis and the radius of gyration (see App. C) is 0.63 in. Thus, the slenderness ratio is

$$L/r = (9)(12)/0.63 = \underline{171.4} \qquad \text{Ans.}$$

The corresponding Euler buckling load is then

$$P_{cr} = \pi^2 EA/(L/r)^2 = \pi^2(30)(10^6)(0.96)/(171.4)^2 = \underline{9680\ \text{lb}} \qquad \text{Ans.}$$

PROBLEMS

Note: In the following problems, assume that all columns are pivot-ended and that Euler's formula (Eq. 10–2) is applicable for steel with L/r greater than 140 and for aluminum or timber with L/r greater than 80.

10-5* A hollow circular steel ($E = 30{,}000$ ksi) column 15 ft long has an inner diameter of 3 in. and an outer diameter of 4 in. Determine

(a) The slenderness ratio.

(b) The Euler buckling load.

(c) The axial stress in the column under the action of the Euler buckling load.

10-6* A meterstick has a rectangular cross section 30 mm wide by 5 mm thick. Determine

(a) The slenderness ratio.

(b) The Euler buckling load. Use $E = 13$ Gpa for the wood.

10-7* A 16-ft column with the cross section shown in Fig. P10–7 is constructed from four rough-cut (full dimension) timbers. The timbers are nailed together so that they act as a unit. Determine

(a) The slenderness ratio.

(b) The Euler buckling load. Use $E = 2000$ ksi for the timber.

(c) The axial stress in the column when the Euler load is applied.

10-8 A 3-m column with the cross section shown in Fig. P10–8 is constructed from two rough-cut (full dimension) timbers. The timbers are nailed together so that they act as a unit. Determine

(a) The slenderness ratio.

(b) The Euler buckling load. Use $E = 13$ GPa for the timber.

(c) The axial stress in the column when the Euler load is applied.

10-9 Determine the maximum allowable compressive load for a 25-ft timber ($E = 1200$ ksi) column having the cross section shown in Fig. P10–9 if a factor of safety of 3 is specified.

FIG. P10–7

FIG. P10–8

FIG. P10–9

FIG. P10–10

10-10*   Determine the maximum allowable compressive load for a 3.5-m aluminum ($E = 70$ GPa) column having the cross section shown in Fig. P10–10 if a factor of safety of 2.25 is specified.

10-11*   A W12 × 50 structural steel section (see Appendix C for cross-sectional properties) is used for a 24-ft column. Determine

    (a)   The slenderness ratio.

    (b)   The Euler buckling load. Use $E = 29,000$ ksi for the steel.

    (c)   The axial stress in the column when the Euler load is applied.

10-12   A WT305 × 46 structural steel section (see Appendix C for cross-sectional properties) is used for a 5-m column. Determine

    (a)   The slenderness ratio.

    (b)   The Euler buckling load. Use $E = 200$ GPa for the steel.

    (c)   The axial stress in the column when the Euler load is applied.

10-13   Determine the maximum compressive load that a WT12 × 47 structural steel column (see Appendix C for cross-sectional properties) can support if it is 30 ft long and a factor of safety of 1.75 is specified. Use $E = 29,000$ ksi.

10-14*   Determine the maximum compressive load that a W254 × 45 structural steel column (see Appendix C for cross-sectional properties) can support if it is 6 m long and a factor of safety of 1.75 is specified. Use $E = 200$ GPa.

10-15*   Design a hollow circular aluminum alloy ($E = 10,500$ ksi) column 12 ft long to carry a maximum compressive load of 2000 lb with a factor of safety of 2. The inner radius of the cross section must be one-half of the outer radius. Select the smallest cross section possible to minimize cost.

FIG. P10–18

10-16* A sand bin with a mass of 70000 kg is to be supported by four square timber columns 5 m long. If the modulus of elasticity of the timber is 13 GPa and a factor of safety of 2.5 is specified, determine the size of the cross section required. Assume that the load will be equally divided among the columns.

10-17 Determine the maximum total compressive load that three rectangular aluminum ($E = 70$ GPa) bars 30 mm by 120 mm by 2.5 m long will carry as columns if

   (a)   Each bar acts independently of the other two.

   (b)   The three bars are welded together to form an H-column.

10-18* Two $L5 \times 3 \times \frac{1}{2}$-in. structural steel angles (see Appendix C for cross-sectional properties) are used for a column that is 15 ft long. Determine the total compressive load required to buckle the two members if

   (a)   They act independently of each other. Use $E = 29,000$ ksi.

   (b)   They are riveted together as shown in Fig. P10–18.

10-19* Two C229 $\times$ 30 structural steel channels (see Appendix C for cross-sectional properties) are used for a column that is 12 m long. Determine the total compressive load required to buckle the two members if

   (a)   They act independently of each other. Use $E = 200$ GPa.

   (b)   They are latticed (see Fig. 10–12a) 150 mm back to back as shown in Fig. P10–19.

FIG. P10–19   FIG. P10–20

10-20 A column 40 ft long is made by riveting three S10 × 25.4 structural steel sections (see Appendix C for cross-sectional properties) together as shown in Fig. P10–20. Determine

(a) The maximum compressive load that this column can support. Use $E = 29{,}000$ ksi.

(b) The maximum compressive load that this column can support if the rivets are removed and the sections act as separate units.

10-4   EFFECTS OF DIFFERENT IDEALIZED END CONDITIONS

The Euler buckling formula, as expressed by either Eq. 10–2 or Eq. 10–3, was derived for a column with pivoted ends. The Euler equation changes for columns with different end conditions such as the four common ones shown in Fig. 10–7.

While it is possible to set up the differential equation with the appropriate boundary conditions to determine the Euler equation for each new case, a more common approach makes use of the concept of an effective length.

Pivot or round ends	Fixed or built-in ends	One free end, one fixed end	One pivot end, one fixed end
(a)	(b)	(c)	(d)

FIG. 10–7

The pivot-ended column, by definition, has zero bending moments at each end. The length L in the Euler equation, therefore, is the distance between successive points of zero bending moment. All that is needed to modify the Euler column formula for use with other end conditions is to replace L by L', where L' is defined as the *effective length* of the column (the distance between two successive inflection points or points of zero moment).

The ends of the column in Fig. 10–7b are built in or fixed; since the deflection curve is symmetrical, the distance between successive points of zero moment (inflection points) is half the length of the column. Thus, the effective length L' of a fixed-end column for use in the Euler column formula is half the true length ($L' = 0.5L$). The column in Fig. 10–7c, being fixed at one end and free at the other end, has zero moment only at the free end. If a mirror image of this column is visualized below the fixed end, however, the effective length between points of zero moment is seen to be twice the actual length of the column ($L' = 2L$). The column in Fig. 10–7d is fixed at one end and pinned at the other end. The effective length of this column cannot be determined by inspection, as could be done in the previous two cases; therefore, it is necessary to solve the differential equation to determine the effective length. This procedure yields $L' = 0.7L$.

A pin-ended column is usually loaded through a pin that, as a result of friction, is not completely free to rotate; therefore, there will always be an indeterminate moment at the ends of a pin-connected column that will reduce the distance between the inflection points to a value less than L. Also, it is impossible to support a column so that all rotation is eliminated, and so the effective length of the column in Fig. 10–7b will be somewhat greater than $L/2$. As a result, it is usually necessary to modify the effective column lengths indicated by the ideal end conditions. The amount of the corrections will depend on the individual application. In summary, *the term L/r in all column formulas in this book is interpreted to mean the effective slenderness ratio L'/r*. In the problems in this book, the length given for a member is assumed to be the effective length unless otherwise noted.

PROBLEMS

Note: In the following problems, the Euler formula (Eq. 10–2) is applicable for steel columns with an effective slenderness ratio L'/r greater than 140 and for aluminum or timber columns with L'/r greater than 80.

10-21   Derive the Euler formula, Eq. 10–2 or 10–3, for an axially loaded, pivot-ended column by solving the differential equation of the elastic curve written with the origin of coordinates at one end of the column.

10-22   Verify the effective length for the fixed-ended column shown in Fig. 10–7b by solving the differential equation of the elastic curve and applying the appropriate boundary conditions.

10-23   Verify the effective length for the fixed-end, free-end column shown in Fig. 10–7c by solving the differential equation of the elastic curve and applying the appropriate boundary conditions.

10-24   A free-body diagram for a fixed-end, pivot-end column is shown in Fig. P10–24. Use the diagram to develop the differential equation of the elastic curve. Verify the effective length for the fixed-end, pivot-end column shown in Fig. 10–7d by solving the differential equation and applying the appropriate boundary conditions.

10-25*   A rigid beam is supported by two fixed-ended columns as shown in Fig. P10–25a. Determine the effective lengths of the columns by solving the differential equation of the elastic curve and applying the appropriate boundary conditions. Assume that buckling occurs as shown in Fig. P10–25b.

(a)   (b)

FIG. P10–25

10-26* A finished 2 × 4 (actual size $1\frac{5}{8}$ in. × $3\frac{1}{2}$ in. × 12 ft long) is used as a fixed-ended column. If the modulus of elasticity for the timber is 2000 ksi and a factor of safety of 2 is specified, determine the maximum safe load for the column.

10-27 An L102 × 76 × 6.4-mm aluminum ($E = 70$ GPa) angle is used for a fixed-end, pinned-end column having an actual length of 2.5 m. Determine the maximum safe load for the column if a factor of safety of 1.75 is specified. See Appendix C for cross-sectional properties; they are the same as those for a steel angle of the same size.

10-28* A W203 × 22 structural steel section (see Appendix C for cross-sectional properties) is used for a fixed-end, free-end column having an actual length of 4 m. Determine the maximum mass the column can support if a factor of safety of 2 is specified. Use $E = 200$ GPa.

10-29* Determine the maximum load that a 2-in. by 3-in. by 8-ft long rectangular aluminum alloy bar ($E = 10,500$ ksi) can support with a factor of safety of 3 if it is used as a fixed-end, pinned-end column.

10-30 A 6-in. by 6-in. by 20-ft long timber ($E = 1900$ ksi) is used as a fixed-end, pinned-end column to support a 40,000-lb load. Determine the factor of safety based on the Euler buckling load.

10-5   THE SECANT FORMULA

Many real columns do not behave as predicted by the Euler formula because of imperfections in the alignment of the loading. In this section, the effect of imperfect alignment is examined by considering an eccentric loading. The pivot-ended column shown in Fig. 10–8 is subjected to compressive forces acting at a distance e from the centerline of the undeformed column. As the loading is increased, the column deflects laterally, as shown in Fig. 10–8b. If

FIG. 10–8

axes are chosen with the origin at the center of the span, as in Art. 10–3, the bending moment at any section is

$$M = P(e + \delta - y)$$

and, if the stress does not exceed the proportional limit and deflections are small, the differential equation for the elastic curve becomes

$$EI \frac{d^2y}{dx^2} = P(e + \delta - y)$$

or

$$\frac{d^2y}{dx^2} + \frac{P}{EI} y = \frac{P}{EI} (e + \delta) \qquad (a)$$

This is the same differential equation as Eq. *a* of Art. 10–3 except that the constant on the right-hand side is $P(e + \delta)/EI$ rather than $P\delta/EI$. The solution, therefore, is of the form

$$y = A \sin px + B \cos px + C \qquad (b)$$

where $p^2 = P/EI$, $C = (e + \delta)$, and *A* and *B* are constants to be determined from the boundary conditions, as was the case in Art. 10–3. Since the slope and deflection of the elastic curve are zero at the origin (that is, when $x = 0$, $y = 0$ and $dy/dx = 0$), the constants *A* and *B* become

$$A = 0 \qquad \text{and} \qquad B = -(e + \delta)$$

The solution to Eq. *a* thus becomes

$$y = (e + \delta)[1 - \cos \sqrt{P/(EI)}x] \qquad (10–4)$$

In this case a relationship can be found between the maximum deflection δ, which is the value of *y* at $x = L/2$, and the load *P*. Thus,

$$\delta = e + \delta - e \cos \sqrt{P/(EI)}(L/2) - \delta \cos \sqrt{P/(EI)}(L/2)$$

or

$$\delta = e \left[\frac{1 - \cos \sqrt{P/(EI)}(L/2)}{\cos \sqrt{P/(EI)}(L/2)} \right] = e[\sec \sqrt{P/(EI)}(L/2) - 1] \qquad (c)$$

Equation *c* indicates that, for a given column in which *E*, *I*, and *L* are fixed and $e > 0$, the column exhibits lateral deflection for even small values of the load *P*. For any value of *e*, the quantity

$$\sec \sqrt{P/(EI)}(L/2)$$

approaches positive or negative infinity as the argument $\sqrt{P/(EI)}(L/2)$ approaches $\pi/2$, $3\pi/2$, $5\pi/2$, . . . , and the deflection δ increases without bound, indicating that the critical load corresponds to one of these angles. If $\pi/2$ is chosen, since this value yields the smallest load, then

$$\sqrt{P/(EI)}(L/2) = \pi/2$$

or

$$\sqrt{P/(EI)} = \pi/L$$

from which

$$P_{cr} = \frac{\pi^2 EI}{L^2} \tag{10--2}$$

which is the Euler formula discussed in Art. 10–3.

Since it has been assumed that the stresses do not exceed the proportional limit (when the elastic curve equation was written), the maximum compressive stress can be obtained by superposition of the axial stress and the maximum bending stress. The maximum bending stress occurs on a section at the midspan of the column where the bending moment assumes its largest value, $M_{max} = P(e + \delta)$. Thus, the maximum stress is

$$\sigma_{max} = \frac{P}{A} + \frac{M_{max}c}{I} = \frac{P}{A} + \frac{P(e + \delta)c}{Ar^2} \tag{d}$$

in which r is the radius of gyration of the column cross section. From Eq. c,

$$e + \delta = e \sec \sqrt{\frac{P}{EI}} \frac{L}{2}$$

Therefore,

$$\sigma_{max} = \frac{P}{A} \left(1 + \frac{ec}{r^2} \sec \sqrt{\frac{P}{EI}} \frac{L}{2} \right)$$

or

$$\frac{P}{A} = \frac{\sigma_{max}}{1 + \frac{ec}{r^2} \sec \left(\frac{L}{2r} \sqrt{\frac{P}{EA}} \right)} \tag{10--5}$$

Equation 10–5 is known as the secant formula and relates the average unit load P/A to the dimensions of the column, the properties of the column material, and the eccentricity e. The term L/r is the same slenderness ratio found in the Euler buckling formula, Eq. 10–3. For columns with different end conditions, the effective slenderness ratio L'/r should be used, as discussed in Art. 10–4. The quantity ec/r^2 is called the *eccentricity ratio* and is seen to depend on the eccentricity of the load and the dimensions of the column. If the column was loaded axially, e would presumably be zero, and the maximum stress would be equal to P/A. It is virtually impossible, however, to eliminate all eccentricity that might result from various factors such as the initial crookedness of the column, minute flaws in the material, and a lack of uniformity of the cross section as well as accidental eccentricity of the load. A committee of the American Society of Civil Engineers[2] made an extensive study of the results of a great many column tests and came to the conclusion that a value of 0.25 for ec/r^2 would give results with the secant formula that would be in good agreement with experimental tests on axially loaded columns of structural steel in ordinary structural sizes. A more complete discussion of eccentrically loaded columns is included in Art. 10–8. When Eq. 10–5 is used, σ_{max} is the proportional limit, although the yield point of structural steel may be used for convenience, and P is the load at which inelastic action would start to occur. This load should be divided by a factor of safety for design purposes. Note that P/A occurs on both sides of the equation, and thus it is not directly proportional to σ. Consequently, the load resulting from reducing σ to some proportion of the proportional limit will not be reduced by the same amount. In order to obtain a given factor of safety, the factor must be applied to the load each time it occurs in the formula and not to the value of σ; that is, in Eq. 10–5 kP_w should be substituted for P, where P_w is the working load (maximum safe load) and k is the factor of safety.

In order to make effective use of Eq. 10–5, curves showing P/A and L/r can be drawn for various values of ec/r^2 for any given material. Figure 10–9 is such a set of curves for a material with $\sigma = 40{,}000$ psi and $E = 30(10^6)$ psi (or $\sigma = 280$ MPa and $E = 210$ GPa). For these material properties, graphic solutions can be obtained from Fig. 10–9. Digital computers can also be programmed to directly solve the secant formula using iterative techniques.[3]

The outer envelope of Fig. 10–9, consisting of the horizontal line $\sigma = 40$ ksi and the Euler curve, corresponds to zero eccentricity. The Euler curve is truncated at 40 ksi since this is the maximum allowable stress for the material. With Young's modulus equal to 30,000 ksi, the truncation occurs when $L/r = 86$. From the data presented in Fig. 10–9, it is seen that eccentricity of loading plays a significant role in reducing the working load (maximum safe load) in the short and intermediate column ranges (slenderness

[2] Final Report of Special Committee on Steel Column Research, *Trans. Am. Soc. of Civil Engrs.*, Vol. 98, 1933.

[3] For a program for the secant formula, see *Numerical Methods in Fortran IV Case Studies,* W. S. Dorn and D. D. McCracken, Wiley, New York, 1972, pp. 43–49.

FIG. 10–9

ratios less than 150 for the steel in Fig. 10–9). For large slenderness ratios, the curves for the various eccentricity ratios tend to merge with the Euler curve. Consequently, the Euler formula can be used to analyze columns with large slenderness ratios. For a given problem, *the slenderness ratio must be computed* to determine whether or not the Euler equation is valid.

The quantities E, I, and L can be eliminated from Eq. c by using Eq. 10–2, in which case the maximum deflection becomes

$$\delta = e[\sec(\pi/2)\sqrt{P/P_{cr}} - 1] \qquad (e)$$

From this equation it would appear that the deflection δ becomes infinite as P approaches P_{cr}; however, under these conditions, the slope of the deflected shape of the column is no longer sufficiently small to be neglected in the curvature expression. As a result, accurate deflections can only be obtained by using the nonlinear form of the differential equation of the elastic curve. Graphs of Eq. e for various values of e are shown in Fig. 10–10. From these curves it can be seen that the deflection is extremely small as e approaches zero until the load is near P_{cr}, at which time the deflection increases very rapidly. In Fig. 10–10 the curve shown in dashed lines for $e = 0$ is a plot of

FIG. 10–10

the exact solution.[4] This limiting case represents an axially loaded column, and the remaining curves illustrate how the deflection under a supposedly axial load can be initiated by a small accidental eccentricity. The curve for $e = 0$ indicates that a load only slightly greater than P_{cr} would probably cause the column to collapse because the proportional limit of the material would have been exceeded.[5] Note the similarity between this curve and that of Fig. 10–3.

[4] See *Theory of Elastic Stability,* S. Timoshenko and J. Gere, Second Edition McGraw-Hill, New York, 1961, pp. 77–82.

[5] The curve is valid for any deflection only until the proportional limit of the material is exceeded, at which time the curve drops.

PROBLEMS

Note: In the following problems, use the effective slenderness ratio for the particular end conditions and assume that Euler's equation is applicable for steel with L/r greater than 140 and for aluminum or timber with L/r greater than 80.

10-31*   The American Society of Civil Engineers recommends using a value of 0.25 for the eccentricity for structural steel columns. What eccentricity does this imply for

   (a)   A solid circular bar of diameter d?

   (b)   A thin-walled tube of thickness t and diameter d?

10-32*   Use Fig. 10–9 to estimate the maximum safe compressive load for a pivot-ended steel strut having a 1 × 2-in. cross section and a 25-in. length. Use a factor of safety of 2.5 and assume $ec/r^2 = 0.20$. What is the corresponding value of the eccentricity?

10-33   Use Fig. 10–9 to estimate the maximum safe compressive load for a fixed-ended solid steel bar having a 50-mm diameter and a 2.50-m length. The load acts 3 mm from the axis of the bar. A factor of safety of 2.5 is specified.

10-34*   Compute the value of L/r corresponding to $P/A = 30$ ksi and $ec/r^2 = 0.30$ for a steel rod with $E = 30,000$ ksi and $\sigma_{PL} = 40$ ksi. Check this point to see if it is in reasonable agreement with Fig. 10–9.

10-35   A W203 × 22 steel wide flange section (see Appendix C for cross-sectional properties) is used for a free-end, fixed-end column that is 1.25 m long. Use Fig. 10–9 to estimate the maximum safe load this column could support if $ec/r^2 = 0.20$ and a factor of safety of 2 is specified.

10-36*   A hollow circular pivot-ended steel column 10 ft long has an inner diameter of 3 in. and an outer diameter of 4 in. When loaded in compression, the column buckles at a load of 121 kips. Use Fig. 10–9 to determine the value of the eccentricity ratio and the corresponding eccentricity of the applied load.

10-37*   A pivot-ended steel strut has a 15-mm by 30-mm rectangular cross section and a 450-mm length. The strut buckles about the long axis of the cross section at a load of 54 kN. Use Fig. 10–9 to determine the eccentricity ratio and the corresponding eccentricity of the applied load.

10-38   A 4-in. diameter solid steel shaft ($E = 30,000$ ksi and $\sigma_{PL} = 40$ ksi) is to support an axial compressive load of 80 kips with a factor of safety of 2.5. Determine the maximum permissible length if $ec/r^2 = 0.20$. Check the computed length with the results shown in Fig. 10–9.

10-39*   A fixed-ended solid steel bar ($E = 210$ GPa and $\sigma_{PL} = 280$ MPa) has a diameter of 50 mm and a length of 2 m. Use Fig. 10–9 to determine the maximum safe compressive load if the load acts 3 mm from the axis of the bar and a factor of safety of 2.5 is specified.

10-40   A W10 $\times$ 22 steel section 12 ft long was used as a pivot-ended column. The column failed under a load of 105 kips. Use Fig. 10–9 to estimate the probable eccentricity of the load.

10-41*   An S305 $\times$ 47 steel section 4.5 m long is used as a fixed-ended, pivot-ended column. If the column carries a load of 225 kN and $ec/r^2 = 0.30$, determine the factor of safety by using the results of Fig. 10–9.

10-42   A W14 $\times$ 82 steel section 20 ft long is used for a fixed-ended column. Use the secant formula to determine the allowable load P if it is assumed that the eccentricity ratio is 0.25, the modulus of elasticity is 30,000 ksi, the proportional limit is 36 ksi, and the factor of safety is 2.

10-43   A W356 $\times$ 64 steel section 2 m long is used for a fixed-ended, pivot-ended column. Use the secant formula to determine the allowable load P if it is assumed that the eccentricity ratio is 0.25, the modulus of elasticity is 200 GPa, the proportional limit is 250 MPa, and the factor of safety is 2.5.

10-6 EMPIRICAL COLUMN FORMULAS—CENTRIC LOADING

Referring to Fig. 10–9, one observes that the envelope of the family of curves is the horizontal line $P/A = 40$ ksi up to L/r of 86 and the plot of the Euler formula (Eq. 10–3) for L/r greater than 86. This curve represents the failure unit loads for *ideal* columns of any length under centric loading. However, for real columns (which will always have built-in eccentricities), the true curve will be an S-shaped curve similar to those shown in Fig. 10–9. Numerous column experiments indicate that the Euler formula (Eq. 10–2 or 10–3) is reliable for designing axially loaded columns, provided the slenderness ratio is within the range in which the eccentricity has relatively little effect. This range is called the *slender* range, which in Fig. 10–9 would begin in the region of L/r equal to 120–140 (the region of transition depends on the material). At the low end of the slenderness scale, the column would behave essentially as a compression block, and the failure unit load would be the compressive strength of the material (for metals, usually the elastic strength σ_y). The extent of this range, the *compression block* range, is a matter of judgment, or is dictated by specifications. In Fig. 10–9, the small eccentricity curves seem to indicate that the range 0 to 40 could be considered the compression block range. The range between the compression block and the slender ranges is known as the *intermediate* range, and the only equation of a rational nature that applies to real columns in this range is the secant formula Eq. 10–5. Since the application of the secant formula to centric loading requires an estimate of the accidental eccentricity ratio, the equation acquires an empirical nature. Furthermore, the application is so involved that simpler empirical formulas have been developed that give results in reasonable agreement with experimental results within the intermediate range. Although a great many equations have been proposed, most of them come under one of three major classifications. These three general formulas are the *Gordon-Rankine,* the *straight-line,* and the *parabolic* formulas, described in the following paragraphs.

The *Gordon-Rankine* formula, one of the first empirical formulas, was developed by Gordon and Rankine, English engineers, during the middle of the nineteenth century. The formula is usually written as

$$\frac{P}{A} = \frac{\sigma_0}{1 + C_1(L/r)^2} \tag{a}$$

in which σ_0 and C_1 are experimentally determined constants. The equation plots as a curve similar to the secant curve.

The *straight-line* formula, proposed by T. H. Johnson about 1890, is written in general form as

$$\frac{P}{A} = \sigma_0 - C_2 \frac{L}{r} \tag{b}$$

where, again, σ_0 and C_2 are experimentally determined material-dependent constants.

At about the same time that T. H. Johnson proposed the straight-line formula, Professor J. B. Johnson proposed a *parabolic* formula of the general form

$$\frac{P}{A} = \sigma_0 - C_3 \left(\frac{L}{r}\right)^2 \tag{c}$$

where σ_0 and C_3 are the experimentally determined constants. This latter formula with the proper choice of constants can be made to agree quite closely in the intermediate range with the secant formula with certain prescribed numerical constants. See, for example, the close correlation between curves ② and ⑦ in Fig. 10–11.

For design purposes, the entire range of unit loads for a given material is covered by an appropriate set of specifications known as a *column code*.

The code will specify the empirical formula for the intermediate range and the limits of the intermediate range; it may also specify the Euler formula in the slender range. If the code is written for *safe* (or *allowable*) loads, the factor of safety k will be specified for the Euler formula and will either be specified or be included in the constants for the empirical formula. A few representative codes are listed in Table 10–1, in which the Euler formula is specified for the slender range, except for code 5. Some codes, such as 4 and 5, are for ultimate (buckling) loads. When these codes are used to determine safe loads, the values obtained from the formulas must be divided by an appropriate factor of safety. The data in Table 10–1 are taken or adapted from handbooks published by the American Institute of Steel Construction (A.I.S.C.), the American Association of State Highway Officials (A.A.S.H.O.), the Armed Forces Supply Support Center (Mil-Hdbk-5), the Aluminum Company of America (Alcoa Handbook), the Aluminum Association (Aluminum Construction Manual), and the Forest Products Laboratory (U. S. D. A. Handbook #72). A graphic comparison of selected codes is provided in Fig. 10–11.

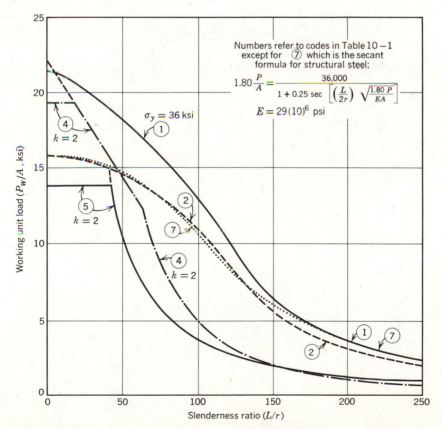

FIG. 10–11

TABLE 10–1 Some Representative Column Codes

Code No.	Source	Material and Modulus of Elasticity	Intermediate Range Formulas and Limitations $\frac{L}{r}$ is the effective ratio $\frac{L'}{r}$	Slender Range Factor of Safety
1	A.I.S.C. Specification, 1970	Steel with a yield point σ_y, $E = 29{,}000$ ksi or $E = 200$ GPa	$\dfrac{P_w}{A} = \dfrac{1}{k}\left[\sigma_y - \dfrac{\sigma_y}{2C_c^2}\left(\dfrac{L}{r}\right)^2\right]$ $C_c^2 = \dfrac{2\pi^2 E}{\sigma_y}$ $k = \dfrac{5}{3} + \dfrac{3(L/r)}{8C_c} - \dfrac{(L/r)^3}{8C_c^3}$ $0 \le \dfrac{L}{r} \le C_c$	23/12
2	Adapted from A.A.S.H.O. Specification, 1973	A36 Structural steel, $E = 29{,}000$ ksi or $E = 200$ GPa	$\dfrac{P_w}{A} = 16{,}000 - \dfrac{1}{2}\left(\dfrac{L}{r}\right)^2$, psi[a] $\dfrac{P_w}{A} = 110 - 0.00345\left(\dfrac{L}{r}\right)^2$, MPa[b] k included $0 \le \dfrac{L}{r} \le 130$	2.24[c]
3	Aluminum Construction Manual	Aluminum alloy 2014-T6, $E = 10{,}600$ ksi or $E = 73$ GPa	$\dfrac{P_w}{A} = 31{,}330 - 238\left(\dfrac{L}{r}\right)$, psi $\dfrac{P_w}{A} = 216 - 1.64\left(\dfrac{L}{r}\right)$, MPa[b] k included $11.3 \le \dfrac{L}{r} \le 55$	1.95

4	Adapted from Alcoa Structural Handbook, 1960	Aluminum alloy 2024-T4 $E = 10,600$ ksi or $E = 73$ GPa	$\dfrac{P_u}{A} = 44,800 - 313\left(\dfrac{L}{r}\right)$, psi $\dfrac{P_u}{A} = 309 - 2.16\left(\dfrac{L}{r}\right)$, MPa[b] Ultimate load; use k as specified $18.5 \le \dfrac{L}{r} \le 64$	As specified
5	Adapted from Mil-Hdbk-5, 1966	Magnesium alloy ZK60A-T5 $E = 6500$ ksi or $E = 45$ GPa	$\dfrac{P_u}{A} = \dfrac{7.6(10^6)}{(L/r)^{1.5}}$, psi $\dfrac{P_u}{A} = \dfrac{52.4(10^3)}{(L/r)^{1.5}}$, MPa[b] Ultimate load; use k as specified $43 < \dfrac{L}{r}$	No Euler equation
6	Adapted from U.S.D.A. Wood Handbook #72, 1955 For square or round solid cross sections	Douglas fir (coast type) $E = 1600$ ksi or $E = 11$ GPa	$\dfrac{P_w}{A} = 1450 - 16.56(10^{-6})\left(\dfrac{L}{r}\right)^4$, psi $\dfrac{P_w}{A} = 10.0 - 114.2(10^{-9})\left(\dfrac{L}{r}\right)^4$, MPa[b] k included $38 \le \dfrac{L}{r} \le 73.5$ Use 1450 psi or 10.0 MPa for $\dfrac{L}{r} < 38$	3

[a] The formula in the AASHO specification is written in terms of the actual column length; changing the constant C_3 to 1/2 adjusts the formula for use with the effective length.

[b] Converted to SI units.

[c] Not in the AASHO specification that specifies the secant formula for the slender range.

In order to provide a smooth transition from the intermediate to the slender range, the constants may be adjusted so that the intermediate curve becomes tangent to the Euler curve, thus defining the upper limit of the intermediate range. For safe loads, the factor of safety to be included in the Euler formula can be adjusted to make this curve tangent to any of the formulas for safe loads in the intermediate range. This is done by equating the two expressions for P/A and equating the two expressions for the first derivatives of P/A with respect to L/r. The resulting two equations can be solved for the factor of safety to be used in the Euler formula and the L/r at the transition from intermediate to slender range. This technique will be demonstrated later. The practice of making the intermediate and Euler curves tangent is not universal; many column codes plot as curves similar to that for curve ④ of Fig. 10–11.

The formulas in Table 10–1 are representative of a large number of column equations that have been incorporated in various design codes. The slenderness ratio in this table is always the effective slenderness ratio L'/r. If no end conditions are specified in the problems presented later, the stated length is the effective length. Note that the use of high-strength materials will increase the allowable (P/A) value for short columns but will have little, if any, effect on the strength of long columns, since the critical load (the Euler load) depends on Young's modulus, not on the elastic strength of the material. Note also that the use of fixed or restrained ends, which has the effect of reducing the length of the column, materially increases the strength of slender columns but has much less influence on short compression members.

All of the discussion so far has been concerned with *primary* instability, in which the column deflects as a whole into a smooth curve. No discussion of compression loading is complete without reference to *local instability* in which the member fails locally by crippling of thin sections. Thin open sections such as angles, channels, and H-sections are particularly sensitive to crippling failure. The design of such members to avoid crippling failure is usually governed by specifications controlling the width-thickness ratios of outstanding flanges. Closed section members of thin material (thin-walled tubes, for example) must also be examined for crippling failure when the members are short (see Mil-Hdbk-5).

EXAMPLE 10–3

Two steel C10 × 25 channels are latticed 5 in. back to back, as shown in Figs. 10–12a and b, to form a column. Determine the maximum allowable axial load for effective lengths of 25 ft and 40 ft. Use Code 1 for structural steel (see Appendix A for properties).

FIG. 10–12

Back of channel

Lacing bars cause channels to act as a unit

5.00″

(a)   *(b)*

SOLUTION

Both I_x and I_y (see Fig. 10–12) or r_x and r_y must be known in order to determine the minimum radius of gyration. Properties of the channel section are given in App. C. The value of I_y for two channels is obtained by using the parallel axis theorem; thus,

$$I_y = 2I_c + 2A\ d^2 = 2Ar_c^2 + 2A\ d^2$$

and

$$r_y = \sqrt{I_y/(2A)} = \sqrt{\frac{2A(r_c^2 + d^2)}{2A}} = \sqrt{r_c^2 + d^2}$$

The above expression indicates that the radius of gyration for the two channels is the same as that for one channel, a fact obtained directly from the definition of radius of gyration. Then,

$$r_y = \sqrt{(0.676)^2 + (2.50 + 0.617)^2} = 3.19 \text{ in.}$$

which is less than the tabular value of 3.52 in. for r_x. This means that the column tends to buckle with respect to the y axis, and the slenderness ratios are

$$\frac{L}{r} = \frac{12(25)}{3.19} = 94.0 \quad \text{and} \quad \frac{12(40)}{3.19} = 150.5$$

From code 1,

$$C_c^2 = 2\pi^2(29)(10^3)/36 = 15{,}900 \qquad C_c = 126$$

The slenderness ratios above indicate that the 25-ft column is in the intermediate range; hence,

$$k\frac{P}{A} = 36 - \frac{36(94^2)}{2(15,900)} = 26.0$$

The factor of safety is

$$k = \frac{5}{3} + \frac{3(94/126)}{8} - \frac{(94/126)^3}{8} = 1.89$$

Hence, the safe load for the 25-ft column is

$$P = \frac{26.0(2)(7.35)}{1.89} = \underline{202 \text{ kips}}$$
Ans.

The 40-ft column having a slenderness ratio of 150.5 is in the slender range; hence,

$$P = \frac{\pi^2 EA}{k(L/r)^2} = \frac{\pi^2(29)(10^3)(2)(7.35)}{(23/12)(150.5^2)} = \underline{96.7 \text{ kips}}$$
Ans.

EXAMPLE 10–4

Determine the dimensions necessary for a 500-mm rectangular strut to carry an axial load of 6.75 kN with a factor of safety of 2. The material is aluminum alloy 2024-T4, and the width of the strut is to be twice the thickness. Use Code 4.

SOLUTION

The code is represented by two different equations that depend on the value of L/r, which in turn will depend on the equation used. Thus, it will be necessary to assume that one of the equations applies and use it to obtain the dimensions of the column, after which the value of L/r must be calculated and used to check the validity of the equation used. Assume L/r is less than 64, in which case the straight-line equation is valid. The cross section is

FIG. 10–13

shown in Fig. 10–13, and the least moment of inertia I_x is equal to $bt^3/12$, the area A is equal to bt or $2t^2$, and the least radius of gyration is

$$r = \left(\frac{bt^3/12}{bt}\right)^{1/2} = \frac{t}{\sqrt{12}}$$

The slenderness ratio is

$$\frac{L}{r} = \frac{0.500\sqrt{12}}{t}$$

and, when this value and the expression for the area are substituted in the straight-line formula of code 4, it becomes

$$k\left(\frac{P}{A}\right) = \frac{2(6.75)(10^3)}{2t^2} = 309(10^6) - 2.16(10^6)\left(\frac{0.500\sqrt{12}}{t}\right)$$

from which

$$t = 13.70(10^{-3})\text{ m} = 13.70\text{ mm}$$

The value of L/r for this thickness is

$$\frac{L}{r} = \frac{0.500\sqrt{12}}{0.0137} = 126.4$$

which is greater than 64 and indicates that the straight-line formula is not valid. The problem must be solved again using the Euler equation. Thus,

$$\frac{P}{A} = \frac{6.75(10^3)}{2t^2} = \frac{\pi^2 E}{2(L/r)^2} = \frac{(\pi^2)(73)(10^9)}{2[(0.500)\sqrt{12}/t]^2}$$

from which

$$t^4 = 28.1(10^{-9})\text{ m}^4 \quad \text{and} \quad t = 12.95(10^{-3})\text{ m} = 12.95\text{ mm}$$

The value of L/r is 133.7 for this thickness which confirms the use of the Euler formula. The dimensions of the cross section are

$$t = \underline{12.95\text{ mm}} \quad \text{and} \quad b = \underline{25.9\text{ mm}} \qquad \text{Ans.}$$

EXAMPLE 10–5
Select the lightest wide flange steel section listed in Appendix C to support a centric load of 150 kips as a 15-ft column. Use code 2.

SOLUTION
If L/r was small, P/A would be 16,000 psi, and the area would be

$$A = 150,000/16,000 = 9.38\text{ sq. in.}$$

A column should be selected from Appendix C with an area greater than 9.38 sq. in. for the first trial. In this case, try a W8 × 40 section for which A is 11.7 sq. in. and r_{min} is 2.04 in. The value of L/r for this column is

$$\frac{L}{r} = \frac{180}{2.04} = 88.2$$

and the load it can support is

$$P = A\left[16,000 - \frac{(L/r)^2}{2}\right] = 11.7\left[16,000 - \frac{88.2^2}{2}\right] = 141,700 \text{ lb}$$

This load is less than the design load; therefore, a column with either a larger area, a larger radius of gyration, or both must be investigated. As a second trial value, use a W10 × 45 section for which A is 13.3 sq. in. and r is 2.01 in. For this section L/r is equal to 90, and the load is

$$P = 13.3(16,000 - 90^2/2) = 158,940 \text{ lb}$$

Since the 45-lb column is stronger than necessary and the 40-lb column is not strong enough, any other sections investigated should weigh between 40 and 45 lb/ft. The only wide flange section in Appendix C that might satisfy the requirements is a W14 × 43 section for which A is 12.6 sq. in. and r is 1.89 in. The slenderness ratio and allowable load for this column are

$$(L/r) = 95.2 \qquad \text{and} \qquad P = 144,530 \text{ lb}$$

which is less than the design load. Thus, the lightest column is a

<u>W10 × 45 section</u>   Ans.

When a rolled section is to be selected to support a specified load, it is usually necessary to make several trial solutions since there is no direct relationship between areas and radii of gyration for different structural shapes. The best section is usually the section with the least area (smallest mass) that will support the load. A minimum area can be obtained by assuming $L/r = 0$. The load-carrying capacity of various sections with areas larger than this minimum can then be calculated, using the proper column formula, to determine the lightest one that will carry the specified load.

The above is not necessarily the best procedure. Different designers have different approaches to the trial-and-error procedure, and for certain problems one approach may be better than another. The important point to make here is that the problem of design involving rolled shapes (other than simple geometric shapes) is, in general, solved by trial and error.

PROBLEMS

Note: Refer to Appendix A for any required material properties.

10-44*   Determine the maximum total compressive load that three rectangular bars 1 by 4 in. by 5 ft long will carry as columns if

(a)   The three bars act as independent axially loaded members.

(b)   The three bars are welded together to form an H-column.

Use Code No. 1 and structural steel for this problem.

10-45*   Determine the maximum compressive load that a W254 × 89 steel section can support as a 9-m column

(a)   If it is unbraced throughout its total length.

(b)   If it is braced such that the effective length for bending about the y axis is reduced to 6 m.

Use Code 1 and low-alloy steel (σ_y = 350 MPa) for this problem.

10-46*   Four L3 × 3 × $\frac{1}{2}$-in. structural steel angles 10 ft long are used as a column. Determine the maximum allowable load

(a)   If the angles are not connected and each acts as an independent axially loaded member.

(b)   If the angles are fastened together as shown in Fig. P10–46.

Use Code 2 for this problem.

10-47*   A 7-m column is composed of two structural steel C254 × 45 channels. Determine the maximum allowable load

(a)   If the channels are not connected and each acts as an independent axially loaded member.

(b)   If the channels are latticed 150 mm back to back as shown in Fig. P10–19.

Use Code 2 for this problem.

10-48   Two L5 × $3\frac{1}{2}$ × $\frac{1}{2}$-in. structural steel angles 12 ft long are used as a column. Determine the maximum allowable load

(a)   If the angles are not connected and each acts as an independent axially loaded member.

(b)   If the angles are fastened together as shown in Fig. P10–18.

Use Code 2 for this problem.

FIG. P10–46

FIG. P10–49

10-49   The compression member *AB* of the truss shown in Fig. P10–49 is a W254 × 67 steel section with the *x–x* axis lying in the plane of the truss. The member is continuous from *A* to *B*. Consider all connections to be the equivalent of pivot ends. Determine

(a)   The maximum safe load for member *AB*.

(b)   The maximum safe load for member *AB* if the bracing member *CD* is removed.

Use Code 2 for this problem.

10-50   Solve Prob. 10-49 with the assumption that the joints at *A* and *B* provide sufficient restraint to reduce the effective length of member *AB* to 10 m. Assume connection *C* offers no restraint to bending.

10-51*   A 30-ft plate and angle column consists of four L5 × $3\frac{1}{2}$ × $\frac{1}{2}$-in. steel angles riveted to a 10 × $\frac{1}{2}$-in. steel plate as shown in Fig. P10–51. Determine the maximum safe load

(a)   If the column is pinned at both ends.

(b)   If the column is fixed at the base and pinned at the top.

Use Code 2 for this problem.

10-52*   A column of aluminum alloy 2024-T4 is composed of two L127 × 127 × 19.1-mm angles riveted together as shown in Fig. P10–52. The length between end connections is 3 m, and the end connections are such that there

FIG. P10–51   FIG. P10–52

is no restraint to bending about the *y* axis; but restraint to bending about the *x* axis reduces the effective length to 2.1 m. Determine the maximum permissible load if a factor of safety of 2 is specified. The cross-sectional properties of aluminum and steel angles are the same; see Appendix C. Use Code 4 for this problem.

10-53   An L5 × 5 × ¾-in. aluminum alloy 2014-T6 angle will be used as a fixed-ended, pivot-ended column to support a load of 120 kips. The cross-sectional properties of steel and aluminum angles are the same; see Appendix C. Determine the maximum permissible length permitted by Code 3.

10-54   A ZK60A-T5 magnesium alloy strut 1.5 m long must transmit an axial compressive load of 110 kN. Restraint at the ends of the strut makes the effective length one-third greater than that for completely fixed ends. The strut is to be rectangular in cross section with one dimension fixed at 60 mm. If a factor of safety of 2 is specified for the strut, determine the other dimension required to satisfy Code 5.

10-55*   A sand bin weighing 200 tons is to be supported by four square timber columns 15 ft long. Assume that the load is equally divided among the columns and that the column load is axial. If Douglas fir timbers are to be used, determine the size required to satisfy Code 6.

10-56*   Select the lightest pair of steel channels listed in Appendix C that can be used to form a 7.5-m column to carry an axial load of 500 kN. The channels are to be spaced and latticed so that resistance to bending is the same for the two axes of symmetry. Specify the size and spacing to satisfy Code 2.

10-57   Select the lightest steel W-section listed in Appendix C that can be used as a 12-ft column to carry an axial load of 200 kips according to Code 1. Use 1.92 as the factor of safety for both ranges of column behavior.

FIG. P10–58

Section A–A

10-58* The alloy steel machine part ($\sigma_y = 62$ ksi and $E = 29,000$ ksi) shown in Fig. P10–58 carries an axial compressive load. The pins at the ends offer no restraint to bending about the axis of the pin, but the restraint about the perpendicular axis reduces the effective length to 36 in. Determine the maximum load permitted by Code No. 1 if a factor of safety of 1.92 is specified for both ranges of column behavior.

10-59* A connecting rod made of SAE 4340 heat-treated steel has the cross section shown in Fig. P10–59. The pins at the ends of the rod are parallel to the x axis and are 1750 mm apart. Assume that the pins offer no restraint to bending about the x axis but provide essentially complete fixity for bending about the y axis. Determine the maximum axial compressive load permitted by Code No. 1.

10-60 Four C152 × 12 structural steel channels 10 m long are used to fabricate a column with the cross section shown in Fig. P10–60. The column is fixed at the base and pinned at the top. The pin at the top offers no restraint to bending about the y axis; but for bending about the x axis, the pin provides restraint sufficient to reduce the effective length to 6 m. Determine the maximum axial compressive load permitted by Code No. 2.

FIG. P10–59

FIG. P10–60

10-7 NONDIMENSIONAL COLUMN FORMULAS—CENTRIC LOADING

Instead of working with sundry codes for particular materials, it has been suggested that, if only one or two general types of curves would be satisfactory for all materials, the equations of the curves could be presented in dimensionless form to apply to columns made of any material. The usual technique is to divide all values of P/A by a stress parameter σ_0, which is usually (but not always) taken as the elastic strength σ_y of a material, and all values of L/r by a slenderness ratio parameter, usually the slenderness ratio, which when applied to the Euler equation will give unity for the ratio $(P/A)/\sigma_0$ (a one-to-one ratio). The resulting curve is illustrated by Fig. 10–14 where, for convenience, the parameters are labeled R and S. The general equation for the Euler part of the curve is

$$R = \frac{1}{S^2} \qquad (S \geqslant S_t) \qquad (10\text{--}6)$$

and the general equation for the parabola, which is tangent to the Euler curve, is

$$R = 1 - C_3 S^2 \qquad (a)$$

FIG. 10–14

In order to find the values of C_3 and S_t, the technique previously described will be employed as follows. Equating the two expressions above yields

$$1 - C_3 S_t^2 = 1/S_t^2$$

and equating the first derivatives of these expressions yields

$$-2C_3 S_t = -2/S_t^3$$

The solution of these two equations for C_3 and S_t yields $C_3 = 0.250$ and $S_t = \sqrt{2} = 1.414$. Hence, the equation of the parabola is

$$R = 1 - 0.250S^2 \qquad (0 \le S \le 1.414) \qquad (10\text{--}7)$$

In a similar fashion, using the general straight-line equation,

$$R = 1 - C_2 S \qquad\qquad (b)$$

and proceeding as before, the constants $C_2 = 0.385$ and $S_t = \sqrt{3} = 1.732$. Hence, the equation of the straight line tangent to the Euler curve is

$$R = 1 - 0.385S \qquad (0 \le S \le 1.732) \qquad (10\text{--}8)$$

To recapitulate, the nondimensional curves presented here apply to any material, provided that the behavior of the column can be reliably predicted by one of the two curves illustrated. Note that for a straight-line formula (Eq. b) the value of σ_0 is often an experimentally determined stress larger than the compressive strength σ_y. In such cases, the straight-line curve in Fig. 10–14 will have a horizontal segment $R_0 = \sigma_y/\sigma_0$ at the low L/r end, so that P/A will never exceed σ_y (see, for example, curve ④ in Fig. 10–11). Note also that if the best straight line for a given material is not tangent to the Euler curve, Eq. 10–8 may not be reliable.

EXAMPLE 10–6
Solve Exam. 10–3 by using nondimensional column Eqs. 10–6 and 10–7 for structural steel. Use the factors of safety specified for Code No. 1 in Exam. 10–3.

SOLUTION
In Exam. 10–3 it was shown that the column will tend to buckle with respect to the y axis and that the slenderness ratios for the 25-ft column and the 40-ft column are 94.0 and 150.5, respectively. The quantity $\pi \sqrt{E/\sigma_0}$ used to establish the slenderness ratio parameter S is evaluated as

$$\pi \sqrt{E/\sigma_y} = \pi \sqrt{29(10^3)/36} = 89.16$$

For the 25-ft column, $S = 94/89 = 1.055$ which is less than 1.414; therefore, the parabolic formula applies and

$$R = 1 - 0.25(1.055^2) = 0.722 = (P/A)/\sigma_y$$

from which

$$P = 0.722(36)(2)(7.33)/1.89 = \underline{202 \text{ kips}} \qquad \text{Ans.}$$

For the 40-ft column, $S = 150.5/89 = 1.69$ which is greater than 1.414; hence, the Euler formula applies and

$$R = \frac{1}{S^2} = \frac{1}{1.69^2} = 0.350 = \frac{P/A}{\sigma_y}$$

from which

$$P = 0.350(36)(2)(7.33)/1.92 = \underline{96.3 \text{ kips}} \qquad \text{Ans.}$$

PROBLEMS

> *Note: Use the nondimensional column formulas (Eqs. 10–6, 10–7, and 10–8) for the following problems. Use the elastic strength for σ_0. Refer to Appendix A for any required material properties.*

10-61* Design a hollow circular column 50 in. long to carry a load of 15 kips with a factor of safety of 2. The material is aluminum alloy 6061-T6 and the ratio of inside to outside diameters is to be 0.80. Assume that the behavior of aluminum is predicted by the straight-line formula in the intermediate column range.

10-62* Design a square column 1.5 m long to carry a load of 25 kN with a factor of safety of 3. The material is annealed 18-8 stainless steel. Assume that the behavior of stainless steel is predicted by the parabolic formula in the intermediate column range.

10-63 A cast steel compression member 10 ft long has the cross section shown in Fig. P10–63. The member can be considered fixed at the bottom and free at the top. Cast steel has an elastic strength of 36 ksi and a modulus of elasticity of 29,000 ksi. Determine the maximum allowable load if a factor

FIG. P10–63

FIG. P10–64

of safety of 3 is specified. Assume that the behavior of cast steel is predicted by the parabolic formula in the intermediate column range.

10-64   Two L152 × 89 × 12.7-mm angles of aluminum alloy 2014-T4 are welded together as shown in Fig. P10–64 to form a 4.75-m pin-ended column. The pins provide no restraint to bending about the x axis but reduce the effective length to 3.25 m for bending about the y axis. The sectional properties of one angle are given on the figure where C is the centroid of one angle. Determine the maximum allowable load if a factor of safety of 2 is specified. Assume that the behavior of aluminum is predicted by the straight-line formula in the intermediate column range.

10-65*   A 6-ft strut having the cross section shown in Fig. P10–65 is made of T-1 steel that has an elastic strength of 90 ksi and a modulus of elasticity of 30,000 ksi. The strut is fixed at both ends and a factor of safety of 2 is specified. Determine the maximum allowable compressive load. Assume

FIG. P10–65

FIG. P10–66

that the behavior of T-1 steel is predicted by the parabolic formula in the intermediate column range.

10-66   A strut of aluminum alloy 2024-T4 having the cross section shown in Fig. P10-66 is to carry an axial compressive load of 500 kN with a factor of safety of 2. The strut is 1.25 m long and is fixed at the bottom and pivoted at the top. Determine the dimension d of the cross section

(a)   By using Code No. 4.

(b)   By using Eqs. 10-6 and 10-8.

10-67   An 8-ft strut having the cross section shown in Fig. P10-67 is made of annealed 18-8 stainless steel. The strut is fixed at the bottom and pivoted at the top. Determine the maximum load that the column can support if a factor of safety of 2 is specified

(a)   By using Code No. 1.

(b)   By using Eqs. 10-6 and 10-7.

10-68   Select a structural steel wide-flange section to serve as a column 4 m long that is to carry an axial load of 1000 kN with a factor of safety of 23/12.

(a)   By using Code No. 1.

(b)   By using Eqs. 10-6 and 10-7.

10-8   ECCENTRICALLY LOADED COLUMNS

Although a given column will support the maximum load when the load is applied centrically, it is sometimes necessary to apply an eccentric load to a column. For example, a beam supporting a floor load in a building may in turn be supported by an angle riveted or welded to the side of a column, as shown in Fig. 10-15. Frequently, the floorbeam is framed into the column with a stiff connection, and the column is then subject to a bending moment due to the continuity of floorbeam and column. The eccentricity of the load, or the bending moment, will increase the stress in the column or reduce its

FIG. 10–15

load-carrying capacity. Three methods will be presented for computing the allowable load on a column subjected to an eccentric load (or an axial load combined with a bending moment).

The Secant Formula   The secant formula (Eq. 10–5) was derived on the assumption that the applied load had an initial eccentricity e; this value can be substituted in the equation to determine the failure load (the load to cause incipient inelastic action). As mentioned previously, there is usually a small amount of unavoidable eccentricity that must be approximated when using this formula for centric loads. This unavoidable eccentricity should be added to the intentional eccentricity, particularly if the intentional eccentricity is small. The use of the secant formula is quite time-consuming, unless a series of curves, such as those in Fig. 10–9, is available.[6] The secant formula is specified by the American Railway Engineering Association and the American Association of State Highway Officials.

Modified Column Formulas   The second solution is based on the arbitrary specification that the sum of the direct and bending stresses ($P/A + Mc/I$) shall not exceed the average unit load prescribed by the appropriate

---

[6] See, for example, App. C of the A.A.S.H.O. Standard Specifications for Highway Bridges.

column formula. For example, when this approach is used with the parabolic formula, the equation becomes

$$\frac{P}{A} + \frac{Mc}{I} = \sigma_0 - C_3 \left(\frac{L}{r}\right)^2 \qquad (a)$$

In effect, this merely replaces the low L/r end of the secant curve with a parabola (see Fig. 10–11), and if the constants in the parabola are properly chosen, the agreement between the two curves is quite good. Since the parabola will deviate very rapidly from the secant curve beyond the point of reverse curvature, Eq. a is limited to the intermediate range. This leaves only the secant formula valid for columns in the slender range with eccentric loading. Although it is possible to use other types of intermediate formulas for the right side of Eq. a, the use of the parabola only is recommended here, since it is an attempt to approximate a formula known to be rational.

The application of the secant formula or Eq. a to an eccentrically loaded column such as that of Fig. 10–15 must be accompanied by an investigation of the stability of the column as an axially loaded member. When the load is applied in the plane containing the axis about which the moment of inertia is least (the y axis of Fig. 10–15), the column might fail by buckling about either axis, and both possibilities should be checked.

1. It can fail by buckling about the axis of least radius of gyration (the y axis in Fig. 10–15) because L/r is largest about this axis. In this case, all formulas developed for centric loading apply[7] and the radius of gyration is r_y.

2. The column can fail by buckling about the axis of maximum radius of gyration (the x axis) due to the combined effect of the bending moment and the direct load. In this case, the values required are r_x, I_x, and the distance c, as shown in Fig. 10–15. Note that c is the maximum distance to the compression side from the centroidal axis. The smaller of the two loads is, of course, the limiting load.

The Interaction Formula   One of the modern expressions for treating combining loads is known as the interaction formula, of which several

---

[7] The load is assumed to be axial; that is, the eccentricity in the y direction does not decrease resistance to buckling about the y axis of Fig. 10–15. For columns with I-sections in the slender range, it can be shown that such an assumption may be unwarranted. See "Lateral Buckling of I-Section Columns with Eccentric End Loads in Plane of the Web," B. Johnston, *Journal of Applied Mechanics,* December 1941, p. A-176.

types are in use. The analysis for compression members subjected to bending and direct stress may be derived as follows. It is assumed that the stress in the column can be written as

$$\frac{P}{A} + \frac{Mc}{I} = \sigma_w$$

and, when this expression is divided by σ_w, it becomes

$$\frac{P/A}{\sigma_w} + \frac{Mc/I}{\sigma_w} = 1 \qquad\qquad (b)$$

When considering eccentrically loaded columns, the value of σ_w will, in general, be different for the two terms. In the first term P/A represents an axial unit load on a column; therefore, the value of σ_w should be the average working unit load on an axially loaded column as obtained by an empirical formula such as those presented in articles 10-6 and 10-7 or the Euler formula. In the second term, Mc/I represents the bending stress induced in the member as a result of the eccentricity of the load or an applied bending moment; therefore, the corresponding value of σ_w should be the allowable flexural stress. Since the two values of σ_w are different, one recommended form of Eq. b is the following interaction formula[8]

$$\frac{P/A}{\sigma_a} + \frac{Mc/I}{\sigma_b} = 1 \qquad\qquad (c)$$

in which P/A = the average unit load on the eccentrically loaded column, σ_a = the *allowable* average unit load for an axially loaded column (note that the greatest value of L/r should be used to calculate σ_a), Mc/I = the bending stress in the column, and σ_b = the *allowable* bending stress.[9]

When Eq. c is used to select the most economic section, it will usually not be possible to obtain a section that will exactly satisfy the equation. Any section that makes the sum of the terms on the left side of the equation less than unity is considered safe, and the *safe* section giving the largest sum (less than unity) is the most efficient section.

EXAMPLE 10–7

A 6-m wide flange steel column with the following properties A = 18 210 mm^2, r_{min} = 68.8 mm, r_{max} = 196 mm, and c = 230.7 mm. supports an

---

[8] The 1970 AISC specifications permit this formula if $(P/A)/\sigma_a \leq 0.15$; otherwise, the second term of the equation is expanded to provide for secondary moments due to lateral displacement of any cross section.

[9] The 1970 AISC specifications gives $\sigma_b = 0.66\sigma_y$ with a reduced value for compression flanges in some cases. See *The Steel Construction Manual*.

eccentric load applied 125 mm from the center of the section measured along the web. Use code 1 with $\sigma_y = 290$ MPa, $C_c^2 = 13\,610$, and $k = 1.90$ and 1.76 for maximum and minimum L/r, respectively. Determine the maximum safe load

(a)   According to the modified column formula.
(b)   According to the interaction formula with σ_b equal to 190 MPa.

SOLUTION
(a)   Assume first that the column tends to buckle about the axis parallel to the web. Note that there is no eccentricity about this axis. The value of L/r is

$$L/r = 6(10^3)/68.8 = 87.2$$

which is less than C_c of 116.7. The average unit load is

$$P/A = \frac{290}{1.90}\left[1 - \frac{(87.2)^2}{2(13610)}\right] = 110.0 \text{ MPa}$$

For bending about the axis perpendicular to the web, the radius of gyration is 196 mm (L/r will be much less than 87.2), and the modified column formula becomes

$$\frac{P}{A} + \frac{Mc}{I} = \frac{\sigma_y}{k}\left[1 - \frac{(L/r)^2}{2C_c^2}\right]$$

The bending moment is Pe, and when numerical values are substituted in the equation, it becomes

$$\frac{P}{A} + \frac{P(0.125)(0.2307)}{A(0.196)^2} = \frac{290}{1.76}\left[1 - \frac{(6/0.196)^2}{2(13610)}\right]$$

from which

$$P/A = 90.9 \text{ MPa}$$

Since this last unit load is less than that for buckling about the web, the maximum allowable load is

$$P = 90.9(10^6)(18210)(10^{-6}) = 1.655(10^6) \text{ N} = \underline{1.655 \text{ MN}} \qquad \text{Ans.}$$

(b)   The interaction formula for eccentrically loaded columns is

$$\frac{P/A}{\sigma_a} + \frac{Mc/I}{\sigma_b} = \frac{P/A}{\sigma_a} + \frac{Pec/(Ar^2)}{\sigma_b} = 1$$

The value of σ_a was found to be 110.0 MPa in part a, and when numerical data are substituted in the expression, it becomes

$$\frac{P}{A}\left[\frac{1}{110.0} + \frac{(0.125)(0.2307)/(0.196)^2}{190}\right] = 1$$

which gives

$$P/A = 76.7 \text{ MPa}$$

The allowable load using this specification is

$$P = 76.7(10^6)(18210)(10^{-6}) = 1.397(10^6) \text{ N} = \underline{1.397 \text{ MN}} \qquad \text{Ans.}$$

The two results are about 18 percent apart in this case.

PROBLEMS

Note: Use the table of codes of Art. 10–6 modified for eccentricity unless otherwise specified. Refer to Appendix C for properties. For σ_0 in Eq. 10–7, use the elastic strength unless specified otherwise.

10-69*   A hollow square steel member with outside and inside dimensions of 4 in. and 3 in. (walls are 1/2 in. thick) functions as a column 15 ft long. Determine the maximum load that the column can carry if the load is applied with a known eccentricity of 5/8 in. along an axis of symmetry parallel to a pair of the sides. Use Fig. 10–9 and a factor of safety of 2. Assume an accidental eccentricity ratio of 0.20.

10-70*   A hollow square steel member with outside and inside dimensions of 100 mm and 80 mm (walls are 10 mm thick) functions as a column 3.75 m long. The load is applied with a known eccentricity of 15 mm along one of the diagonals of the square, and the end conditions are such that for bending about the other diagonal, the effective length is reduced to 2.25 m. Determine the maximum load that the column can carry with a factor of safety of 2. Use Fig. 10–9 and assume an accidental eccentricity ratio of 0.20.

10-71*   A 2-in. diameter steel strut is subjected to an eccentric compressive load as shown in Fig. P10-71. The effective length for bending about the x axis is 60 in.; but for bending about the y axis, the end conditions reduce the effective length to 40 in. Determine the maximum load the strut can carry

FIG. P10–71

FIG. P10–74

$r_{x-x} = 61.2$ mm
$r_{y-y} = 36.7$ mm
$A = 3420$ mm^2

with a factor of safety of 1.5. Use Fig. 10–9 and assume an accidental eccentricity ratio of 0.10.

10-72* A W203 × 46 structural steel section is used for a 6-m column. Determine the maximum load the column can carry if it is applied at a point on the centerline of the web 100 mm from the axis of the column. Use Code No. 2.

10-73 A W8 × 31 structural steel section is used for a 30-ft column. Determine the maximum load the column can carry if it is applied at a point on the centerline of the web 3 in. from the axis of the column. Use Code No. 2.

10-74 The structural steel T-section shown in Fig. P10-74 is to be used as a compression member with an effective length of 7 m. Determine

    (a)   The maximum permissible axial compressive load.

    (b)   The maximum permissible bending moment, applied to the ends of the member in the yz plane (the z axis is the axis of the member), that can be superimposed on the load of part (a) if the moment induces compression in the flange.

Use Code No. 1 with $k = 1.92$ for both ranges.

10-75* Two C8 × 18.75 structural steel sections 30 ft long are latticed 3 in. back to back as shown in Fig. P10–75 to form a column. Determine the maximum load the column can carry if the load is applied at point A of the cross section. Use Code No. 2.

10-76* A 2014-T6 aluminum alloy compression member with an effective length of 1.5 m has the T-cross section shown in Fig. P10–76. Determine

    (a)   The maximum permissible axial compressive load.

FIG. P10–75

FIG. P10–76

(b) The maximum permissible bending moment in the yz plane that can be superimposed as shown on the axial load of part (a).

Use Code No. 3.

10-77 Two L6 $\times$ 3$\frac{1}{2}$ $\times$ $\frac{3}{8}$-in. structural steel angles are welded together, as shown in Fig. P10–77, to form a column with an effective length of 18 ft. Determine

(a) The maximum permissible axial compressive load.

(b) The maximum and minimum values for the distance d when the load of part (a) is applied.

Use Code No. 2.

10-78 A W406 $\times$ 100 structural steel section is used as a 9-m column. The column is fixed at the base but can be considered free at the top for bending

FIG. P10–77

with respect to the x axis (see Appendix C) and pivot-ended at the top for bending with respect to the y axis. The load is applied at a point on the centerline of the web 75 mm from the axis of the column. Determine the maximum permissible value for the eccentric compressive load. Use Code No. 2.

10-79* A W8 × 31 structural steel compression member has an effective length of 23 ft and is subjected to an axial load P and a bending moment of 12 ft-kips with respect to the x axis of the cross section (see Appendix C). Determine the maximum safe load P. Use Code No. 1 and the interaction formula with an allowable flexural stress of $0.66\sigma_y$.

10-80* Two C203 × 17 structural steel channels are latticed 100 mm back-to-back with the flanges turned out to form a column with an effective length of 9.75 m. The load is applied at a point on the axis of symmetry parallel to the backs of the channels 50 mm from the axis of the column. Determine the maximum allowable load. Use Code No. 2 and the interaction formula with an allowable flexural stress of 140 MPa.

10-81 A W16 × 67 structural steel section is used for a 15-ft column that supports an eccentric load of 150,000 lb. The load is applied at a distance e from the axis of the column along the centerline of the web. Determine the maximum permissible value of e. Use Code No. 2 and the interaction formula with an allowable flexural stress of 18,000 psi.

10-82* A 1-m strut with a rectangular cross section supports a compressive load P at the point on the cross section indicated on Fig. P10–82. The material is SAE 4340 heat-treated steel. Determine the maximum allowable value of P and the value of dimension b if the strut is to have equal resistance

FIG. P10–82

FIG. P10–83

FIG. P10–84

$I_{cc} = 106.25(10^6)$ mm^4
$I_{yy} = 125.00(10^6)$ mm^4

to buckling about either principal axis. Use Eqs. 10–6 and 10–7 and a factor of safety of 2.

10-83* Determine the maximum allowable load for a 12-ft Douglas fir column with the cross section shown in Fig. P10–83 if the load is applied at the point indicated on Fig. P10–83. Use Eqs. 10–6 and 10–7 with a factor of safety of 3.

10-84 Determine the maximum force F that can be applied with a factor of safety of 5 to the gray cast-iron frame shown in Fig. P10–84. Use Eqs. 10–6 and 10–7 for the compressive strength, and since gray cast iron is a brittle material, use the ultimate compressive strength for σ_0. Note that the flange may be in tension; therefore, the load may be limited by the tensile strength of the iron.

10-9 SPECIAL COLUMN PROBLEMS

Although the secant formula and its various empirical approximations are adequate for treating a wide variety of engineering problems involving columns, there are certain types of problems involving column action for which the secant formula does not apply. One example is the class of problems known as beam columns—members subjected to lateral loading in addition to compressive axial loading. Unless the problem has been treated previously,[10] it is necessary to revert to the differential equation of the elastic curve to find a solution for the particular problem. It may be difficult to develop an explicit solution for P/A as a function of L/r (as was done in Arts. 10–3 and 10–5); in fact, such a solution may be unnecessary. Much of engineering design is done by iteration; hence, if an expression for the maximum deflection in a beam column can be found, the maximum moment can

[10] There is considerable literature available on this subject. See, for example, Chap. 1, *Theory of Elastic Stability,* S. P. Timoshenko and J. M. Gere, Second Edition, McGraw-Hill, New York, 1961.

be determined and the maximum normal stress is $P/A + Mc/I$, where M is a function of the compressive as well as the transverse loads. With an assumed trial section for the member, the maximum stress can be computed to check the adequacy of the section. The following example will illustrate one method of solving the differential equation.

EXAMPLE 10–8
Develop the expression for the maximum deflection of the beam column of Fig. 10–16a. The column ends are free to rotate, but are restrained against lateral displacement.

SOLUTION
The free-body diagram is shown in Fig. 10–16b, with the y axis chosen positive downward to facilitate the solution. For this choice of axes, the curvature indicated is negative, and the differential equation is

$$EI \frac{d^2y}{dx^2} = - Py - \frac{nP}{2} x + \frac{nP}{2L} x^2$$

The boundary conditions are $y = 0$ at $x = 0$ and, from symmetry, $dy/dx = 0$ at $x = L/2$. Rearranging terms and letting $P/(EI) = k^2$, yields

$$\frac{d^2y}{dx^2} + k^2y = \frac{nk^2}{2L} x^2 - \frac{nk^2}{2} x \qquad (a)$$

The complementary solution for this equation is

$$y_c = A \sin kx + B \cos kx$$

and the particular solution is of the form

$$y_p = Cx^2 + Dx + E$$

$w = nP/L$

L

P

P

R

δ

$R = \frac{wL}{2} = \frac{nP}{2}$

x

y

(a)

(b)

FIG. 10–16

When this last equation is substituted into Eq. a and coefficients of like powers of x are matched, the constants are found to be

$$C = \frac{n}{2L}, \quad D = -\frac{n}{2} \quad \text{and} \quad E = -\frac{n}{Lk^2}$$

Combining the solutions gives

$$y = y_c + y_p = A \sin kx + B \cos kx + \frac{nx^2}{2L} - \frac{nx}{2} - \frac{n}{Lk^2} \qquad (b)$$

Substituting the boundary conditions gives

$$y|_{x=0} = B - \frac{n}{Lk^2} = 0$$

$$\left.\frac{dy}{dx}\right|_{x=L/2} = Ak \cos \frac{kL}{2} - Bk \sin \frac{kL}{2} = 0$$

from which

$$B = \frac{n}{Lk^2} \quad \text{and} \quad A = \frac{n}{Lk^2} \tan \frac{kL}{2}$$

The maximum deflection δ is found by substituting these constants into Eq. b and assigning x the value $L/2$; thus,

$$\delta = \frac{n}{Lk^2} \left(\sec \frac{kL}{2} - 1 \right) - \frac{nL}{8}$$

or in terms of P and the distributed load w

$$\delta = \frac{wEI}{P^2} \left(\sec \frac{L}{2} \sqrt{\frac{P}{EI}} - 1 - \frac{PL^2}{8EI} \right) \qquad \text{Ans.}$$

PROBLEMS

10-85   Develop the expression for the maximum deflection in the beam column of Fig. P10–85 in terms of P, E, I, L, and n.

10-86   Develop the expression for the maximum deflection in the beam column of Fig. P10–86 in terms of P, E, I, L, and n.

10-87   Develop the equation of the elastic curve for the beam column of Fig. P10–87 in terms of n, P, L, E, I, x, and y.

FIG. P10–85

FIG. P10–86

FIG. P10–87

FIG. P10–88

10-88* Determine the critical load for the column of Fig. P10–88 in terms of E, I, and L.

10-89 Develop the equation of the elastic curve for the beam column of Fig. P10–89 in terms of n, P, L, E, I, x, and y.

10-10 INELASTIC COLUMN THEORY

The column formulas discussed in Arts. 10–3, 10–5, and 10–6 were based on the assumption that the action was elastic; that is, the stresses were assumed to be below the proportional limit. The Euler formula can thus be used for an *ideal, axially* loaded column for any value of P/A less than the proportional limit of the material.

FIG. P10–89

The first theory for predicting the buckling loads of columns loaded in the inelastic range was proposed by F. R. Engesser about 1890 and is known as the tangent-modulus theory. It is based on the assumption that the Euler formula can be used in the inelastic range, provided that Young's modulus is replaced by the tangent modulus of the material. The tangent modulus is defined as the slope of the stress-strain diagram for a material at *a particular* stress, and it is thus a function of the stress (and strain) for stresses greater than the proportional limit. For stresses less than the proportional limit, the tangent modulus, of course, is the same as Young's modulus. For an ideal elastoplastic material the tangent modulus becomes zero at the proportional limit since the stress-strain diagram becomes horizontal. If E_T represents the tangent modulus, the buckling load, according to the tangent-modulus theory, is

$$\frac{P}{A} = \frac{\pi^2 E_T}{(L/r)^2} \qquad (10\text{--}9)$$

In order to use Eq. (10–9), values of E_T corresponding to various values of P/A must be obtained from a stress-strain diagram. It is difficult to determine precise values of the slope at specific points on the curve; therefore, it is usually desirable to draw a supplementary curve showing E_T plotted against P/A and use this curve to obtain values for E_T. When corresponding values of E_T and P/A are available, values of L/r can be computed from Eq. (10–9) and used to plot a graph indicating the load-carrying capacity of columns when loaded in the plastic region of the stress-strain diagram. The Euler formula was derived on the assumption that the modulus of elasticity was constant over the cross section of the column, whereas the tangent-modulus formula assumes a variable modulus; therefore, the latter formula cannot be considered to provide a rigorous solution. If the assumption is made that for any given value of the load the *variation* of strain across any section of the column is small, the slope of the stress-strain diagram (the tangent modulus) may be considered a constant for the material at all points in any given section. Since the strains will vary linearly (plane section remains plane), the result will be a linear stress distribution, thus justifying the use of the differential equation for bending. On the basis of this reasoning, it appears that the modified Euler (tangent-modulus) formula should give reasonable results, particularly for materials for which the tangent modulus does not vary too abruptly.

The shape of the tangent-modulus curve is influenced to a considerable degree by the shape of the stress-strain diagram near the proportional limit. Three idealized stress-strain diagrams are shown in Fig. 10–17a for three hypothetical materials with identical diagrams to the proportional limit. Material A is elastoplastic, B has a diagram represented by two straight lines, and C has a gradual curve at the knee. The variation of the tangent modulus

FIG. 10–17

Unit stress, σ

E_T

$\sigma - \epsilon$

E_T

Ⓐ Ⓑ Ⓒ

Unit strain, ϵ
Tangent modulus, E_T

(a)

Unit load on column, P/A

Slenderness ratio, L/r

(b)

with the stress is shown on the same figure, with the modulus represented by horizontal distances. The buckling loads, as predicted by the tangent modulus, are shown in Fig. 10–17b. Note that the buckling load predicted by the tangent modulus formula (Eq. 10–9) is the same as that predicted by the Euler formula (Eq. 10–3) in the elastic range and that this load is usually greater than loads obtained by actual experiment. However, when sufficient care is exercised to assure a centric load, the buckling loads indicated by the tangent-modulus formula are fairly close to experimental results[11] for columns loaded in the elastic and plastic range.

The application of the Euler-Engesser equation (Eq. 10–9) in the inelastic range may require an iterative solution. Such a solution can be obtained with the aid of the digital computer if the time and expense are justified. However, many problems can be solved using successive approximations with a precision consistent with ordinary engineering applications. The following paragraphs outline three of the possible approaches to such solutions.

1. The actual compression stress-strain diagram can be approximated by a series of straight lines. One then solves a given problem by assuming the stress range is entirely within a particular segment of the diagram, using the value of E_T for this segment in Eq. 10–9, solving for σ_{cr} (or P_{cr}/A), and checking this value against the values of σ for which the curve segment is valid. If the computed value of σ_{cr} does not lie in the range selected, the process is repeated using a different segment of the diagram. Since Eq. 10–9 is identical with Eq. 10–3 in the elastic range, the curve segments to be considered include the elastic range. Note that at the junction of two segments, there are two different values of E_T, leading to an abrupt discontinuity in the P/A vs. L/r curve, as indicated for material B in Fig. 10–17b. The Euler-Engesser equation is not valid at such points of discontinuity, since the derivation is based on the assumption that the variation of the tangent modulus over the cross section of the member is small.

2. The nonlinear part of the actual stress-strain diagram can be approximated by one or more continuous curves, each of which is represented by a differentiable function of stress and strain. Within certain limits, the tangent modulus ($E_T = d\sigma/d\varepsilon$) is a continuous function of σ, which can be substituted in Eq. 10–9, resulting in an expression for σ_{cr} as a function of L/r (where σ_{cr} is P/A).

---

[11] "Evaluation of Aeroplane Metals," J. A. Van den Broek, *Journal of the Royal Aeronautical Society,* 1946.

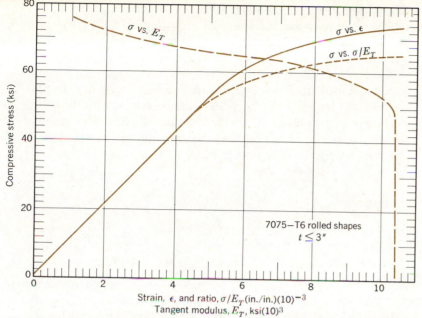

Stress versus strain and stress vs. tangent modulus for aluminum **FIG. 10–18** alloy 7075-T6 rolled shapes.

Here again, one should assume a certain segment of the curve is applicable, then check the value of the computed stress against the range of stresses for which the curve segment is applicable. Note that although the stress-strain functions may provide a reasonably continuous curve, the derivatives of these functions may not be continuous through the change from one function to another. In other words, at the junction of two curves, there can be two different values for E_T, thus invalidating Eq. 10–9, as discussed in the paragraph above.

3.   A solution can be obtained by making use of a stress vs. tangent modulus curve, such as the one shown in Fig. 10–18. The solution can be an iterative process, by assuming values for E_T, solving for σ_{cr}, and checking with the diagram. Some time might be saved by solving Eq. 10–9 for σ/E_T for the given problem and reading the corresponding value of σ_{cr} from a σ vs. σ/E_T curve (such as the one shown in Fig. 10–18) if such a curve is available. This latter curve is usually not provided; hence, it would be necessary to construct the σ vs. σ/E_T curve from the given curves or data. However, in actual design practice, it is more practical to construct a P/A vs. L/r curve by solving Eq. 10–9 for values of L/r

corresponding to values of σ and E_T selected from given curves or data. This curve can then be used as follows: if for a given problem, L/r is known and the critical buckling load is wanted, the value of P/A corresponding to the L/r can be read directly from the curve. If the problem is one of selecting a section to carry a specified load, the trial-solution approach is necessary. In other words, the techniques of Art. 10–6 are followed, with the formulas replaced by graphs. Since there are usually no abrupt discontinuities in the σ vs. E_T curves, the ambiguity referred to in the above paragraphs does not, in general, exist.

PROBLEMS

10-90*   A compression strut with a length of 30 in. between end fittings is composed of two L2 × 2 × $\frac{1}{4}$-in. angles (see Appendix C for cross-sectional properties) fastened together to form a T-section. End conditions at the fittings reduce the effective length to 21 in. Determine the maximum permissible axial load if the strut is made from the aluminum alloy of Fig. 10–18 and a factor of safety of 2 is specified.

10-91   A single L2 × 2 × $\frac{1}{4}$-in. angle (see Appendix C for cross-sectional properties) is to support an axial compressive load of 30 kips with a factor of safety of 2. The angle is made from the aluminum alloy of Fig. 10–18. Determine the maximum allowable total length if the end fittings reduce the effective length to 3/4 of the total length.

10-92*   A compression strut is fabricated from an aluminum alloy 2024-T4 tube having an outside diameter of 30 mm, a wall thickness of 2.00 mm, and a length between fittings of 250 mm. End conditions at the fittings reduce the effective length to 170 mm. The cross-sectional area of the tube is 175.9 mm^2 and the principal radius of gyration is 9.925 mm. The material has the stress-strain diagram shown in Fig. P10–92. Determine the axial buckling load.

10-93   Determine, for the strut of Prob. 10–92, the length (between fittings) at which buckling would probably occur under an axial compressive load of 51 kN.

10-94*   A 3-in. diameter tube with a wall thickness of 0.25 in. is to be used as a compression strut with a length of 50 in. between fittings. The cross-sectional area of the tube is 2.160 sq. in. and the principal radius of gyration is 0.976 in. End conditions at the fittings reduce the effective length to 34 in.

Determine the axial buckling load if the strut is made from an aluminum alloy with the stress-strain diagram shown in Fig. P10–94.

10-95   A compression strut with a length of 16 in. between end fittings is to have a solid rectangular cross section $2.00 \times b$ in. The strut must carry an axial working load of 160 kips with a factor of safety of 1.5. End conditions at the fittings reduce the effective length to $0.7L$. Determine the dimension b if the strut is made from an aluminum alloy with the stress-strain diagram shown in Fig. P10–94.

FIG. P10–94

CHAPTER 11

REPEATED LOADING

11-1 INTRODUCTION

The discussion in previous chapters of this book has been limited to static loading, for which the behavior of a structure under load can be predicted on the basis of the static properties of the material. However, many machine and structural elements are subjected to a type of loading known as *repeated loading,* under which the load is applied and removed, or the magnitude (and frequently the direction) is changed, many thousands of times during the life of the member. For example, the diesel engine rocker arms of Fig. 11–1, as well as the valves, push rods, connecting rods, and crankshaft, are subjected to repeated loading. In many instances, the repeated loading is accidental, as a result of vibrations induced by unpredictable or uncontrollable circumstances.

If the applied load changes from any magnitude in one direction to the same magnitude in the opposite direction, the loading is termed *completely*

FIG. 11–1 Fatigue failures of exhaust valve rocker levers. (*Courtesy: Cleveland Diesel Engine Division. General Motors Corp.*)

reversed, whereas if the load changes from one magnitude to another (the direction does not necessarily change), the load is said to be a *fluctuating load.*

The physical effect of a repeated load on a material is different from that of a static load, the failure always being a fracture (or brittle type of failure) regardless of whether the material is brittle or ductile. Under repeated loading, a small crack forms in a region of high localized stress, and very high stress concentration accompanies the crack. As the load fluctuates, the crack opens and closes and progresses across the section. Frequently, this crack propagation continues until there is insufficient cross section left to carry the load and the member ruptures, the failure being given the very descriptive term *fatigue failure.* It is unfortunate that fatigue failures occur at stresses well below the static *elastic* strength of the material.[1]

Figures 11–2, 11–3, and 11–4 show characteristic fatigue failures. Figure 11–2 is a photograph of a torsional fatigue failure at the front end of a crankshaft. This failure started in an area of fretting corrosion (point *A*) under a hub assembled at this point. From here the fracture progressed to points *B* and *C*, before complete separation occurred. Figure 11–3*b* shows

---

[1] For some ductile metals in the fully annealed condition (low elastic strength), the endurance limit may be slightly higher than the elastic strength.

(Courtesy of Prof. T. J. Dolan, University of Illinois) FIG. 11–2

the fatigue failure of one of the turbine blades of Fig. 11-3*a*. Here, the failure started at the trailing edge (point *A*) and progressed to *BC*, the final fracture showing as the dark area.

The failure of one of the compressor blades, such as *B*, of Fig. 11–4*a* is shown in Fig. 11–4*b*. The fracture appears to have started at *A* and continued, as indicated by the light area. The material to fracture last is identified by the dark area.

The importance of repeated loading was recognized in the first half of the nineteenth century. It is thought[2] that J. V. Poncelot (1788–1867), a French professor of engineering mechanics, mentioned the failure of a spring by fatigue in a book published in 1839. Also, W. J. M. Rankine (1820–1872), a Scotch engineer, in a paper published in 1843 discusses the fracture of railroad locomotive axles. Through the years 1843–1870 the railroad wheel and axle assembly occupied much attention and evoked considerable discussion, including the debate regarding whether the metal crystallized in service. The greatest single contribution to the investigation is that of A. Wöhler (1819–1914), a German engineer who designed fatigue-testing ma-

---

[2] *History of Strength of Materials*, S. Timoshenko, McGraw-Hill, New York, 1953.

FIG. 11–3

chines and conducted many fatigue tests that included the effect of stress concentration due to changes in cross section. The argument that the iron changed from a fibrous to a crystalline structure under the repeated loading apparently was successfully refuted during this same period. Although the idea of change in the metal structure under repeated stressing persists to this day with some individuals, metallurgical investigation has shown that the

FIG. 11–4

crystalline structure of solid metals undergoes no extensive change under such conditions.

Much attention has been devoted to the question of what causes fatigue failure, and although the question is not completely answered, the electron microscope and X-ray techniques have done much to promote an understanding of the mechanism of fatigue damage. Apparently, in a region of high localized stress (usually at or near the surface), localized slip bands form, and under repeated stressing these bands increase in number and width, rapidly at first but more slowly later. As the damage (slip) progresses, microscopic cracks are formed, and eventually these grow and join to form visible cracks.

Although the mechanism of progressive damage seems to be understood, the fundamental reasons for such action are not known. One explanation advanced is that strain hardening accompanies the local inelastic deformation (slip) and that the restraint of adjacent crystals prevents the localized strain from becoming zero as the applied load becomes zero. This results in the localized stress becoming higher with each cycle and the localized material becoming more brittle. For some combination of stress and material, it would be possible for the magnitude of the localized stress eventually to reach the static ultimate strength of the material, and a local crack would result. Another theory suggests that certain dislocations (defects in atomic structure) are displaced (localized slip) with relative ease and, under cyclic loading, migrate through the material until a number of them coalesce at some region where further movement is inhibited. Such a collection of dislocations in a region at or near the surface may be responsible for the action described in the preceding paragraph.

Although it is possible for fatigue failure to occur in a region of compressive stress, the usual failures occur under tensile stress and at stress levels considerably lower than those in a compressive stress failure.

11-2   RESISTANCE TO FAILURE UNDER COMPLETELY REVERSED LOADING

The index of resistance to failure under completely reversed repeated loading is known as the *endurance limit,* defined as *the maximum completely reversed stress to which a material may be subjected millions of times without fracture.* The magnitude of the endurance limit is determined from repeated load (or fatigue) tests. One commonly used machine for determining endurance limit is the R. R. Moore rotating beam machine shown schematically in Fig. 11–5. Observe that the loading simulates that imposed on a railroad car axle between the wheels, the specimen being subjected to pure bending and the flexural stresses being completely reversed, each one-half revolution. There are many other types of machines available, including rotating cantilever beams, nonrotating simple and cantilever beams, and axially loaded members. For the last three, the load may be applied by means of a crank and connecting rod, one machine being indicated diagrammatically in Fig. 11–6. For such experimental setups, the loading may be varied so that fluctuating loads of various ranges may be imposed.

The testing technique is to subject a series of identical specimens to loads of different magnitudes and note the number of cycles of stress (or load) *n* necessary to fracture the specimen. The data are plotted on a stress-

FIG. 11–5

cycle ($\sigma - n$) diagram, and the endurance limit can be determined from the diagram. In order to space the points on the diagram far apart for clarity, it is customary to plot the data on semilogarithmic paper, the stress being plotted to a linear scale and the cycles to a logarithmic scale, as indicated in Fig. 11–7. For some materials (particularly the ferrous materials) the curve seems to break sharply, as in curves A and B of Fig. 11–7, indicating that if the stress has the magnitude of σ_{EA}, material A could probably be subjected to an unlimited number of cycles without failure.[3] On the other hand, the alloys of aluminum and magnesium exhibit curves such as C and D of Fig. 11–7, which are apparently asymptotic to the n axis. For such materials, the stress corresponding to some arbitrary number of cycles, such as $5(10^8)$ for the aluminum alloys, is taken as the endurance limit.

It must be observed that the very nature of the fatigue failure—depen-

FIG. 11–6

[3] The data for Fig. 11–7 were obtained from polished specimens. Some endurance limits in Appendix A are for the material in the as-rolled condition and therefore have lower values.

FIG. 11–7

FIG. 11–8

dent as it is on local imperfections or peculiar configuration of grain structure as well as on the precision necessary for machining identical specimens—would dictate that the results obtained from a large number of
specimens might show considerable scatter. The results could be better
represented by a series of curves representing different probabilities of failure or as a band such as that in Fig. 11–8.[4] The customary $\sigma - n$ diagram
(such as shown in Fig. 11–7) is quite likely an approximate 50 percent probability of failure curve; that is, a specimen tested at a particular maximum
stress will have a 50–50 chance of failure when subjected to the number of
cycles indicated by the intersection of the line representing the test stress
with the 50 percent curve.

If a machine part is to be designed for a limited life, a strength index
higher than the endurance limit can be used. Theoretically, one can estimate
the number of cycles of stress to which the member will probably be subjected during its lifetime, enter the $\sigma - n$ diagram with this value, and read
the corresponding stress from the stress scale. If the member was stressed to
this value, one could expect a fracture failure at the end of the expected
lifetime, and the design of the machine part could be based on this value
reduced by an appropriate factor of safety. In this connection, however, the
statistical nature of the $\sigma - n$ data should be emphasized since the life of any
particular manufactured part may be much less than that indicated by the
ordinary $\sigma - n$ diagram.

11-3   DESIGN FACTORS AFFECTING
FATIGUE STRENGTH OF STRUCTURAL
ELEMENTS

Since the fatigue failure starts at a point of high localized stress, any discontinuity, accidental or intended, is very apt to initiate such a failure. The
accidental discontinuities must be taken care of by the factor of safety;
however, the discontinuities due to deliberate design must be included in
stress computations by an appropriate stress concentration factor (see Art.
1–10). This factor modifies the nominal equations for stress computations;
thus,

$$\sigma = \frac{KP}{A} \quad \tau = \frac{KT\rho}{J} \quad \sigma = \frac{KMy}{I}$$

where K is a factor greater than unity and may be obtained from the curves

[4] "Effect of Stress Amplitude on Statistical Variability in Fatigue Life of 75S-
T6 Aluminum Alloy," G. M. Sinclair and T. J. Dolan, *Trans. A.S.M.E.,* Vol. 75,
1953.

of Figs. 11–9, 11–10, and 11–11.[5] In general, the factors may be based on either the gross or the net section (the minimum cross section at the discontinuity), and it is necessary for the designer to know whether the factors at his disposal apply to the gross or the net section. The factors K_t of Figs. 11–9, 11–10, and 11–11 are all based on the net section except those of curve ④ of Fig. 11–9, which are to be applied to the gross section. Sometimes the solution of a problem is expedited by the use of the factor K_g based on the gross section, and for this purpose conversion expressions are given with the various curves.

Instead of the theoretical stress concentration factor, a similar experimental factor known as the *endurance limit reduction factor* may be used. The endurance limit reduction factor is determined from repeated load tests and is the ratio of the endurance limit as determined from standard test specimens to the endurance limit as determined from specimens containing a particular discontinuity. The magnitude of this factor depends on the size of the specimen, on the kind of material, and on the care exercised in machining of the test specimen. In general, the endurance limit reduction factor is smaller than the theoretical stress concentration factor; however, the larger the specimen and the stronger and more fine-grained the material, the less will be the difference between the two factors. Naturally, if a designer has at his disposal an appropriate endurance limit reduction factor (that is, the factor was obtained from a specimen of comparable size and material to that of the proposed design), such a factor should be used. In case the size and material limitations are not met or are unknown, the stress concentration factor should be used since it will be larger than the other factor and the resulting design will be more conservative.

The remarks in Art. 1–10 concerning the use of the stress concentration factor in design will bear repeating here. Stress concentration is ignored in a ductile material subjected to static loading. However, if the loading is repeated, stress concentration must not be disregarded, regardless of whether the material is brittle or ductile. Example 11–1 will serve to illustrate this

---

[5] Curves 1, 2, 5, 6, and 8: "Photoelastic Studies in Stress Concentration," M. M. Frocht, *Mechanical Engineering*, August 1936.

Curve 3: "Stress Concentration Factors Around a Central Circular Hole in a Plate Loaded Through Pin in the Hole," M. M. Frocht and H. N. Hill, *Trans. A.S.M.E. Journal*, March 1940.

Curve 4: "Plane Stress and Plane Strain in Bipolar Co-ordinates," G. B. Jeffery, *Philosophical Transactions of the Royal Society of London*, Vol. 221, 1921.

Curve 7: "How Holes and Notches Affect Flat Springs," A. M. Wahl, *Machine Design*, May 1940.

Curve 9: "Stress Concentration Factors and Their Effect on Design," G. H. Neugebauer, *Product Engineering*, Feb. 1943.

Curve 10: "Torsional Stress Concentrations in Shafts of Circular Cross-Section and Variable Diameter," L. S. Jacobsen, *Trans. A.S.M.E.*, Vol. 47, 1925.

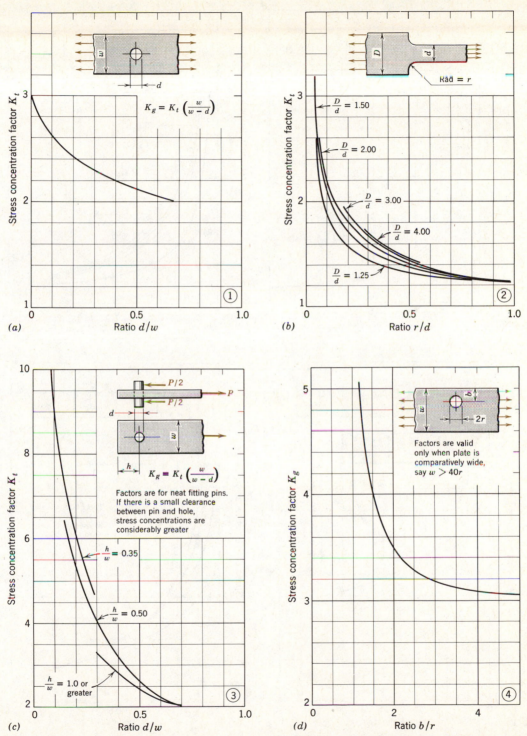

Stress concentration factors for axial loading of flat bars. FIG. 11–9

FIG. 11-10 Stress concentration factors for flexural loading.

Stress concentration factors for torsional loading. FIG. 11–11

point. For the problems in this chapter, unless otherwise indicated, assume that the details of construction are such that stress concentration is negligible at supports and points of application of loads.

EXAMPLE 11–1

The simple beam AB of Fig. 11–12a is constructed of aluminum alloy 2024-T4 and has a square cross section 2 in. wide by 2 in. deep with a pair of notches (detail in insert) at the center. Determine the maximum permissible magnitude for the load P

(a) If applied as a completely reversed repeated load with a factor of safety of 3.

(b) If applied as a steady load with a factor of safety of 2 with respect to failure by slip.

SOLUTION

(a) The critical stress is the maximum tensile stress, which occurs at the surface and is given by the flexure formula modified for stress concentration as necessary. The free-body and bending moment diagrams of Fig. 11–12b indicate a maximum moment at C; however, the stress concentration at the notches may produce a greater flexural stress at D than at C. Hence, it is necessary to obtain the value for K from curve ⑧ of Fig. 11–10 as follows: $h/(2r)$ is 4, d/r is 1/2, and r/b is 0.143, from which K_t is about 1.9, which is the factor based on the section 1.75 in. wide by 2 in. deep. In this instance, the work is expedited by converting K_t to K_g, the factor based on the gross section, since it will then be possible to compare KM at sections C and D. Thus, K_g is 1.9(2)/1.75 or 2.17. The values for KM are 1(3/2)P or 1.5P ft-lb at C and 2.17P ft-lb at D. This analysis indicates that D is the criti-

FIG. 11–12

cal section, and when the endurance limit from Appendix A with a factor of safety k of 3 is used, the expression for the stress is

$$\frac{\sigma}{k} = \frac{KMc}{I} = \frac{18,000}{3} = \frac{2.17P(12)(1)}{2(2^3)/12}$$

from which

$$P = \underline{307 \text{ lb}} \qquad\qquad \text{Ans.}$$

(b)   Since the material has an ultimate elongation of 19 percent (see Appendix A), it is a ductile material, and under static loading the stress concentration is to be neglected. Comparing M/I at C and D, yields

$$\frac{1.5P}{2(8)/12} > \frac{P}{1.75(8)/12}$$

Therefore, the critical section is now at C. Since the mode of failure is slip, the elastic strength will be used; thus

$$\frac{48,000}{2} = \frac{(3P/2)(12)(1)}{2(2^3)/12}$$

from which

$$P = \underline{1780 \text{ lb}} \qquad\qquad \text{Ans.}$$

PROBLEMS

11-1*   Estimate the probable service life of a machine part made of 1020 steel (see Fig. 11–7) if the part is subjected to a completely reversed repeated stress of

(a)   45 ksi at 250 cycles per min.

(b)   40 ksi at 300 cycles per min.

(c)   38 ksi at 350 cycles per min.

(d)   32 ksi at 500 cycles per min.

11-2*   Estimate the probable service life of an aluminum alloy 2024-T4 (see Fig. 11–7) panel in an airplane if the panel is subjected to a completely reversed repeated stress (due to vibration) of

(a)   295 MPa at 100 cycles per min.

FIG. P11–5

FIG. P11–6

(b)   210 MPa at 200 cycles per min.

(c)   180 MPa at 150 cycles per min.

(d)   100 MPa at 250 cycles per min.

11-3*   Determine a suitable working stress with a factor of safety of 1.25 for a machine part made of T-1 steel (see Fig. 11–7) if the part is subjected to a completely reversed repeated load and the estimated service life of the part is

(a)   100,000 cycles.

(b)   500,000 cycles.

(c)   5,000,000 cycles.

11-4*   A machine part made of 1020 steel (see Fig. 11–7) is subjected to a completely reversed repeated stress of 150 MPa. Determine the factor of safety if the required service life of the part is

(a)   100,000 cycles.

(b)   500,000 cycles.

(c)   5,000,000 cycles.

11-5*   The machine part shown in Fig. P11–5 is 1.0 in. thick and is made of structural steel (see Appendix A). Determine the maximum safe axial load if

(a)   The load is a completely reversed repeated load (use a factor of safety of 3).

(b)   The load is a static load (use a factor of safety of 2 with respect to failure by slip).

FIG. P11–7

FIG. P11–8

11-6   The machine part shown in Fig. P11–6 is 30 mm thick and is made of 0.4 percent carbon hot-rolled steel (see Appendix A). Determine the maximum safe axial load if

    (a)   The load is a completely reversed repeated load (use a factor of safety of 3).

    (b)   The load is a static load (use a factor of safety of 2 with respect to failure by slip).

11-7   The bar of Fig. P11–7 is 1/4 in. thick and is made of annealed 18-8 stainless steel (see Appendix A). Determine the maximum safe axial load if

    (a)   The load is completely reversed and repeated and a factor of safety of 2.8 is specified.

    (b)   The load is static and a factor of safety of 2 with respect to failure by slip is specified.

11-8*   The structural steel (see Appendix A) plate of Fig. P11–8 is 15 mm thick and is subjected to a tensile load of 150 kN uniformly distributed across the width of the plate. Determine the minimum allowable width of the plate if a factor of safety of 2.5 with respect to failure by fracture or slip is specified and

    (a)   The load is completely reversed and repeated.

    (b)   The load is static.

11-9*   The cantilever spring shown in Fig. P11–9 is made of heat-treated SAE 4340 steel (see Appendix A) and is 2 in. wide. If P is a completely reversed repeated load of 45 lb, determine the maximum safe values for the distances a and b (b in combination with the maximum value for a). A factor of safety of 3 is specified.

11-10   The cantilever beam shown in Fig. P11–10 is made of 0.4 percent

FIG. P11–9

FIG. P11–10

FIG. P11–11

FIG. P11–12

carbon hot-rolled steel (see Appendix A). If a factor of safety of 2.5 with respect to failure by fracture or slip is specified, determine the minimum permissible thickness h of the bar when

    (a)   The load P is a static load of 1500 N.

    (b)   The load P is a completely reversed and repeated load of 1200 N.

11-11   The circular shaft shown in Fig. P11–11 is made of structural steel (see Appendix A). Determine the maximum safe value of the load P if

    (a)   The load P is a static load and a factor of safety of 2.2 with respect to failure by slip is specified.

    (b)   The load P is a completely reversed repeated load and a factor of safety of 2.8 with respect to failure by fracture is specified.

11-12*   The notched cantilever beam shown in Fig. P11–12 is 60 mm thick and is made of aluminum alloy 6061-T6 (see Appendix A). Determine the maximum completely reversed repeated load P that may be applied if a factor of safety of 3 with respect to failure by fracture is specified.

11-13*   A 4-in. diameter shaft subjected only to bending has a U-shaped groove 0.4-in. deep (radius of groove is 0.2 in.) turned in it at the section of maximum moment. The shaft is made of structural steel (see Appendix A). Determine the maximum bending moment that may be applied to the shaft if

    (a)   The load is static and a factor of safety of 2 with respect to failure by slip is specified.

    (b)   The load is a completely reversed repeated load and a factor of safety of 2.5 with respect to failure by fracture is specified.

11-14   A simple beam is 100 mm wide by 50 mm deep and has a span of 3 m. The beam is loaded with a completely reversed repeated concentrated load P at the center of the span. The beam has two vertical U-shaped grooves opposite each other on the edges of the beam 0.5 m from one end. The grooves are 12.5 mm deep and have a radius of 7.5 mm. Determine the maximum safe value for the load P if a factor of safety of 3 with respect to failure by fracture is specified. The endurance limit for the material is 275 MPa.

11-15   A 4-in. diameter shaft is to have one end turned down to a 3-in. diameter. The shaft is to be made of 1.2 percent carbon steel normalized (use 25 ksi for the endurance limit in torsion). If the shaft is to be subjected to a completely reversed repeated torque of 1850 ft-lb and a factor of safety of 3 is specified, determine the minimum radius needed at the junction between the two sections of the shaft.

11-16*   A 120-mm diameter shaft has one end turned down to a 60-mm diameter. The fillet at the junction between the two sections has a 7.5-mm radius. The shaft is made of 3.5 percent nickel steel, hardened and tempered (use 220 MPa for the endurance limit in torsion). A factor of safety of 3 is specified. Determine

   (a)   The maximum completely reversed repeated torque that can be applied.

   (b)   The maximum completely reversed repeated torque that can be applied after a U-shaped groove 12 mm wide by 18 mm deep (the radius at the bottom of the groove is 6 mm) is introduced in the 120-mm section of the shaft.

11-17*   The bar shown in Fig. P11-7 is made of heat-treated 4130 steel with an endurance limit of 75 ksi, a yield strength of 135 ksi, and an ultimate strength of 160 ksi. Determine the minimum thickness for the bar if it must initially support a static load of 125 kips with a factor of safety of 2 with respect to failure by rupture and later support a completely reversed and repeated load of 50 kips with a factor of safety of 3.

11-18   The bar shown in Fig. P11-6 is made of structural steel (see Appendix A) and must support a static load of 175 kN with a factor of safety of 3, with respect to failure by rupture, during erection of a structure. After the erection is completed, the bar will be subjected to a completely reversed and repeated load of 45 kN. Determine the minimum allowable thickness of the bar if it is to have a factor of safety of 2.5.

11-4   RESISTANCE TO FAILURE UNDER FLUCTUATING LOAD

The maximum stress necessary to cause fracture under a fluctuating load is usually larger than the endurance limit. Considerable data have been accumulated from tests performed in machines similar to that shown schemati-

FIG. 11–13

FIG. 11–14

cally in Fig. 11–6, in which the stress in the specimen may vary from some minimum to a different maximum value. However, the task of experimentally determining the ranges of stress that will cause fatigue failure for all possible (or reasonable) combinations of maximum and minimum stresses and for the numerous materials in use today would be unending. Hence, some method of predicting the strength based on available data is essential. The approach to the problem is made through the interpretation of data from a limited number of tests in various ranges and is based on the concept that a fluctuating stress can be considered the sum of a steady stress and a completely reversed (or alternating) stress, as indicated for normal stresses in Fig. 11–13, where σ_a, the steady stress, is one-half the sum of σ_{max} and σ_{min}, and where σ_r, the alternating stress, is one-half the difference between these two stresses (compressive stresses are negative). The steady and alternating stresses may not be the same kind of stress; for example, one could have a steady torsional shearing stress combined with an alternating flexural stress. However, for the restricted scope of this book, the discussion will be limited to the case in which the steady and alternating stresses at a point are both normal (or both shearing) stresses on the same plane.

The data from tests can be plotted on a diagram, and a curve can be drawn to represent the strength of the material for various combinations of steady and alternating stresses. Two curves that have received widespread attention are shown in Fig. 11–14, in which σ_E is the endurance limit, σ_Y is the static elastic[6] tensile strength, and σ_U is the static ultimate tensile strength. Any point with coordinates (σ_a, σ_r), such as A on the diagram, would represent a particular combination of steady and repeated stresses, and if the point is above the straight line (Soderberg[7] or Goodman,[8] depend-

[6] Yield point, yield strength, or proportional limit.

[7] See "Working Stresses," C. R. Soderberg, *Trans. A.S.M.E.*, Vol. 55, APM-55-16, 1933, p. 131.

[8] Goodman's diagram does not resemble the one of Fig. 11–14. However, since the two diagrams yield the same results, it has become customary to identify the upper line of Fig. 11–14 as the Goodman criterion.

ing on which line is accepted as the failure criterion), failure is presumed to result. Most of the available test data plot above the two lines shown, indicating that both curves are conservative.

When a design problem involves machine elements that will not function properly if the static elastic strength of the material is exceeded (a slip failure), the Soderberg diagram should be regarded as the criterion of failure. The reason for using the more conservative diagram is that σ_{max} might exceed the static elastic strength on the first application of the maximum load, whereupon the usefulness of the mechanism would be impaired or destroyed. If fracture is the only mode of failure to be considered, slip is of no concern, and the less conservative Goodman diagram should be used. Needless to say, the approximate criteria of strength are to be used only if fluctuating strength data from mechanical tests are unavailable. For designs that warrant the expense, fatigue tests should be conducted on structural components loaded to simulate actual service conditions. When the approximate failure criteria are applied, if the endurance limit is unknown *and the material is ferrous,* one-half of the ultimate tensile strength will give a reasonable value for the endurance limit. The use of the Soderberg and Goodman diagrams is illustrated by Exams. 11–2 and 11–3. The choice of diagrams is usually indicated by a statement concerning the mode of failure. For example, a factor of safety with respect to failure by slip or fracture means that both slip (due to the application of the maximum load) and fracture (due to the alternating load) must be avoided; hence, the Soderberg criterion should be used.

EXAMPLE 11–2

A machine element of heat-treated SAE 4340 steel is subjected to stresses that fluctuate from 350 MPa *T* to 70 MPa *C*. Determine the factor of safety with respect to failure by slip or fracture.

SOLUTION

Failure by slip or fracture indicates that the Soderberg diagram should be used. The diagram is shown in Fig. 11–15, for which the properties were

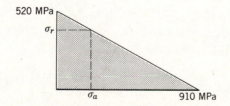

FIG. 11–15

obtained from Appendix A. The first step in the solution is to compute the ratio of σ_r to σ_a as follows:

$$\sigma_{a1} = \frac{350 + (-70)}{2} = 140 \text{ MPa } T$$

and

$$\sigma_{r1} = \frac{350 - (-70)}{2} = 210 \text{ MPa } T \text{ and } C$$

from which

$$\sigma_r = \frac{3\sigma_a}{2}$$

The equation of the line in Fig. 11–15 is

$$\sigma_r = 520 - \left(\frac{520}{910}\right)\sigma_a$$

and since

$$\sigma_r = \frac{3\sigma_a}{2}$$

$$\frac{3\sigma_a}{2} = 520 - \left(\frac{520}{910}\right)\sigma_a$$

from which

$$\sigma_a = 251 \text{ MPa}$$

This value represents the limit of the steady stress. Because of the fixed ratio between σ_a and σ_r, 251 MPa can be used as a strength index; hence, the factor of safety is

$$k = \frac{251 \text{ MPa}}{140 \text{ MPa}} = \underline{1.79} \qquad\qquad \text{Ans.}$$

Observe that this value is obtained regardless of whether one uses the steady, the alternating, the maximum, or the minimum stress as the strength index; for example,

$$\sigma_{\max} = \sigma_a + \sigma_r = \sigma_a + \frac{3\sigma_a}{2} = \frac{5\sigma_a}{2}$$

and the factor of safety is

$$k = \frac{(251)}{(140)} = \frac{(3/2)(251)}{(3/2)(140)} = \frac{(5/2)(251)}{(5/2)(140)} = 1.79$$

EXAMPLE 11–3
The load on the beam of Exam. 11–1 fluctuates from $3P$ downward to $2P$ upward. Determine the maximum permissible magnitude for P if a factor of safety of 3 with respect to failure by fracture is specified.

SOLUTION
Since the only criterion of failure is fracture,[9] the Goodman diagram will be used with σ_E of 18 ksi and σ_U of 68 ksi. Instead of these strength values being used, they may be divided by the factor of safety, thus converting the failure diagram to a working stress diagram. Whether the factor of safety is introduced in the diagram or later is optional. To get the ratio of stresses,

$$P_a = \frac{3P + (-2P)}{2} = \frac{P}{2}$$

and

$$P_r = \frac{3P - (-2P)}{2} = \frac{5P}{2}$$

from which

$$P_r = 5P_a$$

Therefore, since the stresses will be below the proportional limit,

$$\sigma_r = 5\sigma_a$$

and when the Goodman diagram is used with working stresses instead of strengths,

$$\sigma_{r_w} = \frac{18}{3} - \frac{18}{68}\sigma_{a_w} = 5\sigma_{a_w}$$

from which

$$\sigma_{a_w} = 1.14 \text{ ksi}$$

The loading is repeated; therefore the stress concentration will not be ignored, and the analysis of part (a) of Exam. 11–1 is valid. Thus,

---

[9] So far as the static strength is concerned, fracture here denotes that the maximum static load is reached when the maximum flexural stress is equal to the ultimate tensile strength. Actually, under a load of this magnitude, the beam would not collapse since the material is ductile.

$$\frac{\sigma_a}{k} = \frac{KMc}{I} = 1140 = \frac{2.17P_a(12)(1)}{2(2^3)/12}$$

from which

$$P_a = 58.4$$

and

$$P = 2P_a = \underline{117 \text{ lb}} \qquad \text{Ans.}$$

Observe that the correct solution is obtained whether σ_a, σ_r, σ_{max}, or σ_{min} is used, as long as the corresponding value for the load is used. For example, if σ_r is used, the above equation becomes

$$1140(5) = 2.17(5P_a)(9)$$

giving the same result as above.

   When this result is compared with Exam. 11–1, the fluctuating load is seen to vary from -234 lb to $+350$ lb, while a completely reversed load varies from -307 to $+307$ lb.

PROBLEMS

11-19*   Determine the factor of safety with respect to failure by slip or fracture for a machine part for which the stress varies from 10 ksi T to 6 ksi C. The material is aluminum alloy 2024-T4 (see Appendix A).

11-20*   Determine the factor of safety with respect to failure by slip or fracture for a machine part for which the stress varies from 250 MPa T to 90 MPa C. The material is heat-treated SAE 4340 steel (see Appendix A).

11-21*   A machine component is subjected to an axial load that varies from 3000 lb compression to 5000 lb tension. Determine the combination of maximum and minimum stresses to which the member could probably be subjected millions of times without failure by slip or fracture if the proposed material is hardened and tempered 0.93 percent carbon spring-steel with an endurance limit of 56 ksi, a proportional limit of 85 ksi, and an ultimate strength of 97 ksi.

11-22*   A machine component is subjected to a repeated load that varies from 10 kN tension to 30 kN tension. The material is a hardened and tempered 3.5 percent nickel steel with an endurance limit of 435 MPa, a proportional limit of 595 MPa, and an ultimate strength of 815 MPa. Determine the

FIG. P11–25   FIG. P11–26

combination of maximum and minimum stresses to which the member could probably be subjected millions of times without failure by slip or fracture.

11-23 A machine element is subjected to a bending load that varies from 5000 lb tension to 3000 lb compression. The material is structural steel (see Appendix A). Determine the magnitudes of the steady and alternating components of a suitable working stress if a factor of safety of 2 with respect to failure by slip or fracture is specified.

11-24 A machine component is subjected to an axial load that varies from 24 kN tension to 12 kN compression. The material is annealed 18-8 stainless steel (see Appendix A). Determine the magnitudes of the steady and alternating components of a suitable working stress if a factor of safety of 2 with respect to failure by fracture is specified.

11-25* Determine the maximum and minimum values for the fluctuating axial load P of Fig. P11–25 if the magnitude of the load in tension is twice the magnitude in compression. The material is structural steel (see Appendix A) and a factor of safety of 3 with respect to failure by slip or fracture is specified.

11-26 The bar shown in Fig. P11–26 is made of annealed 18-8 stainless steel (see Appendix A) and is subjected to an axial load that fluctuates from $4P$ in tension to P in compression. Determine the maximum permissible value for P if a factor of safety of 3 with respect to failure by fracture is specified. The bar is 10 mm thick.

11-27 The bar shown in Fig. P11–27 is made of structural steel (see Appendix A) and is subjected to an axial load that fluctuates from $3P$ in tension to $2P$ in compression. Determine the maximum permissible value for P if a factor of safety 2.5 with respect to failure by fracture is specified. The bar is 1/2 in. thick.

11-28* A simple beam of structural steel (see Appendix A) is 150 mm wide by 50 mm deep and has a span of 3 m. A pair of symmetrical vertical notches for which the stress concentration factor is 1.5 based on the gross section are present at the center of the span. A concentrated fluctuating load is ap-

FIG. P11–27

0.42" dia. hole   0.46" rad.

FIG. P11–29

plied 1 m from one of the supports. Determine the maximum and minimum values of the load if the magnitude in one direction is 3/2 that in the other direction. A factor of safety of 2 with respect to failure by fracture or slip is specified.

11-29* The connecting rod shown in Fig. P11–29 has a rectangular cross section that is 2 in. thick. The rod is made of normalized 0.57 percent carbon steel that has an endurance limit of 33 ksi, a proportional limit of 35 ksi, and an ultimate strength of 72 ksi. Determine the maximum completely reversed and repeated load P that can be applied if a factor of safety of 3 with respect to failure by fracture or slip is specified. Note that the tension on the net section at the pin fluctuates from zero to a maximum.

11-30 The strut shown in Fig. P11–30 has a rectangular cross section and is made of 0.4 percent carbon hot-rolled steel (see Appendix A). The strut is subjected to a completely reversed and repeated load of 7.5 kN. Determine the minimum thickness required if a factor of safety of 2 with respect to failure by fracture or slip is specified. Note that the tension on the net section at the pin fluctuates from zero to a maximum.

11-31 The cantilever beam shown in Fig. P11–31 is 2 in. wide and is made of heat-treated spring steel that has an endurance limit of 56 ksi, a proportional limit of 85 ksi, and an ultimate strength of 97 ksi. The load at C varies from 3P downward to P upward. Determine the maximum permissible mag-

FIG. P11–30

FIG. P11–31

FIG. P11–32   FIG. P11–33

nitude for *P* if a factor of safety of 2 with respect to failure by fracture or slip is specified.

11-32* The steel cantilever spring shown in Fig. P11–32 is deflected repeatedly from zero to 7.5 mm by means of the rotating cam. The spring has a rectangular cross section 6 mm thick by 36 mm wide. The material is cold-rolled 18-8 stainless steel (see Appendix A). Determine the factor of safety with respect to failure by fracture.

11-33* The notched cantilever beam shown in Fig. P11–33 is 2 in. thick and is made of aluminum alloy 2024-T4 (see Appendix A). If the load fluctuates from 5*P* to the right to *P* to the left and a factor of safety of 3 with respect to failure by fracture or slip is specified, determine the maximum permissible value for *P*.

11-34 The machine part shown in Fig. P11–34 is 15 mm thick and is made of structural steel (see Appendix A). Determine the maximum safe axial load *P* if the magnitude of the load in tension is twice the magnitude in compression and a factor of safety of 2.5 with respect to failure by fracture or slip is specified.

11-35 The unbalanced disk of Fig. P11–35 weighs 300 lb. When the disk

FIG. P11–34   FIG. P11–35

FIG. P11–36

rotates, the eccentricity of the mass center induces an effective force of 150 lb directed outward from the shaft that is attached to and rotates with the disk. Neglect the weight of the shaft and any torsional moment; assume the end bearings are simple supports and consider bending only in a vertical plane. Determine the minimum permissible diameter for the shaft if it is made of heat-treated SAE 4340 steel (see Appendix A) and a factor of safety of 3 with respect to failure by fracture or slip is specified.

11-36*   The large flywheel shown in Fig. P11–36 has a mass of 4500 kg and is supported by and rotates about shaft AB that is fixed at B. A misalignment e of the shaft and the mass center of the wheel (see insert) causes the bottom of the shaft to move in a circular path during rotation of the wheel. Assume that the wheel rotates about its mass center that remains directly below the center of the shaft at support B. Neglect the mass of the shaft and bearing friction. The shaft is made of 0.4 percent carbon hot-rolled steel (see Appendix A) and a factor of safety of 2.5 with respect to failure by fracture or slip is specified. Determine the maximum permissible value for e.

CHAPTER 12

DYNAMIC LOADING

12-1 INTRODUCTION

When the motion of a body is changed (accelerated), the force necessary to produce this acceleration is called a *dynamic force* or *load*. For example, the force an elbow in a pipeline exerts on the fluid in the pipe to change its direction of flow, the pressure on the wings of an airplane pulling out of a dive, the collision of two automobiles, and a man jumping on a diving board are all examples of dynamic loading. A suddenly applied load is called an *impact load*. The last two of the preceding examples are considered impact loads. Under impact loading, if there is elastic action, the loaded system will vibrate until equilibrium is established. A dynamic load may be expressed in terms of mass times the acceleration of the mass center, in terms of the rate of change of the momentum, or in terms of the change of the kinetic energy of the body.

For many problems in mechanics of materials, the energy approach, introduced in Art. 3–4, is most effective. The magnitude of the dynamic load is often expressed in terms of the kinetic energy delivered to the loaded system; hence, it is often referred to as an *energy load*. For example, a particle of mass m moving with a speed v possesses a kinetic energy $mv^2/2$; if this particle is stopped by a body, the energy absorbed by the body (the loaded system) is some fractional part of $mv^2/2$, the balance of the energy being converted into sound, heat, and permanent deformation of the striking particle.

In the loaded system, dynamic loading produces stresses and strains, the magnitude and distribution of which will depend not only on the usual parameters encountered previously but also on the velocity of propagation of the strain waves through the material of which the system is composed. This latter consideration, although very important when loads are applied with high velocities, may often be neglected when the velocity of application of the load is low. The velocities are considered to be low when the loading time permits the material to act in the same manner as it does under static loading; that is, the relations between stress and strain and between load and deflection are essentially the same as those already developed for static loading.

12-2   ENERGY OF CONFIGURATION

Since dynamic loading is conveniently considered to be the transfer of energy from one system to another, the concept of energy of configuration (strain energy) as an index of resistance to failure is important enough to be discussed at some length. In Art. 3–4 the concept of strain energy was introduced by considering the work done by a slowly applied axial load P in elongating a bar of uniform cross section by an amount δ_2. From the load-deformation diagram of Fig. 3–7b it was observed that the work done in elongating an axially loaded bar is

$$W_k = \int_0^{\delta_2} P \, d\delta \qquad (a)$$

Since the work done on the bar must equal the strain energy stored in the bar, the expression for strain energy in terms of axial stress and axial strain became

$$W_k = U = \int_0^{\varepsilon_2} (\sigma)(A)(L) \, d\varepsilon = AL \int_0^{\varepsilon_2} \sigma \, d\varepsilon \qquad (b)$$

When the stress remains below the elastic limit of the material, Hooke's law applies and $d\varepsilon$ may be expressed as $d\sigma/E$. Equation b then becomes

$$U = \left(\frac{AL}{E}\right) \int_0^{\sigma_1} \sigma \, d\sigma$$

or

$$U = AL \left(\frac{\sigma_1^2}{2E} \right)$$ (12–1)

When the energy load U applied to a resisting structural member comes from a body falling from rest, the kinetic energy of the falling weight W equals its change in potential energy, which is $W(h + \delta)$, where h is the distance of free fall and δ the deflection of the resisting member due to the impact. For the special case of an axially loaded bar, the maximum axial stress and the maximum elongation of the bar can be determined by equating the kinetic energy of the falling body to the strain energy in the bar at the instant of maximum deflection if it is assumed that all of the kinetic energy of the body is effective in straining the bar. The loss of energy during impact is discussed in the next article.

If instead of axial loading, one considers a differential element of a material in a region where the stresses vary with respect to the position of the element, there will, in general, be three normal and three shearing stresses acting on the element. The energy intensity expression for this general case was developed in Art. 3–4 (see Eq. 3–11b).

When the state of stress in a body can be represented by one nonzero normal stress, Eq. 12–1 can be written as

$$U = \int_{\text{volume}} \frac{\sigma^2}{2E} \, dx \, dy \, dz$$ (c)

Equation c is valid for a beam subjected to pure bending, and if the energy (usually small) resulting from the shearing stresses is neglected, the expression applies to any beam with transverse loading. The stress in Eq. c must be expressed as a function of one or more of the other variables before integration.

For the case of pure bending in a beam of constant cross section A and length L, the value of σ_x is

$$\sigma_x = \frac{My}{I}$$

and the strain energy (when x is measured along the axis of the beam) is

$$U = \frac{1}{2E} \int \int \left(\frac{My}{I} \right)^2 dA \, dx = \frac{1}{2E} \int_0^L \frac{M^2}{I^2} \, dx \int_{\text{area}} y^2 \, dA$$

which reduces to

$$U = \int_0^L \frac{M^2}{2EI} \, dx$$ (12–2)

The procedure leading to Eq. c for a normal stress can be applied in exactly the same manner to a shearing stress to give

$$U = \int_{\text{volume}} \frac{\tau^2}{2G} \, dx \, dy \, dz \tag{d}$$

where τ must be expressed as a function of x, y, and z before integration. For the case of a circular shaft of length L and cross section A subjected to a torque T, the expression for τ is

$$\tau = \frac{T\rho}{J}$$

and the energy becomes

$$U = \frac{1}{2G} \iint \left(\frac{T\rho}{J}\right)^2 dA \, dx = \frac{1}{2G} \int_0^L \frac{T^2}{J^2} \, dx \int_{\text{area}} \rho^2 \, dA$$

which gives

$$U = \int_0^L \frac{T^2 \, dx}{2GJ} \tag{12-3}$$

One of the important concepts disclosed by Eqs. c and d is that the energy-absorbing capacity of a member (that is, the resistance to failure) is a function of the *volume* of material available, in contrast to the resistance to failure under static loading, which is a function of cross-sectional area or section modulus. Hence, if the length of the bar of Fig. 3–7 is doubled, the capacity for absorbing energy is doubled (end conditions neglected), whereas the static load capacity remains unchanged (except for the additional load imposed by the added weight of the bar).

The integral $\int \sigma \, d\varepsilon$ of Eq. b represents the area under the stress-strain curve, and if evaluated from zero to the elastic limit of the material, it yields the maximum strain energy per unit volume that a material will absorb without inelastic deformation (modulus of resilience). The area under the entire curve from zero to rupture denotes energy per unit volume necessary to produce rupture of the material (modulus of toughness).

PROBLEMS

Note: In the following problems, the mass of the loaded members is to be neglected, and unless stated otherwise, all of the applied energy is to be assumed to be absorbed elastically by the loaded members. Neglect stress concentrations.

12-1*   A solid circular steel ($E = 30,000$ ksi) bar 50 in. long is to be used as a tension member subject to axial energy loads. The allowable axial tensile stress is 25 ksi. Determine the minimum diameter required for an axial energy load of 100 ft-lb.

12-2*   A rolled bronze ($E = 100$ GPa) bar with a cross-sectional area of 2500 mm^2 is to be used as a tension member subject to axial energy loads. The allowable axial tensile stress is 70 MPa. Determine the minimum length of bar required for an axial energy load of 200 N · m.

12-3*   A bar with cross-sectional area A and length L is made of a homogeneous material with specific weight γ and modulus of elasticity E. Determine the total elastic strain energy stored in the bar (as a result of its own weight) if it hangs vertically while suspended from one end. Express the result in terms of γ, L, A, and E.

12-4*   Derive an expression for the total elastic strain energy stored in a cantilever beam with a circular cross section if it is supporting a concentrated load P at the free end of span L. Neglect the energy resulting from shearing stresses. Express the result in terms of P, L, E, and I.

12-5   A rectangular beam of width b, depth $2c$, and span L is simply supported at the ends and is subjected to end couples that produce a constant bending moment for the full length of the beam. Compare the total elastic strain energy stored in this beam with the total elastic strain energy stored in an axially loaded bar of the same size and at the same maximum tensile stress level.

12-6*   A rectangular beam of width b, depth $2c$, and span L is simply supported at the ends and carries a concentrated load P at midspan. Compare the total elastic strain energy stored in this beam with the total elastic strain energy stored in an axially loaded bar of the same size and at the same maximum tensile stress level. Neglect the energy resulting from the shearing stresses.

12-7   A rectangular cantilever beam of width b, depth $2c$, and span L carries a concentrated load P at the free end. Compare the total elastic strain energy stored in this beam with the total elastic strain energy stored in an axially loaded bar of the same size and at the same maximum tensile stress level. Neglect the energy resulting from shearing stresses.

12-8*   A rectangular cantilever beam of width b, depth $2c$, and span L carries a uniformly distributed load w over its entire length. Compare the total elastic strain energy stored in this beam with the total elastic strain energy stored in an axially loaded bar of the same size and at the same maximum tensile stress level. Neglect the energy resulting from shearing stresses.

12-9 A rectangular cantilever beam of width b, depth $2c$, and span L has a couple M applied at the free end. Compare the total elastic strain energy stored in this beam with the total elastic strain energy stored in an axially loaded bar of the same size and at the same maximum tensile stress level.

12-10* A solid circular steel shaft of diameter d and length L is subjected to a constant torque T. Compare the total elastic strain energy stored in this shaft with the total elastic strain energy stored in an axially loaded bar of the same size and at the same maximum stress level.

12-11* A 9-ft steel ($E = 30,000$ ksi) beam is simply supported at the ends. A concentrated load P gradually applied at the center of the span produces a maximum fiber stress of 12,000 psi. Determine the total elastic strain energy stored in the beam in terms of I and C. Neglect the energy resulting from shearing stresses.

12-12* A 25-mm diameter steel ($E = 200$ GPa) rod 900 mm long is supported at the top end and fitted with a loading flange (stop) at the bottom end as shown in Fig. P12–12. A collar W slides freely on the rod. Determine the maximum mass that the collar may have without exceeding the proportional limit (250 MPa) of the rod if

(a) The collar is dropped from a height of 600 mm.

(b) The collar is suddenly applied to the stop (dropped from a very small height).

(c) The collar is slowly applied to the stop (static load).

12-13 A weight of 30 lb is dropped from a height of 4 ft onto the center of a small rigid platform as shown in Fig. P12–13. The two steel ($E = 30,000$ ksi) rods supporting the platform have 1 by 2 in. rectangular cross sections and are 8 ft long. Determine

FIG. P12–12

FIG. P12–13

FIG. P12–14   FIG. P12–15

(a)   The maximum tensile stress developed in the rods.

(b)   The maximum deflection of the platform.

12-14   A flat piece of steel ($E = 200$ GPa) 10 mm thick and 2 m long has a width of 25 mm for 0.5 m of its length and a width of 50 mm for the remaining 1.5 m. A 2-kg collar W is dropped onto a loading flange at the end of the bar as shown in Fig. P12–14. Determine the maximum height h from which the collar can be dropped if the axial tensile stress in the rod is not to exceed 200 MPa.

12-15*   A timber ($E = 1200$ ksi) cantilever beam 6 in. wide, 2 in. deep, and 5 ft long is subjected to an impact load as shown in Fig. P12–15. The block W weighs 10 lb and is dropped 2 in. onto the end of the beam. Determine the maximum fiber stress produced in the beam.

12-16*   A timber ($E = 8.2$ GPa) beam 150 mm wide, 100 mm deep, and 3 m long is simply supported at the ends. From what maximum height can a 14-kg mass be dropped onto the center of the beam without causing the maximum fiber stress to exceed 10 MPa?

12-17   An S12 × 50 structural steel section (see Appendix C) is used as a simply supported beam 10 ft long. What weight falling on the center of the span from a height of 20 in. will produce a maximum fiber stress of 16 ksi in the beam? The modulus of elasticity of structural steel is 29,000 ksi. The web of the beam is vertical.

12-18   A timber ($E = 8.2$ GPa) cantilever beam is 150 mm wide, 75 mm deep, and 1.5 m long. Determine the height from which a 12-kg mass should be dropped upon the free end of the beam in order to produce a maximum deflection three times as great as that produced by the same mass if it is gradually applied.

12-19*   A weight of 10 lb is dropped from a height of 8 in. onto the free end of a steel (E = 30,000 ksi) cantilever beam. The beam is 2 in. wide, 3 in. deep, and 30 in. long. Determine

    (a)   The maximum deflection produced by the falling weight.

    (b)   The static load needed to produce the same deflection.

    (c)   The maximum fiber stress caused by the falling weight.

12-20*   A hollow steel (G = 80 GPa) shaft is 4.5 m long and has an outside diameter of 200 mm and an inside diameter of 150 mm. Determine the torsional energy load required to produce a maximum shearing stress of 85 MPa in the shaft.

12-21   A 10-ft steel (E = 30,000 ksi and G = 12,000 ksi) shaft 3 in. in diameter is subjected to an energy load of 100 ft-lb. Determine

    (a)   The maximum normal stress in the shaft when the load is applied as an axial energy load.

    (b)   The maximum shearing stress in the shaft when the load is applied as a torsional energy load.

12-22*   A steel (G = 80 GPa) shaft 100 mm in diameter and 1.5 m long is subjected to a torsional energy load of 200 J. Determine

    (a)   The maximum shearing stress produced in the shaft.

    (b)   The angle of rotation of one end of the shaft with respect to the other end.

12-23*   A torsional energy load of 20 ft-lb is applied to the steel (G = 12,000 ksi) shaft shown in Fig. P12–23. Determine

    (a)   The maximum shearing stress in the shaft.

    (b)   The maximum angle of rotation at the right end of the shaft.

    (c)   The maximum shearing stress in the shaft if its diameter is reduced to a uniform 2 in. over the complete length of the shaft.

FIG. P12–23

12-24   An elevator cage with a mass of 1400 kg is being lowered at a rate of 1.5 m/s. The elevator cable has an effective cross-sectional area of 650 mm^2 and an effective modulus of elasticity of 170 GPa. If the cage is suddenly stopped by the cable drum (not the brakes on the side of the cage) after descending 60 m, determine the maximum tensile stress developed in the cable. Neglect the mass of the cable.

12-3   THE EQUIVALENT STATIC LOAD

As noted in Art. 12–1, stresses and strains produced by dynamic loading depend on the velocity of propagation of the strain wave through the material. However, for many engineering applications (when velocities of the applied load are not too high), one may neglect such complications and assume that the material behaves as it does under slowly applied loads. This will be the approach taken in this book, and with such a simplifying assumption there is no need to evaluate Eqs. *c* or *d* of Art. 12–2. Instead, one may apply an *equivalent static load* (that is, a slowly applied load that will produce the same maximum deflection as that due to the dynamic load) and *assume that the material behaves identically under the two loadings.* With this assumption, the stress-strain diagram (Fig. 3–7c) for any point in the loaded system, as well as the strain distribution, will be identical under both loads.

As an illustration, consider the steel cantilever beam of Fig. 12–1a on which is dropped the weight W, as indicated. The variation of the end deflection of the beam with respect to time is shown in Fig. 12–1b where Δ, equal to δ_{max}, is the maximum deflection resulting from the falling weight transferring part of its kinetic energy ($mv^2/2$) to the beam and where δ_{stat} is the static deflection resulting when the beam and weight finally reach equilibrium. The static load P, applied as in Fig. 12–1c, produces the same maximum deflection Δ; the deflection curve and stress-strain diagrams are assumed to be the same in both cases. Hence, the strain energy in the beam is the same in both cases, and since the strain energy is equal to the effective applied energy (work done on the beam), the following relation may be stated:

$$\begin{pmatrix} \text{effective} \\ \text{energy} \\ \text{applied} \end{pmatrix} = \begin{pmatrix} \text{work done by} \\ \text{equivalent} \\ \text{static load} \end{pmatrix} \qquad (a)$$

The effective energy is that portion of the applied energy that produces the maximum deflection. In addition to energy lost in the form of sound, heat, and local distortion, some is absorbed by the supports.

From Art. 12–2, the strain energy in terms of the static load equals $\int P\, d\delta$ or $P_{avg}(\delta_{max})$. Equation *a* may then be written as

FIG. 12–1

$$U = P_{avg}(\Delta)$$

and if stress and strain are proportional (load-deflection diagram is a straight line), P_{avg} is equal to one-half P_{max}. Hence, for a material possessing a linear stress-strain relationship,

$$U = P\Delta/2 \tag{12-4}$$

The right-hand side of Eq. 12–4 involves the relations between load, stress, and strain previously encountered under static loading and applies to any type of loading. If the effective energy applied, U, is due to a change of velocity of a translating body, the left side of Eq. 12–4 becomes

$$\eta m v^2/2$$

where η is the effective portion of the applied energy, m is the mass of the body, and v is the velocity of the moving body just before impact. If the effective energy applied comes from a body falling from rest and being stopped by the resisting structure, the left-hand side of Eq. 12–4 comes from the change in potential energy of the body and is

$$\eta m g h$$

where g is the local acceleration due to gravity and h is the total vertical distance through which the body falls.

For torsional energy loading, Eq. a can be written as

$$U = T\theta/2 \tag{12-5}$$

where T is the slowly applied torque that will produce the same maximum angle of twist θ, as the suddenly applied torque if the maximum shearing stress is less than the proportional limit.

A term sometimes encountered in working with dynamic loading is load factor (or impact factor), defined as the *ratio of the equivalent static load to the magnitude of the force the loading mass would exert on the loaded system if slowly applied*. For example, the equivalent static loading on the wings of an airplane pulling out of a dive divided by the gravitational force mg on the plane (loading on horizontal tail surfaces neglected) is the load factor.

Example 12–1 and 12–2 illustrate the application of Eq. 12–4.

EXAMPLE 12–1

The upper 900 mm of the steel (E = 200 GPa) bar AB in Fig. 12–2a is 50 mm wide by 12 mm thick, and the lower 600 mm is 25 mm wide by 12 mm thick. When the collar W is released from rest from a height of 500 mm onto the end

FIG. 12–2

A of the bar, the axial stress in the bar (neglect stress concentration) is not to exceed 100 MPa. Assume that 80 percent of the energy of the falling collar is absorbed elastically by member AB. Determine

(a)   The maximum permissible mass for collar W.
(b)   The maximum permissible mass for W if the bar AB is 25 mm wide through its entire length.

SOLUTION

(a)   Figure 12–2b indicates the static load applied to the bar and the resulting deflections. When Eq. 12–4 is used, the following expression is obtained:

$$\eta mgh = (0.8)(mg)(0.500 + \Delta) = P\Delta/2 \qquad (b)$$

The quantities P and Δ can be calculated directly from the given data. From Fig. 12–2b,

$$0.012(0.050)(\sigma_2) = 0.012(0.025)(\sigma_1) \qquad \text{or} \qquad \sigma_1 = 2\sigma_2$$

This expression indicates that σ_1 (equal to 100 MPa) is the limiting stress and that σ_2 must be 50 MPa. The load P is

$$P = \sigma_1 A_1 = 100(10^6)(0.012)(0.025) = 30(10^3)N = 30 \text{ kN}$$

and the deflection Δ is

$$\Delta = \delta_2 + \delta_1 = 0.900\varepsilon_2 + 0.600\varepsilon_1 = (0.900\sigma_2 + 0.600\sigma_1)/200(10^9)$$

$$= \frac{0.900(50)(10^6) + 0.600(100)(10^6)}{200(10^9)}$$

$$= 0.525(10^{-3})\text{m} = 0.525 \text{ mm}$$

Substituting these values of P and Δ into Eq. *b* gives the expression

$$0.8(m)(9.81)(0.500 + 0.000525) = 30(10^3)(0.525)(10^{-3})/2$$

from which

$$m = \underline{2.00 \text{ kg}} \qquad\qquad \text{Ans.}$$

(b)   If the bar is 25 mm wide for its entire length,

$$\Delta = \frac{(0.900 + 0.600)(100)(10^6)}{200(10^9)} = 0.750(10^{-3})\text{m} = 0.750 \text{ mm}$$

and

$$0.8(m)(9.81)(0.500 + 0.000750) = 30(10^3)(0.750)(10^{-3})/2$$

from which

$$m = \underline{2.86 \text{ kg}} \qquad\qquad \text{Ans.}$$

In view of the development of Art. 12–2, which indicates that the energy-absorbing capacity of a member is a function of the amount of material present, the results in parts (a) and (b) seem paradoxical, as the strength of the member is increased by removing some material. The apparent discrepancy is probably best explained by referring to the stress-strain diagram of Fig. 12–3. For part (a) of the problem, the energy absorbed per unit volume by the lower 600 mm of the bar is represented by the area *ocd*, which gives

$$u = \frac{1}{2}\left[\frac{100(10^6)}{200(10^9)}\right](100)(10^6) = 25(10^3)N \cdot \text{m/m}^3 = 25 \text{ kJ/m}^3$$

The energy absorbed per unit volume by the upper 900 mm of the bar is represented by area *oab* of Fig. 12–3 and is seen to be only one-fourth as much because both the stress and the strain are one-half as much as in the

FIG. 12-3

lower 600 mm. When one-half of the volume of the upper 900 mm is removed, the energy absorbed by each unit volume of the remaining material is quadrupled (area *ocd* as compared to area *oab*), resulting in a net gain in energy-absorbing capacity. For this reason, bolts designed to carry dynamic loads are sometimes turned down so that the diameter of the shank of the bolt is the same as the diameter at the root of the threads. In this way the stress distribution in the bolt is approximately uniform, and the capacity of the bolt to absorb energy is a maximum.

EXAMPLE 12-2

The simply supported 6061-T6 aluminum alloy beam *A* of Fig. 12-4*a* is 3 in. wide by 1 in. deep. The center support is a helical spring with a modulus of 100 lb/in. The spring is initially unstressed and in contact with the beam. When the 50-lb block drops 2 in. onto the top of the beam, 90 percent of the energy is absorbed elastically by the system consisting of beam *A* and spring *B*. Determine

(a)   The maximum flexural stress in the beam for this loading.
(b)   The load factor.
(c)   The proportion of the effective energy absorbed by each of the components *A* and *B*.

SOLUTION

A deflection diagram for the system and free-body diagrams for the component parts are shown in Figs. 12-4*b*, *c*, and *d*. Since the action is assumed to be elastic and the materials obey Hooke's law, the load-deflection relation will be linear, and Eq. 12-4 is valid; hence,

$$50(2 + \Delta)(0.9) = P\Delta/2 \qquad (c)$$

FIG. 12–4

(a)

(b)

(c)

(d)

The fiber stress is wanted; therefore, the forces P and Q will be evaluated, and from these forces and Fig. 12–4c the fiber stress can be determined. From Figs. 12–4c and d and Appendix D,

$$\delta_A = \frac{(P - Q)L^3}{48EI} = \frac{(P - Q)(10^3)(12^3)}{48(10^7)(3)(1^3)/12} = \frac{(P - Q)(144)}{(10^4)}$$

and

$$\delta_B = \frac{Q}{100}$$

From Fig. 12–4b,

$$\delta_A = \delta_B = \Delta$$

or

$$\frac{(P - Q)(144)}{10,000} = \frac{Q}{100}$$

from which

$$Q = \frac{36P}{61}$$

Then,

$$\Delta = \frac{Q}{100} = \frac{36P}{6100}$$

and Eq. *c* becomes

$$50 \left(2 + \frac{36P}{6100}\right)0.9 = \left(\frac{1}{2}\right)\frac{36P^2}{6100}$$

or

$$P^2 - 90P = 30,500$$

from which

$$P = 225.3 \quad \text{and} \quad -135.3$$

The positive root is the magnitude of P necessary to produce the maximum downward deflection of the beam, whereas the negative root is the force necessary to produce the maximum upward deflection (see Fig. 12–1*b*) of the vibrating system.

When the value of 225.3 lb for P is used,

$$Q = \frac{36(225.3)}{61} = 133.0 \text{ lb}$$

and

$$P - Q = 92.3 \text{ lb}$$

(a) From Fig. 12–4*c*, R is 46.15 lb, and the maximum bending moment occurs at the center of the beam and is 46.15(5)(12) in.-lb; therefore, the maximum flexural stress is

$$\sigma = \frac{Mc}{I} = \frac{46.15(60)(0.50)}{3(1^3)/12}$$

$$= \underline{5540 \text{ psi } T \text{ and } C} \qquad \text{Ans.}$$

which is well below the proportional limit, and therefore the action is elastic.

(b) The load factor is

$$\frac{225.3}{50} = \underline{4.51} \qquad\qquad \text{Ans.}$$

(c) The resultant equivalent static load at the center of A is $(P - Q)$ and is 92.3 lb, and the energy absorbed elastically by the beam is

$$U_A = \frac{(92.3)(\Delta)}{2}$$

The elastic energy absorbed by the helical spring is

$$U_B = \frac{(133.3)(\Delta)}{2}$$

The total energy absorbed elastically by the system is

$$U = \frac{(225.3)(\Delta)}{2}$$

Therefore, the proportions absorbed by each component are

$$\frac{92.3}{225.3} \,(100) = \underline{41 \text{ percent for } A} \qquad\qquad \text{Ans.}$$

and

$$\frac{133.0}{225.3} \,(100) = \underline{59 \text{ percent for } B} \qquad\qquad \text{Ans.}$$

To recapitulate, Eq. 12–4 is applicable only to systems having a linear relationship between load and deformation, and only when the velocity of application of the dynamic load is not too high.

PROBLEMS

Note: In the following problems, the mass of the loaded members is to be neglected, and unless stated otherwise, all of the applied energy is to be assumed to be absorbed elastically by the loaded members. Neglect stress concentrations.

12-25*   A load of 1000 lb compresses a given railroad car spring 0.1 in. What static load would be required to produce the same deflection as would be produced by a load of 350 lb dropped from a height of 1 ft?

12-26*   A mass of 24 kg is dropped from a height of 100 mm onto a helical spring that has a modulus of 18 kN/m. Determine

    (a)   The maximum deflection of the spring.

    (b)   The static load that would produce the same deflection.

12-27*   A hot-rolled Monel (see Appendix A) bar with a cross-sectional area of 3.5 in.^{2} is to be used as a tension member subject to axial energy loads. The allowable axial tensile stress is 10 ksi. Determine the minimum length of bar required for an axial energy load of 50 ft-lb when eight-tenths of the energy is effective in producing stress in the bar.

12-28*   A flat bar of steel ($E = 200$ GPa) 50 mm thick and 1.8 m long has a width of 150 mm for 0.6 m of its length and a width of 100 mm for the remaining 1.2 m. Determine the maximum axial stress developed in the bar by an axial energy load of 160 J applied at one end of the bar. Assume that four-fifths of the energy produces stress in the bar.

12-29   A collar W is dropped 10 in. onto the end of the steel ($E = 30,000$ ksi) bar shown in Fig. P12–29. If 80 percent of the energy is effective in producing stress in the bar and if the maximum stress is not to exceed 18 ksi, determine the maximum allowable weight of the collar if

    (a)   The bar has the cross section shown in Fig. P12–29.

FIG. P12–29

(b) The bar has a uniform cross section of 2 sq. in. throughout its entire length.

12-30   An S305 × 61 steel section (see Appendix C) is used as a simply supported beam 3.0 m long. What mass falling on the center of the span from a height of 100 mm will produce a maximum fiber stress of 100 MPa in the beam if three-fourths of the energy is effective in producing stress? The web of the beam is vertical. Use $E = 200$ GPa.

12-31*   A weight of 80 lb is dropped from a height of 2 in. onto the free end of a steel ($E = 30,000$ ksi) cantilever beam. The beam is 2 in. wide, 3 in. deep, and 30 in. long. If only one-half of the energy of the falling weight is effective in producing stress in the beam, determine

(a) The maximum fiber stress produced by the falling weight.

(b) The maximum deflection due to the falling weight.

(c) The static load needed to produce the same deflection.

12-32*   A timber ($E = 8.2$ GPa) beam 150 mm wide, 100 mm deep, and 3 m long is simply supported at the ends. A 14-kg mass is dropped onto the center of the beam from a height of 250 mm. If 80 percent of the energy of the falling body is effective in producing stress in the beam, determine

(a) The maximum fiber stress produced by the falling body.

(b) The maximum deflection due to the falling body.

(c) The static load needed to produce the same deflection.

12-33   A steel ($G = 12,000$ ksi) shaft 15 ft long consists of a 5-ft solid section with a diameter of 10 in. and a 10-ft hollow section with an outside diameter of 10 in. and an inside diameter of 8 in. Determine

(a) The torsional energy load required to produce a maximum shearing stress of 10,000 psi in the shaft.

(b) The angle of twist produced by the energy load of part (a).

12-34*   A torsional energy load of 35 J is applied to the steel ($G = 80$ GPa) shaft shown in Fig. P12–34. Determine

(a) The maximum shearing stress produced by the energy load.

(b) The torsional energy load required to produce the same maximum shearing stress if the shaft is turned to a uniform 50-mm diameter except for a 50-mm length at the support. Neglect stress concentrations.

12-35*   The beam of Fig. P12–35 is 3 in. wide and 1 in. deep. Each of the supporting coil springs has a modulus of 50 lb per in. When the block W is

FIG. P12–34

75 mm

$U = 35$ J

600 mm

50 mm

450 mm

FIG. P12–35

W

1"

h

5'   5'

slowly applied to the center of the beam, the ends of the beam move downward 0.10 in. and the center of the beam moves downward an additional 0.12 in. Determine

(a)   The weight of block W.

(b)   The height h from which the block W must be dropped to produce a fiber stress in the beam of 2400 psi.

12-36   The bronze ($E = 80$ GPa) beam of Fig. P12–36 is 75 mm wide by 25 mm deep. Each of the supporting coil springs has a modulus of 9 kN/m. From what height should the block W, with a mass of 4.5 kg, be dropped in order to produce a total deflection at the center of the beam equal to four times the deflection produced by the same mass when it is slowly applied to the beam?

12-37   A vertical energy load is applied to the coil spring (modulus = 100 lb/in.) of the beam-spring system shown in Fig. P12–37. If the spring itself absorbs 72 in.-lb of energy, determine the energy absorbed by the timber ($E = 1200$ ksi) beam.

12-38*   The beam of Fig. P12–38 is made of an aluminum alloy ($E = 70$ GPa). The 25-kg block B is dropped from a height of 12 mm onto the top of

W

25 mm

h

1.5 m   1.5 m

FIG. P12–36

U

3"

2"

5'   5'

FIG. P12–37

FIG. P12–38   FIG. P12–39

the coil spring C (modulus = 35 kN/m). The resulting load factor is 2. Determine the percentage of the energy of block B that is absorbed by

    (a)   The coil spring C.

    (b)   The beam A.

12-39* A steel (E = 30,000 ksi) cantilever beam with a box cross section is subjected to an energy load through a coil spring (modulus = 2500 lb/in.) as shown in Fig. P12–39. The cross section is 8 in. wide by 6 in. deep and has a cross-sectional moment of inertia of 96 in.[4]. Determine the maximum height from which the 250-lb weight W can be dropped without exceeding a fiber stress of 15 ksi in the beam. Assume that 80 percent of the energy of the falling weight is absorbed elastically by the beam-spring combination.

12-40 When the 35-kg block B of Fig. P12–40 was dropped from a point 50 mm above the top of the coil spring C (modulus = 26 kN/m), the point D on the steel (E = 200 GPa) cantilever beam was observed to deflect 50 mm downward. Determine

    (a)   The percentage of the energy in block B that was absorbed by the beam-spring combination.

    (b)   The load factor.

12-41* The beam AB shown in Fig. P12–41 has a flexural rigidity EI and is securely fastened to the rigid arm C. The modulus of the helical spring is $6EI/L^3$. The structure rotates about a shaft at B. Determine the height h, in

FIG. P12–40

FIG. P12–41

terms of W, L, E, and I, from which W must be dropped to have a load factor of 4. Assume 75 percent of the energy of W is effective in producing elastic deformation in the beam-spring system.

12-42   The structure shown in Fig. P12–42 consists of a solid steel ($G = 12,000$ ksi) rod AB rigidly connected to the steel ($E = 30,000$ ksi) beam BC. The rod is fixed at A and supported by a smooth bearing D at B. The 100-lb weight W is dropped 1 inch onto the beam and 80 percent of the energy of the falling weight is absorbed elastically by the rod and beam. Determine the maximum fiber stress in beam BC and the maximum shearing stress in rod AB.

12-43   Beams A and B of Fig. P12–43 are made of wood ($E = 8$ GPa) and are 50 mm deep by 150 mm wide. When the 12-kg block W is dropped on beam A, 90 percent of the energy is absorbed elastically by beams A and B, and the maximum deflection is 38 mm. Determine

(a)   The height h from which W is dropped.

(b)   The load factor.

(c)   The amount of energy absorbed by each beam.

12-44*   Beams A and B of Fig. P12–44 are made of wood ($E = 1200$ ksi) and have cross sections that are 2 in. deep by 6 in. wide. When the 20-lb block W

FIG. P12–42

FIG. P12–43

is dropped on the beams, nine-tenths of the energy is absorbed elastically by the beams. If the maximum fiber stress developed in beam B is 1200 psi, determine

(a)   The height h from which W is dropped.

(b)   The load factor.

(c)   The amount of energy absorbed by each beam.

12-45   The aluminum alloy ($E = 70$ GPa) beams shown in Fig. P12–45 have cross sections that are 25 mm deep by 100 mm wide. The helical spring C between beams A and B has a modulus of 20 kN/m. The 5-kg block is dropped onto beam A with a load factor of 4. If 80 percent of the energy is absorbed elastically by A, B, and C, determine

(a)   The height h from which the block is dropped.

(b)   The amount of energy absorbed by each component.

12-4   SOME ELEMENTARY DESIGN CONCEPTS

Some design considerations for beams subjected to energy loads are readily obtained from the equivalent static load procedure of Art. 12–3. For example, consider the energy absorbed by a cantilever beam of length L with a concentrated load P at the free end. The strain energy (see Appendix D) is

FIG. P12–44

FIG. P12–45

$$U = P\Delta/2 = \frac{P(PL^3)}{2(3EI)}$$

With the load P replaced by its value in terms of the maximum fiber stress in the beam, σ equals PLc/I, the expression for the energy becomes

$$U = \frac{\sigma^2 IL}{6c^2 E} = \frac{\sigma^2 L}{6E}\left(\frac{I}{c^2}\right)$$

For a rectangular beam I/c^2 is equal to one-third the cross-sectional area, regardless of whether the long or short dimension of the rectangle is parallel to the load. When the parenthetical quantity is replaced by its value in terms of the area of the rectangle, the expression for the energy becomes

$$U = (AL)\frac{\sigma^2}{18E} \tag{a}$$

which shows that for a rectangular cross section the capacity to absorb energy is independent of the moment of inertia of the cross section. Comparison of Eq. a with Eq. 12–1 of Art. 12–2 reveals that the rectangular cantilever beam considered has only one-ninth as much capacity to absorb or store elastic energy as an axially loaded member of the same volume. Too much of the volume of the cantilever beam has low stresses for efficient energy load capacity. The situation can be improved by using a tapered beam whose width varies directly as the distance from the free end.

For this purpose consider the cantilever beam of constant depth h of Fig. 12–5. The maximum fiber stress at any section is

FIG. 12–5

$$\sigma = Mc/I = \frac{Pxh/2}{2zh^3/12} = \frac{6PL}{bh^2}$$

which shows that the maximum fiber stress is independent of the distance x along the beam. The differential equation of the elastic curve for the beam is

$$E\frac{d^2y}{dx^2} = -\frac{Px}{I} = -\frac{12PL}{bh^3}$$

After integration twice and evaluation of the constants of integration, the magnitude of the maximum deflection at the free end of the beam is found to be

$$\Delta = \frac{6PL^3}{bh^3E}$$

The energy U then becomes

$$U = \frac{P\Delta}{2} = \frac{P}{2}\frac{6PL^3}{bh^3E}$$

or in terms of the maximum fiber stress the energy is

$$U = (\text{volume})[\sigma^2/(6E)] \tag{b}$$

Comparison of Eqs. a and b shows that the tapered cantilever beam of Fig. 12–5 can absorb three times as much elastic energy as a cantilever beam of the same volume with a constant rectangular cross section. However, comparison with Eq. 12–1 of Art. 12–2 shows that the beam of Fig. 12–5 is still only one-third as effective as an axially loaded member of the same volume.

For torsion members subjected to energy loads, a hollow shaft will provide higher energy-absorbing capacity per unit volume than a solid circular shaft.

CHAPTER 13

CONNECTIONS

13-1 GENERAL DISCUSSION OF RIVETED CONNECTIONS

One of the most vexing problems confronting the structural designer is the proportioning and detailing of connections to transmit forces from one structural element to another. Among the common types of fastenings are rivets, bolts, welds, pins, and threaded couplings. This article is concerned with riveted (or bolted) connections such as the joints of Fig. 13–1.

A riveted joint can fail as a result of the cross shearing stresses in the rivets, the bearing stresses between the rivets and the plates, the tensile stresses in the plates, or perhaps a combination of these stresses.

Before discussing the analysis of these connections, it is best to define, with the aid of Fig. 13–2, some of the terms used. Figure 13–2a represents a *double-riveted butt joint* with one *cover (or splice) plate*. The term *double-riveted* (or *single-riveted,* and the like) refers to the number of rows of rivets that transfer the total load from one element to another (in this case from main plate to cover plate, or from cover plate to main plate.) A butt joint may also be fabricated with two cover plates, one on each side of the main plates. In some butt joints, both cover plates engage all rows of rivets; in others, one plate may engage only the inner rows and the other plate all rows (see Fig. 13–2b). Figure 13–2c represents a *triple-riveted lap joint*. The various terms used are indicated in the diagrams. Note carefully the designa-

FIG. 13–1

Typical shear, bearing, and tension failures in riveted joints.

(a) Double-riveted lap joint. Shear failure.

(b) Single-riveted lap joint. Bearing failure.

(c) Triple-riveted lap joint. Tension failure in outer row.

(d) Triple-riveted lap joint. Tension failure in inner row.

(e) Triple-riveted butt joint. Tension failure in inner rows of cover plates (top plate on left side of joint and bottom plate on right side).

tion of inner and outer rows in the two different types of joints. The term *pitch* alone always denotes spacing between rivets in a row.

When analyzing long joints such as those in boilers, water tanks, and air compressor tanks, a representative length of the joint, called the *repeating group,* is investigated rather than the entire joint. The repeating group of rivets is the shortest pattern that repeats itself along the joint, and the length of the group is the least common multiple of the rivet pitches in the joint. It is recommended that the rivet pattern be sketched as an aid to obtaining the correct number of rivets (or shear areas) in the repeating group. Such a sketch is given in Fig. 13–3, which is an example of a triple-riveted joint with pitches of 2, 3, and 4 in. It will be observed that there are thirteen rivets in the length of 12 in. This sketch is extended to show that the patterns for two consecutive 12-in. lengths are identical.

The analysis of a riveted or bolted connection based on elastic action[1] is unsatisfactory because the solution is so laborious, and the results of the computations (which are based on certain assumptions) are usually consider-

FIG. 13–2

FIG. 13–3

[1] For a discussion of elastic analysis, see "Work of Rivets in Riveted Joints," A. Hrennikoff, *Trans. Am. Soc. of Civil Engrs.*, Vol. 99, 1934, p. 437. The accompanying discussions include an extensive bibliography on the subject.

FIG. 13–4

ably different from experimental results, which of necessity involve indeterminate variables due to fabrication methods. For a brief exploration into the field of elastic analysis, consider section a–a of the quadruple-riveted lap joint of Fig. 13–4 in which the rivet distortions are exaggerated. If it is assumed that the distortions in rivets 1 and 2 are equal (equivalent to assuming that the forces on the two rivets are equal) and if there is no slip in the joint (the rivets fill the holes completely and elastic action prevails), the axial deformation between rivets 1 and 2 in plates A and B is the same. Therefore, the average axial strain in A is equal to that in B; if Hooke's law applies and the plates are of the same size and material, the tensile force in A equals that in B, equals $P/2$. When this result is applied to the free-body diagrams of Fig. 13–5, the condition of equilibrium indicates that F_1 is equal to $P/2$. Now, when the same reasoning is applied to section b–b and the free-body diagrams of Fig. 13–6 are drawn, it becomes apparent that if F_1 is equal to $P/2$, F_2 must be zero. This statement is not consistent with the original assumption, indicating that if elastic action prevails throughout the joint and the rivets completely fill the holes, a uniform distribution of load to rivets is impossible (except for the special case of the double-riveted joint with balanced plate areas).

One possible elastic analysis of the joint of Fig. 13–4, based on the assumptions of negligible friction, all rivets bearing on the plates immediately on application of load, and the force on any rivet being a constant[2] times the rivet distortion δ, would indicate that for a steel joint having the

---

[2] This constant is experimentally determined and varies considerably with materials, joints, loads, and investigators. For this analysis, the constant $12.5(10^6)$ lb/in. was taken from reference 3.

FIG. 13–5   FIG. 13–6

dimensions shown, at working load rivets 1 and 4 would each carry about 35 percent of the total force P and rivets 2 and 3 would each carry about 15 percent. These results, although closer to the truth than those obtained under the first assumption, are still in error because experiments[3,4,5] show that slip may begin upon application of even a small load, indicating that although friction may be negligible, the rivets do not completely fill the holes (which is to be expected if the cooling of the rivet after driving is considered); therefore, the assumption of no slip is invalidated. Furthermore, bending of the plates was neglected, and as can be seen from Fig. 13–4, the applied forces P cannot become collinear until the plates are bent.

The foregoing discussion will serve to indicate the need for some simplifying assumption or concept regarding the action of a riveted or bolted connection leading to a reliable easily applied method of analysis. Two widely used methods are outlined in the two succeeding articles. For both methods, the following approximations are made:

1. *The nominal size of the rivet or bolt* is used for *shear* and *bearing* stresses in the design of *steel bridges* and *buildings*. In the design of *pressure vessels* and *aircraft structures,* the actual diameter of the hole in the plate is used (the rivet when driven is assumed to fill the hole).
2. The net section for computing the tensile stress in the plates is calculated using the nominal size of the rivet or bolt plus 1/8 in. (3 mm) for *steel bridge* and *building design*. The actual (specified) size of the hole is used to calculate the net section in the design of *aircraft structures and pressure vessels.*

[3] "Riveted and Pin-connected Joints of Steel and Aluminum Alloys," L. S. Moisseiff, E. C. Hartmann, and R. L. Moore, *Trans. Am. Soc. of Civil Engrs.,* Vol. 109, 1944, p. 1359.

[4] "Tension Tests of Large Riveted Joints," R. E. Davis, G. B. Woodruff, and H. E. Davis, *Trans. Am. Soc. of Civil Engrs.,* Vol. 105, 1940, p. 1193.

[5] "Tests of Joints in Wide Plates," W. M. Wilson, J. Mather, and C. O. Harris, *Bulletin No.* 239, Engr. Exp. Sta., Univ. of Ill., 1931.

3.   *Bending* is neglected in all joints. *Friction* is neglected in *riveted* or *ordinary bolted* joints. In *friction type* bolted joints, the high-strength bolts are torqued to produce sufficient friction to prevent the joint from slipping into bearing; hence, bearing stress need not be considered in such joints.

A term frequently encountered in discussions of riveted connections is the *efficiency* of a joint, here defined as *the maximum permissible load on a joint divided by the maximum permissible axial load on the solid plate.* Efficiency may also be defined as *the strength of the joint divided by the strength of the solid plate.* One may speak of the efficiency in shear, bearing, or tension as the permissible load (or strength) in shear, bearing, or tension divided by the safe load capacity (or strength) of the solid plate; however, the *efficiency of the joint* is as defined above.

13-2   ANALYSIS OF RIVETED OR BOLTED JOINTS—UNIFORM SHEAR METHOD

One of the common assumptions made regarding load distribution in riveted and bolted connections is that the cross shearing stress in the rivets or bolts is uniformly distributed over all the shear areas (the cross-sectional areas of the rivets or bolts). After the discussion in Art. 13–1, this may appear to be a somewhat indefensible assumption; however, tests have been conducted that seem to indicate that such a condition is possibly attained with only a small amount of slip in the joint.[6] The tensile stress in the plate is assumed to

[6] "Riveted and Pin-connected Joints of Steel and Aluminum Alloys," L. S. Moisseiff, E. C. Hartmann, and R. L. Moore, *Trans. Am. Soc. of Civil Engrs.,* Vol. 109, 1944, p. 1359. Included in these tests were symmetrical triple-riveted butt joints having $\frac{3}{4}$-in. main plates and two $\frac{3}{8}$-in. cover plates with twelve $\frac{7}{8}$-in. rivets in each half. At design loads (average shearing stress = 13,500 psi), the rivets in the inner row were found to carry 33–42 percent of the load, those in the intermediate row 16–26 percent, and those in the outer row 36–43 percent. However, the ultimate loads for the joints designed to fail in shear were within 2 percent of the predicted loads based on the uniform shearing stress assumption and on shearing strengths obtained from auxiliary control specimens having one and two rivets in each half. These results indicate that although at working loads there is definitely a nonuniform distribution of load to the rivets, the inequality disappears by the time the ultimate load is reached. Joints designed to fail in tension averaged within 4 percent of the loads predicted on the basis of the tensile properties of the plate material (as obtained from control specimens) on the net section. The writers of this paper concluded that "the amount of plastic yielding necessary to level off high stress concentrations and to effect a uniform distribution of load among the rivets must be relatively small." Of course, this means that a reasonably ductile material is necessary (see also references 4 and 5 of Art. 13–1).

TABLE 13–1 Allowable Stresses for Riveted Joints

Specification	Tensile Stress Net Section of Plate (psi)	(MPa)	Shearing Stress in Rivet (psi)	(MPa)	Bearing Stress on Plate (psi)	(MPa)
Am. Soc. of Mech. Engrs. Boiler Construction Code for Unfired Pressure Vessels, 1971; for temps to 650°F						
SA-515 plate, SA-31A rivets	13,700	95	9,000	62	18,000	124
Am. Inst. of Steel Const. for Struct. Steel Bldgs., 1970						
A36 steel, A502-1 rivets, A325 bolts[a]	21,600	149	15,000	103	48,600	335
Am. Assoc. of State Highway Officials for Steel Bridges, 1973						
A36 steel, A502-1 rivets, A325 bolts[a]	20,000	138	13,500	93	40,000	276
Am. Ry. Engrs. Assoc., 1973						
A36 steel, A502-1 rivets, A325 bolts[a]	20,000	138	13,500[b]	93[b]	27,000[c]	186[c]
			20,000	138	36,000	248
Am. Soc. of Civil Engrs. Spec. for Alum. Alloy 2014-T6						
(14S-T6), 1952	22,000	152	10,000[d]	70[d]	36,000[e]	248[e]
			8,000	55		

[a] The shearing stress applies and the bearing stress is not restricted for friction-type joints with A325 bolts.

[b] The smaller value for rivets, the other for A325 bolts in nonfriction joints.

[c] The smaller value for rivets in single shear, the other for rivets in double shear.

[d] The larger value for alloy 2117-T3 driven cold, the smaller value for alloy 6061-T43 driven hot. If the rivet is used in relatively thin plates, the allowable shear stress shall be reduced (see spec.).

[e] If the margin is less than two rivet diameters, the stress shall be reduced (see spec.).

be uniformly distributed over the net area of the plate at each row of rivets. The bearing stress between the rivets and the plates is assumed to be uniformly distributed over the projection of the contact area; that is, the plate thickness multiplied by the rivet diameter.

In order to design or check the safety of riveted joints, working stresses must be specified. Table 13–1 lists allowable stresses from some widely used specifications.

To illustrate the analysis of riveted joints by assuming a uniform shearing stress, the joint of Fig. 13–1d and a boiler joint will be discussed in Exams. 13–1 and 13–2.

EXAMPLE 13–1

The triple-riveted lap joint of Fig. 13–1d is composed of $1\frac{3}{8}$-in. wide by 0.025-in. thick aluminum alloy 2024-T4 plates connected with $\frac{5}{32}$-in. aluminum alloy 2117-T4 rivets. The holes and rivets are essentially the same size for this joint. The joint ruptured under an axial load of 1640 lb. Determine for this loading

    (a)   The average shearing stress in the rivets.
    (b)   The average bearing stress between rivet and plate.
    (c)   The maximum average tensile stress and state where it occurs.

SOLUTION

    (a,b)   From the free-body diagram of Fig. 13–7a and the equilibrium equation $\Sigma F_x = 0$, the force on each rivet is

$$F_s = 1640/5 = 328 \text{ lb}$$

and, from Fig. 13–7c, the average shearing stress is

$$\tau = \frac{328}{(\pi/4)(5/32)^2} = \underline{17{,}100 \text{ psi}} \qquad \text{Ans.}$$

From Figs. 13–7b and c, the average bearing stress is

$$\sigma_b = \frac{328}{0.025(5/32)} = \underline{84{,}000 \text{ psi}} \qquad \text{Ans.}$$

        The specified strengths of these materials in shear and bearing are 27,000 psi and 90,000 psi, respectively;[7] therefore, this joint should not fail in shear or bearing under the given load.

    (c)   The average tensile stress (axial stress) may be maximum at either the outer row where there is only one hole or the inner

[7] ANC-5, Strength of Aircraft Elements, December, 1942. Note that specified strengths are guaranteed minimum values and are usually less than typical values.

FIG. 13–7

(a)

(b)

(c)

(d)

(e)

row where there are three holes and a reduction in the load. The section at the outer row will be checked first with the aid of the free-body diagram of Fig. 13–7d. Summing forces in the horizontal direction gives

$$P = \sigma_t A_t = \sigma_t \left(\frac{11}{8} - \frac{5}{32}\right)(0.025)$$

$$1640 = \sigma_t \left(\frac{39}{32}\right)(0.025)$$

from which

$$\sigma_t = 53,900 \text{ psi } T \text{ in the outer row}$$

The section through the inner row will be checked with the aid of Fig. 13–7e. Note that in order to separate the joint at the inner row, not only must the plate be cut at this section, but the rivet ahead of the section must also be removed (or cut); according to the assumption of uniform shearing stress, the load on this rivet is one-fifth of the total load. Again, summing forces in the horizontal direction gives

$$P = \frac{P}{5} + \sigma_t A_t \qquad \text{or} \qquad \frac{4P}{5} = \sigma_t \left[\frac{11}{8} - 3\left(\frac{5}{32}\right)\right](0.025)$$

$$\frac{4(1640)}{5} = \sigma_t \left(\frac{29}{32}\right)(0.025)$$

from which

$$\sigma_t = \underline{57{,}900 \text{ psi } T \text{ at the inner row}} \qquad \text{**Ans.**}$$

There is no need to check the next row because the situation at the outer row of the other plate is identical to that already checked. The results of these computations indicate that, in tension, the joint is weaker at the inner row than at the outer row; this fact is verified by Fig. 13–1d, which shows the plate ruptured at the inner row. The specified tensile strength of this material is 62,000 psi (see reference 7); therefore, the average stress obtained was about 7 percent below the specified strength.[8]

EXAMPLE 13–2

A 1.25-m diameter boiler is made of 20-mm plate. The longitudinal seam is a double-riveted butt joint with two 14-mm cover plates and with 22-mm diameter rivets in 24-mm holes. The outer rows of rivets have a pitch of 125 mm and the inner rows a pitch of 50 mm. Using the ASME boiler code, determine

    (a)   The maximum allowable internal pressure.

    (b)   The efficiency of the joint.

SOLUTION

    (a)   From the sketch of the rivet pattern shown in Fig. 13–8a, it is evident that the repeating group has two rivets in each outer row and five rivets in each inner row. The free-body diagram of Fig.

---

[8] This tends to verify the conclusions in references 4 and 5 that protecting the minimum net section by increasing the pitch in the outer rows is not as effective as the analysis would indicate. This joint would very likely have carried a larger load if it had been fabricated with two ⅛-in. rivets in each of the three rows.

FIG. 13–8

13–8b shows a section of the main plate with a repeating group of rivets and indicates the fourteen shearing forces F_s, two forces for each rivet. The free-body diagram in Fig. 13–8c indicates that $F_b = 2F_s$ since each rivet is in double shear. The maximum allowable values of F_b and F_s are

$$(F_b)_{max} = \sigma_b A_b = 124(10^6)(0.020)(0.024)$$

$$= 59.5(10^3) \text{ N} = 59.5 \text{ kN}$$

and

$$(F_s)_{max} = \tau A_s = 62(10^6)(\pi/4)(0.024^2)$$

$$= 28.0(10^3) \text{ N} = 28.0 \text{ kN}$$

It is evident that the bearing capacity of the rivets and cover plate is greater than that of the main plate because the thickness of the two cover plates exceeds that of the main plate. Since twice the allowable shearing load is less than the maximum allowable bearing load, the shearing strength of the rivets is the critical factor as far as shear and bearing are concerned. The maximum load that can be carried by the rivets is, from equilibrium (see Fig. 13–8b),

$$P_s = 7(2F_s) = 7(2)(28.0) = 392 \text{ kN}$$

Three different sections must be checked for tensile capacity: (1) the main plate at the outer row of rivets, (2) the main plate at the inner row of rivets with some of the load already transferred to the cover plates, and (3) the cover plates at the inner row of rivets. The summation of forces in the direction of the load for the free-body diagram of Fig. 13–8d gives

$$P_t(\text{outer row}) = \sigma_t A_t = (95)(10^6)[0.250 - 2(0.024)](0.020)$$

$$= 384(10^3) \text{ N} = 384 \text{ kN}$$

The same equilibrium equation for the free-body diagram of Fig. 13–8e for tension in the inner row with $(2/7)P$ already transferred to the cover plates gives

$$P_t(\text{inner row}) = [0.250 - 5(0.024)](0.020)(95)(10^6) + \left(\frac{2}{7}\right) P_t$$

from which

$$P_t = 346(10^3) \text{ N} = 346 \text{ kN}$$

as the maximum allowable load for tension at the inner row of the main plate. The summation of forces in the direction of the load for the free-body diagram of Fig. 13–8f, which shows the tension in the cover plates at the inner row, gives

$$P_t(\text{cover}) = 2[0.250 - 5(0.024)](0.014)(95)(10^6)$$

$$= 346(10^3) \text{ N} = 346 \text{ kN}$$

The 346-kN load that can be carried by the main plate at the inner row of rivets and by the cover plates is the smallest of the four loads computed and is therefore the maximum allowable load on the 250-mm portion of the joint investigated.

The free-body diagram of Fig. 13–8g shows a 250-mm length of one-half of the boiler where p is the internal pressure in the boiler. The applicable equation of equilibrium gives

FIG. 13–9

$$p(0.250)(1.250) = 2(346)(10^3)$$

or

$$p = 2.21(10^6) \text{ N/m}^2 = \underline{2.21 \text{ MPa}} \qquad \text{Ans.}$$

(b)   The efficiency of the joint is the allowable load on the joint divided by the allowable load on the main plate or

$$\eta = \frac{346(10^3)}{(0.020)(0.250)(95)(10^6)} = 0.728 = \underline{72.8 \text{ percent}} \quad \text{Ans.}$$

A type of failure not yet considered here is a *marginal failure,* frequently called *tearout.* If a joint is designed to satisfy the usual specifications, the margins will be wide enough ($1\frac{1}{2}$ to 2 rivet diameters is usually specified) to prevent tearout failures. Should a narrow margin be needed, the shear tearout investigation should be made as indicated in Fig. 13–9. For the problems in this book, the margins will be assumed to be wide enough to preclude marginal failures, and the transverse pitch great enough to prevent failure along a diagonal (or zigzag) line between rivets in different rows.

PROBLEMS

Note: In the following problems, when the size of the hole is given, use the hole diameter for shear, bearing, and tension investigations. When the size of the rivet or bolt is given and no hole diameter is specified, use the nominal diameter of the rivet (or bolt) for shear and bearing investigations, and the nominal diameter plus $\frac{1}{8}$ in. (3 mm) for the size of the hole in tension investigations.

FIG. P13–1

FIG. P13–3

Transverse section

FIG. P13–2

Section A–A

13-1*   A bolted joint from a truss is shown in Fig. P13–1. Member A consists of two 75 × 8-mm bars; members B and C each consist of two L76 × 76 × 6.4-mm angles. The diameter of the bolts is 20 mm. The total tensile force P_1 = 90 kN. Determine

(a)   The highest average tensile stress in member A.

(b)   The maximum average bearing stress on the connection of B.

(c)   The average shearing stress in the bolts at C.

13-2*   A 10-in. steel channel is riveted to a gusset plate as shown in Fig. P13–2. The cross-sectional area of the channel is 4.47 in.² and the web is 1/4 in. thick. The diameter of the rivets is 3/4 in. Determine

(a) The average shearing stress in the rivets.

(b) The average bearing stress between rivets and channel web.

(c) The highest average tensile stress in the channel.

13-3 A compression member in a pin-connected truss is shown in Fig. P13–3. The member consists of two C254 × 37 steel channels (see Appendix C) latticed together. The bearing area on the pin is increased by riveting two 10-mm pin plates (a and b) to each channel web with 20-mm diameter rivets as shown. The total load on the member (two channels) is 980 kN. If the bearing stress on the pin is uniformly distributed, determine the average shearing stress in the rivets on the plane of contact between

(a) Pin plate b and the channel web.

(b) Pin plates a and b.

13-4* A long steel conduit 6 ft in diameter is subjected to an internal pressure of 90 psi. It is made of 3/8-in. steel plates joined with a longitudinal, double-riveted lap joint. The rivets are 3/4 in. in diameter in 13/16-in. holes. The pitch in each row is 3 in. Determine

(a) The average shearing stress in the rivets.

(b) The average bearing stress on the plates.

(c) The highest average tensile stress in the plates.

13-5* The triple-riveted lap joint shown in Fig. P13–5 is used to join two 12-mm plates. The rivets are 25 mm in diameter and are spaced 150 mm in the outer rows and 75 mm in the inner row. When the tensile stress is 100 MPa in the gross section of the plate, determine

(a) The average shearing stress in the rivets.

(b) The average bearing stress on the plates.

(c) The highest average tensile stress and its location.

FIG. P13–5

13-6   A riveted lap splice in the stressed skin of an airplane wing is shown in Fig. P13-6. The material is aluminum alloy. The rivets are 5/32 in. in diameter with 0.159-in. holes. Determine

(a)   The average shearing stress in the rivets.

(b)   The maximum average bearing stress in either sheet.

(c)   The maximum average tensile stress in either sheet.

13-7*   A cylindrical compressed air tank with an inside diameter of 900 mm is made of 10-mm steel plate and is subjected to an internal pressure of 2 MPa. The longitudinal seam is a double-riveted butt joint with two 6-mm cover plates. The 20-mm diameter rivets in 22-mm holes are spaced at 60 mm in the inner rows and 120 mm in the outer rows. Determine

(a)   The average shearing stress in the rivets.

(b)   The highest average bearing stress in any plate.

(c)   The highest average tensile stress in any plate.

13-8   A cylindrical pressure vessel with an inside diameter of 36 in. is made of 1/2-in. steel plate. The longitudinal seam is a double-riveted butt joint with two 5/16-in. cover plates. The 3/4-in. diameter rivets in 13/16-in. holes are spaced at 2.5 in. in the inner rows and 5 in. in the outer rows. When the vessel is subjected to an internal pressure of 250 psi, determine

(a)   The average shearing stress in the rivets.

(b)   The highest average bearing stress in any plate.

(c)   The highest average tensile stress in any plate.

13-9   Two 11-mm steel plates are connected by a triple-riveted butt joint with two 6-mm cover plates. The 15-mm diameter rivets are spaced at 100 mm in the outer rows, 50 mm in the intermediate rows, and 75 mm in the inner rows. The joint is loaded in tension and each shear area carries a total force of 13 kN. Determine the highest average tensile stress in the joint.

FIG. P13–6

13-10*   Two 7/16-in. steel plates are connected with a triple-riveted butt joint with two 1/4-in. cover plates. The 5/8-in. rivets are spaced at 4 in. in the outer rows, 2 in. in the intermediate rows, and 3 in. in the inner rows. The joint is loaded in tension and each shear area carries a total force of 3000 lb. Determine the highest average tensile stress in the joint.

13-11*   A boiler with a diameter of 1.00 m is made of 12-mm plate. The longitudinal seam is a double-riveted lap joint having 25-mm diameter rivets in 27-mm holes at a pitch of 90 mm in each row. Using the ASME boiler code, determine

   (a)   The maximum allowable internal pressure.

   (b)   The efficiency of the riveted joint.

13-12   A cylindrical pressure tank with a diameter of 40 in. is made of 5/16-in. steel plate. The longitudinal seam is a triple-riveted butt joint with one 3/8-in. cover plate and 3/4-in. diameter rivets in 13/16-in. holes at pitches of 2.5 in. in the inner and intermediate rows and 5 in. in the outer rows. Using the AISC specification, determine

   (a)   The maximum allowable internal pressure.

   (b)   The efficiency of the riveted joint.

13-13*   A triple-riveted lap joint is to be made of steel plates and 20-mm rivets at pitches of 120 mm in the outer rows and 60 mm in the inner row. Using the AISC building code, determine the minimum permissible thickness of the plate for the maximum allowable load on the joint.

13-14   Using the AASHO specification for steel bridges, determine the efficiency of the riveted joint described in Prob. 13-10.

13-3   ANALYSIS OF RIVETED AND BOLTED JOINTS—ULTIMATE STRENGTH METHOD

A widely used method for the design and investigation of riveted connections is based on the concept that if the load necessary to produce excessive inelastic action in a joint (ultimate load) can be predicted, a safe working load can then be determined by introducing an appropriate factor of safety. The determination of stresses at design loads (working loads) is of no consequence; hence, the question of distribution of load to the various elements of a connection at design loads merits no consideration.

The ultimate strength of the joint is determined by considering the various methods and combinations of failure that could occur. Any rivet can fail in either shear or bearing, and these possibilities are usually investigated first. When a joint contains some rivets in single shear and some in double shear, it is possible that the shearing strength of the material could be the controlling factor for the rivets in single shear and that the bearing strength could limit the load carried by the rivets in double shear.

The joint can fail in tension at the outer row of rivets in a lap joint, at the outer row of rivets in the main plate of a butt joint, or at the inner row of rivets in the cover plate of a butt joint. Failure can also result from a tensile failure of the main plate, at a row other than the outer row, combined with

TABLE 13–2   Ultimate Strengths for Some Selected Materials

Specification	Tensile Strength of Plate		Bearing Strength of Plate		Shearing Strength of Rivet	
	(psi)	(MPa)	(psi)	(MPa)	(psi)	(MPa)
Am. Soc. of Mech. Engrs. Boiler Construction Code for Power Boilers, 1971						
SA 515-grade 55 plate	55,000	380	95,000	655		
SA 31, Grade A rivets					44,000	305
MIL-HDKB-5 Met. Matls. and Elements for Flight Vehicle Struct., Feb. 1966						
2024-T3 sheet	64,000[a]	440[a]	124,000[b]	850[b]		
2117-T3 rivets					30,000[a,c]	200[a,c]
2024-T31 rivets					41,000[a,c]	280[a,c]
Structural steel (1025)	55,000	380	90,000[b]	620[b]	35,000	240
Stainless steel, AISI 301						
Annealed	75,000	520	150,000[b]	1,040[b]		
Sheet or strip, $\frac{1}{4}$ hard	125,000	860	250,000[b]	1,720[b]		
Monel rivets					49,000	340

[a] Probability (not guaranteed) values.

[b] For ratio of margin to rivet diameter = 2 and ratio of rivet diameter to sheet thickness <5.5.

[c] For ratio of rivet diameter to sheet thickness ≤3 for single shear and 1.5 for double shear. See the specification for reduction factors.

rivet failure (shear or bearing), which would have to occur if the joint was to separate. All possible combinations of tension failure with shear or bearing failure must be visualized and checked to determine the weakest combination. The smallest of the computed loads is the strength of the joint.

Table 13–2 lists a few representative values of ultimate strengths that may be used in the analysis of riveted connections. Although the preceding discussion concerned ultimate loads on joints, the same analysis can be made using working stresses, instead of ultimate strengths, to obtain the maximum working load on the joint. Such a procedure amounts to dividing by the factor of safety before calculating the joint strength instead of afterward.

The ultimate strength method of analyzing riveted joints is illustrated in Exams. 13–3 and 13–4.

EXAMPLE 13–3

The triple-riveted butt joint of Fig. 13–1e is composed of $1\frac{3}{8} \times 0.040$-in. main plates spliced with two 0.025-in. cover plates and $\frac{5}{32}$-in. diameter rivets. The diameter of the hole is essentially the same as that of the rivet. The plate material is aluminum alloy 2024-T4, and the rivets are 2117-T6. If the guaranteed[9] ultimate strengths of these materials are 27,000 psi shear 90,000 psi bearing, and 62,000 psi tension, determine the ultimate load for the specimen.

SOLUTION

The modes of failure to be investigated are: shearing of all rivets, bearing on all rivets, and tension at any net section in combination with shear or bearing failure of any rivets assisting the net section.

Ultimate shearing and bearing loads will be computed first with the aid of Figs. 13–10a and b (note that in Fig. 13–10a there are six additional forces F_s, not shown, on the opposite side of the plate) as follows:

$$2F_s = 2(27,000)\left(\frac{\pi}{4}\right)\left(\frac{5}{32}\right)^2 = 1035 \text{ lb shear per rivet}$$

$$F_b = 90,000\left(\frac{5}{32}\right)(0.040) = 562 \text{ lb bearing per rivet}$$

The cover plates will carry more load in bearing than the main plate since the bearing area is greater. The results above indicate that the shearing load is not critical; therefore, as far as shear and bearing are concerned, the ultimate load on the joint is

$$P_b = 6F_b = 6(562) = 3370 \text{ lb}$$

[9] ANC-5, Strength of Aircraft Elements, December 1942. Note that guaranteed values are minimum acceptable values and are usually less than typical values.

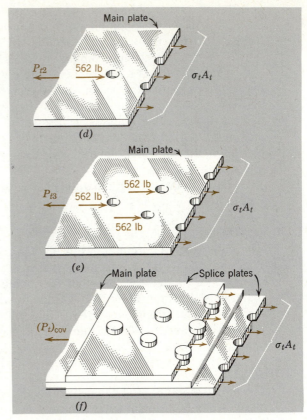

FIG. 13-10

Next, the ultimate tensile load will be determined with the aid of the free-body diagrams of Figs. 13–10c, d, e, and f. The load for each rivet assisting the net section in the main plate (Figs. 13–10d and e) is 562 lb, since the previous shear and bearing analysis indicates that the rivet will fail in bearing before the ultimate shear load is attained. Summing forces in the horizontal direction on each of the free-body diagrams in order yields the following:

(c) $P_{t1} = \sigma_t A_t = 62{,}000 \left(\dfrac{11}{8} - \dfrac{5}{32} \right) (0.040) = 3020 \text{ lb}$

(d) $P_{t2} = 562 + 62{,}000 \left[\dfrac{11}{8} - 2 \left(\dfrac{5}{32} \right) \right] (0.040) = 3200 \text{ lb}$

(e) $P_{t3} = 562(3) + 62{,}000 \left[\dfrac{11}{8} - 3 \left(\dfrac{5}{32} \right) \right] (0.040) = 3930 \text{ lb}$

(f) $(P_t)_{\text{cov}} = 62{,}000 \left(\dfrac{11}{8} - \dfrac{5}{32} \right) (0.025)(2) = \underline{2810 \text{ lb}}$ **Ans.**

FIG. 13–11

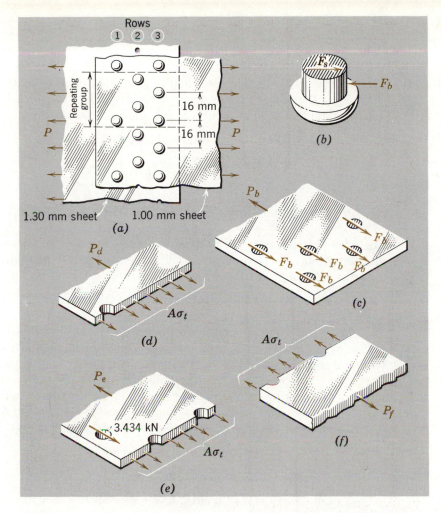

As the load on the cover plates is the least of the five loads computed, the joint might be expected to fracture at a load of 2810 lb. The average of two such joints, one of which is shown in Fig. 13–1e, loaded to fracture was 2790 lb.

EXAMPLE 13–4

Figure 13–11a represents a riveted lap splice in the stressed skin of an airplane wing. The sheet material is 2024-T3, and the 4-mm diameter rivets in 4.04-mm holes are 2024-T31. Use the MIL-HDBK-5 specifications with the shearing strength reduced to 0.964 times the value given (because the rivet diameter to plate thickness ratio is greater than 3) and determine

(a) The maximum permissible load per linear meter of joint if a factor of safety of 1.5 is specified.

(b) The efficiency of the joint.

SOLUTION

(a) From Fig. 13–11*a* the length of a repeating group of rivets is seen to be 32 mm with one rivet in the left row and two rivets in each of the other rows. Each rivet is in single shear, and the free-body diagram in Fig. 13–11*b* indicates that $F_b = F_s$. The bearing load is limited by the lower plate since the thickness of this plate is smaller. The failure loads per rivet in shear and bearing are

$$(F_s)_{max} = (\pi/4)(0.00404)^2(280)(10^6)(0.964)$$
$$= 3.46(10^3) \text{ N} = 3.46 \text{ kN}$$

and

$$(F_b)_{max} = (0.001)(0.00404)(850)(10^6) = 3.434(10^3) \text{ N} = 3.434 \text{ kN}$$

There are five rivets in each repeating group, and the free-body diagram in Fig. 13–11*c* indicates that the failure load in the joint, as far as shear and bearing are concerned, is

$$P_b = 5(3.434)(10^3) = 17.17(10^3) \text{ N} = 17.17 \text{ kN}$$

There are three possibilities of joint failure involving tensile failure of the plates to be investigated. The joint might fail by tearing the lower plate at row 1; the joint might fail by tearing the lower plate at row 2 with bearing failure in row 1; or the joint might fail by tearing the top plate at row 3. These three possibilities are shown in the free-body diagrams in Figs. 13–11*d*, *e*, and *f*. The equations of equilibrium for these free-body diagrams give

From (*d*) $P_d = 0.001(0.032 - 0.00404)(440)(10^6)$

$$= 12.3(10^3) \text{ N} = 12.3 \text{ kN}$$

From (*e*) $P_e = 3434 + 0.001[0.032 - 2(0.00404)](440)(10^6)$

$$= 13.96(10^3) \text{ N} = 13.96 \text{ kN}$$

From (*f*) $P_f = 0.00130[0.032 - 2(0.00404)](440)(10^6)$

$$= 13.68(10^3) \text{N} = 13.68 \text{ kN}$$

It is not necessary to check other failure combinations because the plates have as much or more areas at all other rows and more rivets would have to fail for other combinations. The controlling load on a repeating group is the 12.3-kN load from Fig. 13–11*d*. With a factor of safety of 1.5, the load per meter becomes

$$P = \frac{12.3(10^3)}{0.032} \frac{1}{1.5} = 256(10^3) \text{ N/m} = \underline{256 \text{ kN/m}} \qquad \textbf{Ans.}$$

(b) The efficiency of the joint is

$$\eta = \frac{12.3(10^3)}{0.001(0.032)(440)(10^6)} = 0.874 = \underline{87.4 \text{ percent}} \quad \text{Ans.}$$

PROBLEMS

> *Note: In the following problems, when the size of the hole is given, use the hole diameter for shear, bearing, and tension investigations. When the size of the rivet or bolt is given and no hole diameter is specified, use the nominal diameter of the rivet (or bolt) for shear and bearing investigations, and the nominal diameter plus $\frac{1}{8}$ in. (3 mm) for the size of the hole in tension investigations.*

13-15*   A boiler with a diameter of 1.00 m is made of 13-mm plate. The longitudinal seam is a double-riveted lap joint having 19-mm diameter rivets in 21-mm holes at a pitch of 64 mm in each row. Using the ultimate strengths from the ASME boiler code and a factor of safety of 5, determine

(a)   The maximum allowable internal pressure.

(b)   The efficiency of the riveted joint.

13-16*   The triple-riveted lap joint shown in Fig. P13–16 connects two 3/8-in. structural steel plates with 3/4-in. diameter rivets. Using the ultimate strengths from the MIL-HDBK-5 specifications and a factor of safety of 3, determine

(a)   The maximum allowable load.

(b)   The efficiency of the riveted joint.

FIG. P13–16

13-17   A triple-riveted lap joint is used to join two 6-mm plates with 16-mm diameter rivets in 18-mm holes. The pitch in the outer rows is 125 mm and in the inner row is 50 mm. Using the ultimate strengths from the ASME boiler code, determine the efficiency of the joint.

13-18*   A triple-riveted lap joint is to be made of structural steel plates and 5/8-in. diameter rivets at pitches of 4 in. in the outer rows and 2 in. in the inner row. Using the MIL-HDBK-5 specifications and a factor of safety of 3, determine the minimum permissible thickness of the plate for the maximum allowable load on the rivets.

13-19   A 1.20-m diameter boiler with an internal steam pressure of 1.25 MPa is made of 10-mm plate. The double-riveted longitudinal butt joint is made with two 6-mm splice plates and 18-mm diameter rivets in 20-mm holes at pitches of 60 mm in the inner rows and 160 mm in the outer rows. Determine the factor of safety of the joint by using the ultimate strength values from the ASME boiler code.

13-20*   A 50-in. diameter boiler is made of 3/4-in. plate. The longitudinal seam is a double-riveted butt joint with two 9/16-in. cover plates and 3/4-in. diameter rivets in 13/16-in. holes. The outer rows of rivets have a pitch of 6 1/4 in. and the inner rows a pitch of 2 1/2 in. Using the ultimate strength values from the ASME boiler code and a factor of safety of 5, determine the maximum allowable internal pressure.

13-21*   The triple-riveted lap joint shown in Fig. P13–21 connects two 10-mm structural steel plates with 20-mm diameter rivets. Using the ultimate strength values from the MIL-HDBK-5 specifications and a factor of safety of 3, determine

FIG. P13–21

FIG. P13–22

(a) The maximum load that a 600-mm length of joint can support.

(b) The efficiency of the joint.

13-22 Two 6 by 5/8-in. AISI annealed stainless steel plates are spliced with two 3/8-in. cover plates and 3/4-in. Monel rivets as shown in Fig. P13–22. Using the ultimate strength values from the MIL-HDBK-5 specifications and a factor of safety of 2, determine the maximum allowable load P.

13-23 The double-riveted butt joint with unequal-length cover plates, shown in Fig. P13–23, is used to join two 15-mm structural steel plates. The rivets are 20 mm in diameter with 22-mm holes. Using the ultimate strength values from the ASME boiler code and a factor of safety of 3, determine

(a) The maximum load that a 500-mm length of joint can support.

(b) The efficiency of the joint.

13-24 Two 10 by 3/4-in. steel plates are to be connected by a riveted butt joint with two cover plates and 1-in. diameter rivets. The joint is to carry an axial load of 65,000 lb. Using the ultimate strength values from the ASME boiler code and a factor of safety of 5, determine

(a) The minimum allowable number of rivets.

(b) The minimum allowable number of rows of rivets.

10 mm plate   15 mm plate

3 @ 60 = 180 mm

FIG. P13–23

FIG. P13–25

(c) The best distribution of rivets in the rows.

(d) The minimum allowable thickness for the cover plates.

13-25   The cover plates shown in Fig. P13–25 are of equal thickness, the rivets are 25 mm in diameter, and the material is structural steel. Using the ultimate strength values from the MIL-HDBK-5 specifications and a factor of safety of 3, determine

(a) The maximum permissible load on the joint.

(b) The minimum permissible thickness for the cover plates.

13-4   WELDED CONNECTIONS

Connections between metal plates, angles, pipes, and other structural elements are frequently made by welding. Fusion welds are made by melting portions of the materials to be joined with an electric arc, a gas flame, or with Thermit. In fusion welds, additional welding material is usually added to the melted metal to fill the space between the two parts to be joined or to form a fillet. A gas shield is provided when welding certain metals to prevent rapid oxidation of the molten metal. A good welded joint will usually develop the full strength of the material being joined unless the high temperature necessary for the process changes the properties of the materials.

Metals may also be joined by resistance welding in which a small area or spot is heated under high localized pressure. The material is not melted with this type of welding. Other joining methods for metals include brazing and soldering, in which the joining metal is melted but the parts to be connected are not melted. Such connections are usually much weaker than the materials being connected. Fusion welding is the most effective method when high strength is an important factor, and it will be discussed more in detail.

Figure 13–12 illustrates the two general types of fusion-welded joints. The joints illustrated in Figs. 13–12*a*, *b*, and *c* are called *butt* joints and are made with groove welds. The weld may be square for fairly thin materials, or there may be one or two vees for thicker metals. When approved welding techniques and materials are used and complete penetration of the material is obtained, the strength of the joint is assumed to be the same as that of the material being joined.

A fillet weld is illustrated by Fig. 13–12*d* and is used to join a plate, angle, or other structural shape to another structural member. The end weld may be omitted in some applications. Ideally, the fillet should be slightly concave, rather than convex or straight as shown, to reduce possible stress concentrations at reentrant corners.

From an analysis similar to that for the riveted connection (see Art. 13–1), it can be shown that the shearing stress in the side weld cannot be uniformly distributed along the weld as long as the action is elastic, and

Butt joint
Square groove weld
(*a*)

Butt joint in pipe
Single vee groove weld
(*b*)

Butt joint
Double vee groove weld
(*c*)

End weld
Side weld
Fillet weld
(*d*)

FIG. 13–12

FIG. 13–13

tests[10] indicate the distribution is somewhat as indicated in Fig. 13–13. However, if the material is ductile (the usual condition for welded connections), a slight amount of inelastic action at the ends of the welds will result in essentially a uniform shearing stress distribution; therefore, in the design of welded connections, a uniform distribution is assumed. Fillet welds are designed on the assumption that failure will occur by shearing the minimum section of the weld, which is called the throat of the weld (section AB in Fig. 13–14a) whether the weld is a side (longitudinal) weld or an end (transverse) weld. The strength of side fillet welds can be determined from Figs. 13–14a and b. Although it is desirable to make fillet welds slightly concave, calculations are usually based on the assumption that the weld surface is straight and makes an angle of 45° with the two legs. The maximum allowable load P that can be carried by a side fillet weld with a length L and legs t (it is common practice to designate the size of fillet welds by the length of leg t) can be obtained from the free-body diagram in Fig. 13–14b and is

$$P = 0.707tL\tau$$

where τ is the allowable shearing stress permitted in the weld material. Some allowable shearing stresses for steel welds using A233 class E60XX electrodes are as follows: AISC, 1970, 18 ksi (124 MPa); AASHO, 1973, 12.4 ksi (85 MPa); AREA, 1973, 12.4 ksi (85 MPa).

FIG. 13–14

(a)

(b)

[10] "Distribution of Shear in Welded Connections," H. W. Troelsch, *Trans. Am. Soc. of Civil Engrs.*, Vol. 99, 1934, p. 409.

Tests indicate that the strength of a transverse fillet weld is about 30 percent greater than that of a side fillet weld with the same dimensions. It is common practice, however, to assume that the strength of a transverse weld is the same as that of a side weld with the same dimensions. This practice will be followed in this book.

The following example illustrates the analysis of a centrically loaded fillet weld.

EXAMPLE 13–5

An L8 × 6 × $\frac{1}{2}$ steel angle is to be welded to a flat plate with the long side of the angle against the plate as shown in Figs. 13–15a and b.

(a) Determine the minimum lengths L_1 and L_2 that will cause the angle to carry the maximum allowable axial load. The allowable tensile stress for the material in the angle is 18,000 psi, and the allowable shearing stress in the weld material is 13,600 psi. Each leg of the weld is 0.50 in.

(b) Solve part (a) if a transverse weld is placed across the end of the angle.

(a)

(b)

(c)

(d)

FIG. 13–15

SOLUTION

(a) The maximum longitudinal force in the angle will be developed when the tensile stress is uniformly distributed and will act through the centroid of the cross section of the angle. The area of the cross section and the location of its centroid can be obtained from the table of properties of sections of structural shapes in Appendix C. The area is 6.75 sq. in., and the location of the centroid is shown on Fig. 13–15c, which is a free-body diagram of a part of the angle. The forces F_1 and F_2 are assumed to act at the edges of the angle. The force P is

$$P = \sigma A = 18,000(6.75) = 121,500 \text{ lb}$$

The equation $\Sigma M_A = 0$, applied to the free-body diagram, gives

$$(\uparrow+) \qquad \Sigma M_A = 8F_2 - 2.47P = 0$$

from which

$$F_2 = 37,500 \text{ lb as shown}$$

Similarly,

$$F_1 = 84,000 \text{ lb as shown}$$

The shear area of the throat that must carry the load F_2 is $L_2(0.707)(0.50)$, and the shearing stress multiplied by this area must equal the applied load. Thus,

$$F_2 = 37,500 = \tau A = 13,600 L_2(0.707)(0.50)$$

from which

$$L_2 = \underline{7.80 \text{ in.}} \qquad\qquad \text{Ans.}$$

In the same manner

$$L_1 = \underline{17.45 \text{ in.}} \qquad\qquad \text{Ans.}$$

(b) When a weld is placed across the end of the angle, an additional force must be shown on the free-body diagram, as in Fig. 13–15d. The force F'_3 is

$$F'_3 = \tau A = 13,600(8)(0.707)(0.50) = 38,500 \text{ lb.}$$

Note that the transverse weld is assumed to carry the same load per inch of weld as the side weld. The equation $\Sigma M_A = 0$ applied to Fig. 13–15d gives

$$\overset{\curvearrowleft}{(+)} \qquad \Sigma M_A = 8F_2' + 4(38,500) - 2.47P = 0$$

from which

$$F_2' = 18,250 \text{ lb} = 13,600L_2'(0.707)(0.50)$$

as in part (a) of the solution. This equation gives

$$L_2' = \underline{3.80 \text{ in.}} \qquad\qquad \text{Ans.}$$

Similarly,

$$L_1' = \underline{13.45 \text{ in.}} \qquad\qquad \text{Ans.}$$

PROBLEMS

13-26* Design a fillet-welded joint, for attaching an L4 × 4 × 3/8 steel angle (see Appendix C) to a gusset plate, that will develop the full working tensile strength of the angle. Use a tensile working stress of 15,000 psi in the angle and a maximum allowable force per linear inch of weld of 3000 lb. Weld along the end of the angle in order to shorten the gusset plate.

13-27* Design a fillet-welded joint, for attaching an L127 × 127 × 19.1 steel angle (see Appendix C) to a gusset plate, that will develop the full working tensile strength of the angle. Use a tensile working stress of 100 MPa in the angle and a maximum allowable force per linear millimeter of weld of 1000 N. Weld along the end of the angle in order to shorten the gusset plate.

13-28* Design a fillet-welded joint, for attaching an L6 × 4 × 3/8 steel angle (see Appendix C) to a gusset plate with the 6-in. side against the plate, that will develop the full tensile working strength of the angle. Use a tensile working stress of 24 ksi in the angle and an allowable shearing stress in the 5/16-in. weld of 13.6 ksi. Weld along the end of the angle in order to shorten the gusset plate.

13-29 Design a fillet-welded joint, for attaching an L127 × 76 × 12.7 steel angle (see Appendix C) to a gusset plate with the 127-mm side against the plate, that will develop the full tensile working strength of the angle. Use a tensile working stress of 165 MPa in the angle and an allowable shearing stress in the 11-mm weld of 85 MPa. Weld along the end of the angle in order to shorten the gusset plate.

13-30 A cylindrical pressure tank is to be constructed by welding 3/8-in. thick plates on the ends of a 6-ft length of steel pipe. The pipe has an outside

FIG. P13–31

diameter of 14 in. and a thickness of 0.20 in. Determine the size of fillet weld needed to develop the full tensile working strength of the pipe. The working stress in the pipe is 20,000 psi, and the allowable force per inch of weld is 10,600t where t is the size of the weld.

13-31*   An improvised tension member is to be made by splicing two lengths of L152 × 102 × 12.7 steel angle (see Appendix C). Since the angles are too short to permit groove welding, a joint composed of plate A and bar B fillet welded to the angles as shown in Fig. P13–31 is proposed. The working stress for the angles is 125 MPa, and the maximum allowable load on each millimeter of weld is 700 N. Determine the length of each weld a, b, and c required to develop the full axial tensile capacity of the angles with no bending about any axis in the joint.

13-32   The flanges of a steel cantilever beam AB of Fig. P13–32 are to be connected to the column with 1 × 14-in. butt welds. The web connection is to bc 3/8 × 2-in. intermittent fillet welds on both sides of the web. Recall that the 80-kip load can be replaced with a vertical force at A and a couple. Assume that the couple is resisted only by the flange welds and that the vertical force is resisted only by the web welds. Determine

FIG. P13–32

(a) The average tensile stress in the flange weld (on the vertical plane).

(b) The number of 3/8 × 2-in. welds to be used on each side of the web (use AISC specifications).

13-33 Prove that the use of a transverse end weld has the effect of shortening each of the side welds by equal amounts when a structural angle is welded to a plate in such a manner that the maximum tensile strength of the angle is developed.

13-5 CONNECTIONS UNDER TORSIONAL LOADING

Power-transmitting shafts are frequently connected by flange couplings (see, for example, Fig. 5–2), and the design of the coupling is, of course, important. The flange coupling is not the only device called on to transmit a torque. Many structural connections are subjected to torsional loading, often combined with centric loading. The latter case is taken up in the next article. This article is limited to pure torsion.

When a torque is applied to the plate C of Fig. 13–16a that is riveted to another structural element, the plate will rotate (a slight amount for elastic action) about some point such as O in the figure. If the assumption is made that any straight line in the plate (such as OB) before the torque is applied remains straight after application of the torque, the deformations of the rivets, and hence the average cross shearing strain in each rivet, will be proportional to the distance from O. If, in addition, stress and strain are proportional, the rivet cross shearing stress will be proportional to the distance from O. This means that for rivets (or bolts) of the same diameter, the force on each rivet is proportional to the distance from O and is in the direction of the strain, that is, perpendicular to the radius from O.

The equations of equilibrium can be used to locate the center of rotation O. The resultant of the resisting forces F_1, F_2, $\cdots$ in Fig. 13–16b is a couple; therefore, $\sum_{i=1}^{n} F_i$ is equal to zero in any direction, where F_i is one of the n resisting forces. The summation of forces in the y direction gives

FIG. 13-16

(a)   (b)

$$\Sigma F_y = F_1 \cos \theta_1 + F_2 \cos \theta_2 + \cdots = \sum_{i=1}^{n} F_i \cos \theta_i = 0$$

Also,

$$F_i = \tau_i A_i = (\tau_i/\rho_i)(\rho_i A_i) = k\rho_i A_i$$

because the cross shearing stress is proportional to the radius ρ. The summation thus becomes

$$\Sigma F_y = k \sum_{i=1}^{n} \rho_i \cos \theta_i A_i = k \sum_{i=1}^{n} x_i A_i = k\bar{x}A_t = 0$$

where $\bar{x}$ is the x coordinate of the centroid of the rivet areas and A_t is the sum of the rivet areas.

This expression proves that $\bar{x}$ is zero because neither k nor A_t can be zero. The summation of forces in the x direction will indicate that $\bar{y}$ must also be zero, which means that the center of rotation of the plate and the centroid of the rivet areas coincide. The following is an example of a riveted joint loaded in pure torsion.

EXAMPLE 13-6

The plates and rivets of Fig. 13-16a are of structural steel, the rivets are all 12 mm in diameter, and the applied torque is 2.7 kN · m. Determine the magnitude of the maximum cross shearing stress in the rivets.

SOLUTION

The free-body diagram of Fig. 13-16b will be used for this problem. The centroid of the rivet areas, which is the same as the center of rotation, is located by applying the principle of moments of areas with respect to the x' and y' axes. The coordinates of the centroid are

$$\bar{x}' = 50 \text{ mm} \quad \text{and} \quad \bar{y}' = 50 \text{ mm}$$

The radial distances from 0 are

$$\rho_1 = 79.1 \text{ mm}, \rho_2 = 90.1 \text{ mm}, \rho_3 = 50.0 \text{ mm}, \rho_4 = 70.7 \text{ mm}, \rho_5 = 90.1 \text{ mm}$$

The summation of moments of forces with respect to the O axis gives

$$\circlearrowleft+ \quad \Sigma M_0 = 0.0791 F_1 + 0.0901 F_2 + 0.0500 F_3$$
$$+ 0.0707 F_4 + 0.0901 F_5 - 2700 = 0$$

Since the rivet areas are all equal, the forces on the rivets are proportional to the radial distances from O; thus,

$$\frac{F_1}{0.0791} = \frac{F_2}{0.0901} = \cdots$$

The maximum forces are F_2 and F_5; if the forces are written in terms of F_2, they become

$$F_1 = \frac{0.0791 F_2}{0.0901}, \quad F_3 = \frac{0.0500 F_2}{0.0901}, \quad F_4 = \frac{0.0707 F_2}{0.0901}, \quad F_5 = F_2$$

When these values are substituted in the moment equation, it becomes

$$0.0791 \left(\frac{0.0791 F_2}{0.0901}\right) + 0.0901 F_2 + 0.0500 \left(\frac{0.0500 F_2}{0.0901}\right)$$
$$+ 0.0707 \left(\frac{0.0707 F_2}{0.0901}\right) + 0.0901(F_2) = 2700$$

from which $F_2 = 8110$ N. The maximum cross section shearing stress is

$$\tau_{max} = \frac{F_2}{A} = \frac{8110}{(\pi/4)(0.012)^2} = 71.7(10^6) \text{ N/m}^2 = \underline{71.7 \text{ MPa}} \qquad \text{Ans.}$$

PROBLEMS

13-34* A hollow circular shaft is subjected to a torque of 10 kN · m. The inside diameter is 75 mm and the outside diameter is 100 mm. Determine

   (a) The maximum shearing stress in the shaft.

   (b) The number of 20-mm bolts needed for a coupling on a bolt circle of 140-mm diameter for a cross shearing stress in the bolts of 70 MPa.

13-35* A shaft coupling has six 1/2-in. diameter bolts on a 4-in. diameter bolt circle. Determine the maximum horsepower the coupling can transmit at 300 rpm if the cross shearing stress in the bolts is limited to 8 ksi.

13-36* Six bolts, placed as shown in Fig. P13–36, are used to transmit a torque of 55 kN · m from one plate to another. Determine

    (a)   The magnitude of the force on bolt A.

    (b)   The magnitude of the force on bolt B.

13-37 The riveted structural connection in Fig. P13–37 is subjected to a torque as shown. If all rivets are the same size, determine

    (a)   Which rivet transmits the largest force.

    (b)   The magnitude of the force of part (a).

13-38 A riveted structural connection is subjected to the torque shown in Fig. P13–38. If all rivets are the same size, determine

    (a)   Which rivet transmits the largest force.

    (b)   The magnitude of the force of part (a).

13-39* Two plates are riveted together as shown in Fig. P13–39. Determine the cross shearing stress in each of the nine 1/2-in. diameter rivets.

13-40 A motor supplies the torque Q to the driving pulley B, which is mounted on the bronze ($G = 41$ GPa) shaft of Fig. P13–40. Pulleys A and C are for power takeoff with 150 kW being taken off at C and the remainder at A. The allowable shearing stress in the hollow steel ($G = 80$ GPa) segment AB is 100 MPa. The angle of twist at A with respect to C is not to exceed 0.01 rad counterclockwise looking from A toward C. The bronze and steel seg-

FIG. P13–36

FIG. P13–37

FIG. P13–38

FIG. P13–39

FIG. P13–40

ments are rigidly connected by a rigid flange coupling with a bolt circle of 200-mm diameter. When the shaft is rotating at 200 rpm, determine

(a)   The maximum permissible power that can be supplied to pulley B.

(b)   The number of 18-mm diameter bolts required for the flange coupling with an allowable cross shearing stress of 70 MPa in the bolts.

13-41*   A hollow steel ($G = 12,000$ ksi) shaft has diameters of 6 in. and 4 in. and an allowable shearing stress of 9000 psi. Determine

(a)   The horsepower transmitted at 25 rad per sec.

(b)   The magnitude of the angle of twist in a length of 10 ft.

(c)   The required number of 3/4-in. diameter bolts for a flange coupling in which two-thirds of the bolts are on a 12-in. diameter bolt circle and one-third are on a 9-in. diameter circle. The allowable cross shearing stress in the bolts is 12,000 psi.

13-6   ECCENTRIC LOAD ON A RIVETED OR BOLTED JOINT

A common and important combination of torsional and centric loads occurs when a riveted or bolted joint is subjected to an eccentric load. The presentation in Arts. 13–1 through 13–3 was limited to centric loads on riveted or bolted joints. The present discussion will be limited to eccentric loads in the plane of the shear areas. In Art. 13–5 bolted or riveted connections subjected to torsional loads were analyzed. The proof that a riveted joint rotates or tends to rotate about the centroid of the rivet areas when the joint is subjected to a couple was given in Art. 13–5.

Any load on a riveted or bolted joint that does not act through the centroid of the rivet areas is defined as an eccentric load. Fortunately, any eccentric load can be replaced by a centric load and a couple in the plane of the shear areas. Furthermore, the force acting on any given shear area can be readily obtained by superposition of the component forces due to the torsional and centric loads. The following example illustrates the procedure.

EXAMPLE 13–7

Determine the cross shearing stress in the rivet of the joint of Fig. 13-17a carrying the greatest load. The diameter of each rivet is $\frac{3}{4}$ in.

SOLUTION

By symmetry the x coordinate of the centroid of the rivet areas is $\bar{x} = 5.00$ in. By the principal of moments the y coordinate of the centroid of the rivet areas is

FIG. 13-17

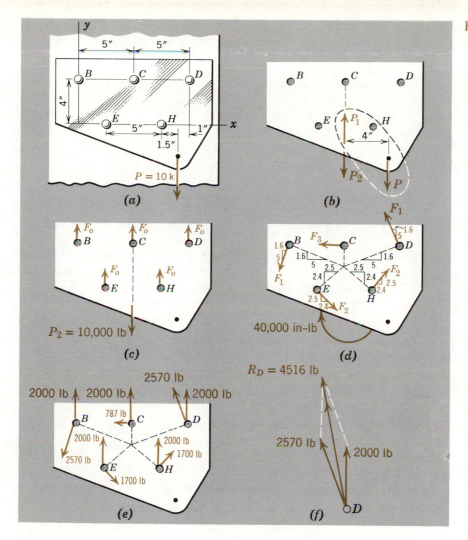

$$\bar{y} = \frac{\Sigma M_x}{\Sigma A} = \frac{(3A)4}{5A} = 2.40 \text{ in.}$$

where A is the cross-sectional area of one rivet.

Two equal opposite centric loads P (labeled P_1 and P_2) are introduced as shown in Fig. 13–17b. By this means the eccentric load P is replaced by a centric load P_2 of 10,000 lb and a clockwise couple $P_1 P$ of 40,000 in.-lb, as shown in Fig. 13–17b. An upward resisting force of F_O, equal to 2000 lb, is required on each shear area, as shown in Fig. 13–17c, to hold the centric load P_2 in equilibrium. The five resisting forces shown in Fig. 13–17d are required to hold the 40,000-in.-lb couple PP_1 in equilibrium. As shown in Art. 13–5,

$$\frac{F_1}{\sqrt{27.56}} = \frac{F_2}{\sqrt{12.01}} = \frac{F_3}{1.6}$$

and

$$2F_1\sqrt{27.56} + 2F_2\sqrt{12.01} + 1.6F_3 = 40,000$$

The simultaneous solution of these equations gives

$$F_1 = 2570 \text{ lb} \qquad F_2 = 1700 \text{ lb} \qquad \text{and} \qquad F_3 = 787 \text{ lb, all as shown}$$

The component forces acting on each rivet as shown in Fig. 13–7e reveal that rivet D is carrying the greatest load. The two resisting forces on rivet D are added vectorally to give the resultant resisting force R_D of 4516 lb, as shown in Fig. 13–17f. The cross shearing stress on rivet D is

$$\tau = \frac{4516}{\pi(3/8)^2} = \underline{10,240 \text{ psi}} \qquad\qquad \text{Ans.}$$

PROBLEMS

13-42*   A gusset plate carrying a load of 15 kN is held by three 14-mm diameter bolts, as shown in Fig. P13–42. Determine

    (a)   Which bolt carries the greatest load.

    (b)   The cross shearing stress in the bolt of part (a).

FIG. P13–42

FIG. P13–43

FIG. P13–44

13-43*   Determine the cross shearing stress in bolt *A* of the bolted connection shown in Fig. P13–43. All bolts are 1 in. in diameter.

13-44*   The diameter of each rivet in the joint shown in Fig. P13–44 is 25 mm. Determine the cross shearing stresses in rivets *A* and *B*.

13-45   The bolts in the joint shown in Fig. P13–45 each have a diameter of 5/8 in. Determine the maximum cross shearing stress in any of the bolts.

13-46   Each of the rivets of the joint shown in Fig. P13–46 has a diameter of 26 mm. Determine the cross shearing stress in the rivet carrying the greatest load.

FIG. P13–45

FIG. P13–46

FIG. P13–47

FIG. P13–48

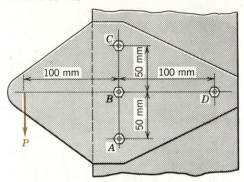

13-47*   The diameter of each rivet in the joint shown in Fig. P13–47 is 3/4 in. Determine the cross shearing stresses in the top and bottom rivets.

13-48*   The bolts in the joint shown in Fig. P13–48 are 12 mm in diameter. If the allowable cross shearing stress in the bolts is 70 MPa, determine the maximum permissible value for the load.

13-49   A load P is applied to the plate riveted to a support as shown in Fig. P13–49. Determine the maximum allowable value of P if the maximum cross shearing stress in the rivets is 12,000 psi. All rivets are 1 in. in diameter.

13-50   The bolted joint shown in Fig. P13–50 is to be designed to support the 2400-lb eccentric load. The diameter of the bolts is 1/2 in. and the allowable cross shearing stress in the bolts is 8000 psi. Determine the minimum permissible spacing a.

FIG. P13–49

FIG. P13–50

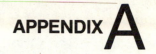

APPENDIX A

1 AVERAGE PROPER-TIES OF SELECTED MATERIALS
2 STRESS-STRAIN DIAGRAMS FOR TYPICAL METALS

TABLE A–1a Average Properties of Selected Engineering Materials (English System of Units)

Exact values may vary widely with changes in composition, heat treatment, and mechanical working. More precise information can be obtained from manufacturers.

Materials	Specific Weight (lb/in.³)	Elastic Strength[a]			Ultimate Strength			Endurance Limit[c] (ksi)	Modulus of Elasticity (1000 ksi)	Modulus of Rigidity (1000 ksi)	Percent Elongation in 2 in.	Coefficient of Thermal Expansion (10⁻⁶/°F)
		Tension (ksi)	Comp. (ksi)	Shear (ksi)	Tension (ksi)	Comp. (ksi)	Shear (ksi)					
Ferrous metals												
Wrought iron	0.278	30	b		48	b	25	23	28		30[d]	6.7
Structural steel	0.284	36	b		66	b		28	29	11.0	28[d]	6.6
Steel, 0.2% C hardened	0.284	62	b		90	b			30	11.6	22	6.6
Steel, 0.4% C hot-rolled	0.284	53	b		84	b		38	30	11.6	29	
Steel, 0.8% C hot-rolled	0.284	76	b		122	b			30	11.6	8	
Cast iron—gray	0.260	25			25	100		12	15		0.5	6.7
Cast iron—malleable	0.266	32	b		50	b			25		20	6.6
Cast iron—nodular	0.266	70			100				25		4	6.6
Stainless steel (18-8) annealed	0.286	36	b		85	b		40	28	12.5	55	9.6
Stainless steel (18-8) cold-rolled	0.286	165	b		190	b		90	28	12.5	8	9.6
Steel, SAE 4340, heat-treated	0.283	132	145		150	b	95	76	29	11.0	19	
Nonferrous metal alloys												
Aluminum, cast, 195-T6	0.100	24	25		36	b	30	7	10.3	3.8	5	
Aluminum, wrought, 2014-T4	0.101	41	41	24	62	b	38	18	10.6	4.0	20	12.5
Aluminum, wrought, 2024-T4	0.100	48	48	28	68	b	41	18	10.6	4.0	19	12.5
Aluminum, wrought, 6061-T6	0.098	40	40	26	45	b	30	13.5	10.0	3.8	17	12.5

Material											
Magnesium, extrusion, AZ80X	0.066	35	26	49	b	21	19	6.5	2.4	12	14.4
Magnesium, sand cast, AZ63-HT[e]	0.066	14	14	40	b	19	14	6.5	2.4	12	14.4
Monel, wrought, hot-rolled	0.319	50	b	90	b	40		26	9.5	35	7.8
Red brass, cold-rolled	0.316	60		75				15	5.6	4	9.8
Red brass, annealed	0.316	15	b	40	b			15	5.6	50	9.8
Bronze, cold-rolled	0.320	75		100				15	6.5	3	9.4
Bronze, annealed	0.320	20	b	50	b			15	6.5	50	9.4
Titanium alloy, annealed	0.167	135	b	155	b			14	5.3	13	9.4
Invar, annealed	0.292	42	b	70	b			21	8.1	41	0.6
Nonmetallic materials											
Douglas fir, green[e]	0.022	4.8	3.4		3.9	0.9		1.6			
Douglas fir, air dry[e]	0.020	8.1	6.4		7.4	1.1		1.9			
Red oak, green[e]	0.037	4.4	2.6		3.5	1.2		1.4			
Red oak, air dry[e]	0.025	8.4	4.6		6.9	1.8		1.8			1.9
Concrete, medium strength	0.087		1.2		3.0			3.0			6.0
Concrete, fairly high strength	0.087		2.0		5.0			4.5			6.0

[a] Elastic strength may be represented by proportional limit, yield point, or yield strength at a specified offset (usually 0.2 percent for ductile metals).

[b] For ductile metals (those with an appreciable ultimate elongation), it is customary to assume the properties in compression have the same values as those in tension.

[c] Rotating beam.

[d] Elongation in 8 in.

[e] All timber properties are parallel to the grain.

TABLE A–1b Average Properties of Selected Engineering Materials (International System of Units) Exact values may vary widely with changes in composition, heat treatment, and mechanical working. More precise information can be obtained from manufacturers.

Materials	Density (Mg/m³)	Elastic Strength[a]			Ultimate Strength			Endurance Limit[c] (MPa)	Modulus of Elasticity (GPa)	Modulus of Rigidity (GPa)	Percent Elongation in 50 mm	Coefficient of Thermal Expansion (10⁻⁶/°C)
		Tension (MPa)	Comp. (MPa)	Shear (MPa)	Tension (MPa)	Comp. (MPa)	Shear (MPa)					
Ferrous metals												
Wrought iron	7.70	210	b		330	b	170	160	190		30[d]	12.1
Structural steel	7.87	250	b		450	b		190	200	76	28[d]	11.9
Steel, 0.2% C hardened	7.87	430	b		620	b			210	80	22	11.9
Steel, 0.4% C hot-rolled	7.87	360	b		580	b		260	210	80	29	
Steel, 0.8% C hot-rolled	7.87	520	b		840	b			210	80	8	
Cast iron—gray	7.20				170	690		80	100		0.5	12.1
Cast iron—malleable	7.37	220	b		340	b			170		20	11.9
Cast iron—nodular	7.37	480			690				170		4	11.9
Stainless steel (18-8) annealed	7.92	250	b		590	b		270	190	86	55	17.3
Stainless steel (18-8) cold-rolled	7.92	1140	b		1310	b		620	190	86	8	17.3
Steel, SAE 4340, heat-treated	7.84	910	1000		1030	b	650	520	200	76	19	
Nonferrous metal alloys												
Aluminum, cast 195-T6	2.77	160	170		250	b	210	50	71	26	5	
Aluminum, wrought, 2014-T4	2.80	280	280	160	430	b	260	120	73	28	20	22.5
Aluminum, wrought, 2024-T4	2.77	330	330	190	470	b	280	120	73	28	19	22.5
Aluminum, wrought, 6061-T6	2.71	270	270	180	310	b	210	93	70	26	17	22.5

Material										
Magnesium, extrusion, AZ80X	1.83	240	180	340	b	140	130	45	16	12	25.9
Magnesium, sand cast, AZ63-HT	1.83	100	96	270	b	130	100	45	16	12	25.9
Monel, wrought, hot-rolled	8.84	340	b	620	b		270	180	65	35	14.0
Red brass, cold-rolled	8.75	410	b	520	b			100	39	4	17.6
Red brass, annealed	8.75	100	b	270	b			100	39	50	17.6
Bronze, cold-rolled	8.86	520	b	690				100	45	3	16.9
Bronze, annealed	8.86	140	b	340	b			100	45	50	16.9
Titanium alloy, annealed	4.63	930	b	1070	b			96	36	13	
Invar, annealed	8.09	290	b	480	b			140	56	41	1.1
Nonmetallic materials										
Douglas fir, green[e]	0.61	33	23	27		6.2		11			
Douglas fir, air dry[e]	0.55	56	44	51		7.6		13			
Red oak, green[e]	1.02	30	18	24		8.3		10			3.4
Red oak, air dry[e]	0.69	58	32	48		12.4		12			
Concrete, medium strength	2.41	8		21				21			
Concrete, fairly high strength	2.41	14		34				31			

[a] Elastic strength may be represented by proportional limit, yield point, or yield strength at a specified offset (usually 0.2 percent for ductile metals).

[b] For ductile metals (those with an appreciable ultimate elongation), it is customary to assume the properties in compression have the same values as those in tension.

[c] Rotating beam.

[d] Elongation in 200 mm.

[e] All timber properties are parallel to the grain.

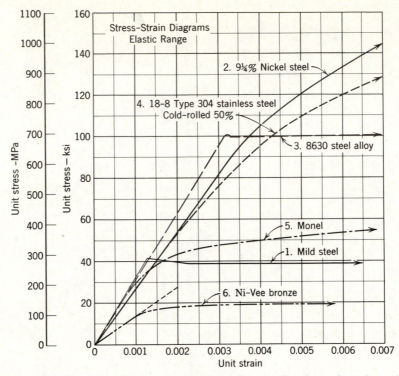

FIG. A–1 Stress-strain diagrams in the elastic range for a selection of engineering materials. (*Courtesy of the International Nickel Company.*)

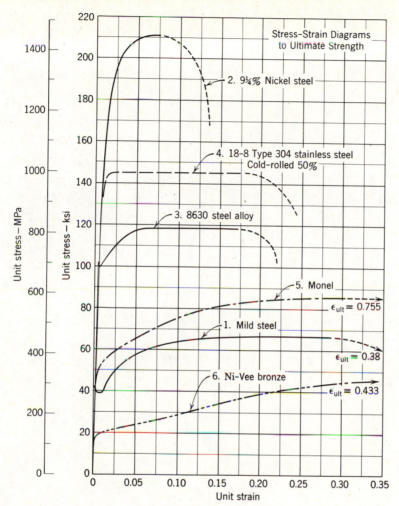

Stress-strain diagrams to ultimate strength for a selection of engineering FIG. A–2 materials. (*Courtesy of the International Nickel Company.*)

1 PROPERTIES OF SELECTED AREAS
2 SECOND MOMENT OF PLANE AREAS

Rectangle	Segment of 2nd-Degree Parabola
$A = bh$ $I_x = bh^3/12$ $I_y = b^3h/12$	$y = kx^2 = (h/b^2)x^2$ $A = 2bh/3$ $\bar{x} = 3b/8$ $\bar{y} = 3h/5$
Circle	Spandrel of 2nd-Degree Parabola
$A = \pi R^2 = \pi d^2/4$ $I_x = I_y = \pi R^4/4$ $= \pi d^4/64$ $J_0 = \pi R^4/2 = \pi d^4/32$	$y = kx^2 = (h/b^2)x^2$ $A = bh/3$ $\bar{x} = 3b/4$ $\bar{y} = 3h/10$
Semicircle	Spandrel of 3rd-Degree Parabola
$A = \pi R^2/2$ $I_x = \pi R^4/8$ $I_y = \pi R^4\left(\frac{1}{8} - \frac{8}{9\pi^2}\right)$	$y = kx^3 = (h/b^3)x^3$ $A = bh/4$ $\bar{x} = 4b/5$ $\bar{y} = 2h/7$
Right Triangle	Spandrel of nth-Degree Parabola
$A = bh/2$ $I_x = bh^3/36$ $I_y = hb^3/36$	$y = kx^n = (h/b^n)x^n$ $A = \dfrac{bh}{n+1}$ $\bar{x} = \dfrac{n+1}{n+2}b$ $\bar{y} = \dfrac{n+1}{2(2n+1)}h$

TABLE B–1   Properties of Selected Areas

$A_{TOT}\bar{Y}_{TOT} = A_1\bar{Y}_1 + A_2\bar{Y}_2$

$A_{TOT} = 20 in^2$

$A_1 = 8 in^2$

$A_2 = 12 in^2$

$Y_1 = 8 + 1 = 9''$

$Y_2 = 4''$

$\bar{Y} = \dfrac{4(12) + 8(9)}{20} = 6''$

$I_{NA} = I'_{x_1} + A_1 d_1^2 + I'_{x_2} + A_2 d_2^2 + I'_{x_3} + A_3 d_3^2$

$I_{NA} = \frac{1}{12}(16)(6^3) + (1.5)(6)(3^2) + \frac{1}{12}(1.5)(2^3) + (1.5)(2)(1^2) + \frac{1}{12}(4)(2^3) + (2)(4)(3^2)$

$M_E = \frac{\sigma_y}{c} I_{NA}$

— M_P - NEED A NEW $\mathcal{Q}$. USE $\bar{F}_T = F_C \Rightarrow A_T \sigma_y = A_C \sigma_y$

— FULLY PLASTIC MOMENT $F_t = \left(\frac{36000 + 0}{2}\right)(1.5)(a) + (36000)(1.5)(10-2a)$

$F_C = F_C$

$F_C = \left(\frac{36000 + 0}{2}\right)(4)(a) - \left(\frac{36000}{a}\right)\left(\frac{1}{2}\right)(2.5)(a-2)$

$M_{FULL} = M_T + M_C$

$M_t = F_{t_1}(\text{MOMENT ARM}) + F_{t_2}(\text{MOMENT ARM})$

$M_C = F_{C_1}(MA_1) - F_{C_2}(MA_2)$

SECOND MOMENT OF PLANE AREAS

B–1   Definitions   The second moment of the area in Fig. B–1 with respect to the x axis is defined as the integral $\int_A y^2\, dA$ where the integration is carried out over the area A. The quantity is called the second moment of the area to distinguish it from the first moment about the axis, which is defined as $\int_A y\, dA$. The second moment is frequently referred to as *moment of inertia* of the area because of the similarity of the defining equation to that used to express the moment of inertia of the mass of a body and the terms *second moment* and *moment of inertia* are used interchangeably.

From the definition in the preceding paragraph, it may be noted that a similar equation can be written for the second moment of the area with respect to any axis in the plane of the area. The symbol I is used for the second moment with respect to any axis in the plane of the area and is sometimes called the rectangular moment of inertia; thus,

$$I_x = \int_A y^2\, dA \qquad I_y = \int_A x^2\, dA \qquad\qquad (B-1)$$

When the reference axis is normal to the plane of the area, for example through O in Fig. B–1, the integral is called the polar second moment or polar moment of inertia J_O and can be written as

$$J_O = \int_A r^2\, dA = \int_A (x^2 + y^2)\, dA = \int_A x^2\, dA + \int_A y^2\, dA$$

Thus,

$$J_O = I_y + I_x \qquad\qquad (B-2)$$

Note that the x and y axes can be any two mutually perpendicular axes intersecting at O.

From the definition, the second moment of an area is the product of a distance squared multiplied by an area and thus is always positive and has the dimension of a length raised to the fourth power (L^4). Common units are mm^4 and $in.^4$.

FIG. B–1

B–2 **Parallel Axis Theorem**   When the second moment of an area has been determined with respect to a given axis, the second moment with respect to a parallel axis can be obtained by means of the *parallel axis theorem* (sometimes called the transfer formula), provided one of the axes passes through the centroid of the area.

The second moment of the area in Fig. B–2 about the b axis is

$$I_b = \int_A (y + d)^2 \, dA = \int_A y^2 \, dA + 2d \int_A y \, dA + d^2 \int_A dA$$

$$= I_x + 2d \int_A y \, dA + Ad^2 \qquad\qquad (a)$$

The integral $\int_A y \, dA$ is the first moment of the area with respect to the x axis. If the x axis passes through the centroid of the area, the first moment is zero and Eq. a becomes

$$I_b = I_c + Ad^2 \qquad\qquad (B\text{–}3)$$

where I_c is the second moment of the area with respect to an axis parallel to b and passing through the centroid C, and d is the distance between the two axes. In a similar manner it can be shown that

$$J_B = J_c + Ad_1^2 \qquad\qquad (B\text{–}4)$$

The parallel-axis theorem states that the *second moment of an area with respect to any axis is equal to the second moment of the area with respect to a parallel axis through the centroid of the area added to the product of the area and the square of the distance between the two axes.* The theorem demonstrates that the second moment of an area with respect to an axis through the centroid of the area is less than that for any parallel axis. Thus, solving Eq. B–3 for I_c gives

$$I_c = I_b - Ad^2$$

FIG. B–2

B–3   Second Moments of Areas by Integration   When determining the second moment of an area by integration, it is first necessary to select an element of area. If all parts of the element are the same distance from the axis, the second moment can be determined directly from the definition (see Art. B–1). When any other element is selected, the second moment of the *element* with respect to the axis must either be known or be obtainable from a known result by the parallel-axis theorem.

Either single or double integration may be involved, depending on the element selected. When double integration is used, all parts of the element will be the same distance from the moment axis, and the second moment of the element can be written directly. Special care must be taken in establishing the limits for the two integrations to see that the correct area is included. If a strip element is selected, the second moment can usually be obtained by a single integration, but the element must be properly selected in order for its second moment about the reference axis to be either known or readily calculated. Either Cartesian or polar coordinates can be used for some problems. The choice of a coordinate system and the selection of an element is a matter of either personal preference or previous experience. The following examples illustrate the procedure for determining the second moments of areas by integration.

EXAMPLE B–1
A rectangle has a base *b* and a height *h*. Determine the second moment of the area with respect to

(a)   The base of the rectangle.
(b)   An axis through the centroid parallel to the base.
(c)   An axis through the centroid normal to the area (polar second moment).

SOLUTION
(a)   The rectangular area is shown in Fig. B–3, together with an element of area parallel to the reference axis. The second moment of the area of the element about the *x* area is

FIG. B–3

$$dI_x = y^2 \, dA = y^2 b \, dy$$

and the second moment of the entire area is

$$I_x = b \int_0^h y^2 \, dy = \frac{bh^3}{3} \qquad \text{Ans.}$$

(b) The parallel-axis theorem can be used to determine the second moment about an axis through c parallel to the x axis. Thus,

$$I_{xc} = I_x - Ad^2 = \frac{bh^3}{3} - (bh)\left(\frac{h}{2}\right)^2 = \frac{bh^3}{12} \qquad \text{Ans.}$$

(c) The second moment with respect to the vertical axis through c can be obtained from the preceding solution by interchanging b and h; that is,

$$I_{yc} = \frac{hb^3}{12}$$

and the polar second moment of the area about an axis at c is

$$J_c = I_{xc} + I_{yc} = \frac{bh^3}{12} + \frac{hb^3}{12} = \frac{bh}{12}\,(h^2 + b^2) \qquad \text{Ans.}$$

EXAMPLE B–2
Determine the second moment of a circular area with a radius R with respect to a diameter of the circle.

SOLUTION
Figure B–4 shows a circular area with a radius R. The x axis is the moment

FIG. B–4

axis. Polar coordinates are convenient for this problem. The second moment of the element about the x axis is

$$dI_x = (r \sin \theta)^2 \, dr(r \, d\theta) = r^3 \sin^2\theta \, dr \, d\theta$$

The second moment of the entire area is

$$I_x = \int_0^{2\pi} \int_0^R r^3 \sin^2\theta \, dr \, d\theta = \int_0^{2\pi} \frac{R^4}{4} \sin^2\theta \, d\theta$$

$$= \frac{R^4}{4} \int_0^{2\pi} \frac{1 - \cos 2\theta}{2} \, d\theta = \frac{R^4}{4} \left[\frac{\theta}{2} - \frac{\sin 2\theta}{4} \right]_0^{2\pi} = \frac{\pi R^4}{4} \qquad \text{Ans.}$$

EXAMPLE B–3

The equation of the parabolic curve in Fig. B–5 is $cy = x^2$ where c is a constant. Determine the second moment of the shaded area with respect to the x and y axes.

SOLUTION

The constant in the equation can be obtained by substituting $x = b$ and $y = h$ in the equation, which gives $c = b^2/h$ and the equation becomes

$$b^2y = hx^2$$

A horizontal element (Fig. B–5) is selected for this example. The area of the element is

$$dA = x \, dy = bh^{-1/2}y^{1/2}dy$$

FIG. B–5

and the second moment of the element about the x axis is

$$dI_x = y^2 \, dA = bh^{-1/2}y^{5/2}dy$$

The second moment of the area is

$$I_x = bh^{-1/2} \int_0^h y^{5/2}dy = bh^{-1/2} \left[\frac{2}{7} y^{7/2} \right]_0^h = \frac{2}{7} bh^3 \qquad \text{Ans.}$$

Since all parts of the element are *not* the same distance from the y axis, the horizontal element can be used if its second moment with respect to the y axis is known. The element is a rectangle of height x and width dy and the reference axis is the base of the rectangle. From Exam. B–1, the second moment of the element about the y axis is

$$dI_y = \frac{1}{3} (dy)x^3 = \frac{1}{3} \left(\sqrt{\frac{b^2y}{h}} \right)^3 dy = \frac{b^3y^{3/2}}{3h^{3/2}} \, dy$$

and the second moment of the area is

$$I_y = \frac{b^3}{3h^{3/2}} \int_0^h y^{3/2}dy = \frac{2}{15} b^3h \qquad \text{Ans.}$$

EXAMPLE B–4
Determine the second moment of the area in Fig. B–6 with respect to the x axis.

SOLUTION
An element of area parallel to the x axis is selected as shown in Fig. B–6. The area of the element is $dA = x \, dy$ where

$$x = a + \frac{b - a}{h} y$$

FIG. B–6

and the second moment of the element with respect to the x axis is

$$dI_x = y^2\, dA = \left(ay^2 + \frac{b-a}{h}\, y^3\right) dy$$

The second moment of the area about the x axis is

$$I_x = \int_0^h \left(ay^2 + \frac{b-a}{h}\, y^3\right) dy = \underline{\frac{(a+3b)h^3}{12}} \qquad \text{Ans.}$$

B–4   Radius of Gyration of Areas   It is sometimes convenient to express the second moment of area (moment of inertia) as a function of the area and a length. Since the second moment of an area has the dimensions of length to the fourth power, it can be expressed as the product of the area multiplied by a length squared. *The radius of gyration of an area with respect to a specified axis is defined as the length that when squared and multiplied by the area, will give the second moment of the area with respect to the specified axis.* The definition can be expressed by the equations

$$Ak_b^2 = I_b \qquad \text{and} \qquad Ak_O^2 = J_O \qquad (B–5)$$

where k_b and k_O are the radii of gyration with respect to the b axis and the polar axis through O, respectively. The radius of gyration is *not* the distance from the reference axis to a fixed point in the area (such as the centroid), but it is a useful property of the area and the specified axis.

The radius of gyration of an area with respect to any axis is always greater than the distance from the axis to the centroid of the area. From Fig. B–7, the moment of inertia of the area with respect to the b axis is

$$I_b = I_c + Ad^2 = Ak_b^2 = Ak_c^2 + Ad^2$$

or

$$k_b^2 = k_c^2 + d^2$$

from which

$$k_b = (k_c^2 + d^2)^{1/2}$$

Area $= A$

FIG. B–7

This equation indicates that k_b will always be greater than the distance d from the b axis to the centroid. If the distance d is increased, the value of k_b will also be increased, but the increase in k_b will be less than the increase in d, indicating that k_b is *not* the distance from the b axis to a fixed point in the area.

The radius of gyration of an area is a convenient means of expressing a property of an area and is often useful in formulas developed in mechanics—for example, formulas giving the strength of columns.

B-5   Second Moments of Composite Areas   It is often necessary to calculate the second moment of an irregularly shaped area. When such an area is made up of a series of simple shapes such as rectangles, triangles, and circles, the second moment can be determined by use of the parallel-axis theorem rather than by integration. The second moment of the irregular area, the *composite area,* with respect to any axis is equal to the sum of the second moments of the separate parts of the area with respect to the specified axis. When an area such as a hole is removed from a larger area, its second moment must be subtracted from the second moment of the larger area to obtain the resulting second moment.

It is useful to have formulas available giving second moments of areas frequently encountered in routine work. Table B-1 lists properties of several geometric shapes frequently encountered. Tables listing second moments of the cross-sectional areas of various structural shapes are found in engineering handbooks and in data books prepared by industrial organizations such as the American Institute of Steel Construction and the Aluminum Company of America. Properties of a few selected shapes are included in Appendix C for use in solving problems.

EXAMPLE B-5

Determine the second moment of the shaded area in Fig. B-8 with respect to the x axis.

FIG. B-8

SOLUTION

The shaded area can be divided into a 10×6-in. rectangle A and a 10×6-in. triangle B with a 4-in. radius semicircle C removed. The location of the centroids of the three areas, as determined from Table B–1, is shown to the right of the figure. The second moments of the three areas with respect to the x axis can be obtained, by using information from Table B–1, as follows:

For area A,

$$I_x = \frac{bh^3}{3} = \frac{10(6)^3}{3} = 720 \text{ in.}^4$$

For area B,

$$I_x = I_c + Ad^2 = \frac{bh^3}{36} + Ad^2$$

$$= \frac{10(6)^3}{36} + \frac{10(6)}{2}(8)^2 = 1980 \text{ in.}^4$$

For area C, although I_c for the semicircle is given in Table B–1, the development of the expression here will expand the coverage of the subject matter. Since the area is half of a circle, its second moment with respect to a diameter is half that of a full circle, and from the parallel-axis theorem

$$I_c = I_{\text{dia}} - Ad^2 = \frac{1}{2} \frac{\pi R^4}{4} - \frac{\pi R^2}{2}\left(\frac{4R}{3\pi}\right)^2$$

$$= \frac{\pi(4)^4}{8} - \frac{8(4)^4}{9\pi} = 28.10 \text{ in.}^4$$

and

$$I_x = I_c + Ad^2 = 28.10 + \frac{\pi(4)^2}{2}(4.698)^2 = 582.8 \text{ in.}^4$$

The second moment of the composite area is

$$I_x = 720 + 1980 - 583 = \underline{2117 \text{ in.}^4} \qquad \text{Ans.}$$

EXAMPLE B–6

A column (cross section in Fig. B–9) is constructed of a W254 $\times$ 89 wide flange section and a C381 $\times$ 50 standard channel. Determine the second moments and radii of gyration of the cross-sectional area with respect to horizontal and vertical axes through the centroid of the section.

FIG. B–9

SOLUTION

The significant dimensions of the section are shown on the figure. Centroidal distances are shown to the right. Properties of the structural shapes can be obtained from Appendix C. In Fig. B–9, the x axis passes through the centroid of the wide-flange section and the x' axis passes through the centroid of the channel. The x_c axis is located using the principle of moments as applied to areas. The total area is

$$A = A_{WF} + A_C = 11\ 355 + 6425 = 17\ 780 \text{ mm}^2$$

and the moment of the area about the x axis is

$$M_x = \Sigma A y_c = 11\ 355(0) + 6425(120.2) = 17\ 780(y_c)$$

from which the distance from the x axis to the centroid, y_c, is 43.4 mm. The second moment of each area with respect to the x_c axis is

for the wide-flange section,

$$I_{x_c} = I_c + Ad^2 = 142(10^6) + 11\ 355(43.4)^2 = 163.4(10^6) \text{ mm}^4$$

for the channel,

$$I_{x_c} = I_c + Ad^2 = 3.38(10^6) + 6425(120.2 - 43.4)^2 = 41.3(10^6) \text{ mm}^4$$

and for the composite area,

$$I_{x_c} = 163.4(10^6) + 41.3(10^6) = \underline{204.7(10^6) \text{ mm}^4} \qquad \text{Ans.}$$

The radius of gyration about the x_c axis is

$$k_{x_c} = \left(\frac{I}{A}\right)^{1/2} = \left(\frac{204.7(10^6)}{17\ 780}\right)^{1/2} = \underline{107.3 \text{ mm}} \qquad \text{Ans.}$$

The y axis passes through the centroid of each of the areas; thus, the parallel axis theorem is not needed to obtain I_y; that is,

$$I_y = 48.3(10^6) + 131(10^6) = \underline{179.3(10^6) \ mm^4} \qquad \text{Ans.}$$

and

$$k_y = \left(\frac{I}{A}\right)^{1/2} = \left(\frac{179.3(10^6)}{17\ 780}\right)^{1/2} = \underline{100.4 \ mm} \qquad \text{Ans.}$$

B–6   Products of Inertia of Areas   The product of inertia dI_{xy} of the element of area dA in Fig. B–10 with respect to the x and y axes is defined as the product of the two coordinates of the element multiplied by the area of the element; the product of inertia of the total area A about the x and y axes is the sum of the products of inertia of the elements of the area; thus,

$$I_{xy} = \int_A xy \ dA \qquad (B–6)$$

The dimensions of the product of inertia are the same as for the moment of inertia (second moment), but since the product xy can be either positive or negative, the product of inertia can be positive, negative, or zero as contrasted with the second moment, which is always positive.

The product of inertia of an area with respect to any two orthogonal axes is zero when either of the axes is an axis of symmetry. This statement can be demonstrated by means of Fig. B–11, which is symmetrical with respect to the x axis. The products of inertia of the elements dA and dA' on opposite sides of the axis of symmetry will be equal in magnitude and opposite in sign and will thus cancel each other in the summation. Therefore, the resulting product of inertia will be zero.

FIG. B–10

FIG. B–11

The parallel-axis theorem for products of inertia can be derived from Fig. B–12 in which the x' and y' axes pass through the centroid C and are parallel to the x and y axes. The product of inertia with respect to the x and y axes is

$$I_{xy} = \int_A xy \, dA = \int_A (x_c + x')(y_c + y') \, dA$$

$$= x_c y_c \int_A dA + x_c \int_A y' \, dA + y_c \int_A x' \, dA + \int_A x' \, y' \, dA$$

The second and third integrals in the preceding equation are zero since x' and y' are centroidal axes and the last integral is the product of inertia with respect to the centroidal axes. Consequently, the product of inertia is

$$I_{xy} = I_{x'y'} + x_c y_c A \tag{B–7}$$

The parallel-axis theorem for products of inertia can be stated as follows: *The product of inertia of an area with respect to any two perpendicular axes x and y is equal to the product of inertia of the area with respect to a pair of centroidal axes parallel to the x and y axes added to the product of the area and the two centroidal distances from the x and y axes.*

The product of inertia is useful in determining principal axes of inertia, as discussed in the following article. The determination of the product of inertia is illustrated in the next two examples.

EXAMPLE B–7

Determine the product of inertia of the triangular area shown in Fig. B–13 with respect to

(a)   The x and y axes.

(b)   A pair of centroidal axes parallel to the x and y axes.

FIG. B–12

FIG. B–13

SOLUTION

(a) The horizontal element in Fig. B–13 has a length x and a height dy. From symmetry, the product of inertia of the element with respect to axes through the centroid of the element parallel to the x and y axes is zero. Thus, the product of inertia of the element with respect to the x and y axes is

$$dI_{xy} = \frac{x}{2} y \, dA = \frac{x}{2} y(x \, dy)$$

From similar triangles, x is equal to $(b/h)y$ and the product of inertia of the element becomes

$$dI_{xy} = \frac{b^2}{2h^2} y^3 \, dy$$

The product of inertia of the triangular area is

$$I_{xy} = \frac{b^2}{2h^2} \int_0^h y^3 \, dy = \frac{b^2 h^2}{8} \qquad \text{Ans.}$$

(b) The product of inertia with respect to the centroidal axes can be obtained from the parallel-axis theorem; thus,

$$I_{x'y'} = I_{xy} - A x_c y_c = \frac{b^2 h^2}{8} - \frac{bh}{2} \left(\frac{b}{3}\right)\left(\frac{2h}{3}\right) = \frac{b^2 h^2}{72} \qquad \text{Ans.}$$

EXAMPLE B–8
Determine the product of inertia of the quadrant of the circle in Fig. B–14 with respect to the x and y axes.

SOLUTION
The horizontal element selected is shown in Fig. B–14. As in Exam. B–7, the product of inertia of the element with respect to its centroidal axes is zero and its product of inertia with respect to the x and y axes is

FIG. B–14

$$dI_{xy} = \frac{x}{2}(y)(x\,dy) = \frac{y}{2}x^2\,dy$$

Since $R^2 = x^2 + y^2$, the preceding expression becomes

$$dI_{xy} = \frac{y}{2}(R^2 - y^2)\,dy$$

and the product of inertia of the area is

$$I_{xy} = \frac{1}{2}\int_0^R (R^2y - y^3)\,dy = \frac{R^4}{8} \qquad \text{Ans.}$$

B–7   Principal Moments of Inertia   The moment of inertia of the area A in Fig. B–15 with respect to the x' axis through O will, in general, vary with the angle θ. The x and y axes used to obtain Eq. B–2 were any pair of orthogonal axes in the plane of the area passing through O; therefore,

$$J_O = I_x + I_y = I_{x'} + I_{y'}$$

where x' and y' are any pair of orthogonal axes through O. Since the sum of $I_{x'}$ and $I_{y'}$ is a constant, $I_{x'}$ will be maximum and the corresponding $I_{y'}$ will be minimum for one particular value of θ.

The set of axes for which the second moments are maximum and minimum are called the *principal axes* of the area through point O and are

FIG. B–15

designated as the u and v axes. The second moments of the area with respect to these axes are called the *principal moments of inertia* of the area and are designated I_u and I_v. There is only one set of principal axes for any point in an area unless all axes have the same second moment, such as the diameters of a circle.

A convenient way to determine the principal moments of inertia of an area is to express $I_{x'}$ as a function of I_x, I_y, I_{xy}, and θ and then set the derivative of $I_{x'}$ with respect to θ equal to zero to obtain the value of θ giving the maximum and minimum second moments. From Fig. B–15,

$$dI_{x'} = y'^2 \, dA = (y \cos \theta - x \sin \theta)^2 \, dA$$

and

$$
\begin{aligned}
I_{x'} &= \cos^2\theta \int_A y^2 \, dA - 2 \sin \theta \cos \theta \int_A xy \, dA + \sin^2\theta \int_A x^2 \, dA \\
&= I_x \cos^2\theta - 2I_{xy} \sin \theta \cos \theta + I_y \sin^2\theta \\
&= \frac{1}{2}(I_x + I_y) + \frac{1}{2}(I_x - I_y)\cos 2\theta - I_{xy} \sin 2\theta
\end{aligned}
\tag{B–8}
$$

The angle 2θ for which $I_{x'}$ is maximum can be obtained by setting the derivative of $I_{x'}$ with respect to θ equal to zero; thus,

$$\frac{dI_{x'}}{d\theta} = -2(1/2)(I_x - I_y)\sin 2\theta - 2I_{xy} \cos 2\theta = 0$$

from which

$$\tan 2\beta = -\frac{I_{xy}}{(I_x - I_y)/2} \tag{B–9}$$

where β represents the two values of θ that locate the principal axes u and v. Equation B–9 gives two values of 2β that are 180° apart, and thus two values of β that are 90° apart. The principal moments of inertia can be obtained by substituting these values of β in Eq. B–8. From Eq. B–9,

$$\cos 2\beta = \mp \frac{(I_x - I_y)/2}{\{[(I_x - I_y)/2]^2 + I_{xy}^2\}^{1/2}}$$

and

$$\sin 2\beta = \pm \frac{I_{xy}}{\{[(I_x - I_y)/2]^2 + I_{xy}^2\}^{1/2}}$$

When these expressions are substituted in Eq. B–8, the principal moments of inertia reduce to

$$I_{u,v} = \frac{I_x + I_y}{2} \pm \sqrt{\left(\frac{I_x - I_y}{2}\right)^2 + I_{xy}^2} \qquad \text{(B–10)}$$

The product of inertia of the element of area in Fig. B–15 with respect to the x' and y' axes is

$$dI_{x'y'} = x'\, y'\, dA = (x \cos\theta + y \sin\theta)(y \cos\theta - x \sin\theta)\, dA$$

and the product of inertia of the area is

$$I_{x'y'} = (\cos^2\theta - \sin^2\theta) \int_A xy\, dA + \sin\theta \cos\theta \int_A (y^2 - x^2)\, dA$$

$$= I_{xy} \cos 2\theta + \frac{1}{2}(I_x - I_y)\sin 2\theta \qquad \text{(B–11)}$$

The product of inertia $I_{x'y'}$ will be zero when

$$\tan 2\theta = -\frac{I_{xy}}{(I_x - I_y)/2}$$

which is the same as Eq. B–9, which locates the principal axes. Consequently, *the product of inertia is zero with respect to the principal axes.* Since the product of inertia is zero with respect to any axis of symmetry, it follows that any axis of symmetry must be a principal axis for any point on the axis of symmetry.

The following example illustrates the procedure for determining second moments of areas with respect to the principal axes.

FIG. B–16

EXAMPLE B–9
Determine the maximum and minimum second moments of the area of the
unequal-leg angle in Fig. B–16 with respect to axes through the centroid of
the area.

SOLUTION
The area is divided into two rectangles A and B, and the location of the
centroid of each area is indicated by the dimensions. The centroid of the
composite area is at C. The values of I_x and I_y for the two areas are obtained
by applying the parallel-axis theorem to the expression for I_c from Table
B–1; thus,

$$I_x = \frac{bh^3}{12} + Ay_c^2$$

$$I_x = I_{Ax} + I_{Bx} = \frac{1(8)^3}{12} + 1(8)(1.346)^2 + \frac{5(1)^3}{12} + 5(1)(2.154)^2$$

$$= 80.8 \text{ in.}^4$$

$$I_y = I_{Ay} + I_{By} = \frac{8(1)^3}{12} + 8(1)(1.154)^2 + \frac{1(5)^3}{12} + 1(5)(1.846)^2$$

$$= 38.8 \text{ in.}^4$$

The products of inertia are determined as indicated in Art. B–6; thus,

$$I_{xy} = I_{x'y'} + Ax_cy_c = 0 + Ax_cy_c$$

$$I_{xy} = I_{Axy} + I_{Bxy} = 1(8)(-1.154)(1.346) + 5(1)(1.846)(-2.154)$$

$$= -32.3 \text{ in.}^4$$

From Eq. B–9,

$$\tan 2\beta = -\frac{-32.3}{(80.8 - 38.8)/2} = 1.538$$

$$2\beta = 57.0° \quad \text{or} \quad 237°$$

Therefore,

$$\beta = 28.5° \quad \text{or} \quad 118.5°$$

From Eq. B–8 with $\beta = 28.5°$, the maximum second moment is

$$I_u = 80.8(0.879)^2 - 2(-32.3)(0.477)(0.879) + 38.8(0.477)^2$$

$$= \underline{98.3 \text{ in.}^4} \qquad \qquad \text{Ans.}$$

With $\beta = 118.5°$, the minimum second moment is

$$I_v = 80.8(-0.477)^2 - 2(-32.3)(0.879)(-0.477) + 38.8(0.879)^2$$

$$= 21.3 \text{ in.}^4 \qquad\qquad\qquad \text{Ans.}$$

The principal moments of inertia can also be determined by means of Eq. B–10.

$$I_{u,v} = \frac{80.8 + 38.8}{2} \pm \sqrt{\left(\frac{80.8 - 38.8}{2}\right)^2 + (-32.3)^2}$$

$$= \underline{98.3 \text{ in.}^4} \quad \text{and} \quad \underline{21.3 \text{ in.}^4} \qquad \text{Ans.}$$

B–8 Mohr's Circle for Second Moments of Areas   The use of Mohr's circle for determining principal stresses was discussed in Art. 1–8. A comparison of Eq. 1–4a and 1–4b with Eq. B–8 indicates that a similar procedure can be used to obtain the principal moments of inertia of an area.

Figure B–17 illustrates the use of Mohr's circle for second moments. Assume that I_x is greater than I_y and that I_{xy} is positive. Moments of inertia are plotted along the horizontal axis, and products of inertia are plotted along the vertical axis. Second moments are always positive and are plotted to the right of the origin. Products of inertia can be either positive or negative; positive values are plotted above the horizontal axis. The distance OA' along the I axis is equal to I_x and $A'A$, parallel to the I_{xy} axis, is equal to I_{xy}.

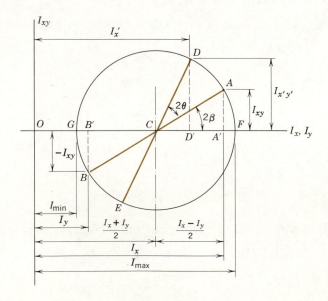

FIG. B–17

Similarly, OB' is laid off equal to I_y and $B'B$ is equal to $-I_{xy}$. The line AB intersects the I axis at C and AB is the diameter of Mohr's circle for second moments. Each point on the circle represents $I_{x'}$ and $I_{x'y'}$ for one particular orientation of the x' and y' axes. To demonstrate this statement, the diameter DE is drawn at an angle 2θ counterclockwise from line AB. From the figure

$$OD' = OC + CD \cos(2\beta + 2\theta)$$

which, since CD is equal to CA, reduces to

$$OD' = OC + CA \cos 2\beta \cos 2\theta - CA \sin 2\beta \sin 2\theta$$

Since $CA \cos 2\beta$ and $CA \sin 2\beta$ are equal to CA' and $A'A$, respectively, the distance OD' becomes

$$OD' = \frac{1}{2}(I_x + I_y) + \frac{1}{2}(I_x - I_y)\cos 2\theta - I_{xy} \sin 2\theta = I_{x'}$$

from Eq. B–8. In a similar manner,

$$D'D = CD \sin(2\beta + 2\theta)$$

$$= \frac{1}{2}(I_x - I_y)\sin 2\theta + I_{xy} \cos 2\theta = I_{x'y'}$$

from Eq. B–11.

Since the horizontal coordinate of each point on the circle represents a particular value of $I_{x'}$, the maximum second moment is represented by OF and its value is

$$I_{max} = OF = OC + CF = OC + CA$$

$$= \frac{I_x + I_y}{2} + \sqrt{\left(\frac{I_x - I_y}{2}\right)^2 + I_{xy}^2}$$

which agrees with Eq. B–10. Figure B–18 represents an area and a set of axes for which the data used to construct Fig. B–17 are valid. In deriving

FIG. B–18

Eq. B–8, the angle between the x and x' axes was θ. In obtaining the same equation from Mohr's circle, the angle between the radii to points (I_x, I_{xy}), line CA, and $(I_x, I_{x'y'})$, line CD, is 2θ. Thus, all angles on Mohr's circle are twice the corresponding angles for the given area.

EXAMPLE B–10
Solve Exam. B–9 by means of Mohr's circle.

SOLUTION
Values of I_x, I_y, and I_{xy} must be computed in the same manner as in Exam. B–9. These values are

$$I_x = 80.8 \text{ in.}^4, \qquad I_y = 38.8 \text{ in.}^4, \qquad I_{xy} = -32.3 \text{ in.}^4$$

The circle in Fig. B–19 is constructed as indicated from these values. Point A is located at I_x and I_{xy} (which is negative) and B is at I_y and $-I_{xy}$. Point C is the center of the circle and is 59.8 units from O. The radius of the circle is

$$CA = \sqrt{(21.0)^2 + (32.3)^2} = 38.5 \text{ in.}^4$$

and the principal moments of inertia are

$$I_{max} = OC + CF = 59.8 + 38.5 = \underline{98.3 \text{ in.}^4} \qquad \text{Ans.}$$

$$I_{min} = OC - CG = 59.8 - 38.5 = \underline{21.3 \text{ in.}^4} \qquad \text{Ans.}$$

FIG. B–19

Twice the angle from the x axis to the principal axis with the maximum moment of inertia is shown on the sketch as $2\beta_{max}$ and

$$\beta_{max} = \frac{1}{2} \tan^{-1} \frac{32.3}{21.0} = 28.5° \text{ counterclockwise from the } x \text{ axis.}$$

The angle from the x axis to the principal axis with the minimum moment of inertia is β_{min} and is

$$\beta_{min} = \beta_{max} + 90° = 118.5° \text{ counterclockwise from the } x \text{ axis.}$$

These values were obtained analytically from the geometry of Mohr's circle. They could also have been measured directly from the figure; however, the accuracy of the results obtained by scaling the distances from the figure would depend on the scale used and the care employed in constructing the figure.

PROPERTIES OF ROLLED STEEL SHAPES

Wide Flange Shapes:
Properties for Designing
(*W*-Shapes)
(U.S. Customary Units)

Desig-nation	Area (in.²)	Depth (in.)	Flange Width (in.)	Flange Thickness (in.)	Web Thickness (in.)	Axis X–X I (in.⁴)	Axis X–X S (in.³)	Axis X–X r (in.)	Axis Y–Y I (in.⁴)	Axis Y–Y S (in.³)	Axis Y–Y r (in.)
W36 × 230	67.6	35.90	16.470	1.260	0.760	15000	837	14.9	940	114	3.73
× 160	47.0	36.01	12.000	1.020	0.650	9750	542	14.4	295	49.1	2.50
W33 × 201	59.1	33.68	15.745	1.150	0.715	11500	684	14.0	749	95.2	3.56
× 152	44.7	33.49	11.565	1.055	0.635	8160	487	13.5	273	47.2	2.47
× 130	38.3	33.09	11.510	0.855	0.580	6710	406	13.2	218	37.9	2.39
W30 × 132	38.9	30.31	10.545	1.000	0.615	5770	380	12.2	196	37.2	2.25
× 108	31.7	29.83	10.475	0.760	0.545	4470	299	11.9	146	27.9	2.15
W27 × 146	42.9	27.38	13.965	0.975	0.605	5630	411	11.4	443	63.5	3.21
× 94	27.7	26.92	9.990	0.745	0.490	3270	243	10.9	124	24.8	2.12
W24 × 104	30.6	24.06	12.750	0.750	0.500	3100	258	10.1	259	40.7	2.91
× 84	24.7	24.10	9.020	0.770	0.470	2370	196	9.79	94.4	20.9	1.95
× 62	18.2	23.74	7.040	0.590	0.430	1550	131	9.23	34.5	9.80	1.38
W21 × 101	29.8	21.36	12.290	0.800	0.500	2420	227	9.02	248	40.3	2.89
× 83	24.3	21.43	8.355	0.835	0.515	1830	171	8.67	81.4	19.5	1.83
× 62	18.3	20.99	8.240	0.615	0.400	1330	127	8.54	57.5	13.9	1.77
W18 × 97	28.5	18.59	11.145	0.870	0.535	1750	188	7.82	201	36.1	2.65
× 76	22.3	18.21	11.035	0.680	0.425	1330	146	7.73	152	27.6	2.61
× 60	17.6	18.24	7.555	0.695	0.415	984	108	7.47	50.1	13.3	1.69
W16 × 100	29.4	16.97	10.425	0.985	0.585	1490	175	7.10	186	35.7	2.52
× 67	19.7	16.33	10.235	0.665	0.395	954	117	6.96	119	23.2	2.46
× 40	11.8	16.01	6.995	0.505	0.305	518	64.7	6.63	28.9	8.25	1.57
× 26	7.68	15.69	5.500	0.345	0.250	301	38.4	6.26	9.59	3.49	1.12
W14 × 120	35.3	14.48	14.670	0.940	0.590	1380	190	6.24	495	67.5	3.74
× 82	24.1	14.31	10.130	0.855	0.510	882	123	6.05	148	29.3	2.48
× 43	12.6	13.66	7.995	0.530	0.305	428	62.7	5.82	45.2	11.3	1.89
× 30	8.85	13.84	6.730	0.385	0.270	291	42.0	5.73	19.6	5.82	1.49
W12 × 96	28.2	12.71	12.160	0.900	0.550	833	131	5.44	270	44.4	3.09
× 65	19.1	12.12	12.000	0.605	0.390	533	87.9	5.28	174	29.1	3.02
× 50	14.7	12.19	8.080	0.640	0.370	394	64.7	5.18	56.3	13.9	1.96
× 30	8.79	12.34	6.520	0.440	0.260	238	38.6	5.21	20.3	6.24	1.52
W10 × 60	17.6	10.22	10.080	0.680	0.420	341	66.7	4.39	116	23.0	2.57
× 45	13.3	10.10	8.020	0.620	0.350	248	49.1	4.33	53.4	13.3	2.01
× 30	8.84	10.47	5.810	0.510	0.300	170	32.4	4.38	16.7	5.75	1.37
× 22	6.49	10.17	5.750	0.360	0.240	118	23.2	4.27	11.4	3.97	1.33
W8 × 40	11.7	8.25	8.070	0.560	0.360	146	35.5	3.53	49.1	12.2	2.04
× 31	9.13	8.00	7.995	0.435	0.285	110	27.5	3.47	37.1	9.27	2.02
× 24	7.08	7.93	6.495	0.400	0.245	82.8	20.9	3.42	18.3	5.63	1.61
× 15	4.44	8.11	4.015	0.315	0.245	48.0	11.8	3.29	3.41	1.70	0.876
W6 × 25	7.34	6.38	6.080	0.455	0.320	53.4	16.7	2.70	17.1	5.61	1.52
× 16	4.74	6.28	4.030	0.405	0.260	32.1	10.2	2.60	4.43	2.20	0.967
W5 × 16	4.68	5.01	5.000	0.360	0.240	21.3	8.51	2.13	7.51	3.00	1.27
W4 × 13	3.83	4.16	4.060	0.345	0.280	11.3	5.46	1.72	3.86	1.90	1.00

Courtesy of The American Institute of Steel Construction.

Wide Flange Shapes: Properties for Designing (W-Shapes) (Converted to SI Units)

Designation	Area (mm^2)	Depth (mm)	Flange Width (mm)	Flange Thickness (mm)	Web Thickness (mm)	Axis X–X I $(10^6$ $mm^4)$	Axis X–X S $(10^3$ $mm^3)$	Axis X–X r (mm)	Axis Y–Y I $(10^6$ $mm^4)$	Axis Y–Y S $(10^3$ $mm^3)$	Axis Y–Y r (mm)
W914 × 342	43610	912	418	32.0	19.3	6245	13715	378	391	1870	94.7
× 238	30325	915	305	25.9	16.5	4060	8880	366	123	805	63.5
W838 × 299	38130	855	400	29.2	18.2	4785	11210	356	312	1560	90.4
× 226	28850	851	294	26.8	16.1	3395	7980	343	114	775	62.7
× 193	24710	840	292	21.7	14.7	2795	6655	335	90.7	620	60.7
W762 × 196	25100	770	268	25.4	15.6	2400	6225	310	81.6	610	57.2
× 161	20450	758	266	19.3	13.8	1860	4900	302	60.8	457	54.6
W686 × 217	27675	695	355	24.8	15.4	2345	6735	290	184	1040	81.5
× 140	17870	684	254	18.9	12.4	1360	3980	277	51.6	406	53.8
W610 × 155	19740	611	324	19.1	12.7	1290	4230	257	108	667	73.9
× 125	15935	612	229	19.6	11.9	985	3210	249	39.3	342	49.5
× 92	11750	603	179	15.0	10.9	645	2145	234	14.4	161	35.1
W533 × 150	19225	543	312	20.3	12.7	1005	3720	229	103	660	73.4
× 124	15675	544	212	21.2	13.1	762	2800	220	33.9	320	46.5
× 92	11805	533	209	15.6	10.2	554	2080	217	23.9	228	45.0
W457 × 144	18365	472	283	22.1	13.6	728	3080	199	83.7	592	67.3
× 113	14385	463	280	17.3	10.8	554	2395	196	63.3	452	66.3
× 89	11355	463	192	17.7	10.5	410	1770	190	20.9	218	42.9
W406 × 149	18970	431	265	25.0	14.9	620	2870	180	77.4	585	64.0
× 100	12710	415	260	16.9	10.0	397	1915	177	49.5	380	62.5
× 60	7615	407	178	12.8	7.7	216	1060	168	12.0	135	39.9
× 39	4950	399	140	8.8	6.4	125	629	159	3.99	57.2	28.4
W356 × 179	22775	368	373	23.9	15.0	574	3115	158	206	1105	95.0
× 122	15550	363	257	21.7	13.0	367	2015	154	61.6	480	63.0
× 64	8130	347	203	13.5	7.7	178	1025	148	18.8	185	48.0
× 45	5710	352	171	9.8	6.9	121	688	146	8.16	95.4	37.8
W305 × 143	18195	323	309	22.9	14.0	347	2145	138	112	728	78.5
× 97	12325	308	305	15.4	9.9	222	1440	134	72.4	477	76.7
× 74	9485	310	205	16.3	9.4	164	1060	132	23.4	228	49.8
× 45	5670	313	166	11.2	6.6	99.1	633	132	8.45	102	38.6
W254 × 89	11355	260	256	17.3	10.7	142	1095	112	48.3	377	65.3
× 67	8580	257	204	15.7	8.9	103	805	110	22.2	218	51.1
× 45	5705	266	148	13.0	7.6	70.8	531	111	6.95	94.2	34.8
× 33	4185	258	146	9.1	6.1	49.1	380	108	4.75	65.1	33.8
W203 × 60	7550	210	205	14.2	9.1	60.8	582	89.7	20.4	200	51.8
× 46	5890	203	203	11.0	7.2	45.8	451	88.1	15.4	152	51.3
× 36	4570	201	165	10.2	6.2	34.5	342	86.7	7.61	92.3	40.9
× 22	2865	206	102	8.0	6.2	20.0	193	83.6	1.42	27.9	22.3
W152 × 37	4735	162	154	11.6	8.1	22.2	274	68.6	7.12	91.9	38.6
× 24	3060	160	102	10.3	6.6	13.4	167	66.0	1.84	36.1	24.6
W127 × 24	3020	127	127	9.1	6.1	8.87	139	54.1	3.13	49.2	32.3
W102 × 19	2470	106	103	8.8	7.1	4.70	89.5	43.7	1.61	31.1	25.4

American Standard Beams:
Properties for Designing
(*S*-Shapes)
(U.S. Customary Units)

Desig-nation	Area (in.²)	Depth (in.)	Flange Width (in.)	Flange Thickness (in.)	Web Thickness (in.)	Axis X–X I (in.⁴)	Axis X–X S (in.³)	Axis X–X r (in.)	Axis Y–Y I (in.⁴)	Axis Y–Y S (in.³)	Axis Y–Y r (in.)
S24 × 121	35.6	24.50	8.050	1.090	0.800	3160	258	9.43	83.3	20.7	1.53
× 106	31.2	24.50	7.870	1.090	0.620	2940	240	9.71	77.1	19.6	1.57
× 100	29.3	24.00	7.245	0.870	0.745	2390	199	9.02	47.7	13.2	1.27
× 90	26.5	24.00	7.125	0.870	0.625	2250	187	9.21	44.9	12.6	1.30
× 80	23.5	24.00	7.000	0.870	0.500	2100	175	9.47	42.2	12.1	1.34
S20 × 96	28.2	20.30	7.200	0.920	0.800	1670	165	7.71	50.2	13.9	1.33
× 86	25.3	20.30	7.060	0.920	0.660	1580	155	7.89	46.8	13.3	1.36
× 75	22.0	20.00	6.385	0.795	0.635	1280	128	7.62	29.8	9.32	1.16
× 66	19.4	20.00	6.255	0.795	0.505	1190	119	7.83	27.7	8.85	1.19
S18 × 70	20.6	18.00	6.251	0.691	0.711	926	103	6.71	24.1	7.72	1.08
× 54.7	16.1	18.00	6.001	0.691	0.461	804	89.4	7.07	20.8	6.94	1.14
S15 × 50	14.7	15.00	5.640	0.622	0.550	486	64.8	5.75	15.7	5.57	1.03
× 42.9	12.6	15.00	5.501	0.622	0.411	447	59.6	5.95	14.4	5.23	1.07
S12 × 50	14.7	12.00	5.477	0.659	0.687	305	50.8	4.55	15.7	5.74	1.03
× 40.8	12.0	12.00	5.252	0.659	0.462	272	45.4	4.77	13.6	5.16	1.06
× 35	10.3	12.00	5.078	0.544	0.428	229	38.2	4.72	9.87	3.89	0.980
× 31.8	9.35	12.00	5.000	0.544	0.350	218	36.4	4.83	9.36	3.74	1.00
S10 × 35	10.3	10.00	4.944	0.491	0.594	147	29.4	3.78	8.36	3.38	0.901
× 25.4	7.46	10.00	4.661	0.491	0.311	124	24.7	4.07	6.79	2.91	0.954
S8 × 23	6.77	8.00	4.171	0.426	0.441	64.9	16.2	3.10	4.31	2.07	0.798
× 18.4	5.41	8.00	4.001	0.426	0.271	57.6	14.4	3.26	3.73	1.86	0.831
S7 × 20	5.88	7.00	3.860	0.392	0.450	42.4	12.1	2.69	3.17	1.64	0.734
× 15.3	4.50	7.00	3.662	0.392	0.252	36.7	10.5	2.86	2.64	1.44	0.766
S6 × 17.25	5.07	6.00	3.565	0.359	0.465	26.3	8.77	2.28	2.31	1.30	0.675
× 12.5	3.67	6.00	3.332	0.359	0.232	22.1	7.37	2.45	1.82	1.09	0.705
S5 × 14.75	4.34	5.00	3.284	0.326	0.494	15.2	6.09	1.87	1.67	1.01	0.620
× 10	2.94	5.00	3.004	0.326	0.214	12.3	4.92	2.05	1.22	0.809	0.643
S4 × 9.5	2.79	4.00	2.796	0.293	0.326	6.79	3.39	1.56	0.903	0.646	0.569
× 7.7	2.26	4.00	2.663	0.293	0.193	6.08	3.04	1.64	0.764	0.574	0.581
S3 × 7.5	2.21	3.00	2.509	0.260	0.349	2.93	1.95	1.15	0.586	0.468	0.516
× 5.7	1.67	3.00	2.330	0.260	0.170	2.52	1.68	1.23	0.455	0.390	0.522

Courtesy of The American Institute of Steel Construction.

American Standard Beams:
Properties for Designing
(*S*-Shapes)
(Converted to SI Units)

Desig-nation	Area (mm²)	Depth (mm)	Flange		Web Thick-ness (mm)	Axis X–X			Axis Y–Y		
			Width (mm)	Thick-ness (mm)		I (10⁶ mm⁴)	S (10³ mm³)	r (mm)	I (10⁶ mm⁴)	S (10³ mm³)	r (mm)
S610 × 180	22970	622.3	204.5	27.7	20.3	1315	4225	240	34.7	339	38.9
× 158	20130	622.3	199.9	27.7	15.7	1225	3935	247	32.1	321	39.9
× 149	18900	609.6	184.0	22.1	18.9	995	3260	229	19.9	216	32.3
× 134	17100	609.6	181.0	22.1	15.9	937	3065	234	18.7	206	33.0
× 119	15160	609.6	177.8	22.1	12.7	874	2870	241	17.6	198	34.0
S508 × 143	18190	515.6	182.9	23.4	20.3	695	2705	196	20.9	228	33.8
× 128	16320	515.6	179.3	23.4	16.8	658	2540	200	19.5	218	34.5
× 112	14190	508.0	162.2	20.2	16.1	533	2100	194	12.4	153	29.5
× 98	12520	508.0	158.9	20.2	12.8	495	1950	199	11.5	145	30.2
S457 × 104	13290	457.2	158.8	17.6	18.1	358	1690	170	10.0	127	27.4
× 81	10390	457.2	152.4	17.6	11.7	335	1465	180	8.66	114	29.0
S381 × 74	9485	381.0	143.3	15.8	14.0	202	1060	146	6.53	91.3	26.2
× 64	8130	381.0	139.7	15.8	10.4	186	977	151	5.99	85.7	27.2
S305 × 74	9485	304.8	139.1	16.7	17.4	127	832	116	6.53	94.1	26.2
× 61	7740	304.8	133.4	16.7	11.7	113	744	121	5.66	84.6	26.9
× 52	6645	304.8	129.0	13.8	10.9	95.3	626	120	4.11	63.7	24.1
× 47	6030	304.8	127.0	13.8	8.9	90.7	596	123	3.90	61.3	25.4
S254 × 52	6645	254.0	125.6	12.5	15.1	61.2	482	96.0	3.48	55.4	22.9
× 38	4815	254.0	118.4	12.5	7.9	51.6	408	103	2.83	47.7	24.2
S203 × 34	4370	203.2	105.9	10.8	11.2	27.0	265	78.7	1.79	33.9	20.3
× 27	3490	203.2	101.6	10.8	6.9	24.0	236	82.8	1.55	30.5	21.1
S178 × 30	3795	177.8	98.0	10.0	11.4	17.6	198	68.3	1.32	26.9	18.6
× 23	2905	177.8	93.0	10.0	6.4	15.3	172	72.6	1.10	23.6	19.5
S152 × 26	3270	152.4	90.6	9.1	11.8	10.9	144	57.9	0.961	21.3	17.1
× 19	2370	152.4	84.6	9.1	5.9	9.20	121	62.2	0.758	17.9	17.9
S127 × 22	2800	127.0	83.4	8.3	12.5	6.33	99.8	47.5	0.695	16.6	15.7
× 15	1895	127.0	76.3	8.3	5.4	5.12	80.6	52.1	0.508	13.3	16.3
S102 × 14	1800	101.6	71.0	7.4	8.3	2.83	55.6	39.6	0.376	10.6	14.5
× 11	1460	101.6	67.6	7.4	4.9	2.53	49.8	41.7	0.318	9.41	14.8
S76 × 11	1425	76.2	63.7	6.6	8.9	1.22	32.0	29.2	0.244	7.67	13.1
× 8.5	1075	76.2	59.2	6.6	4.3	1.05	27.5	31.2	0.189	6.39	13.3

Standard Channels:
Properties for Designing
(*C*-Shapes)
(U.S. Customary Units)

Desig-nation	Area (in.²)	Depth (in.)	Flange		Web Thick-ness (in.)	Axis X–X			Axis Y–Y			
			Width (in.)	Average Thick-ness (in.)		I (in.⁴)	S (in.³)	r (in.)	I (in.⁴)	S (in.³)	r (in.)	$\bar{x}$ (in.)
*C18 × 58	17.1	18.00	4.200	0.625	0.700	676	75.1	6.29	17.8	5.32	1.02	0.862
× 51.9	15.3	18.00	4.100	0.625	0.600	627	69.7	6.41	16.4	5.07	1.04	0.858
× 45.8	13.5	18.00	4.000	0.625	0.500	578	64.3	6.56	15.1	4.82	1.06	0.866
× 42.7	12.6	18.00	3.950	0.625	0.450	554	61.6	6.64	14.4	4.69	1.07	0.877
C15 × 50	14.7	15.00	3.716	0.650	0.716	404	53.8	5.24	11.0	3.78	0.867	0.798
× 40	11.8	15.00	3.520	0.650	0.520	349	46.5	5.44	9.23	3.37	0.886	0.777
× 33.9	9.96	15.00	3.400	0.650	0.400	315	42.0	5.62	8.13	3.11	0.904	0.787
C12 × 30	8.82	12.00	3.170	0.501	0.510	162	27.0	4.29	5.14	2.06	0.763	0.674
× 25	7.35	12.00	3.047	0.501	0.387	144	24.1	4.43	4.47	1.88	0.780	0.674
× 20.7	6.09	12.00	2.942	0.501	0.282	129	21.5	4.61	3.88	1.73	0.799	0.698
C10 × 30	8.82	10.00	3.033	0.436	0.673	103	20.7	3.42	3.94	1.65	0.669	0.649
× 25	7.35	10.00	2.886	0.436	0.526	91.2	18.2	3.52	3.36	1.48	0.676	0.617
× 20	5.88	10.00	2.739	0.436	0.379	78.9	15.8	3.66	2.81	1.32	0.692	0.606
× 15.3	4.49	10.00	2.600	0.436	0.240	67.4	13.5	3.87	2.28	1.16	0.713	0.634
C9 × 20	5.88	9.00	2.648	0.413	0.448	60.9	13.5	3.22	2.42	1.17	0.642	0.583
× 15	4.41	9.00	2.485	0.413	0.285	51.0	11.3	3.40	1.93	1.01	0.661	0.586
× 13.4	3.94	9.00	2.433	0.413	0.233	47.9	10.6	3.48	1.76	0.962	0.669	0.601
C8 × 18.75	5.51	8.00	2.527	0.390	0.487	44.0	11.0	2.82	1.98	1.01	0.599	0.565
× 13.75	4.04	8.00	2.343	0.390	0.303	36.1	9.03	2.99	1.53	0.854	0.615	0.553
× 11.5	3.38	8.00	2.260	0.390	0.220	32.6	8.14	3.11	1.32	0.781	0.625	0.571
C7 × 14.75	4.33	7.00	2.299	0.366	0.419	27.2	7.78	2.51	1.38	0.779	0.564	0.532
× 12.25	3.60	7.00	2.194	0.366	0.314	24.2	6.93	2.60	1.17	0.703	0.571	0.525
× 9.8	2.87	7.00	2.090	0.366	0.210	21.3	6.08	2.72	0.968	0.625	0.581	0.540
C6 × 13	3.83	6.00	2.157	0.343	0.437	17.4	5.80	2.13	1.05	0.642	0.525	0.514
× 10.5	3.09	6.00	2.034	0.343	0.314	15.2	5.06	2.22	0.866	0.564	0.529	0.499
× 8.2	2.40	6.00	1.920	0.343	0.200	13.1	4.38	2.34	0.693	0.492	0.537	0.511
C5 × 9	2.64	5.00	1.885	0.320	0.325	8.90	3.56	1.83	0.632	0.450	0.489	0.478
× 6.7	1.97	5.00	1.750	0.320	0.190	7.49	3.00	1.95	0.479	0.378	0.493	0.484
C4 × 7.25	2.13	4.00	1.721	0.296	0.321	4.59	2.29	1.47	0.433	0.343	0.450	0.459
× 5.4	1.59	4.00	1.584	0.296	0.184	3.85	1.93	1.56	0.319	0.283	0.449	0.457
C3 × 6	1.76	3.00	1.596	0.273	0.356	2.07	1.38	1.08	0.305	0.268	0.416	0.455
× 5	1.47	3.00	1.498	0.273	0.258	1.85	1.24	1.12	0.247	0.233	0.410	0.438
× 4.1	1.21	3.00	1.410	0.273	0.170	1.66	1.10	1.17	0.197	0.202	0.404	0.436

Courtesy of The American Institute of Steel Construction.
* Not part of the American Standard Series.

Standard Channels:
Properties for Designing
(*C*-Shapes)
(Converted to SI Units)

Desig-nation	Area (mm²)	Depth (mm)	Flange			Axis X–X			Axis Y–Y			
			Width (mm)	Thick-ness (mm)	Web Thick-ness (mm)	I (10⁶ mm⁴)	S (10³ mm³)	r (mm)	I (10⁶ mm⁴)	S (10³ mm³)	r (mm)	$\bar{x}$ (mm)
C457 × 86	11030	457.2	106.7	15.9	17.8	281	1230	160	7.41	87.2	25.9	21.9
× 77	9870	457.2	104.1	15.9	15.2	261	1140	163	6.83	83.1	26.4	21.8
× 68	8710	457.2	101.6	15.9	12.7	241	1055	167	6.29	79.0	26.9	22.0
× 64	8130	457.2	100.3	15.9	11.4	231	1010	169	5.99	76.9	27.2	22.3
C381 × 74	9485	381.0	94.4	16.5	18.2	168	882	133	4.58	61.9	22.0	20.3
× 60	7615	381.0	89.4	16.5	13.2	145	762	138	3.84	55.2	22.5	19.7
× 50	6425	381.0	86.4	16.5	10.2	131	688	143	3.38	51.0	23.0	20.0
C305 × 45	5690	304.8	80.5	12.7	13.0	67.4	442	109	2.14	33.8	19.4	17.1
× 37	4740	304.8	77.4	12.7	9.8	59.9	395	113	1.86	30.8	19.8	17.1
× 31	3930	304.8	74.7	12.7	7.2	53.7	352	117	1.61	28.3	20.3	17.7
C254 × 45	5690	254.0	77.0	11.1	17.1	42.9	339	86.9	1.64	27.0	17.0	16.5
× 37	4740	254.0	73.3	11.1	13.4	38.0	298	89.4	1.40	24.3	17.2	15.7
× 30	3795	254.0	69.6	11.1	9.6	32.8	259	93.0	1.17	21.6	17.6	15.4
× 23	2895	254.0	66.0	11.1	6.1	28.1	221	98.3	0.949	19.0	18.1	16.1
C229 × 30	3795	228.6	67.3	10.5	11.4	25.3	221	81.8	1.01	19.2	16.3	14.8
× 22	2845	228.6	63.1	10.5	7.2	21.2	185	86.4	0.803	16.6	16.8	14.9
× 20	2540	228.6	61.8	10.5	5.9	19.9	174	88.4	0.733	15.7	17.0	15.3
C203 × 28	3555	203.2	64.2	9.9	12.4	18.3	180	71.6	0.824	16.6	15.2	14.4
× 20	2605	203.2	59.5	9.9	7.7	15.0	148	75.9	0.637	14.0	15.6	14.0
× 17	2180	203.2	57.4	9.9	5.6	13.6	133	79.0	0.549	12.8	15.9	14.5
C178 × 22	2795	177.8	58.4	9.3	10.6	11.3	127	63.8	0.574	12.8	14.3	13.5
× 18	2320	177.8	55.7	9.3	8.0	10.1	114	66.0	0.487	11.5	14.5	13.3
× 15	1850	177.8	53.1	9.3	5.3	8.87	99.6	69.1	0.403	10.2	14.8	13.7
C152 × 19	2470	152.4	54.8	8.7	11.1	7.24	95.0	54.1	0.437	10.5	13.3	13.1
× 16	1995	152.4	51.7	8.7	8.0	6.33	82.9	56.4	0.360	9.24	13.4	12.7
× 12	1550	152.4	48.8	8.7	5.1	5.45	71.8	59.4	0.288	8.06	13.6	13.0
C127 × 13	1705	127.0	47.9	8.1	8.3	3.70	58.3	46.5	0.263	7.37	12.4	12.1
× 10	1270	127.0	44.5	8.1	4.8	3.12	49.2	49.5	0.199	6.19	12.5	12.3
C102 × 11	1375	101.6	43.7	7.5	8.2	1.91	37.5	37.3	0.180	5.62	11.4	11.7
× 8	1025	101.6	40.2	7.5	4.7	1.60	31.6	39.6	0.133	4.64	11.4	11.6
C76 × 9	1135	76.2	40.5	6.9	9.0	0.862	22.6	27.4	0.127	4.39	10.6	11.6
× 7	948	76.2	38.0	6.9	6.6	0.770	20.3	28.4	0.103	3.82	10.4	11.1
× 6	781	76.2	35.8	6.9	4.6	0.691	18.0	29.7	0.082	3.31	10.3	11.1

Angles Equal Legs:
Properties for Designing
(*L*-Shapes)
(U.S. Customary Units)

Size and Thickness (in.)	Weight (lb/ft)	Area (in.^{2})	Axis X–X or Y–Y				Axis Z–Z
			I (in.^{4})	S (in.^{3})	r (in.)	x or y (in.)	r (in.)
L8 × 8 × 1	51.0	15.0	89.0	15.8	2.44	2.37	1.56
× 7/8	45.0	13.2	79.6	14.0	2.45	2.32	1.57
× 3/4	38.9	11.4	69.7	12.2	2.47	2.28	1.58
× 5/8	32.7	9.61	59.4	10.3	2.49	2.23	1.58
× 1/2	26.4	7.75	48.6	8.36	2.50	2.19	1.59
L6 × 6 × 1	37.4	11.0	35.5	8.57	1.80	1.86	1.17
× 7/8	33.1	9.73	31.9	7.63	1.81	1.82	1.17
× 3/4	28.7	8.44	28.2	6.66	1.83	1.78	1.17
× 5/8	24.2	7.11	24.2	5.66	1.84	1.73	1.18
× 1/2	19.6	5.75	19.9	4.61	1.86	1.68	1.18
× 3/8	14.9	4.36	15.4	3.53	1.88	1.64	1.19
L5 × 5 × 7/8	27.2	7.98	17.8	5.17	1.49	1.57	0.973
× 3/4	23.6	6.94	15.7	4.53	1.51	1.52	0.975
× 5/8	20.0	5.86	13.6	3.86	1.52	1.48	0.978
× 1/2	16.2	4.75	11.3	3.16	1.54	1.43	0.983
× 3/8	12.3	3.61	8.74	2.42	1.56	1.39	0.990
L4 × 4 × 3/4	18.5	5.44	7.67	2.81	1.19	1.27	0.778
× 5/8	15.7	4.61	6.66	2.40	1.20	1.23	0.779
× 1/2	12.8	3.75	5.56	1.97	1.22	1.18	0.782
× 3/8	9.8	2.86	4.36	1.52	1.23	1.14	0.788
× 1/4	6.6	1.94	3.04	1.05	1.25	1.09	0.795
L3$\frac{1}{2}$ × 3$\frac{1}{2}$ × 1/2	11.1	3.25	3.64	1.49	1.06	1.06	0.683
× 3/8	8.5	2.48	2.87	1.15	1.07	1.01	0.687
× 1/4	5.8	1.69	2.01	0.794	1.09	0.968	0.694
L3 × 3 × 1/2	9.4	2.75	2.22	1.07	0.898	0.932	0.584
× 3/8	7.2	2.11	1.76	0.833	0.913	0.888	0.587
× 1/4	4.9	1.44	1.24	0.577	0.930	0.842	0.592
L2$\frac{1}{2}$ × 2$\frac{1}{2}$ × 1/2	7.7	2.25	1.23	0.724	0.739	0.806	0.487
× 3/8	5.9	1.73	0.984	0.566	0.753	0.762	0.487
× 1/4	4.1	1.19	0.703	0.394	0.769	0.717	0.491
L2 × 2 × 3/8	4.7	1.36	0.479	0.351	0.594	0.636	0.389
1/4	3.19	0.938	0.348	0.247	0.609	0.592	0.391
1/8	1.65	0.484	0.190	0.131	0.626	0.546	0.398

Courtesy of The American Institute of Steel Construction.

Angles Equal Legs:
Properties for Designing
(*L*-Shapes)
(Converted to SI Units)

| Size and Thickness (mm) | Mass (kg/m) | Area (mm²) | Axis *X–X* or *Y–Y* | | | | Axis *Z–Z* |
			I (10^6 mm⁴)	S (10^3 mm³)	r (mm)	x or y (mm)	r (mm)
L203 × 203 × 25.4	75.9	9675	37.0	259	62.0	60.2	39.6
× 22.2	67.0	8515	33.1	229	62.2	58.9	39.9
× 19.1	57.9	7355	29.0	200	62.7	57.9	40.1
× 15.9	48.7	6200	24.7	169	63.2	56.6	40.1
× 12.7	39.3	5000	20.2	137	63.5	55.6	40.4
L152 × 152 × 25.4	55.7	7095	14.8	140	45.7	47.2	29.7
× 22.2	49.3	6275	13.3	125	46.0	46.2	29.7
× 19.1	42.7	5445	11.7	109	46.5	45.2	29.7
× 15.9	36.0	4585	10.1	92.8	46.7	43.9	30.0
× 12.7	29.2	3710	8.28	75.5	47.2	42.7	30.0
× 9.5	22.2	2815	6.61	57.8	47.8	41.7	30.2
L127 × 127 × 22.2	40.5	5150	7.41	84.7	37.8	39.9	24.7
× 19.1	35.1	4475	6.53	74.2	38.4	38.6	24.8
× 15.9	29.8	3780	5.66	63.3	38.6	37.6	24.8
× 12.7	24.1	3065	4.70	51.8	39.1	36.3	25.0
× 9.5	18.3	2330	3.64	39.7	39.6	35.3	25.1
L102 × 102 × 19.1	27.5	3510	3.19	46.0	30.2	32.3	19.8
× 15.9	23.4	2975	2.77	39.3	30.5	31.2	19.8
× 12.7	19.0	2420	2.31	32.3	31.0	30.0	19.9
× 9.5	14.6	1845	1.81	24.9	31.2	29.0	20.0
× 6.4	9.8	1250	1.27	17.2	31.8	27.7	20.2
L89 × 89 × 12.7	16.5	2095	1.52	24.4	26.9	26.9	17.3
× 9.5	12.6	1600	1.19	18.8	27.2	25.7	17.4
× 6.4	8.6	1090	0.837	13.0	27.7	24.6	17.6
L76 × 76 × 12.7	14.0	1775	0.924	17.5	22.8	23.7	14.8
× 9.5	10.7	1360	0.732	13.7	23.2	22.6	14.9
× 6.4	7.3	929	0.516	9.46	23.6	21.4	15.0
L64 × 64 × 12.7	11.5	1450	0.512	11.9	18.8	20.5	12.4
× 9.5	8.8	1115	0.410	9.28	19.1	19.4	12.4
× 6.4	6.1	768	0.293	6.46	19.5	18.2	12.5
L51 × 51 × 9.5	7.0	877	0.199	5.75	15.1	16.2	9.88
× 6.4	4.75	605	0.145	4.05	15.5	15.0	9.93
× 3.2	2.46	312	0.079	2.15	15.9	13.9	10.1

Angles Unequal Legs:
Properties for Designing
(*L*-Shapes)
(U.S. Customary Units)

Size and Thickness	Weight (lb/ft)	Area (in.^{2})	Axis X–X				Axis Y–Y				Axis Z–Z	
			I (in.^{4})	S (in.^{3})	r (in.)	y (in.)	I (in.^{4})	S (in.^{3})	r (in.)	x (in.)	r (in.)	Tan α
L9 × 4 × 5/8	26.3	7.73	64.9	11.5	2.90	3.36	8.32	2.65	1.04	0.858	0.847	0.216
1/2	21.3	6.25	53.2	9.34	2.92	3.31	6.92	2.17	1.05	0.810	0.854	0.220
L8 × 6 × 1	44.2	13.0	80.8	15.1	2.49	2.65	38.8	8.92	1.73	1.65	1.28	0.543
× 3/4	33.8	9.94	63.4	11.7	2.53	2.56	30.7	6.92	1.76	1.56	1.29	0.551
× 1/2	23.0	6.75	44.3	8.02	2.56	2.47	21.7	4.79	1.79	1.47	1.30	0.558
L8 × 4 × 1	37.4	11.0	69.6	14.1	2.52	3.05	11.6	3.94	1.03	1.05	0.846	0.247
× 3/4	28.7	8.44	54.9	10.9	2.55	2.95	9.36	3.07	1.05	0.953	0.852	0.258
× 1/2	19.6	5.75	38.5	7.49	2.59	2.86	6.74	2.15	1.08	0.859	0.865	0.267
L7 × 4 × 3/4	26.2	7.69	37.8	8.42	2.22	2.51	9.05	3.03	1.09	1.01	0.860	0.324
× 1/2	17.9	5.25	26.7	5.81	2.25	2.42	6.53	2.12	1.11	0.917	0.872	0.335
× 3/8	13.6	3.98	20.6	4.44	2.27	2.37	5.10	1.63	1.13	0.870	0.880	0.340
L6 × 4 × 3/4	23.6	6.94	24.5	6.25	1.88	2.08	8.68	2.97	1.12	1.08	0.860	0.428
× 1/2	16.2	4.75	17.4	4.33	1.91	1.99	6.27	2.08	1.15	0.987	0.870	0.440
× 3/8	12.3	3.61	13.5	3.32	1.93	1.94	4.90	1.60	1.17	0.941	0.877	0.446
L6 × 3½ × 1/2	15.3	4.50	16.6	4.24	1.92	2.08	4.25	1.59	0.972	0.833	0.759	0.344
× 3/8	11.7	3.42	12.9	3.24	1.94	2.04	3.34	1.23	0.988	0.787	0.767	0.350
L5 × 3½ × 3/4	19.8	5.81	13.9	4.28	1.55	1.75	5.55	2.22	0.977	0.996	0.748	0.464
× 1/2	13.6	4.00	9.99	2.99	1.58	1.66	4.05	1.56	1.01	0.906	0.755	0.479
× 3/8	10.4	3.05	7.78	2.29	1.60	1.61	3.18	1.21	1.02	0.861	0.762	0.486
× 1/4	7.0	2.06	5.39	1.57	1.62	1.56	2.23	0.830	1.04	0.814	0.770	0.492
L5 × 3 × 1/2	12.8	3.75	9.45	2.91	1.59	1.75	2.58	1.15	0.829	0.750	0.648	0.357
× 3/8	9.8	2.86	7.37	2.24	1.61	1.70	2.04	0.888	0.845	0.704	0.654	0.364
× 1/4	6.6	1.94	5.11	1.53	1.62	1.66	1.44	0.614	0.861	0.657	0.663	0.371
L4 × 3½ × 1/2	11.9	3.50	5.32	1.94	1.23	1.25	3.79	1.52	1.04	1.00	0.722	0.750
× 3/8	9.1	2.67	4.18	1.49	1.25	1.21	2.95	1.17	1.06	0.955	0.727	0.755
× 1/4	6.2	1.81	2.91	1.03	1.27	1.16	2.09	0.808	1.07	0.909	0.734	0.759
L4 × 3 × 1/2	11.1	3.25	5.05	1.89	1.25	1.33	2.42	1.12	0.864	0.827	0.639	0.543
× 3/8	8.5	2.48	3.96	1.46	1.26	1.28	1.92	0.866	0.879	0.782	0.644	0.551
× 1/4	5.8	1.69	2.77	1.00	1.28	1.24	1.36	0.599	0.896	0.736	0.651	0.558
L3½ × 3 × 1/2	10.2	3.00	3.45	1.45	1.07	1.13	2.33	1.10	0.881	0.875	0.621	0.714
× 3/8	7.9	2.30	2.72	1.13	1.09	1.08	1.85	0.851	0.897	0.830	0.625	0.721
× 1/4	5.4	1.56	1.91	0.776	1.11	1.04	1.30	0.589	0.914	0.785	0.631	0.727
L3½ × 2½ × 1/2	9.4	2.75	3.24	1.41	1.09	1.20	1.36	0.760	0.704	0.705	0.534	0.486
× 3/8	7.2	2.11	2.56	1.09	1.10	1.16	1.09	0.592	0.719	0.660	0.537	0.496
× 1/4	4.9	1.44	1.80	0.755	1.12	1.11	0.777	0.412	0.735	0.614	0.544	0.506
L3 × 2½ × 1/2	8.5	2.50	2.08	1.04	0.913	1.00	1.30	0.744	0.722	0.750	0.520	0.667
× 3/8	6.6	1.92	1.66	0.810	0.928	0.956	1.04	0.581	0.736	0.706	0.522	0.676
× 1/4	4.5	1.31	1.17	0.561	0.945	0.911	0.743	0.404	0.753	0.661	0.528	0.684
L3 × 2 × 1/2	7.7	2.25	1.92	1.00	0.924	1.08	0.672	0.474	0.546	0.583	0.428	0.414
× 3/8	5.9	1.73	1.53	0.781	0.940	1.04	0.543	0.371	0.559	0.539	0.430	0.428
× 1/4	4.1	1.19	1.09	0.542	0.957	0.993	0.392	0.260	0.574	0.493	0.435	0.440
L2½ × 2 × 3/8	5.3	1.55	0.912	0.547	0.768	0.813	0.514	0.363	0.577	0.581	0.420	0.614
× 1/4	3.62	1.06	0.654	0.381	0.784	0.787	0.372	0.254	0.592	0.537	0.424	0.626

Angles Unequal Legs:
Properties for Designing
(*L*-Shapes)
(Converted to SI Units)

Size and Thickness	Mass (kg/m)	Area (mm²)	Axis *X–X*				Axis *Y–Y*				Axis *Z–Z*	
			I (10⁶ mm⁴)	*S* (10³ mm³)	*r* (mm)	*y* (mm)	*I* (10⁶ mm⁴)	*S* (10³ mm³)	*r* (mm)	*x* (mm)	*r* (mm)	Tan α
L229 × 102 × 15.9	39.1	4985	27.0	188	73.7	85.3	3.46	43.4	26.4	21.8	21.5	0.216
× 12.7	31.7	4030	22.1	153	74.2	84.1	2.88	35.6	26.7	20.6	21.7	0.220
L203 × 152 × 25.4	65.8	8385	33.6	247	63.2	67.3	16.1	146	43.9	41.9	32.5	0.543
× 19.1	50.3	6415	26.4	192	64.3	65.0	12.8	113	44.7	39.6	32.8	0.551
× 12.7	34.2	4355	18.4	131	65.0	62.7	9.03	78.5	45.5	37.3	33.0	0.558
L203 × 102 × 25.4	55.7	7095	29.0	231	64.0	77.5	4.83	64.6	26.2	26.7	21.5	0.247
× 19.1	42.7	5445	22.9	179	64.8	74.9	3.90	50.3	26.7	24.2	21.6	0.258
× 12.7	29.2	3710	16.0	123	65.8	72.6	2.81	35.2	27.4	21.8	22.0	0.267
L178 × 102 × 19.1	39.0	4960	15.7	138	56.4	63.8	3.77	49.7	27.7	25.7	21.8	0.324
× 12.7	26.6	3385	11.1	95.2	57.2	61.5	2.72	34.7	28.2	23.3	22.1	0.335
× 9.5	20.2	2570	8.57	72.8	57.7	60.2	2.12	26.7	28.7	22.1	22.4	0.340
L152 × 102 × 19.1	35.1	4475	10.2	102	47.8	52.8	3.61	48.7	28.4	27.4	21.8	0.428
× 12.7	24.1	3065	7.24	71.0	48.5	50.5	2.61	34.1	29.2	25.1	22.1	0.440
× 9.5	18.3	3230	5.62	54.4	49.0	49.3	2.04	26.2	29.7	23.9	22.3	0.446
L152 × 89 × 12.7	22.8	2905	6.91	69.5	48.8	52.8	1.77	26.1	24.7	21.2	19.3	0.344
× 9.5	17.4	2205	5.37	53.1	49.3	51.8	1.39	20.2	25.1	20.0	19.5	0.350
L127 × 89 × 19.1	29.5	3750	5.79	70.1	39.4	44.5	2.31	36.4	24.8	25.3	19.0	0.464
× 12.7	20.2	2580	4.16	49.0	40.1	42.2	1.69	25.6	25.7	23.0	19.2′	0.479
× 9.5	15.5	1970	3.24	37.5	40.6	40.9	1.32	19.8	25.9	21.9	19.4′	0.486
× 6.4	10.4	1330	2.24	25.7	41.1	39.6	0.928	13.6	26.4	20.7	19.6	0.492
L127 × 76 × 12.7	19.0	2420	3.93	47.7	40.4	44.5	1.07	18.8	21.1	19.1	16.5	0.357
× 9.5	14.6	1845	3.07	36.7	40.9	43.2	0.849	14.6	21.5	17.9	16.6	0.364
× 6.4	9.82	1250	2.13	25.1	41.1	42.2	0.599	10.1	21.9	16.7	16.8	0.371
L102 × 89 × 12.7	17.7	2260	2.21	31.8	31.2	31.8	1.58	24.9	26.4	25.4	18.3	0.750
× 9.5	13.5	1725	1.74	24.4	31.8	30.7	1.23	19.2	26.9	24.3	18.5	0.755
× 6.4	9.22	1170	1.21	16.9	32.3	29.5	0.870	13.2	27.2	23.1	18.6	0.759
L102 × 76 × 12.7	16.5	2095	2.10	31.0	31.8	33.8	1.01	18.4	21.9	21.0	16.2	0.543
× 9.5	12.6	1600	1.65	23.9	32.0	32.5	0.799	14.2	22.3	19.9	16.4	0.551
× 6.4	8.63	1090	1.15	16.4	32.5	31.5	0.566	9.82	22.8	18.7	16.5	0.558
L89 × 76 × 12.7	15.2	1935	1.44	23.8	27.2	28.7	0.970	18.0	22.4	22.2	15.8	0.714
× 9.5	11.8	1485	1.13	18.5	27.7	27.4	0.770	13.9	22.8	21.1	15.9	0.721
× 6.4	8.04	1005	0.795	12.7	28.2	26.4	0.541	9.65	23.2	19.9	16.0	0.727
L89 × 64 × 12.7	14.0	1775	1.35	23.1	27.7	30.5	0.566	12.5	17.9	17.9	13.6	0.486
× 9.5	10.7	1360	1.07	17.9	27.9	29.5	0.454	8.70	18.3	16.8	13.6	0.496
× 6.4	7.29	929	0.749	12.4	28.4	28.2	0.323	6.75	18.7	15.6	13.8	0.506
L76 × 64 × 12.7	12.6	1615	0.866	17.0	23.2	25.4	0.541	12.2	18.3	19.1	13.2	0.667
× 9.5	9.82	1240	0.691	13.3	23.6	24.3	0.433	9.52	18.7	17.9	13.3	0.676
× 6.4	6.70	845	0.487	9.19	24.0	23.1	0.309	6.62	19.1	16.8	13.4	0.684
L76 × 51 × 12.7	11.5	1450	0.799	16.4	23.5	27.4	0.280	7.77	13.9	14.8	10.9	0.414
× 9.5	8.78	1115	0.637	12.8	23.9	26.4	0.226	6.08	14.2	13.7	10.9	0.428
× 6.4	6.10	768	0.454	8.88	24.3	25.2	0.163	4.26	14.6	12.5	11.0	0.440
L64 × 51 × 9.5	7.89	1000	0.380	8.96	19.5	20.7	0.214	5.95	14.7	14.8	10.7	0.614
× 6.4	5.39	684	0.272	6.24	19.9	20.0	0.155	4.16	15.0	13.6	10.8	0.626

Structural Tees Cut from W Shapes:
Properties for Designing
(*WT*-Shapes)
(U.S. Customary Units)

Desig-nation	Area (in.^{2})	Depth of Tee (in.)	Flange Width (in.)	Flange Thickness (in.)	Stem Thickness (in.)	Axis X–X I (in.^{4})	S (in.^{3})	r (in.)	y (in.)	Axis Y–Y I (in.^{4})	S (in.^{3})	r (in.)
WT18 × 115	33.8	17.950	16.470	1.260	0.760	934	67.0	5.25	4.01	470	57.1	3.73
× 80	23.5	18.005	12.000	1.020	0.650	740	55.8	5.61	4.74	147	24.6	2.50
WT15 × 66	19.4	15.155	10.545	1.000	0.615	421	37.4	4.66	3.90	98.0	18.6	2.25
× 54	15.9	14.915	10.475	0.760	0.545	349	32.0	4.69	4.01	73.0	13.9	2.15
WT12 × 52	15.3	12.030	12.750	0.750	0.500	189	20.0	3.51	2.59	130	20.3	2.91
× 47	13.8	12.155	9.065	0.875	0.515	186	20.3	3.67	2.99	54.5	12.0	1.98
× 42	12.4	12.050	9.020	0.770	0.470	166	18.3	3.67	2.97	47.2	10.5	1.95
× 31	9.11	11.870	7.040	0.590	0.430	131	15.6	3.79	3.46	17.2	4.90	1.38
WT9 × 38	11.2	9.105	11.035	0.680	0.425	71.8	9.83	2.54	1.80	76.2	13.8	2.61
× 30	8.82	9.120	7.555	0.695	0.415	64.7	9.29	2.71	2.16	25.0	6.63	1.69
× 25	7.33	8.995	7.495	0.570	0.355	53.5	7.79	2.70	2.12	20.0	5.35	1.65
× 20	5.88	8.950	6.015	0.525	0.315	44.8	6.73	2.76	2.29	9.55	3.17	1.27
WT8 × 50	14.7	8.485	10.425	0.985	0.585	76.8	11.4	2.28	1.76	93.1	17.9	2.51
× 25	7.37	8.130	7.070	0.630	0.380	42.3	6.78	2.40	1.89	18.6	5.26	1.59
× 20	5.89	8.005	6.995	0.505	0.305	33.1	5.35	2.37	1.81	14.4	4.12	1.57
× 13	3.84	7.845	5.500	0.345	0.250	23.5	4.09	2.47	2.09	4.80	1.74	1.12
WT7 × 60	17.7	7.240	14.670	0.940	0.590	51.7	8.61	1.71	1.24	247	33.7	3.74
× 41	12.0	7.155	10.130	0.855	0.510	41.2	7.14	1.85	1.39	74.2	14.6	2.48
× 34	9.99	7.020	10.035	0.720	0.415	32.6	5.69	1.81	1.29	60.7	12.1	2.46
× 24	7.07	6.985	8.030	0.595	0.340	24.9	4.48	1.87	1.35	25.7	6.40	1.91
× 15	4.42	6.920	6.730	0.385	0.270	19.0	3.55	2.07	1.58	9.79	2.91	1.49
× 11	3.25	6.870	5.000	0.335	0.230	14.8	2.91	2.14	1.76	3.50	1.40	1.04
WT6 × 60	17.6	6.560	12.320	1.105	0.710	43.4	8.22	1.57	1.28	172	28.0	3.13
× 48	14.1	6.355	12.160	0.900	0.550	32.0	6.12	1.51	1.13	135	22.2	3.09
× 36	10.6	6.125	12.040	0.670	0.430	23.2	4.54	1.48	1.02	97.5	16.2	3.04
× 25	7.34	6.095	8.080	0.640	0.370	18.7	3.79	1.60	1.17	28.2	6.97	1.96
× 15	4.40	6.170	6.520	0.440	0.260	13.5	2.75	1.75	1.27	10.2	3.12	1.52
× 8	2.36	5.995	3.990	0.265	0.220	8.70	2.04	1.92	1.74	1.41	0.706	0.773
WT5 × 56	16.5	5.680	10.415	1.250	0.755	28.6	6.40	1.32	1.21	118	22.6	2.68
× 44	12.9	5.420	10.265	0.990	0.605	20.8	4.77	1.27	1.06	89.3	17.4	2.63
× 30	8.82	5.110	10.080	0.680	0.420	12.9	3.04	1.21	0.884	58.1	11.5	2.57
× 15	4.42	5.235	5.810	0.510	0.300	9.28	2.24	1.45	1.10	8.35	2.87	1.37
× 6	1.77	4.935	3.960	0.210	0.190	4.35	1.22	1.57	1.36	1.09	0.551	0.785
WT4 × 29	8.55	4.375	8.220	0.810	0.510	9.12	2.61	1.03	0.874	37.5	9.13	2.10
× 20	5.87	4.125	8.070	0.560	0.360	5.73	1.69	0.988	0.735	24.5	6.08	2.04
× 12	3.54	3.965	6.495	0.400	0.245	3.53	1.08	0.999	0.695	9.14	2.81	1.61
× 9	2.63	4.070	5.250	0.330	0.230	3.41	1.05	1.14	0.834	3.98	1.52	1.23
× 5	1.48	3.945	3.940	0.205	0.170	2.15	0.717	1.20	0.953	1.05	0.532	0.841
WT3 × 10	2.94	3.100	6.020	0.365	0.260	1.76	0.693	0.774	0.560	6.64	2.21	1.50
× 6	1.78	3.015	4.000	0.280	0.230	1.32	0.564	0.861	0.677	1.50	0.748	0.918
WT2 × 6.5	1.91	2.080	4.060	0.345	0.280	0.526	0.321	0.524	0.440	1.93	0.950	1.00

Courtesy of The American Institute of Steel Construction.

Structural Tees Cut from W Shapes:
Properties for Designing
(*WT*-Shapes)
(Converted to SI Units)

Desig-nation	Area (mm²)	Depth of Tee (mm)	Flange Width (mm)	Flange Thickness (mm)	Stem Thickness (mm)	Axis X–X I (10⁶ mm⁴)	Axis X–X S (10³ mm³)	Axis X–X r (mm)	Axis X–X y (mm)	Axis Y–Y I (10⁶ mm⁴)	Axis Y–Y S (10³ mm³)	Axis Y–Y r (mm)
WT457 × 171	21805	455.9	418.3	32.0	19.3	389	1098	133	102	196	936	94.7
× 119	15160	457.3	304.8	25.9	16.5	308	914	142	120	61.2	403	63.5
WT381 × 98	12515	384.9	267.8	25.4	15.6	175	613	118	99.1	40.8	305	57.2
× 80	10260	378.8	266.1	19.3	13.8	145	524	119	102	30.4	228	54.6
WT305 × 77	9870	305.6	323.9	19.1	12.7	78.7	328	89.2	65.8	54.1	333	73.9
× 70	8905	308.7	230.3	22.2	13.1	77.4	333	93.2	75.9	22.7	197	50.3
× 63	8000	306.1	229.1	19.6	11.9	69.1	300	93.2	75.4	19.6	172	49.5
× 46	5875	301.5	178.8	15.0	10.9	54.5	256	96.3	87.9	7.16	80.3	35.1
WT229 × 57	7225	231.3	280.3	17.3	10.8	29.9	161	64.5	45.7	31.7	226	66.3
× 45	5690	231.6	191.9	17.7	10.5	26.9	152	68.8	54.9	10.4	109	42.9
× 37	4730	228.5	190.4	14.5	9.0	22.3	128	68.6	53.8	8.32	87.7	41.9
× 30	3795	227.3	152.8	13.3	8.0	18.6	110	70.1	58.2	3.98	51.9	32.3
WT203 × 74	9485	215.5	264.8	25.0	14.9	32.0	187	57.9	44.7	38.8	293	63.8
× 37	4755	206.5	179.6	16.0	9.7	17.6	111	61.0	48.0	7.74	86.2	40.4
× 30	3800	203.3	177.7	12.8	7.7	13.8	87.7	60.2	46.0	5.99	67.5	39.9
× 19	2475	199.3	139.7	8.8	6.4	9.78	67.0	62.7	53.1	2.00	28.5	28.4
WT178 × 89	11420	183.9	372.6	23.9	15.0	21.5	141	43.4	31.5	103	552	95.0
× 61	7740	181.7	257.3	21.7	13.0	17.1	117	47.0	35.3	30.9	239	63.0
× 51	6445	178.3	254.9	18.3	10.5	13.6	93.2	46.0	32.8	25.3	198	62.5
× 36	4560	177.4	204.0	15.1	8.6	10.4	73.4	47.5	34.3	10.7	105	48.5
× 22	2850	175.8	170.9	9.8	6.9	7.91	58.2	52.6	40.1	4.07	47.7	37.8
× 16	2095	174.5	127.0	8.5	5.8	6.16	47.7	54.4	44.7	1.46	22.9	26.4
WT152 × 89	11355	166.6	312.9	28.1	18.0	18.1	135	39.9	32.5	71.6	459	79.5
× 71	9095	161.4	308.9	22.9	14.0	13.3	100	38.4	28.7	56.2	364	78.5
× 54	6840	155.6	305.8	17.0	10.9	9.66	74.4	37.6	25.9	40.6	265	77.2
× 37	4735	154.8	205.2	16.2	9.4	7.78	62.1	40.6	29.7	11.7	114	49.8
× 22	2840	156.7	165.6	11.2	6.6	5.62	45.1	44.5	32.3	4.25	51.1	38.6
× 12	1525	152.3	101.3	6.7	5.6	3.62	33.4	48.8	44.2	0.587	11.6	19.6
WT127 × 83	10645	144.3	264.5	31.8	19.2	11.9	105	33.5	30.7	49.1	370	68.1
× 65	8325	137.7	260.7	25.1	15.4	8.66	78.2	32.3	26.9	37.2	285	66.8
× 45	5690	129.8	256.0	17.3	10.7	5.37	49.8	30.7	22.5	24.2	188	65.3
× 22	2850	133.0	147.6	13.0	7.6	3.86	36.7	36.8	27.9	3.48	47.0	34.8
× 9	1140	125.3	100.6	5.3	4.8	1.81	20.0	39.9	34.5	0.454	9.03	19.9
WT102 × 43	5515	111.1	208.8	20.6	13.0	3.80	42.8	26.2	22.2	15.6	150	53.3
× 30	3785	104.8	205.0	14.2	9.1	2.39	27.7	25.1	18.7	10.2	99.6	51.8
× 18	2285	100.7	165.0	10.2	6.2	1.47	17.7	25.4	17.7	3.80	46.0	40.9
× 13	1695	103.4	133.4	8.4	5.8	1.42	17.2	29.0	21.2	1.66	24.9	31.2
× 7	955	100.2	100.1	5.2	4.3	0.895	11.7	30.5	24.2	0.437	8.72	21.4
WT76 × 15	1895	78.7	152.9	9.3	6.6	0.733	11.4	19.7	14.2	2.76	36.2	38.1
× 9	1150	76.6	101.6	7.1	5.8	0.549	9.24	21.9	17.2	0.624	12.3	23.3
WT51 × 10	1230	52.8	103.1	8.8	7.1	0.219	5.26	13.3	11.2	0.803	15.6	25.4

TABLE OF BEAM DEFLECTIONS AND SLOPES

Case	Load and Support (Length L)	Slope at End ($+$ $\measuredangle$)	Maximum Deflection ($+$ upward)
1		$\theta = -\dfrac{PL^2}{2EI}$ at $x = L$	$y_{max} = -\dfrac{PL^3}{3EI}$ at $x = L$
2		$\theta = -\dfrac{wL^3}{6EI}$ at $x = L$	$y_{max} = -\dfrac{wL^4}{8EI}$ at $x = L$
3		$\theta = -\dfrac{wL^3}{24EI}$ at $x = L$	$y_{max} = -\dfrac{wL^4}{30EI}$ at $x = L$
4		$\theta = +\dfrac{ML}{EI}$ at $x = L$	$y_{max} = +\dfrac{ML^2}{2EI}$ at $x = L$
5		$\theta_1 = -\dfrac{Pb(L^2 - b^2)}{6LEI}$ at $x = 0$ $\theta_2 = +\dfrac{Pa(L^2 - a^2)}{6LEI}$ at $x = L$	$y_{max} = -\dfrac{Pb(L^2 - b^2)^{3/2}}{9\sqrt{3}\,LEI}$ at $x = \sqrt{(L^2 - b^2)/3}$ y_{center} not max $= -\dfrac{Pb(3L^2 - 4b^2)}{48EI}$
6		$\theta_1 = -\dfrac{PL^2}{16EI}$ at $x = 0$ $\theta_2 = +\dfrac{PL^2}{16EI}$ at $x = L$	$y_{max} = -\dfrac{PL^3}{48EI}$ at $x = L/2$
7		$\theta_1 = -\dfrac{wL^3}{24EI}$ at $x = 0$ $\theta_2 = +\dfrac{wL^3}{24EI}$ at $x = L$	$y_{max} = -\dfrac{5wL^4}{384EI}$ at $x = L/2$
8		$\theta_1 = -\dfrac{ML}{6EI}$ at $x = 0$ $\theta_2 = +\dfrac{ML}{3EI}$ at $x = L$	$y_{max} = -\dfrac{ML^2}{9\sqrt{3}\,EI}$ at $x = L/\sqrt{3}$ y_{center} not max $= -\dfrac{ML^2}{16EI}$

TABLE D-1 Beam Deflections and Slopes

THE INTERNATIONAL SYSTEM OF UNITS (SI)

THE INTERNATIONAL SYSTEM OF UNITS

E–1 Introduction The importance of the regulation of weights and measures was recognized as early as 1787 when Article 1, Section 8, of the United States Constitution was written. The metric system was legalized in the United States in 1866; in 1893, the international meter and kilogram became the fundamental standards of length and mass both for metric and customary weights and measures. International standardization began with an International Metric Convention in 1875, which established a permanent International Bureau of Weights and Measures. The National Bureau of Standards represents the United States in this international body.

The original metric system provided a set of units for the measurement of length, area, volume, capacity, and mass based on two fundamental units: the meter and the kilogram. With the addition of a unit of time, practical measurements began to be based on the meter-kilogram-second (MKS) system. In 1960, the Eleventh General Conference on Weights and Measures formally adopted the *International System of Units,* for which the abbreviation is *SI* in all languages, as the international standard. Thirty-six countries, including the United States, participated in this conference.

TABLE E–1

Quantity	Name of Base SI Unit	Symbol
Length	meter	m
Mass	kilogram	kg
Time	second	s
Electric current	ampere	A
Thermodynamic temperature	kelvin	K
Amount of substance	mole	mol
Luminous intensity	candela	cd

TABLE E–2

Quantity	Name of Supplementary SI Unit	Symbol
Plane angle	radian	rad
Solid angle	steradian	sr

TABLE E–3

Quantity	Derived SI Unit	Special Name	Symbol
Area	square meter	—	m^2
Volume	cubic meter	—	m^3
Linear velocity	meter per second	—	m/s
Angular velocity	radian per second	—	rad/s
Linear acceleration	meter per second squared	—	m/s^2
Frequency	(cycle) per second	hertz	Hz
Density	kilogram per cubic meter	—	kg/m^3
Force	kilogram · meter per second squared	newton	N
Moment of force	newton · meter	—	N · m
Pressure	newton per meter squared	pascal	Pa
Stress	newton per meter squared	pascal ·	Pa or N/m^2
Work, energy	newton · meter	joule	J
Power	joule per second	watt	W

E–2 SI Units   The International System of Units adopted by the conference includes three classes of units: (1) base units, (2) supplementary units, and (3) derived units. The system is founded on the seven base units listed in Table E–1.

Certain units of the international system have not been classified under either base units or derived units. These units, listed in Table E–2, are called supplementary units and may be regarded either as base units or as derived units.

Derived units are expressed algebraically in terms of base units and/or supplementary units. Their symbols are obtained by means of the mathematical signs of multiplication and division. For example, the SI unit for velocity is meter per second (m/s) and the SI unit for angular velocity is radian per second (rad/s). For some of the derived units, special names and symbols exist; those of interest in mechanics are listed in Table E–3.

E–3 Multiples of SI Units   Prefixes are used to form names and symbols of multiples (decimal multiples and sub-multiples) of SI units. The choice of the appropriate multiple is governed by convenience and should usually be chosen so that the numerical values will be between 0.1 and 1000. Only one prefix should be used in forming a multiple of a compound SI unit, and prefixes in the denominator should be avoided. Approved prefixes with their names and symbols are listed in Table E–4.

TABLE E–4

Factor by Which Unit is Multiplied	Prefix Name	Symbol
10^{12}	tera	T
10^{9}	giga	G
10^{6}	mega	M
10^{3}	kilo	k
10^{2}	hecto*	h
10	deca*	da
10^{-1}	deci*	d
10^{-2}	centi*	c
10^{-3}	milli	m
10^{-6}	micro	μ
10^{-9}	nano	n
10^{-12}	pico	p
10^{-15}	femto	f
10^{-18}	atto	a

* To be avoided when possible.

E–4   Conversion Between the SI and U.S. Customary Systems

As the use of SI becomes more commonplace in the United States, engineers will be required to be familiar with both SI and the U.S. Customary system in common use today. As an aid to interpreting the physical significance of answers in SI units for those more accustomed to the U.S. Customary system, the following conversion factors are provided:

CONVERSION FACTORS

	U.S. Customary to SI	SI to U.S. Customary
Length:	1 in. = 25.40 mm	1 m = 39.37 in.
	1 ft = 0.3048 m	1 m = 3.281 ft
Area:	1 in.^{2} = 645.2 mm^2	1 m^2 = 1550 in.^{2}
	1 ft^2 = 0.09290 m^2	1 m^2 = 10.76 ft^2
Volume:	1 in.^{3} = 16.39(10^3)mm^3	1 cm^3 = 0.06102 in.^{3}
	1 ft^3 = 0.02832 m^3	1 m^3 = 35.31 ft^3
Moment of Inertia:	1 in.^{4} = 0.4162(10^6) mm^4	1 mm^4 = 2.402(10^{-6}) in.^{4}
Force:	1 lb = 4.448 N	1 N = 0.2248 lb
Distributed Load:	1 lb/ft = 14.59 N/m	1 kN/m = 68.53 lb/ft
Pressure or Stress:	1 psi = 6.895 kPa	1 MPa = 145.0 psi
	1 ksi = 6.895 MPa	1 MPa = 145.0 psi
Bending Moment or Torque:	1 ft · lb = 1.356 N · m	1 N · m = 0.7376 ft · lb
Work or Energy:	1 ft · lb = 1.356 J	1 J = 0.7376 ft · lb
Power:	1 hp = 745.7 W	1 kW = 1.341 hp

ANSWERS

1-35   (a) $\sigma_x = 221$ MPa T
    (b) $\tau = 61.0$ MPa

1-38   (a) $\tau_v = -1200$ psi
        $\tau_h = 1200$ psi
    (b) $\sigma_y = 1950$ psi T

1-39 $\sigma = 1.487$ MPa T
    $\tau = 16.63$ MPa

1-43 $\sigma_n = 1.160$ ksi C
    $\tau_{nt} = 12.79$ ksi

1-44 $\sigma_n = 60.3$ MPa C
    $\tau_{nt} = 95.8$ MPa

1-45 $\sigma_n = 4.37$ ksi T
    $\tau_{nt} = 14.24$ ksi

1-46 $\sigma_n = 163.8$ MPa T
    $\tau_{nt} = 9.85$ MPa

1-49 $\sigma_n = 22.0$ ksi T
    $\sigma_t = 6.04$ ksi T
    $\tau_{nt} = 5.79$ ksi

1-51 $\sigma_{p1} = 16.45$ ksi T
    $\sigma_{p2} = 10.45$ ksi C
    $\sigma_{p3} = 0$
    $\tau_p = 13.45$ ksi
    $\tau_{max} = 13.45$ ksi
    $\theta_p = 24.0°$)

1-52 $\sigma_{p1} = 98.2$ MPa T
    $\sigma_{p2} = 118.2$ MPa C
    $\sigma_{p3} = 0$
    $\tau_p = 108.2$ MPa
    $\tau_{max} = 108.2$ MPa
    $\theta_p = 16.85°$)

1-55 $\sigma_{p1} = 30.4$ ksi T
    $\sigma_{p2} = 6.57$ ksi T
    $\sigma_{p3} = 0$
    $\tau_p = 11.93$ ksi
    $\tau_{max} = 15.22$ ksi
    $\theta_p = 28.5°$)

1-57 $\sigma_{p1} = 23.9$ ksi T
    $\sigma_{p2} = 4.15$ ksi T
    $\sigma_{p3} = 0$
    $\tau_p = 9.85$ ksi
    $\tau_{max} = 11.93$ ksi
    $\theta_p = 33.0°$)

1-59 $\sigma_{p1} = 21.0$ ksi T

    $\sigma_{p2} = 29.0$ ksi C
    $\sigma_{p3} = 0$
    $\tau_{xy} = -15.00$ ksi
    $\tau_p = 25.0$ ksi
    $\tau_{max} = 25.0$ ksi
    $\theta_p = 18.43°$)

1-62 $\sigma_{p1} = 15.21$ ksi T
    $\sigma_{p2} = 0.789$ ksi T
    $\sigma_{p3} = 0$
    $\tau_p = 7.21$ ksi
    $\tau_{max} = 7.61$ ksi
    $\theta_p = 28.2°$)

1-65 $\tau_{xy} = \pm120.0$ MPa
    $\sigma_{p2} = 160.0$ MPa C
    $\sigma_{p3} = 0$
    $\tau_p = 130.0$ MPa
    $\tau_{max} = 130.0$ MPa
    $\theta_p = \pm33.7°$

1-66 $\sigma_y = 8.00$ ksi C
    $\tau_{xy} = \pm12.00$ ksi
    $\sigma_{p2} = 14.00$ ksi C
    $\sigma_{p3} = 0$
    $\tau_p = 15.00$ ksi
    $\tau_{max} = 15.00$ ksi
    $\theta_p = \pm26.6°$

1-68 $\sigma_x = 3.93$ ksi C
    $\sigma_y = 17.78$ ksi C
    $\sigma_{p2} = 24.7$ ksi C
    $\sigma_{p3} = 0$
    $\tau_p = 13.86$ MPa
    $\tau_{max} = 13.86$ MPa

1-71 $\sigma_x = 96.2$ MPa T
    $\sigma_{p1} = 116.2$ MPa T
    $\sigma_{p3} = 0$
    $\tau_p = 54.1$ MPa
    $\tau_{max} = 58.1$ MPa
    $\theta_p = 25.5°$)

1-72 $\sigma_n = 12.01$ ksi T
    $\tau_n = 1.680$ ksi

1-73 $\sigma_n = 143.6$ MPa T
    $\tau_n = 17.68$ MPa

1-75   (a) $\sigma_n = 1.160$ ksi C
        $\tau_{nt} = 12.79$ ksi

(b) $\sigma_{p1} = 16.45$ ksi T
    $\sigma_{p2} = 10.45$ ksi C
    $\sigma_{p3} = 0$
    $\tau_p = 13.45$ ksi
    $\tau_{\max} = 13.45$ ksi
    $\theta_p = 24.0°$⟩

1-76 (a) $\sigma_n = 60.3$ MPa C
    $\tau_{nt} = -95.8$ MPa
  (b) $\sigma_{p1} = 98.2$ MPa T
    $\sigma_{p2} = 118.2$ MPa C
    $\sigma_{p3} = 0$
    $\tau_p = 108.2$ MPa
    $\tau_{\max} = 108.2$ MPa
    $\theta_p = 16.85°$⟩

1-81 (a) $\sigma_n = 22.0$ ksi T
    $\sigma_t = 6.03$ ksi T
    $\tau_{nt} = 5.79$ ksi
  (b) $\sigma_{p1} = 23.8$ ksi T
    $\sigma_{p2} = 4.15$ ksi T
    $\sigma_{p3} = 0$
    $\tau_p = 9.85$ ksi
    $\tau_{\max} = 11.93$ ksi
    $\theta_p = 33.0°$⟩

1-83 $\sigma_{p1} = 15.21$ ksi T
  $\sigma_{p2} = 0.789$ ksi T
  $\sigma_{p3} = 0$
  $\tau_p = 7.21$ ksi
  $\tau_{\max} = 7.61$ ksi
  $\theta_p = 28.2°$⟩

1-86 $\tau_{xy} = \pm120.0$ MPa
  $\sigma_{p2} = 160.0$ MPa C
  $\sigma_{p3} = 0$
  $\tau_p = 130.0$ MPa
  $\tau_{\max} = 130.0$ MPa
  $\theta_p = \pm33.7°$

1-89 (a) $\sigma_n = 153.3$ MPa T
    $\tau_{nt} = 61.6$ MPa
  (b) $\sigma_{p1} = 171.8$ MPa T
    $\sigma_{p2} = 51.8$ MPa C
    $\sigma_{p3} = 0$
    $\tau_p = 111.8$ MPa
    $\tau_{\max} = 111.8$ MPa
    $\theta_p = 13.29°$⟩

1-90 (a) $\sigma_n = 884$ psi T
    $\tau_{nt} = 4400$ psi
  (b) $\sigma_{p1} = 4110$ psi T
    $\sigma_{p2} = 5110$ psi C
    $\sigma_{p3} = 0$
    $\tau_p = 4610$ psi
    $\tau_{\max} = 4610$ psi
    $\theta_p = 6.27°$⟩

1-93 (a) $\sigma_{avg} = 11.50$ ksi C
    $\tau_p = 8.50$ ksi
  (b) $\sigma_x = 19.00$ ksi C
    $\sigma_y = 4.00$ ksi C
    $\tau_{xy} = 4.00$ ksi
  (c) $\tau_{\max} = 10.00$ ksi

1-94 (a) $\sigma_{avg} = 125.0$ MPa T
    $\tau_p = 75.0$ MPa
  (b) $\sigma_x = 185.0$ MPa T
    $\sigma_y = 65.0$ MPa T
    $\tau_{xy} = 45.0$ MPa
  (c) $\tau_{\max} = 100.0$ MPa

1-98 (a) $\sigma_n = 37.8$ MPa T
    $\tau_n = 27.4$ MPa
  (b) $\sigma_{p1} = 60.0$ MPa T
    $\sigma_{p2} = 20.0$ MPa T
    $\sigma_{p3} = 40.0$ MPa C
    $\tau_{\max} = 50.0$ MPa

1-99 (a) $\sigma_n = 14.29$ ksi T
    $\tau_n = 4.17$ ksi
  (b) $\sigma_{p1} = 18.00$ ksi T
    $\sigma_{p2} = 12.00$ ksi T
    $\sigma_{p3} = 6.00$ ksi T
    $\tau_{\max} = 6.00$ ksi

1-102 (a) $\sigma_n = 41.3$ MPa T
    $\tau_n = 53.0$ MPa
  (b) $\sigma_{p1} = 81.3$ MPa T
    $\sigma_{p2} = 4.74$ MPa C
    $\sigma_{p3} = 36.6$ MPa C
    $\tau_{\max} = 59.0$ MPa

1-103 (a) $\sigma_{p1} = 106.2$ MPa T
    $\sigma_{p2} = 10.00$ MPa T
    $\sigma_{p3} = 3.77$ MPa T
    $\tau_{\max} = 51.2$ MPa
  (b) $l = 0.734$

$m = 0.560$
$n = 0.385$
$(\theta_x = 42.8°)$
$(\theta_y = 56.0°)$
$(\theta_z = 67.4°)$

1-105 (a) $\sigma_n = 17.20$ MPa T
$\tau_n = 98.6$ MPa

(b) $\sigma_{p1} = 179.9$ MPa T
$\sigma_{p2} = 27.6$ MPa T
$\sigma_{p3} = 127.5$ MPa C
$\tau_{max} = 153.7$ MPa

(c) $l = 0.702$
$m = 0.288$
$n = -0.651$
$(\theta_x = 45.4°)$
$(\theta_y = 73.2°)$
$(\theta_z = 130.6°)$

CHAPTER 2

2-1 $\varepsilon_{avg} = 0.300$ in./in.
2-2 $\varepsilon_{50}/\varepsilon_{200} = 1.795$
2-3   (a) $\varepsilon = 0.001200$ in./in.
       (b) $\varepsilon = 0.001179$ in./in.
2-4   (a) $\varepsilon = 0.005379$ m/m
       (b) $\varepsilon = 0.005483$ m/m
2-7 $\varepsilon_{avg} = \gamma L/6E$
2-8 $\delta = 228$ mm
2-11 $\gamma_P = 1560$ μrad
2-13 $\gamma_P = 7970$ μrad
2-15 $\varepsilon_{p1} = 1085$ μ
       $\varepsilon_{p2} = -885$ μ
       $\varepsilon_{p3} = 0$
       $\gamma_p = 1970$ μ
       $\gamma_{max} = 1970$ μ
       $\theta_p = 11.98°$
2-16 $\varepsilon_{p1} = 1575$ μ
       $\varepsilon_{p2} = 585$ μ
       $\varepsilon_{p3} = 0$
       $\gamma_p = 990$ μ
       $\gamma_{max} = 1575$ μ
       $\theta_p = 38.0°$
2-19   (a) $\varepsilon_{p1} = 1645$ μ
            $\varepsilon_{p2} = -1045$ μ

$\varepsilon_{p3} = 0$
$\gamma_P = 2690$ μ
$\gamma_{max} = 2690$ μ
$\theta_p = 24.0°$

(b) $\varepsilon_n = -1016$ μ
$\varepsilon_t = 1616$ μ
$\gamma_{nt} = -559$ μ

2-21   (a) $\varepsilon_{p1} = 1643$ μ
            $\varepsilon_{p2} = -243$ μ
            $\varepsilon_{p3} = 0$
            $\gamma_p = 1887$ μ
            $\gamma_{max} = 1887$ μ
            $\theta_p = 29.0°$

       (b) $\varepsilon_n = 1023$ μ
            $\varepsilon_t = 377$ μ
            $\gamma_{nt} = 1773$ μ

2-23 $\varepsilon_{p1} = 1521$ μ
       $\varepsilon_{p2} = 78.9$ μ
       $\varepsilon_{p3} = 0$
       $\gamma_p = 1442$ μ
       $\gamma_{max} = 1521$ μ
       $\theta_p = 28.2°$

2-25 $\gamma_{xy} = \pm 2030$ μ
       $\varepsilon_{p2} = -1360$ μ
       $\gamma_p = 2400$ μ
       $\gamma_{max} = 2400$ μ
       $\theta_p = \pm 28.9°$

2-27 $\varepsilon_y = -640$ μ
       $\gamma_{xy} = \pm 1920$ μ
       $\varepsilon_{p2} = -1120$ μ
       $\theta_p = \pm 26.6°$

2-29 $\varepsilon_x = -294$ μ
       $\varepsilon_y = -1334$ μ
       $\varepsilon_{p2} = -1853$ μ
       $\gamma_p = 2078$ μ
       $\gamma_{max} = 2078$ μ

2-31 $\varepsilon_x = 800$ μ
       $\varepsilon_{p1} = 1280$ μ
       $\gamma_{max} = 2400$ μ
       $\theta_p = 26.6°$

2-32 $\varepsilon_{p1} = 600$ μ
       $\varepsilon_{p2} = -400$ μ
       $\varepsilon_{p3} = \varepsilon_z = -85.7$ μ

$\gamma_p = 1000\ \mu$
$\gamma_{max} = 1000\ \mu$
$\theta_p = 18.43°$

2-33　$\varepsilon_{p1} = 787\ \mu$
$\varepsilon_{p2} = 293\ \mu$
$\varepsilon_{p3} = \varepsilon_z = -463\ \mu$
$\gamma_p = 495\ \mu$
$\gamma_{max} = 1250\ \mu$
$\theta_p = 38.0°$

2-34　$\varepsilon_{p1} = 1281\ \mu$
$\varepsilon_{p2} = -281\ \mu$
$\varepsilon_{p3} = \varepsilon_z = -429\ \mu$
$\gamma_p = 1562\ \mu$
$\gamma_{max} = 1710\ \mu$
$\theta_p = 19.9°$

2-36　(a)　$\varepsilon_x = 950\ \mu$
$\varepsilon_y = -225\ \mu$
$\gamma_{xy} = 275\ \mu$
(b)　$\varepsilon_{p1} = 966\ \mu$
$\varepsilon_{p2} = -241\ \mu$
$\varepsilon_{p3} = \varepsilon_z = -311\ \mu$
$\gamma_p = 1207\ \mu$
$\gamma_{max} = 1277\ \mu$
$\theta_p = 6.59°$

2-38　(a)　$\varepsilon_x = 850\ \mu$
$\varepsilon_y = 550\ \mu$
$\gamma_{xy} = 1559\ \mu$
(b)　$\varepsilon_{p1} = 1494\ \mu$
$\varepsilon_{p2} = -93.8\ \mu$
$\varepsilon_{p3} = \varepsilon_z = -600\ \mu$
$\gamma_p = 1588\ \mu$
$\gamma_{max} = 2094\ \mu$
$\theta_p = 39.6°$

2-40　$\varepsilon_{p1} = 822\ \mu$
$\varepsilon_{p2} = -604\ \mu$
$\varepsilon_{p3} = \varepsilon_z = -93.4\ \mu$
$\gamma_p = 1426\ \mu$
$\gamma_{max} = 1426\ \mu$
$\theta_p = 22.8°$

2-42　$\varepsilon_{p1} = 955\ \mu$
$\varepsilon_{p2} = -648\ \mu$
$\varepsilon_{p3} = \varepsilon_z = -131.6\ \mu$
$\gamma_p = 1603\ \mu$

$\gamma_{max} = 1603\ \mu$
$\theta_p = 13.08°$

2-44　$\varepsilon_a = 906\ \mu$
$\varepsilon_b = -50.0\ \mu$
$\varepsilon_c = -306\ \mu$

CHAPTER 3

3-1　(a)　$E = 16,300$ ksi
　　(b)　$\nu = 0.333$
　　(c)　$\sigma_{PL} = 24.4$ ksi

3-2　(a)　$E = 73.4$ GPa
　　(b)　$\nu = 0.326$
　　(c)　$\sigma_{PL} = 330$ MPa

3-3　(a)　$\nu = 0.350$
　　(b)　$E = 25,000$ ksi
　　(c)　$G = 9260$ ksi

3-5　(a)　$\nu = 0.333$
　　(b)　$E = 15,000$ ksi
　　(c)　$G = 5620$ ksi

3-9　(a)　$\sigma_x = 8185$ psi T
　　　　$\sigma_y = 587$ psi T
　　　　$\tau_{xy} = -760$ psi
　　(b)　$\sigma_{p1} = 8260$ psi T
　　　　$\sigma_{p2} = 512$ psi T
　　　　$\sigma_{p3} = 0$
　　　　$\tau_p = 3874$ psi
　　　　$\tau_{max} = 4130$ psi
　　　　$\theta_p = 5.65°$

3-10　(a)　$\sigma_x = 201$ MPa T
　　　　$\sigma_y = 39.2$ MPa T
　　　　$\tau_{xy} = 32.3$ MPa
　　(b)　$\sigma_{p1} = 207$ MPa T
　　　　$\sigma_{p2} = 33.1$ MPa T
　　　　$\sigma_{p3} = 0$
　　　　$\tau_p = 87.0$ MPa
　　　　$\tau_{max} = 103.6$ MPa
　　　　$\theta_p = 10.9°$

3-13　$\varepsilon_{p1} = 700\ \mu$
$\varepsilon_{p2} = -300\ \mu$
$\varepsilon_{p3} = \varepsilon_z = -171.4\ \mu$
$\gamma_p = 1000\ \mu$
$\gamma_{max} = 1000\ \mu$
$\theta_p = 18.43°$

$\sigma_{p1} = 20.1$ ksi T
$\sigma_{p2} = 2.97$ ksi C
$\sigma_{p3} = 0$
$\tau_p = 11.54$ ksi
$\tau_{max} = 11.54$ ksi
$\theta_p = 18.43°$↰

3-14 $\sigma_x = 17.80$ ksi T
$\sigma_y = 0.659$ ksi C
$\tau_{xy} = 6.92$ ksi
$\sigma_{p1} = 20.1$ ksi T
$\sigma_{p2} = 2.97$ ksi C
$\sigma_{p3} = 0$
$\tau_p = 11.54$ ksi
$\tau_{max} = 11.54$ ksi
$\theta_p = 18.42°$↰

3-17 (a) $\varepsilon_{p1} = 967 \mu$
(b) $\gamma_{max} = 1767 \mu$
(c) $\sigma_{p1} = 13.25$ ksi T
(d) $\tau_{max} = 6.63$ ksi
(e) $\theta_p = 45.0°, 135.0°$

3-20 (a) $u = 4.43$ in.-lb/in.^{3}
(b) $u = 4.43$ in.-lb/in.^{3}

3-21 (a) $u = 41.0$ kN · m/m^3
(b) $u = 41.0$ kN · m/m^3

3-23 (a) $u = 70.0$ kN · m/m^3
(b) $u = 70.0$ kN · m/m^3
(c) $u = 70.0$ kN · m/m^3

3-25 (a) $u = 3.29$ in.-lb/in.^{3}
(b) $u = 3.29$ in.-lb/in.^{3}

3-27 $\Delta L = 16.38$ in.

3-28 $\Delta L = 185.9$ mm
$\Delta D = 5.21$ mm

3-29 $\delta = 0.059$ in. upward

3-31 Add 0.2852 in. to the reading.

3-33 (a) $\sigma_{max} = 19.50$ ksi C
(b) $\tau_{max} = 9.75$ ksi
(c) $P = 234$ kips C

3-34 (a) $\sigma_{max} = 75.0$ MPa T
(b) $\Delta D = 0.0804$ mm decrease

3-36 (a) $\sigma = -20\alpha E$
(b) $\sigma = 31.5$ MPa C

3-38 (a) $k_s = 2.25$
(b) $k_f = 4.13$

3-39 (a) $A = 1160$ mm^2
Use L102 × 89 × 6.4 angles.
(b) $k = 2.02$

3-40 (a) $W_{max} = 2322$ lb
(b) $k_A = 1.25$
$k_B = 1.18$
(c) $k_A = 3.61$
$k_B = 2.94$

3-42 (a) $\sigma_{CD} = 5.31$ ksi T
(b) $\tau_B = 3.42$ ksi
(c) $k_{CD} = 7.73$
(c) $k_B = 11.11$

3-45 (a) $\sigma_{CF} = 25.0$ MPa C
(b) $\tau_C = 76.6$ MPa
(c) $k_{CE} = 11.2$
(d) $k_C = 3.66$

CHAPTER 4

4-1 $P_{max} = 39.3$ kips T

4-2 $P_{max} = 193.8$ kN T

4-3 (a) $\delta = 0.0531$ in.
(b) $\delta = 0.0489$ in.
(c) $k_a = 3.12$
$k_b = 4.34$

4-5 $\delta = 0.1226$ in.

4-8 $\delta = 3.07$ mm

4-10 $\delta = \gamma L^2/6E$

4-12 (a) $k_A = 1.78$
(b) $P = 24.8$ kips
(c) $\delta_C = 0.0375$ in.

4-13 (a) $k_A = 3.20$
(b) $P = 153.4$ kN
(c) $\delta_D = 1.800$ mm

4-15 (a) $\sigma_A = 10.50$ ksi T
$\sigma_B = 34.0$ ksi T
(b) $P = 162.4$ kips
(c) $\delta_C = 0.088$ in.

4-17 (a) $\sigma_A = 166.1$ MPa T
$\sigma_B = 40.0$ MPa T
$\sigma_C = 160.0$ MPa T

(b) $P = 402$ kN

(c) $\delta_D = 4.00$ mm

4-19 (a) $\sigma_A = 291$ MPa T

$\sigma_B = 40.0$ MPa T

$\sigma_C = 240$ MPa T

(b) $P = 626$ kN

$\delta_D = 7.00$ mm

4-22 (a) $\sigma_c = 1222$ psi C

$\sigma_s = 12{,}220$ psi C

(b) $\delta = 0.00978$ in.

4-23 (a) $P_{max} = 2453$ kN

(b) $\delta = 0.375$ mm

4-25 (a) $\sigma_A = 275$ MPa T

$\sigma_B = 119.4$ MPa C

(b) $k_A = 1.56$

$k_B = 2.76$

(c) $\delta = 0.982$ mm

4-27 (a) $\sigma_A = 254$ MPa T

$\sigma_B = 206$ MPa T

(b) $k_A = 1.69$

$k_B = 1.60$

(c) $\delta = 1.354$ mm

4-30 (a) $\sigma_A = 200$ MPa T

$\sigma_B = 150$ MPa T

(b) $k_A = 2.50$

$k_B = 2.67$

(c) $\delta_C = 1.000$ mm

4-32 $P_{max} = 168.5$ kips

4-35 $\sigma_{CE} = 35.5$ MPa T

4-36 (a) $W_E = 7.84$ kips

(b) $\sigma_A = 10.24$ ksi T

4-38 (a) $\sigma_{bolt} = 164.3$ MPa T

$\sigma_{bar} = 60.4$ MPa C

(b) $d = 250.856$ mm

4-39 $\sigma_B = 11.05$ ksi T

$\sigma_C = 2.09$ ksi C

4-41 $P = 46.0$ kN

4-43 $\sigma_A = 27.7$ ksi T

$\sigma_B = 35.5$ ksi T

$\delta_C = 0.0886$ in.

4-46 $\sigma_A = 170.2$ MPa T

4-48 $\delta_A = 0.038$ in.

$\delta_C = 0.012$ in.

4-50 $\sigma_A = 25$ ksi T

$\sigma_B = 30$ ksi T

4-52 $\delta_C = 0.122$ in.

4-53 $\sigma = 960$ psi T

$\tau = 490$ psi

4-54 $F = 210$ kN

4-57 $t = 0.600$ in.

4-58 $p = 1.951$ MPa

4-61 $\sigma = 8.54$ ksi T

$\tau = 3.20$ ksi

4-63 $\sigma_m = 7\gamma r^2/18t$ T

$\sigma_t = \gamma r^2/9t$ T

4-66 $\sigma_a = 496$ psi T

$\sigma_h = 1489$ psi T

4-67 $\sigma_a = 1.653$ MPa T

$\sigma_h = 5.95$ MPa T

4-70 $\sigma_{max} = 25.0$ ksi T

$\tau_{max} = 20.0$ ksi

4-71 (a) $\sigma_{max} = 195.0$ MPa T

(b) $\sigma_t = 146.4$ MPa T

$\sigma_r = 26.4$ MPa C

4-73 $F = 454$ kN/m C

4-76 $t_{min} = 0.633$ in.

4-78 (a) $\sigma_{max} = 13.00$ ksi T

(b) $p_s = 1.875$ ksi

4-79 $t_{min} = 72.5$ mm

CHAPTER 5

5-1 $\tau_{max} = 12.73$ ksi

$\theta = 0.0764$ rad

5-2 $\tau_{max} = 55.9$ MPa

$\theta = 0.0279$ rad

5-3 $T_{max} = 63.6$ in.-kips

5-4 $d_{min} = 53.5$ mm

5-5 (a) $\tau_{AB} = 9.43$ ksi

$\tau_{BC} = 3.77$ ksi

(b) $\theta_{BA} = 0.0252$ rad

$\theta_{CA} = 0.0327$ rad

5-8 (a) $d_{AB} = 63.4$ mm

$d_{BC} = 79.9$ mm

(b) $\theta_{AC} = 0.0483$ rad

5-9 $T_{max} = 15.71$ in.-kips

5-13 $T_{max} = 81.9$ in.-kips

5-16 $T_1 = 15.02$ kN · m
$T_2 = 111.5$ kN · m

5-17 (a) $\tau_{max} = 11.21$ ksi
$\theta_{DB} = 0.00764$ rad ↻——

5-20 (a) $\tau_{max} = 86.6$ MPa
(b) $\theta_3 = 0.0729$ rad ↻——
(c) $\theta_{4.5} = 0.0749$ rad ↻——

5-22 $\theta = 3TL/(16Gt\pi r^3)$

5-23 $\theta = 28TL/(45\pi Gr^4)$

5-25 $\theta = qL^2/(\pi Gc^4)$

5-27 $T_{max} = 1091$ in.-lb

5-28 $T_{max} = 11.96$ kN · m

5-31 (a) $T_{max} = 149.6$ in.-kips
(b) $k_\sigma = 5.77$
$k_\tau = 6.00$

5-33 Power $= 170.3$ hp

5-34 (a) Power $= 286$ kW
(b) $\theta = 0.0600$ rad

5-35 $d_{min} = 4.24$ in.

5-38 $d_{min} = 84.4$ mm

5-41 (a) $\gamma_c = 0.004$ rad
(b) Ratio $= 1.292$

5-42 (a) $T_{max} = 23.0$ kN · m
(b) $\Delta T = 39.5\%$

5-44 $T = 16.08$ kN · m

5-47 (a) $T = 155.5$ in.-kips
(b) $\theta = 0.270$ rad

5-49 (a) $T = 417$ in.-kips
(b) $\theta/L = 0.0885$ rad

5-51 $T = 4.12$ kN · m

5-53 (a) $\tau_{max} = 34.4$ ksi
(b) $\theta = 0.0687$ rad

5-54 (a) $T_{max} = 18.82$ kN · m
(b) $\theta_{1.25} = 0.0547$ rad

5-55 (a) $\tau_{max} = 16.34$ ksi
(b) $\theta = 0.0408$ rad

5-56 (a) $\tau_A = 73.3$ MPa
(b) $\tau_B = 67.8$ MPa
(c) $\theta = 0.01878$ rad

5-58 (a) $\theta = 0.0730$ rad
(b) $\tau_s = 73.0$ MPa
$\tau_b = 109.5$ MPa

5-60 (a) $\tau_s = 93.8$ MPa

$\tau_b = 69.6$ MPa
$\theta = 0.01419$ rad ↻——

5-63 $T_{max} = 2.18$ in.-kips

5-66 $\tau_s = 61.1$ MPa
$\tau_b = 18.47$ MPa

5-67 $T = 748$ in.-kips

5-68 $T = 15.93$ kN · m

5-70 (a) $\tau_s = 120.0$ MPa
$\tau_b = 141.1$ MPa
(b) $\theta = 0.0767$ rad ↻——

5-73 (a) $T_1 = 137.5$ in.-kips
$T_2 = 83.1$ in.-kips
$T_3 = 108.0$ in.-kips
$\theta_1 = 0.0262$ rad
$\theta_2 = 0.0433$ rad
$\theta_3 = 0.0333$ rad

5-74 (a) $T_1 = 8.36$ kN · m
$T_2 = 6.56$ kN · m
(b) $T_3 = 6.30$ kN · m
$\theta_1 = 0.0220$ rad
$\theta_2 = 0.0280$ rad
$\theta_3 = 0.0292$ rad

5-75 (a) $\tau = 7.87$ ksi
(b) $\theta/L = 0.00539$ rad

5-78 $t_{min} = 3.81$ mm

5-80 (a) $\tau_{max} = 74.4$ MPa
(b) $\theta/L = 0.00353$ rad

5-82 (a) $T_{max} = 1004$ in.-kips
(b) $\theta/L = 0.00152$ rad

CHAPTER 6

6-1 $M_r = 51.2$ in.-kips

6-2 (a) $M_r = 8.00$ kN · m
(b) $\Delta M_r = 21.1\%$

6-3 (a) $\sigma_{max} = 10.00$ ksi C
(b) $M_r = 544$ in.-kips

6-4 $\sigma_{max} = 12.80$ MPa T and C

6-7 $F = 38.4$ kips T

6-10 $M_f/M = 84.1\%$

6-11 $M_r = 51.2$ in.-kips

6-12 $M_r = 8.00$ kN · m

6-13 $M_r = 544$ in.-kips

6-14 $\sigma_{max} = 12.80$ MPa T and C

6-17 (a) $\sigma = 3.69$ ksi T

    (b) $\sigma_{max} = 4.22$ ksi
T and C

6-20 (a) $\sigma = 48.7$ MPa C

    (b) $\sigma_{max} = 57.2$ MPa
T and C

6-21 (a) $\sigma_A = 387$ psi C

    (b) $\sigma_B = 646$ psi T

6-24 (a) $\sigma_A = 2.87$ MPa C

    (b) $\sigma_B = 12.45$ MPa T

6-26 $w_{max} = 3.26$ kN/m

6-29 $V = 6000 - 1000x$

    $M = -500x^2 + 6000x$
$- 8000$

6-30 $V = 22 - 8x$

    $M = -4x^2 + 22x - 19$

6-31 $V = 5200 - 800x$

    $M = -400x^2 + 5200x$
$+ 26{,}000$

6-34 (a) $V = 70x - 60$

    $M = 35x^2 - 60x - 90$

    (b) $V = 80$

    $M = 80x - 230$

6-36 (a) $V = w(L + x)$

    $M = \dfrac{w}{2}(x^2 + 2Lx$
$- 2L^2)$

    (b) $V = 2w(2L - x)$

    $M = -w(x^2 - 4Lx$
$+ 4L^2)$

    (c) $V_{max} = 2wL$ at $x = L$

    $M_{max} = -wL^2$ at
$x = 0$ and $x = L$

6-41 (a) $L = 66$ ft

    (b) $V = 970x - 24{,}000$

    $M = 485x^2 - 24{,}000x$
$+ 192{,}000$

6-42 Partial results:

$R_L = 6000$ lb

$R_R = 4000$ lb

$V_{max} = 4000$ lb

$M_{max} = 10{,}000$ ft-lb

6-43 Partial results:

$R_L = 27$ kN

$R_R = 12$ kN

$V_{max} = 18$ kN

$M_{max} = 10.8$ kN · m

6-44 Partial results:

$R_L = 1700$ lb

$R_R = 8300$ lb

$V_{max} = -4300$ lb

$M_{max} = -10{,}000$ ft-lb

6-45 Partial results:

$R_1 = 70$ N/m

$R_2 = 45$ N

$V_{max} = 80$ N

$M_{max} = -115.7$ N · m

6-48 Partial results:

$R_L = 9600$ lb

$R_R = 16{,}800$ lb

$V_{max} = -10{,}800$ lb

$M_{max} = 38{,}400$ ft-lb

6-49 Partial results:

$V_{max} = 12$ kN

$M_{max} = -7.5$ kN · m

$M = -2x^2 + 18x - 30$

6-52 Partial results:

$V_{max} = 5000$ lb

$M_{max} = -12{,}000$ ft-lb

    (a) $M = -250x^2 + 5000x$
$- 12{,}000$

    (b) $M = -500x^2 + 6000x$
$- 12{,}000$

6-55 Partial results:

$R_L = 14.75$ kN

$R_R = 7.25$ kN

$V_{max} = 14.75$ kN

$M_{max} = 15.94$ kN · m

6-56 Partial results:

$V_{max} = 1600$ lb

$M_{max} = 5400$ ft-lb

6-59 Partial results:

$V_{max} = -6$ kN

$M_{max} = 5.5$ kN · m

$M = -1.5x^2 + 6x - 0.5$

6-63 (a) $\sigma_{max} = 64.6$ MPa T

(b) $\sigma_{max} = 93.8$ MPa C

6-64 (a) $\sigma_{max} = 5.44$ ksi T

(b) $\sigma_{max} = 8.17$ ksi C

6-66 $w_{max} = 12.05$ kN/m

6-69 $b = 3.69$ in.

$h = 7.39$ in.

6-70 $b = 114.5$ mm

$h = 229$ mm

6-76 (a) $M_2 = 1620$ ft-lb

$F_1 = 2880$ lb

$F_2 = 4320$ lb

(b) $V_H = 1440$ lb

(c) $\tau_{avg} = 40.0$ psi

6-77 (a) $F_1 = 24.0$ kN

$F_2 = 36.0$ kN

(b) $V_H = 12.00$ kN

(c) $\tau_{avg} = 882$ kPa

6-80 $\tau_{top} = 740$ kPa

$\tau_{bot} = 634$ kPa

6-83 $s = 14.26$ in.

6-84 (a) $\tau_V = 143.7$ psi

(b) $\tau_{max} = 171.3$ psi at NA

6-85 (a) $\tau_V = 707$ kPa

(b) $\tau_{max} = 822$ kPa at NA

6-88 (a) $\sigma_A = 247$ psi T

$\tau_A = 91.0$ psi

(b) $\tau_{max} = 128.7$ psi,

$b-b$ at NA

(c) $\sigma_{max} = 3574$ psi C,

5 ft from a-a

6-91 $P_{max} = 2.19$ kN

6-92 $w_{max} = 1800$ lb/ft

6-96 (a) $\sigma_{max} = 6.73$ ksi T

(b) $\sigma_{max} = 5.05$ ksi C

(c) $\tau_{max} = 579$ psi

$\tau_A = 547$ psi

6-97 (a) $\sigma_{max} = 64.1$ MPa T

(b) $\tau_{max} = 24.0$ MPa

(c) $\tau_A = 19.49$ MPa

6-99 $\sigma_{p1} = 86.1$ MPa T

$\sigma_{p2} = 6.35$ MPa C

$\sigma_{p3} = 0$

$\tau_p = 46.2$ MPa

$\tau_{max} = 46.2$ MPa

$\theta_p = 15.2°$⤴

6-100 $\sigma_{p1} = 21.5$ psi T

$\sigma_{p2} = 335$ psi C

$\sigma_{p3} = 0$

$\tau_p = 178.0$ psi

$\tau_{max} = 178.0$ psi

$\theta_p = 14.2°$⤴

6-102 $P_{max} = 50.1$ kips

6-105 $\sigma_{max} = 75.2$ MPa T and C

$\tau_{max} = 43.7$ MPa

6-108 (a) $\sigma_{max} = 21.5$ ksi

T and C

$\tau_{max} = 12.69$ ksi

(b) $\sigma_{max} = 20.2$ ksi

T and C

$\tau_{max} = 10.11$ ksi

6-109 (a) $\sigma_{max} = 1519$ psi

T and C

(b) $\beta = 75.3°$⤴

6-110 (a) $\sigma_{max} = 70.8$ MPa C

(b) $\beta = 67.4°$⤴

6-111 (a) $\sigma_A = 5.84$ ksi T

(b) $\sigma_{max} = 7.48$ ksi C

(c) $\beta = 30.5°$⤴

6-114 (a) $\sigma_A = 16.46$ MPa C

$\sigma_B = 16.46$ MPa T

$\beta = 45°$⤴

6-117 (a) $\sigma_A = 9.62$ ksi T

$\sigma_B = 8.11$ ksi C

(b) $\beta = 60.3°$⤴

6-118 $\Delta\sigma = 31.3\%$

6-120 $M_p = 4380$ in.-kips

6-121 $M_p = 909$ kN · m

6-123 $M_p/M_e = 1.698$

6-124 $M_p/M_e = 1.136$

6-126 (a) $a = 3.775$ in.

(b) $M = 1558$ in.-kips

(c) $M_p/M_e = 1.714$

6-129 (a) $c = 119.6$ mm

(b) $M = 19.43$ kN · m

6-130 (a) $e = 29.5$ mm

(b) $\tau_{max} = 11.11$ MPa

6-131 (a) $e = 1.706$ in.
    (b) $\tau_{max} = 8.38$ ksi

6-134 (a) $e = 6.98$ mm
    (b) $\tau_{max} = 13.69$ MPa

6-135 (a) $e = 4R/\pi$
    (b) $\tau_A = 318$ psi

6-138 (a) $e = 1.429$ in.
    (b) $\tau_0 = 857$ psi

6-140 $\tau_{max} = 1875$ psi
    $e = 1.50$ in.

6-144 $\sigma_s = 11.31$ ksi T

6-145 $\sigma_w = 9.28$ MPa C

6-146 $\sigma_w = 1000$ psi T
    $\sigma_a = 10,000$ psi T

6-150 $\sigma_w = 5.26$ MPa T and C
    $\sigma_s = 120.9$ MPa T and C

6-152 $\sigma_c = 918$ psi C
    $\sigma_s = 15,330$ psi T

6-153 $\sigma_c = 6.16$ MPa C
    $\sigma_s = 117.5$ MPa T

6-156 $d = 18.02$ in.
    $A_s = 1.605$ in.^{2}

6-158 $M_{max} = 562$ in.-kips

6-161 $\sigma_i = 129.0$ MPa C
    $\sigma_o = 107.2$ MPa T

6-162 $b = 6.24$ in.

6-164 $\sigma_o = 8.01$ ksi C
    $\sigma_i = 7.65$ ksi T

6-166 $\sigma_o = 3.90$ ksi C
    $\sigma_i = 2.55$ ksi T

6-167 $M_{max} = 150.5$ kN · m

CHAPTER 7

7-1 (a) $\rho = 10,700$ in.
    (b) $\delta = 0.242$ in.
    (c) $\sigma_{max} = 8.82$ ksi
          T and C

7-2 (a) $\rho = 640$ m
    (b) $\delta = 0.781$ mm
    (c) $\sigma_{max} = 53.5$ MPa T

7-5 $t_{max} = 0.600$ in.

7-6 $\sigma_{max} = 333$ MPa T and C

7-9 (a) $y = \dfrac{w}{24EI}(-x^4 + 2Lx^3 - 2L^3x + L^4)$
    (b) $\delta = wL^4/24EI \uparrow$

7-10 (a) $y = \dfrac{w}{16EI}(Lx^3 - 4L^2x^2 + 3L^3x)$
    (b) $\delta = 5wL^4/128EI \uparrow$

7-11 (a) $y = \dfrac{w}{96EI}(-4x^4 + 7Lx^3 - 3L^3x)$
    (b) $\delta = 7wL^4/768EI \downarrow$

7-12 (a) $y = \dfrac{w}{48EI}(-2x^4 + 3Lx^3 - 3L^2x^2 + 2L^3x)$
    (b) $\delta = wL^4/96EI \uparrow$

7-15 (a) $y = \dfrac{w}{120EIL}(-x^5 + 5L^4x - 4L^5)$
    (b) $\delta = wL^4/30EI \downarrow$

7-16 (a) $y = \dfrac{w}{120EIL}(-x^5 + 10L^2x^3 - 20L^3x^2)$
    (b) $\delta = 11wL^4/120EI \downarrow$

7-19 (a) $\delta = 0.338$ in. $\uparrow$
    (b) $\delta_{max} = 0.405$ in. $\downarrow$

7-20 (a) $\delta = 10.37$ mm $\downarrow$
    (b) $\delta = 22.0$ mm $\downarrow$

7-23 $\delta_{max} = 0.03208ML^2/EI$

7-24 (a) $\delta_{L/2} = 0.0143PL^3/EI \downarrow$
    (b) $\delta_{max} = 0.01456PL^3/EI \downarrow$

7-27 $\delta = 5wL^4/48EI \downarrow$

7-28 $\delta = 8.82$ mm $\uparrow$

7-31 $\delta = 20PL^3/81EI \downarrow$

7-32 $\delta = wL^4/24EI \uparrow$

7-33 $\delta = 5PL^3/48EI \downarrow$

7-34 $\delta = 2wL^4/3EI \downarrow$

7-37 (a) $\delta_{L/2} = 0.01432PL^3/EI \downarrow$
    (b) $\delta_{max} = 0.01456PL^3/EI \downarrow$

7-40 $\delta = 5wL^4/48EI \downarrow$

7-41 (a) $\delta = 29wL^4/48EI \downarrow$

(b) $\delta = 129wL^4/384EI \downarrow$

7-42 (a) $\delta = 9wL^4/128EI \downarrow$

(b) $\delta_{max} = 0.0394wL^4/EI \uparrow$

7-57 $\delta = 0$

7-58 $\delta = 67PL^3/162EI \downarrow$

7-59 $\delta = 29wL^4/48EI \downarrow$

7-60 $\delta = 23wL^4/24EI \downarrow$

7-63 $\delta = 23wL^4/8EI \downarrow$

7-67 (a) $\delta = 19PL^3/324EI \downarrow$

(b) $\delta = PL^3/48EI \uparrow$

7-70 $\delta = 305wL^4/384EI \downarrow$

7-71 $\delta = 101wL^4/128EI \downarrow$

7-73 $\delta = 0.632$ in. $\downarrow$

7-74 $\delta = 8.82$ mm $\uparrow$

7-77 $\delta = 0.1312$ in. $\downarrow$

7-79 $\delta_{max} = 0.484PL^3/EI \downarrow$

7-80 $\delta_{max} = 5.82PL^3/EI \downarrow$

7-81 $\delta_{max} = 0.0394wL^4/EI \uparrow$

7-82 $\delta_{max} = 0.00915wL^4/EI \downarrow$

7-85 (a) $\delta = 0.317$ in. $\downarrow$

(b) $\delta = 0.0465$ in. $\uparrow$

7-86 (a) $\delta = 26.0$ mm $\downarrow$

(b) $\delta = 26.25$ mm $\downarrow$

7-89 $\delta_{max} = 0.0258PL^3/EI \uparrow$

7-91 $\delta = 2.50$ in. $\downarrow$

7-92 $\delta = 11.14$ mm $\downarrow$

7-93 $\delta = 0.722$ in. $\downarrow$

7-95 $\delta = 5wL^4/32EI \downarrow$

7-97 $\delta = 7wL^4/768EI \downarrow$

7-99 $\delta = 11wL^4/120EI \downarrow$

7-102 $\delta = 23wL^4/12EI \downarrow$

7-104 $\delta = 0$

7-107 $\delta = 16PL^3/1875EI \downarrow$

7-108 $\delta = PL^3/6EI \uparrow$

7-109 $\delta = 0$

7-110 $\delta = 29wL^4/24EI \downarrow$

7-113 $\delta = PL^3/EI \downarrow$

7-116 $\delta = 0.334$ in. $\uparrow$

7-117 $\delta = 11.31$ mm $\downarrow$

7-120 $\delta = 13wL^4/384EI \downarrow$

7-121 $\delta = 49wL^4/240EI \downarrow$

7-122 $\delta_v = \pi PR^3/4EI \downarrow$

7-124 $\delta_v = \pi PR^3/2EI \downarrow$

CHAPTER 8

8-1 $Q = wL^2/4$

8-2 $Q = wL^2/6$

8-4 (a) $P = 14wL/9$

(b) $\delta_A = 46wL^4/27EI \downarrow$

8-6 (a) $R_A = 3wL/8 \uparrow$

(b) $\delta_{max} = 0.00542wL^4/EI \downarrow$

8-8 (a) $P = wL/4$

(b) $\delta_{max} = 0.00542wL^4/EI \downarrow$

8-10 (a) $M = 3wL^2/4$

(b) $R_A = 3wL/2 \uparrow$

$V_B = wL/2 \downarrow$

$M_B = wL^2/4 \,\rangle$

8-11 (a) $R_A = 5wL/12 \uparrow$

$R_B = 7wL/6 \uparrow$

$R_C = 5wL/12 \uparrow$

(b) $\delta_{L/2} = 5wL^4/1152EI \downarrow$

8-14 (a) $R_A = 26.7$ kN $\uparrow$

(b) $\delta_B = 8.99$ mm $\downarrow$

8-15 (a) $R_A = 1960$ lb $\uparrow$

$R_B = 4470$ lb $\uparrow$

$R_C = 3370$ lb $\uparrow$

(b) $\delta = 0.000784$ in. $\uparrow$

8-18 (a) $R_A = 11wL/28 \uparrow$

$R_B = 8wL/7 \uparrow$

$V_C = 13wL/28 \uparrow$

$M_C = wL^2/14 \,\rangle$

(b) $M_B = 3wL^2/28 \,\rangle$

8-19 $R_B = wL \uparrow$

8-22 $R_C = 163.5$ kN $\uparrow$

8-23 $R_L = WL/40 \downarrow$

8-26 (a) $R_L = wL/4 \uparrow$

(b) $\delta_L = 5wL^4/6EI \uparrow$

8-27 (a) $R_R = 43wL/64 \uparrow$

(b) $\delta_R = 17wL^4/6144EI \uparrow$

8-28 (a) $R_A = 3wL/8 \uparrow$

(b) $\delta_{max} = 0.00542wL^4/EI \downarrow$

8-31 (a) $R_A = 38wL/27 \uparrow$

(b) $\delta_{max} = 0.798wL^4/EI \downarrow$

8-32 (a) $V_A = 13wL/16 \uparrow$

$M_A = 11wL^2/48 \,\langle$

(b) $\delta_B = wL^4/48EI \downarrow$

8-35 (a) $M = PL \,\rangle$

8-36 (b) $R_B = 3P \downarrow$

8-36 (a) $R_A = 5wL/12 \uparrow$
$R_B = 7wL/6 \uparrow$
$R_C = 5wL/12 \uparrow$

(b) $\delta_{L/2} = 5wL^4/1152EI \downarrow$

8-39 (a) $R_A = 15.38$ kN $\uparrow$
$R_B = 41.0$ kN $\uparrow$
$R_C = 23.6$ kN $\uparrow$

(b) $\delta_R = 0.1184$ mm $\uparrow$

8-40 (a) $R_A = 9wL/8 \uparrow$
$R_B = 13wL/4 \downarrow$
$R_C = 57wL/8 \uparrow$

(b) $\delta_R = 89wL^4/48EI \downarrow$

8-43 $R_A = 3wL/8 \uparrow$

8-44 $R_B = 7wL/6 \uparrow$

8-46 $R_B = 8wL/7 \uparrow$

8-47 $R_B = wL \uparrow$

8-50 $R_A = 38wL/27 \uparrow$

8-51 $R_L = 2.28$ kips $\uparrow$
$R_C = 3.44$ kips $\uparrow$
$R_R = 2.28$ kips $\uparrow$

8-53 $P = 434$ lb T

8-55 $R_B = 26.4$ kN $\uparrow$

8-59 $\sigma = 69.6$ MPa T

8-61 $\delta_B = 1.875$ in. $\downarrow$

8-63 (a) $Q = wL^2/6$
(b) $\delta = wL^4/24EI \downarrow$

8-64 (a) $M = wL^2/60$
(b) $\delta = wL^4/160EI \uparrow$

8-65 (a) $R_B = 7wL/6 \uparrow$
(b) $\delta_{L/2} = 5wL^4/1152EI \downarrow$

8-68 (a) $V_A = 3wL/8 \uparrow$
$M_A = wL^2/24 \left($
(b) $\delta_C = 7wL^4/288EI \downarrow$

8-70 $R_A = 11wL/28 \uparrow$
$R_B = 8wL/7 \uparrow$

8-71 $R_B = wL \uparrow$

8-74 $R = wL/40 \downarrow$

8-76 $P = 434$ lb T

8-78 $R_B = 1.404P \leftarrow$

8-81 $w = 11.66M_p/L^2$

8-82 $P = 2M_p/L$

8-85 Use a W18 × 60 section.

8-86 (a) $w = 54.5$ kN/m
(b) Use a W457 × 89 section.
(c) 7.19 m to the right of A.

8-88 Use a W16 × 40 section.

8-90 Use a W406 × 60 section.

CHAPTER 9

9-1 $\sigma_{p1} = 2000$ psi T
$\sigma_{p2} = 500$ psi C
$\sigma_{p3} = 0$
$\tau_p = 1250$ psi
$\tau_{max} = 1250$ psi
$\theta_p = 26.6°$⌝

9-2 $\sigma_{p1} = 42.9$ MPa T
$\sigma_{p2} = 28.4$ MPa C
$\sigma_{p3} = 0$
$\tau_p = 35.7$ MPa
$\tau_{max} = 35.7$ MPa
$\theta_p = 39.1°$⌝

9-4 $P = 1623$ kN

9-5 $P = 125.0$ kips

9-7 (a) $\tau = 22.9$ ksi
(b) $\tau = 28.6$ ksi

9-8 (a) $P = 5.62$ kN
(b) $P = 4.72$ kN

9-10 (a) $\tau = 233$ MPa
(b) $\delta = 160.0$ mm

9-13 $k = 202$ lb/in.

9-14 $\sigma_L = 1000$ psi T
$\sigma_R = 1400$ psi C

9-15 $\sigma_A = 5.83$ MPa T
$\sigma_B = 0.833$ MPa T
$\sigma_C = 9.17$ MPa C
$\sigma_D = 4.17$ MPa C

9-18 $\sigma_A = 1100$ psi T
$\sigma_B = 500$ psi T
$\sigma_C = 1500$ psi C
$\sigma_D = 900$ psi C

9-19 $\sigma_{max} = 22.8$ MPa C

9-21 $P_{max} = 316$ kN

9-22 $h = 1.000$ in.

9-25 $\sigma_{max} = 1520$ psi C

9-27 $\sigma_{p1} = 32.3$ ksi T
$\sigma_{p2} = 0.279$ ksi C
$\sigma_{p3} = 0$
$\tau_p = 16.28$ ksi
$\tau_{max} = 16.28$ ksi
$\theta_p = 5.31°$↰

9-30   At point G:
$\sigma_{p1} = 0$
$\sigma_{p2} = 16.56$ MPa C
$\sigma_{p3} = 0$
$\tau_p = 8.28$ MPa
$\tau_{max} = 8.28$ MPa
$\theta_p = 0$
At point H:
$\sigma_{p1} = 7.75$ MPa T
$\sigma_{p2} = 0.01190$ MPa C
$\sigma_{p3} = 0$
$\rho_p = 3.88$ MPa
$\tau_{max} = 3.88$ MPa
$\theta_p = 2.24°$↰

9-32 $P_{max} = 9.63$ kips

9-34 $\sigma_{p1} = 6.74$ ksi T
$\sigma_{p2} = 1.557$ ksi C
$\sigma_{p3} = 0$
$\tau_p = 4.15$ ksi
$\tau_{max} = 4.15$ ksi
$\theta_p = 25.7°$↰

9-35 $\sigma_{p1} = 75.8$ MPa T
$\sigma_{p2} = 31.5$ MPa C
$\sigma_{p3} = 0$
$\tau_p = 53.7$ MPa
$\tau_{max} = 53.7$ MPa
$\theta_p = 32.8°$↰

9-36 $\sigma_{p1} = 23.9$ ksi T
$\sigma_{p2} = 0.955$ ksi C
$\sigma_{p3} = 0$
$\tau_p = 12.41$ ksi
$\tau_{max} = 12.41$ ksi
$\theta_p = 11.31°$↰

9-39 $R_{max} = 17.62$ kN

9-40 $\sigma_{p1} = 0.094$ ksi T

$\sigma_{p2} = 6.21$ ksi C
$\sigma_{p3} = 0$
$\tau_p = 3.15$ ksi
$\tau_{max} = 3.15$ ksi
$\theta_p = 7.02°$↰

9-43 $\sigma_{p1} = 27.5$ MPa T
$\sigma_{p2} = 77.5$ MPa C
$\sigma_{p3} = 0$
$\tau_p = 52.5$ MPa
$\tau_{max} = 52.5$ MPa
$\theta_p = 30.8°$↰

9-45   (a) $\sigma = 65.8$ MPa T
$\tau = 23.5$ MPa
(b) $\sigma_{p1} = 77.9$ MPa T
$\sigma_{p2} = 22.0$ MPa T
$\sigma_{p3} = 1.700$ MPa C
$\tau_p = 28.0$ MPa
$\tau_{max} = 39.8$ MPa
$\theta_p = 24.8°$↰

9-46 $\sigma_{p1} = 1800$ psi T
$\sigma_{p2} = 3200$ psi C
$\sigma_{p3} = 0$
$\tau_p = 2500$ psi
$\tau_{max} = 2500$ psi
$\theta_p = 36.9°$↰

9-49   Failure predicted by:
1. Maximum-shearing-stress theory
2. Maximum-distortion-energy-theory

9-50   Failure predicted by:
1. Maximum-shearing-stress-theory
2. Maximum-distortion-energy-theory

9-52   (a) $k = 1.60$ (Maximum-normal-stress)
$k = 1.60$ (Maximum-shearing-stress)
$k = 2.00$ (Maximum-normal-strain)
$k = 1.83$ (Maximum-distortion-energy)

(b) $k = 1.60$ (Maximum-
normal-stress)
$k = 1.00$ (Maximum-
shearing-stress)
$k = 1.33$ (Maximum-
normal-strain)
$k = 1.14$ (Maximum-
distortion-energy)

9-55 (a) $P = 475$ kips
(b) $P = 339$ kips
(c) $P = 442$ kips
(d) $P = 408$ kips
9-57 (a) $R = 5.84$ kips
(b) $R = 6.11$ kips
9-60 $p = 81.3$ MPa

CHAPTER 10
10-1 $P_{max} = 240$ lb
10-2 $P_{max} = 750$ N
10-3 $P_{max} = 720$ lb
10-5 (a) $L/r = 144$
(b) $P_{cr} = 78.5$ kips
(c) $\sigma = 14.28$ ksi C
10-6 (a) $L/r = 693$
(b) $P_{cr} = 40.1$ N
10-7 (a) $L/r = 92.2$
(b) $P_{cr} = 46.4$ kips
(c) $\sigma = 2.32$ ksi
10-10 $P_w = 164.4$ kN
10-11 (a) $L/r = 146.9$
(b) $P_{cr} = 194.3$ kips
(c) $\sigma = 13.22$ ksi
10-14 $P_w = 218$ kN
10-15 $R = 1.021$ in.
10-16 Use 178×178-mm posts.
10-18 (a) $P_{cr} = 27.8$ kips
(b) $P_{cr} = 82.8$ kips
10-19 (a) $P_{cr} = 27.7$ kN
(b) $P_{cr} = 694$ kN
10-25 $L' = L$
10-26 $P_w = 2.38$ kips
10-28 $m = 2.23$ Mg
10-29 $P_w = 15.30$ kips

10-31 (a) $e = d/32$
(b) $e = d/16$
10-32 $P_w = 19.6$ kips
$e = 0.0333$ in.
10-34 $L/r = 28.5$
10-36 $ec/r^2 = 0.2$
$e = 0.1560$ in.
10-37 $ec/r^2 = 0.43$
$e = 1.075$ mm
10-39 $P_w = 112.3$ kN
10-41 $k = 2.89$
10-44 (a) $P_w = 41.4$ kips
(b) $P_w = 206$ kips
10-45 (a) $P_w = 616$ kN
(b) $P_w = 1302$ kN
10-46 (a) $P_w = 33.3$ kips
(b) $P_w = 128.7$ kips
10-47 (a) $P_w = 59.1$ kN
(b) $P_w = 997$ kN
10-51 (a) $P_w = 97.0$ kips
(b) $P_w = 193.7$ kips
10-52 $P_w = 850$ kN
10-55 $b = 9.41$ in.
10-56 Use C203 × 28 sections.
10-58 $P_w = 85.6$ kips
10-59 $P_w = 712$ kN
10-61 $d_o = 2.34$ in.
10-62 $a = 32.2$ mm
10-65 $P_w = 254$ kips
10-69 $P_w = 41.0$ kips
10-70 $P_w = 212$ kN
10-71 $P_w = 38.3$ kips
10-72 $P_w = 240$ kN
10-75 $P_w = 49.3$ kips
10-76 (a) $P_w = 145.6$ kN
(b) $M = 4.35$ kN · m
10-79 $P = 56.9$ kips
10-80 $P = 137.0$ kN
10-82 $P = 426$ kN
$b = 40.2$ mm
10-83 $P = 12.05$ kips
10-88 $P = 1.553\pi^2 EI/L^2$
10-90 $P_w = 58.3$ kips

10-92 $P_{cr} = 57.4$ kN

10-94 $P_{cr} = 101.0$ kips

CHAPTER 11

11-1   (a) $t = 6.67$ hr

      (b) $t = 12.67$ hr

      (c) $t = 23.8$ hr

      (d) $t =$ unlimited

11-2   (a) $t = 16.67$ hr

      (b) $t = 83.3$ hr

      (c) $t = 556$ hr

      (d) $t =$ unlimited

11-3   (a) $\sigma_w = 66$ ksi (455 MPa)

      (b) $\sigma_w = 59$ ksi (407 MPa)

      (c) $\sigma_w = 54$ ksi (372 MPa)

11-4   (a) $k = 2.07$

      (b) $k = 1.73$

      (c) $k = 1.60$

11-5   (a) $P_{max} = 9.33$ kips

      (b) $P_{max} = 36.0$ kips

11-8   (a) $w = 461$ mm

      (b) $w = 130$ mm

11-9 $a = 6.89$ in.

      $b = 1.95$ in.

11-12 $P_{max} = 1275$ N

11-13   (a) $M_{max} = 57.9$ in.-kips

      (b) $M_{max} = 15.7$ in.-kips

11-16   (a) $T_{max} = 1920$ N $\cdot$ m

      (b) $T_{max} = 1920$ N $\cdot$ m

11-17 $t_{min} = 1.760$ in.

11-19 $k = 2.06$

11-20 $k = 2.41$

11-21 $\sigma_{max} = 60.1$ ksi T

      $\sigma_{min} = 36.1$ ksi C

11-22 $\sigma_{max} = 530$ MPa T

      $\sigma_{min} = 176.7$ MPa T

11-25 $P_{max} = 2.48$ kips T

      $P_{min} = 1.24$ kips C

11-28 $P_{max} = 8.25$ kN

      $P_{min} = 5.50$ kN

11-29 $P = 18.4$ kips

11-32 $k = 5.91$

11-33 $P = 101.6$ lb

11-36 $e = 17.06$ mm

CHAPTER 12

12-1 $d_{min} = 1.713$ in.

12-2 $L_{min} = 3.27$ m

12-3 $U = \gamma^2 A L^3 / 6E$

12-4 $U = P^2 L^3 / 6EI$

12-6 $U_{bar} = 9 U_{beam}$

12-8 $U_{bar} = 15 U_{beam}$

12-10 $U_{shaft} = (1 + \nu) U_{bar}$

12-11 $U = 86.4 I / c^2$

12-12   (a) $m = 11.71$ kg

      (b) $m = 6250$ kg

      (c) $m = 12500$ kg

12-15 $\sigma_{max} = 939$ psi T and C

12-16 $h = 204$ mm

12-19   (a) $\Delta = 0.1039$ in.

      (b) $P = 1559$ lb

      (c) $\sigma = 15.59$ ksi

12-20 $U = 2180$ J

12-22   (a) $\tau_{max} = 73.7$ MPa

      (b) $\theta = 0.0276$ rad

12-23   (a) $\tau_{max} = 16.48$ ksi

      (b) $\theta = 0.01854$ rad

      (c) $\tau_{max} = 10.09$ ksi

12-25 $P = 9520$ lb

12-26   (a) $\Delta_{max} = 65.9$ mm

      (b) $P = 1186$ N

12-27 $L_{min} = 71.3$ in.

12-28 $\sigma_{max} = 80.0$ MPa

12-31   (a) $\sigma_{max} = 15.90$ ksi

             T and C

      (b) $\Delta_{max} = 0.1060$ in.

      (c) $P = 1590$ lb

12-32   (a) $\sigma_{max} = 9.83$ MPa

             T and C

      (b) $\Delta_{max} = 17.98$ mm

      (c) $P = 3.28$ kN

12-34   (a) $\tau_{max} = 100.2$ MPa

      (b) $U = 62.2$ J

12-35   (a) $W = 10$ lb

(b) $h = 0.88$ in.

12-38   (a) $\eta_{sp} = 21.1\%$

(b) $\eta_{beam} = 60.8\%$

12-39 $h = 21.6$ in.

12-41 $h = 20WL^3/3EI$

12-44   (a) $h = 1.80$ in.

(b) $LF = 4.50$

(c) $U_A = 6.00$ in.-lb

$U_B = 48.0$ in.-lb

CHAPTER 13

13-1   (a) $\sigma_t = 108.2$ MPa T

(b) $\sigma_b = 175.8$ MPa C

(c) $\tau = 26.9$ MPa

13-2   (a) $\tau = 8.71$ ksi

(b) $\sigma_b = 20.5$ ksi C

(c) $\sigma_t = 12.10$ ksi T

13-4   (a) $\tau = 9.38$ ksi

(b) $\sigma_b = 15.95$ ksi C

(c) $\sigma_t = 11.84$ ksi T

13-5   (a) $\tau = 91.7$ MPa

(b) $\sigma_b = 150.0$ MPa C

$\sigma_t = 123.0$ MPa T

(outer row)

13-7   (a) $\tau = 47.3$ MPa

(b) $\sigma_b = 163.6$ MPa C

$\sigma_t = 118.4$ MPa T

13-10 $\sigma_t = 18.30$ ksi

13-11   (a) $p_{max} = 1.578$ MPa

(b) $\eta = 69.2\%$

13-13 $t_{min} = 8.96$ mm

13-15   (a) $p_{max} = 1.322$ MPa

(b) $\eta = 66.8\%$

13-16   (a) $P_{max} = 36.1$ kips

(b) $\eta = 75.0\%$

13-18 $t_{min} = 0.240$ in.

13-20 $p_{max} = 204$ psi

13-21   (a) $P_{max} = 402$ kN

(b) $\eta = 52.9\%$

13-26 $L_1 = 2.07$ in.

$L_3 = 8.22$ in.

13-27 $L_1 = 72.7$ mm

$L_3 = 248$ mm

13-28 $L_1 = 6.33$ in.

$L_3 = 16.53$ in.

13-31 $a = 181.8$ mm

$b = 231$ mm

$c = 134.6$ mm

13-34   (a) $\tau_{max} = 74.5$ MPa

(b)   Use 7 bolts.

13-35   Power $= 89.7$ hp

13-36   (a) $F_A = 55.9$ kN

(b) $F_B = 93.2$ kN

13-39 $\tau_A = 9.18$ ksi

$\tau_B = 5.51$ ksi

$\tau_C = 0$

13-41   (a)   Power $= 1160$ hp

(b) $\theta = 0.030$ rad

(c) $N = 12$ bolts

13-42   (a)   Bolt C

(b) $\tau = 81.2$ MPa

13-43 $\tau_A = 10.88$ ksi

13-44 $\tau_A = 54.0$ MPa

$\tau_B = 86.1$ MPa

13-47 $\tau_{top} = 8.15$ ksi

$\tau_{bot} = 0$

13-48 $P_{max} = 11.20$ kN

INDEX

$$\tan 2\theta_p = \frac{\gamma_{xy}}{\epsilon_x - \epsilon_y}$$

$$\epsilon_{p,1,2} = \frac{\epsilon_x + \epsilon_y}{2} \pm \sqrt{\left(\frac{\epsilon_x - \epsilon_y}{2}\right)^2 + \frac{\gamma_{xy}^2}{2}}$$

$$\gamma_p = 2\sqrt{\left(\frac{\epsilon_x - \epsilon_y}{2}\right)^2 + \frac{\gamma_{xy}^2}{2}}$$

$$\tau_{1S} = \frac{V}{It}\, st\left(50 - \frac{20}{52}\frac{s}{2}\right) = \frac{V}{I}\left[50s - \frac{10s^2}{52}\right]$$

$$\tau_{2S} = \tau_1 + \frac{V}{It}\, st\left(30 - \frac{s}{2}\right) = \tau_1 + \frac{V}{I}\left(30s - \frac{s^2}{2}\right)$$

$$F_1 = \int \tau_{1S}\, dA = \frac{Vt}{I}\int_0^{52}\left(50s - \frac{10s^2}{52}\right)ds$$

$$(F_1)_H = F_1\left(\frac{48}{52}\right) = .6459\%$$

$$M_0 = 0$$

$$Ve = 2(30)(F_1)_H \qquad e = 38.94mm$$

$$F_2 = V - 2(F_1)_H$$

$$J = \left[\frac{1}{12}(15.6)(20^3) + (15.6)(20)(40)^2\right] \times 2 + \frac{1}{12}(6)(60)^3$$

ENERGY

$$\frac{M_E}{I_{NA}} = \frac{\sigma_y}{Y} = \frac{E}{R_{CURVATURE}}$$

$$\theta = \frac{\delta L}{C}$$

RESIDUALS

REVERSE APPLIED

$- \text{APPLIED} = \int_0^a 2\pi r^3 (K)\, dr$

$K =$

THEN ADD τ_{LOAD}

$+\ C_{UNLOAD}$ TO FIND

RESID. AT

CERTAIN POINTS.

RESIDUAL ANGLE

OF TWIST

$\gamma_{AT\ OUTER\ FIBER\ LOAD}$

$+\ \gamma_{AT\ OUTER\ FIBER\ UNLOAD}$

$\dfrac{L}{G} = \gamma$

$\dfrac{\gamma_a}{a} = \dfrac{\gamma_b}{b}$

AVERAGE PROPERTIES OF SELECTED ENGINEERING MATERIALS (INTERNATIONAL SYSTEM OF UNITS)

Exact values may vary widely with changes in composition, heat treatment, and mechanical working. More precise information can be obtained from manufacturers.

Materials	Density (Mg/m³)	Elastic Strength[a] Tension (MPa)	Comp. (MPa)	Shear (MPa)	Ultimate Strength Tension (MPa)	Comp. (MPa)	Shear (MPa)	Endurance Limit[c] (MPa)	Modulus of Elasticity (GPa)	Modulus of Rigidity (GPa)	Percent Elongation in 50 mm	Coefficient of Thermal Expansion (10^{-6}/°C)
Ferrous metals												
Wrought iron	7.70	210	b		330	b	170	160	190		30[d]	12.1
Structural steel	7.87	250	b		450	b		190	200	76	28[d]	11.9
Steel, 0.2% C hardened	7.87	430	b	160	620	b			210	80	22	11.9
Steel, 0.4% C hot-rolled	7.87	360	b	190	580	b		260	210	80	29	
Steel, 0.8% C hot-rolled	7.87	520	b	180	840	b			210	80	8	
Cast iron—gray	7.20				170	690		80	100		0.5	12.1
Cast iron—malleable	7.37	220	b		340	b			170		20	11.9
Cast iron—nodular	7.37	480			690	b			170		4	11.9
Stainless steel (18-8) annealed	7.92	250	b		590	b		270	190	86	55	17.3
Stainless steel (18-8) cold-rolled	7.92	1140	b		1310	b		620	190	86	8	17.3
Steel, SAE 4340, heat-treated	7.84	910	1000		1030	b	650	520	200	76	19	
Nonferrous metal alloys												
Aluminum, cast 195-T6	2.77	160	170		250	b	210	50	71	26	5	
Aluminum, wrought, 2014-T4	2.80	280	280		430	b	260	120	73	28	20	22.5
Aluminum, wrought, 2024-T4	2.77	330	330		470	b	280	120	73	28	19	22.5
Aluminum, wrought, 6061-T6	2.71	270	270		310	b	210	93	70	26	17	22.5

$-0.427\ F_{SU} = V_{CENTROID}$

$.398\ F_{SU} = V_{WEB-FLANGE}$

Q5